ROUTLEDGE HANDBOOK OF ECOTOURISM

This handbook presents a timely, broad-ranging, and provocative overview of the essential nature of ecotourism. The chapters will both advance the existing central themes of ecotourism and provide challenging and divergent observations that will thrust ecotourism into new areas of research, policy, and practice.

The volume is arranged around four key themes: sustainability, ethics and identity, change, conflict, and consumption, and environment and learning, with a total of 28 chapters. The first section focuses on sustainability as a core ecotourism criterion, with a primary focus on some of the macro sustainability issues that have an impact on ecotourism. Foremost among these topics is the linkage to the UN's Sustainable Development Goals, which have relevance to ecotourism as one of the greenest or most responsible forms of tourism. The chapters in the second section provide a range of different topics that pull ecotourism research into new directions, including a chapter on enriching indigenous ecotourism through culturally sensitive universalism. The third section includes chapters on topics ranging from persons with disabilities as a neglected body of research in ecotourism, to ecotourism as a form of luxury consumption. The final section emphasises the link between ecotourism and learning about the natural world, including a deeply theoretical chapter on rewilding Europe. With contributions from authors around the world, this handbook gives a global platform to local voices, in both developed and emerging country contexts.

The multidisciplinary and international *Routledge Handbook of Ecotourism* will be of great interest to researchers, students, and practitioners working in tourism and sustainability.

David A. Fennell is a professor in the Department of Geography and Tourism Studies, Brock University, Canada. He is the founding editor-in-chief of the *Journal of Ecotourism*.

ROUTLEDGE HANDBOOK OF ECOTOURISM

Edited by
David A. Fennell

LONDON AND NEW YORK

First published 2022
by Routledge
2 Park Square, Milton Park, Abingdon, Oxon OX14 4RN

and by Routledge
605 Third Avenue, New York, NY 10158

Routledge is an imprint of the Taylor & Francis Group, an informa business

British Library Cataloguing-in-Publication Data
A catalogue record for this book is available from the British Library

Library of Congress Cataloging-in-Publication Data
A catalog record has been requested for this book

ISBN: 978-0-367-43192-1 (hbk)
ISBN: 978-1-032-06723-0 (pbk)
ISBN: 978-1-003-00176-8 (ebk)

DOI: 10.4324/9781003001768

Typeset in Bembo
by MPS Limited, Dehradun

CONTENTS

FIGURES

TABLES

CONTRIBUTORS

Areej Shabib Aloudat is an associate professor at Yarmouk University, Irbid, Jordan. She is appointed as a vice dean for quality assurance and accreditation at the Faculty of Tourism and Hotel Management for five years. She is an author of a number of journal articles and book chapters on tour guiding, tourist behaviour, grounded theory, qualitative research,female entrepreneurs in tourism.

Georgette Leah Burns is an environmental anthropologist who specialises in human interactions with nature. This ranges from research topics on rural development to wildlife tourism. In addition to numerous book chapters and journal articles, Leah is co-editor of *Engaging with Animals: Interpretations of a Shared Existence* (2014). From 2013 to 2015 she resided in Iceland where she was Head of the Department of Rural Tourism at Hólar University College, and Head of Tourism Research the Icelandic Seal Center.

Kadir Çakar is an assistant professor at Mardin Artuklu University, Turkey. His PhD examined the motivations and experiences of travelers visiting Gallipoli Peninsula within the context of dark tourism. His main research areas include qualitative research, dark tourism, tourist destination governance, tourism education, information communication technologies, destination marketing and management, crisis management, and sustainable tourism.

Kellee Caton is associate professor of tourism studies at Thompson Rivers University, where she teaches sociocultural theory and applied philosophy to TRU's wonderful students. Her research program focuses on the humanistic side of tourism with a broad interest in moral philosophy, critical theory, and tourism epistemology and pedagogy. Her work explores how we come to know tourism as a field of academic study, as well as how we come to know the world around us through tourism discourse and practice. Some of her recent projects include explorations of the role of tourism in ideological production in educational and religious tourism contexts, conceptual analyses of the knowledge advancement process in tourism studies, and advocacy projects for the value of humanities content in tourism curricula. Kellee is incoming co-chair of the Critical Tourism Studies international network and a member of the Tourism Education Futures Initiative executive, and she sits on the editorial boards of Annals of Tourism Research and Tourism Analysis.

David D'Acunto holds a PostDoc position at Free University of Bozen-Bolzano, Faculty of Economics and Management, Italy.

Simon Darcy is a professor in management at UTS Business School, he specialises in developing inclusive organisational approaches for diversity groups, including people with disability. Simon's work has spanned tourism, sport management, events, volunteers, transport, employment, entrepreneurship, the built environment and disability services. It is characterised by an evidence-based approach to changing practice in the business, government and not-for-profit sectors. His impact includes contributing significantly to the understanding of the business case for accessible tourism nationally and internationally through the UN World Tourism Organisation and, with Professor Jock Collins, undertaking research that has illuminated entrepreneurship among people with disability – informing development of the IgniteAbility startups program.

Jonathon Day is an associate professor focused on sustainable tourism at Purdue University, United States. His research focuses on sustainable tourism, responsible travel, and strategic destination governance "in the role of business in solving grand challenges through corporate social responsibility programmes and social entrepreneurship."

Cecilia De Bernardi is a lecturer in tourism at Dalarna University. She has a PhD in tourism researcher and two bachelor degrees in communication sciences and tourism research. Her research has previously focused on Indigenous tourism and authenticity, but lately she has worked with Airbnb, nostalgia, science fiction, second homes, resilience, tourism policy, and music festivals.

Fatma Cam Deniz is a joint postdoctoral researcher at the Gallant Lab in UC Berkeley's Helen Wills Neuroscience Institute and the International Computer Science Institute. She is interested in how sensory information is encoded in the brain and uses machine learning approaches to fit computational models to large-scale brain data acquired using functional magnetic resonance imaging (fMRI).

Nils Lindahl Elliot is an independent researcher and writer with a specialism in transdisciplinary research about mediated understandings of science and nature. For two decades, she was a lecturer and researcher in higher education. Her research has received the support of the UK's Arts and Humanities Research Board (now AHRC), the British Academy, and the Economic and Social Research Council (ESRC).

David Fennell conducts research mainly in the areas of ecotourism, tourism ethics, and moral issues tied to the use of animals in the tourism industry, and sustainability. A major thrust of his research involves the use of theory from other disciplines (e.g., biology, philosophy) to gain traction on many of tourism's most persistent issues and problems. Fennell is the founding Editor-in-Chief of the *Journal of Ecotourism*.

Lee A. Fitzgerald is professor of zoology and faculty curator of amphibians and reptiles in the Department of Wildlife and Fisheries Sciences at Texas A&M University. His biological specialty is the evolutionary ecology and conservation biology of amphibians and reptiles (herpetology).

Brian Garrod is a professor in marketing (research), business at Swansea University. He is the author of eight textbooks and more than 50 research articles, focusing mainly on the marketing and management of tourism destinations. He is co-editor-in chief of the *Journal of Destination Marketing & Management* and sits on the editorial board of seven other academic journals. He has undertaken research for a wide range of organisations, including the World Tourism Organization (UNWTO), the Organization for Economic Cooperation and Development (OECD), the European Union Interreg IIc fund, the Welsh Government, Visit Wales, the Countryside Agency, and the Arts and Humanities Research Council (AHRC).

Silvia Giralt-Escobar is a professor at the University of Alcalá, Madrid, Spain. She is a researcher and author of several scientific articles and advisor of doctoral and master's theses.

Sonya Graci is an associate professor at the Ted Rogers School of Hospitality and Tourism Management at Ryerson, University in Toronto, Ontario. Dr. Graci is also the director of the Hospitality and Tourism Research Institute. Dr. Graci has worked on numerous projects around the world related to sustainable tourism development and has focused her attention on community capacity building in Honduras, Indonesia, Canada, Fiji, and China. She has a keen interest in working with Aboriginal communities in developing sustainable forms of tourism. She also has a passion for increasing sustainability in marine environments and has focused much of her research on sustainable tourism development in island states. Dr. Graci is the author of two books and several journal articles and industry publications.

Bryan S. R. Grimwood is an associate professor in recreation and leisure studies and associate chair for graduate studies at University of Waterloo. His areas of research are tourism and Indigenous peoples, tourism ethics and responsibility, northern landscapes, and outdoor experiential education.

Burcin Kalabay Hatipoglu is an assistant professor of human resource management at Boğaziçi University in Istanbul, Turkey. She holds a PhD from the University of New South Wales, Sydney. In the last 10 years, Dr. Hatipoglu has been involved in several sustainable tourism development projects and consulted industry organisations. Her research interests include human capital reporting, education for sustainability, and organisational sustainability.

Huong H. Do is a lecturer of tourism at Hoa Sen University. Huong does research in quantitative social research and qualitative social research. Her most recent publication is *Tourism students' motivational orientations: the case of Vietnam.*

Carter A. Hunt is an associate professor in tourism management and anthropology at The Pennsylvania State University, Pennsylvania, USA. His research interests focus on biodiversity conservation, sustainable community development, and nature-based tourism. He explores these themes through the lenses of environmental anthropology, the anthropology of tourism, and conservation psychology.

Chris E. Hurst is a Ph.D. candidate in the Department of Recreation and Leisure Studies at the University of Waterloo. Her doctoral research examines human-nature relationships in nature-based recreation. She employs a multidisciplinary approach, drawing upon posthuman and social theories, as well as the fields of human geography, decoloniality, leisure, and tourism. Chris is a part of the Canadian research team for the ARCTISEN Project, an international project supporting culturally sensitive tourism in the Arctic. In addition, she works as a research assistant under her supervisor Dr. Bryan S. R. Grimwood, and previously served as the editorial assistant for the *Journal of Leisure Sciences.*

Tümay Imamoğlu is the president of Şile Tourism, Culture and Promotion Association.

Helen Kopnina, Ph.D. Cambridge University, 2002, is currently employed at The Hague University of Applied Science (HHS) in the Netherlands, coordinating the Sustainable Business program, and conducting research within three main areas: sustainability, environmental education, and biological conservation. Helen is the author of over 200 articles and (co)author and (co)editor of 17 books.

Laura Lawton is an associate professor and acting head of the Department of Tourism, Sport and Hotel Management at Griffith University, Australia. She was formally the program director for the Master of Business Programs in the Department. Previously she held appointments at the University of South Carolina and George Mason University, Virginia, USA. She previously taught at Bond University, Gold Coast.

James Malitoni Chilembwe is a lecturer in the faculty of Tourism, Hospitality and Management at Mzuzu University, Mzuzu, Malawi. He is a holder of MSc in international tourism enterprise obtained at Glasgow Caledonian University (GCU) in 2012. He acquired tourism management and teaching training qualifications from Austria in 2009 at the Institute of Tourism and Hospitality Management (ITH). He also holds professional travel industry qualifications obtained from 1999 to 2007 in Switzerland and Canada from the International Air Transport Association (IATA) and United Federation of Travel Agents Association (UFTAA). He is a current Ph.D. candidate (doctor of philosophy) in tourism management at Glasgow Caledonian University (GCU), UK.

Maria Marshall is a professor of agricultural economics at Purdue University. She conducts an applied research, teaching, and extension program in small and family business management, disaster recovery, and entrepreneurship. Dr. Marshall is the director of the North Central Regional Center for Rural Development. She is also the director and founder of the Purdue Institute for Family Business (PIFB).

Ian E. Munanura, Ph.D., is an assistant professor of nature-based tourism in the Department of Forest Ecosystems and Society at Oregon State University. His areas of research interests are ecotourism, integrated social and ecological systems, sustainable rural livelihoods, and community resilience. He has previously led and advised USAID-funded ecotourism projects. He also served as country director for the Wildlife Conservation Society in Rwanda.

Steve Noakes is an adjunct research fellow at Griffith Institute for Tourism, Griffith University, Australia and visiting professor in sustainable tourism and Indigenous culture, adventure tourism, hotel and tourism strategic management at Bond University, Australia. He is also a visiting professor in business economics at UiT The Arctic University of Norway. He has an extensive background as a tourism and hospitality sector entrepreneur, educator, trainer, researcher, and international consultant. He is also a leading contributor throughout Asia and the Pacific on sustainable tourism approaches, including tourism and Indigenous people, international development, and human rights issues.

Joe Pavelka is a professor of ecotourism and outdoor leadership at Mount Royal University, Calgary Alberta Canada. Joe's research focuses on aspects of destination management (both here in Canada and abroad) and marketing and tourism motivation. His past education includes a bachelor of arts in geography (1987) and honours bachelors of outdoor recreation (1987) both from Lakehead University in Thunder Bay, Ontario. He obtained his master's in recreation administration from the University of Alberta (1990) in Edmonton, and his doctorate in geography (2010) from the University of Calgary, studying the resort community evolution from an interdisciplinary approach focusing on amenity migration.

Ige Pirnar works as chair of Department of BA, Yasar University and has her MBA from Bilkent University (1989) and Ph.D. from Ankara University, BA (1998). She has many articles, conference papers in English and Turkish and has eight books in Turkish (three edited and two with co-authors) and one in English. Her areas of expertise are: international business, marketing management, entrepreneurship, hospitality marketing, and international tourism management.

Manuel Ramón Gonzalez Herrera is a professor at the Autonomous University of Ciudad Juárez in the bachelor's degree in tourism. Some of the subjects he has taught to date are tourism planning, tourism and the environment, tourism geography, tourism destination management, tourism research, among others. His most recent research is fundamentally based on the theme of sustainable tourism development management for the minimization of impacts and crisis events in tourism. Although his interest in other lines of research is also predominant.

John B. Read IV is a teaching assistant and Ph.D. student at University of Waterloo. He has a background in fire protection (B.S.) with experience working as a firefighter and emergency medical practitioner in municipal, military, and wildland settings as well as stints as a fire marshal at research facilities in the Antarctic and military installations in Saudi Arabia. Engaging in a career transition, John began working for the National Park Service while pursuing an M.S. with an emphasis in recreation administration. He currently sits as co-chair for the US Association of Polar Early Career Scientists and in his free time enjoys birdwatching and coffee.

Andrew Rylance is a technical expert in environmental economics and sustainable financing. He is currently the UNDP Acting Head of the Environment, Energy and Climate Change portfolio and the UNDP Chief Technical Advisor on Sustainable Financing of Protected Areas in Papua New Guinea.

Edwin Sabuhoro is a Rwandan national, received his law degree from the National University of Rwanda, a master's of science degree in conservation and tourism from the University of Kent at Canterbury, UK and a Ph.D. in parks, recreation and tourism management at Clemson University in the USA. His work to set up community-based projects for local communities and reformed-poachers around the Mountain Gorilla's Volcanoes National Park which has helped in the reduction of poaching in the park, and improved the livelihoods of local communities was recognized nationally and internationally through numerous awards. Among them is the International Union for Conservation of Nature (IUCN) Young Conservationist of the year 2008 Award, 2010 President Obama Young African Leaders Innitiative (YALI), 2015 UN-GRASP-Ian Redmond Conservation Award and has been featured on 2015 CNN African Voices, and was recognised as a 2015 CNN Hero.

Sudipta Kiran Sarkar is a senior lecturer in tourism at Anglia Ruskin University, Cambridge Campus, UK. Before working in the United Kingdom, Sudipta worked as a tourism academic in India, Malaysia, Hong Kong, and South Korea. He has authored and co-authored a number of book chapters, conference, and journal papers in the areas of technology and social media in tourism, ecotourism, destination life cycle models, sustainability and technology in urban tourism, and tourism education. Sudipta's experience also involves teaching and learning activities in various domains of tourism and hospitality, and one of his main pedagogical interest is experiential learning in tourism.

Stephen Schweinsberg is a senior lecturer in sustainable management in the UTS Business School. Stephen's current research interests are around sustainable tourism, tourism's place based setting and academic discourse in a post COVID-19 world. Stephen is on the editorial board for *Tourism Review* and his research has been published in a range of leading journals including *Tourism Management, Annals of Tourism Research* and the *Journal of Sustainable Tourism.*

Valerie A. Sheppard has more than 30 years of experience in branding, business strategy development and execution, and coaching from both line management, teaching, and consulting roles. She is certified professionally qualified to teach at the collegiate level by the AACSB, and has taught both undergraduate and graduate level marketing courses at The University of California Irvine Merage School of Business.

Anna Spenceley is a tourism expert with over 20 years' international experience and an extensive publication record, rich diversity of project experience, and an international network of associates with whom she collaborates. She is the founder STAND Ltd, a boutique consultancy company, specialising in meeting the development needs of international organisations addressing sustainability agendas. Anna is a member of the Independent Advisory Panel of Travalyst, and is a senior research fellow at the School of Tourism and Hospitality at the University of Johannesburg, and an honorary fellow at the University of Brighton. She sits on the editorial board of the *Journal for Sustainable Tourism*, the *Journal of Ecotourism*, and the *Journal of South African National Parks: Koedoe.*

Amanda L. Stronza is an environmental anthropologist and photographer with 30 years of research and conservation work in the Amazon, the Okavango Delta, and other parts of the tropics. As a social scientist, her contributions to conservation come through learning and documenting how things like cultural beliefs, social norms, institutions, and economic incentives shape our relationships with the environment and other species. Her work straddles theory and practice, and she is active in designing and implementing conservation programs that support positive human-wildlife interactions and community-based conservation. She has joint appointments in the Departments of Ecology and Conservation Biology, and Rangeland, Wildlife, and Fisheries Management, and she co-directs the Applied Biodiversity Science Program.

Sandra Sydnor, M.B.A., Ph.D., is an associate professor at Purdue University's School of Hospitality and Tourism Management. Her multi-unit franchisee industry experience with the Burger King® Corporation's restaurants and executive leadership positions in marketing research and advertising agencies fuel her interest in resilience and sustainability science, wellbeing, and service management. Her current investigations include community and industry resilience and regenerative design and the impact of mindfulness behaviors on customer-facing employees and student engagement.

Bastian Thomsen is a DPhil student in anthropology with a research focus in environmental anthropology, anthrozoology, science and technology studies (STS), and multi-species livelihoods. Bastian is an assistant professor of international social entrepreneurship at Boise State University in Idaho, United States. He co-established and co-leads Boise State's Global Scholars Program with his wife, Dr. Jennifer Thomsen, which takes undergraduate and graduate students internationally to work on applied ethnographic service-learning projects with environmental nonprofit organisations.

Prof. Serena Volo is an Associate Professor of Marketing at the Faculty of Economics and Management of the Free University of Bozen-Bolzano, Italy. She is Vice-director of the Competence Centre in Tourism Management and Tourism Economics, TOMTE and has chaired several editions of CBTS, the Consumer Behavior in Tourism Symposium. She is Editor-in-Chief of the International Journal of Culture, Tourism and Hospitality Research published by Emerald. She is also on the editorial board of leading scientific journals in the field of tourism, hospitality and leisure. She is an elected member of the Executive Council of IATE, the International Association for Tourism Economics. Her research interests include consumer behavior in tourism, experience and emotions in tourism, visual research methods and big data, innovation and competitiveness in tourism. She has had work, research and life experiences in Italy, Ireland, England, USA, France, Switzerland.

David Weaver is a professor of tourism research at Department of Tourism, Sport and Hotel Management at Griffith University. He received his Ph.D. in geography from the University of Western Ontario (Canada) in 1986, and has published more than 120 journal articles, book chapters, and books. He maintains an active research agenda in sustainable destination and protected area management, ecotourism, and resident perceptions of tourism. He is a fellow of the International Academy for the Study of Tourism and has delivered numerous invited international keynote addresses on innovative tourism management topics. He has worked with organizations such as UNWTO and PATA as an expert advisor.

Nicholas Wise is an assistant professor in the School of Community Resources and Development at Arizona State University, USA. His current research focuses on placemaking, competitiveness, social regeneration, and community impacts, conducting work across the areas of sport, events, and tourism. He brings an international perspective to his teaching informed by scholarly research focusing on the Dominican Republic, Argentina, Brazil, Croatia, Serbia, and Italy. He has published across several disciplines and collaborates with colleagues across a number of academic disciplines. He is also a senior fellow of the Higher Education Academy (now AdvanceHE).

INTRODUCTION

David A. Fennell

That you are reading this page, and book, is a minor miracle. All of us know the monumental challenges that COVID-19 forced upon us in so many different ways. This holds true for the present volume where the vast majority of authors had to make considerable adjustments regarding how, when, and sometimes where, to get work done. Priorities around family, personal health and wellness, and community health emerged as first priorities as authors adjusted to formidable work-life challenges. The number of emails sent my way over the course of 2020 is proof of the challenges. For example,

> Thanks for your email. I am sorry not to inform you earlier. I was struggling during COVID time. But things are more settled now. I would appreciate if you could wait one month more to complete it …

> Thank you for your emails and apologies for the late reply. We are finalizing the chapter and I plan to send it to you by the 29th. I am really sorry for the prolonged delay these months have been quite challenging in many ways. Hope you and your family are keeping in good health and high spirits …

It was not just sitting at one's desk and finding the time to write amidst all the chaos on confusion, but also collecting data. Primary data collectors were hampered by the inability to actually get into the field to gather the data that served as the foundation for their work. For example

> Despite our best efforts, my co-authors that are at the site had difficulties collecting data from two of our samples, and we are running late in putting together the rest of our chapter. I believe we will be ready to send the final manuscript by next week. I am hoping that we have not caused any inconvenience. Thank you for your understanding. Warm regards …

But enough of COVID-19. Really.

Another challenge, and this one is on me, was the *perception* that ecotourism's most productive days, at least from a research standpoint, were in the past. I say this from the

DOI: 10.4324/9781003001768-101

perspective of a disciplinary challenge, based on the fact that less emphasis is being placed on ecotourism in the jobs that frequently appear on TRINET, and perhaps less of a focus on ecotourism as a major theme of international conferences. After all, tourism conferences try to capture the spirit of the day in nudging their way to some level of global relevance. The 1990s, and even the 2000s, were a time of tremendous growth in ecotourism as university tourism programs sought to fill their rosters with competency in this area. Ecotourism was the flavour of the day. But is it still?

Moving away from disciplinary challenges, in the field and practice of ecotourism critics abound. Opponents argue that in reality ecotourism is simply business as usual. Costs to destinations, people, and the planet far outweigh benefits, as corporations, governments, ecotourists, and other stakeholders pursue their own interests. The trend towards self-interest persists, either individual or organisational, as, for example, national park administrators change policy as a strategic goal to increase visitors to protected areas, while ecotourists support low-impact environmental impacts (Morrison-Saunders, Hughes, Pope, Douglas, & Wessels, 2019), increasingly there are moral concerns about the use of animals in ecotourism when ecotourism ought to be about their protection (Taylor, Hurst, Stinson, & Grimwood, 2020), and community ecotourism development, although the most popular theme in ecotourism research and practice *still* does not always lead to good outcomes for all members of the community (Duffy, 2006).

And enough about challenges.

Rest assured that ecotourism is alive and well. To use one measure as barometer, a rather important one I would argue, downloads of the *Journal of Ecotourism* (the main stand-alone conduit for publishing in this area) are up 18% in 2020 compared to 2019; it has gone from three editions to four editions starting in 2018; and the journal had 65% more submissions in 2020 than in 2019 (Journal of Ecotourism Publishing Report, 2021).

This book is a testament to ecotourism's vitality as one of tourism's greenest and most ethical forms. Pundits contend that it is an effective tool for the conservation of biodiversity, to build community cohesion and improve livelihoods, reduce poverty, and induce learning for in both providers and consumers. In an effort to continue the journey towards a deep understanding of the complexities inherent in ecotourism, the book has two main objectives. The first is to hear from, and about, new voices, many from early-stage academics, new regions, new case studies from developed and lesser develop contexts, and to strike a balance between female and male representation. Accomplished. The second objective is to provide researchers and students with a state-of-the-art and provocative overview of the essential nature of ecotourism through novel areas of research. The hope is that these new areas will have important implications not only for research but also for advancing policy and practice. Four main themes on sustainability; ethics and identities; change, conflict and consumption; and learning and learning, containing 28 chapters, are intended to accomplish this end.

The first chapter in the section on sustainability by Spenceley and Rylance situate their discussion around ecotourism and the United Nation's 17 Sustainable Development Goals (SDGs) for 2015 to 2030. Tourism, the authors, reason, has a role to play in all of these goals, and ecotourism, given its focus on economic development and conservation, can play a leading role, especially in the areas of ecolodge development for benefits at local levels. The UNWTO has an important role to play in linking the SDGs with ecotourism through cooperation, partnerships, incentives, good practices, and raising awareness of business opportunities. The following chapter by Day, Sydnor, Marshall, and Noakes highlights the important role that ecotourism continues to play as a model of cutting-edge trends in sustainability. They argue that the rise of the circular economy, redefinition of growth and social benefits, the emerging focus on regenerative tourism, and regenerative recovery provide opportunities for operational efficiency. The famous Binna Burra Lodge in Lamington National Park, Australia, is the case study used to show the process of recovery and regeneration from the historic bushfires in 2020.

Schweinsberg and Darcy switch the discussion on ecotourism and sustainability to the trouble with transportation, which is one of the most persistent and deep-seated problems hampering ecotourism as a sustainable and moral industry. A business-as-usual approach to transportation will not move the sector forward. Innovative in their message is the importance of developing accessible forms of transportation given the sheer magnitude of people, globally who have disability and access needs. Sheppard's chapter moves the discussion into linking resilience and sustainability in efforts to build a more effective business model for ecotourism operations. Important in this configuration is the need to combine individual (e.g., spiritual, cognitive, behavioural, and emotional) and community (e.g., learning to live with change, nurturing diversity, and combining different kinds of knowledge) resilience enhancing characteristics in building stronger social and ecological systems through ecotourism. Finally, Aloudat tackles the thorny issue of overtourism in the Petra as a popular tourism destination. Perta Protected Area is an example of the long-established view that at times we are loving our tourism attractions and destinations to death. Tour guides feature prominently in Aloudat's research as interpreters of the environment, motivators of environmentally responsible behaviour and conservation values, and as special information givers. Tour guides recognise the formidable impact taking place on Petra Protected Area, with the need for decision makers to factor tour guide experiences into methods designed to offset impacts. Fennell's chapter on personalised interactive real-time tours demonstrates how new technologies can provide opportunities for operators and local people in times of crisis (COVID-19), as well as for those who may not wish to travel (the sustainable citizen) or cannot travel (persons with disabilities).

The second theme in the book, ethics and identities, provides a range of different topics that pull ecotourism research into new directions. The first part of the section focuses on ethics, and the second on identities. Caton, Hurst, and Grimwood employ enchantment and care ethics (the latter of which is fast emerging in importance in tourism studies research), in a discussion of entangled relationships between humans and non-human others. Such entanglement means a rejection of commodification, colonisation, and neoliberalism in favour of an approach to ecotourism that is premised on sensory and sympoietic (complex, self-organising, and boundreyless systems) designed to repair the damaged social and ecological conditions that are so important for the provision of ecotourism. A focus not on cost-benefit analyses, but rather the "desire to be good to one another." This is followed by a chapter on enriching indigenous ecotourism through culturally sensitive universalism by Read and Grimwood. Culturally sensitive universalism is a bridge that spans the fields of global ethics and development ethics and allows an element of moral free space (unique forms of expression in communities) within broader universal ethical frameworks in preserving cultural values. The compelling story of the Franklin Shipwrecks in the Northwest Passage and Inuit empowerment provide an excellent case study by which to investigate these issues local-to-cosmopolitan issues. Thomsen illustrates the value of a posthumanist ethical approach for the entrenched issues and complexities tied to wolf ecotourism. His argument is that agency, rights, and welfare need to bubble to the surface in efforts to mitigate wolf-human conflicts, and that multi-site analyses indicates that a 'one size fits all' will not move the agenda forward.

The indigenous theme continues in the transition from ethics to identities in Theme 2. Graci reminds us that ecotourism, as noted by Read and Grimwood, is often a reflection of both natural and cultural values. Indigenous ownership and operation of ecotourism companies is essential for self-determination and this is represented in two successful Canadian cases studies: Tundra North Tours in Inuvik, Northwest Territories and Spirit Bear Lodge in Klemtu, British Columbia. In these examples are important lessons about authenticity, capacity building, cultural connectedness, education, conservation, intercultural understanding, and economic recovery. The link between ecotourism, nature, culture, and operations is further sustained in work by de Bernardi on the Sámi people of northern Europe in efforts to preserve identity and authenticity. Ecolabels and certification are an important facet of identity, and marketing and promotion of ecotourism programs should protect Sámi culture from unethical use. Furthermore, like so much of the ecotourism literature on Indigenous people, Sámi people need to be involved in decision making.

In Chapter 12 by Chilembwe, identity is represented though enhancement of stewardship from volunteer ecotourism at a wildlife rescue, rehabilitation, and education centre in Malawi. Chilembwe found that there needs to be symmetry between the ethics and values of local people and volunteers, which is often difficult because of the short time that volunteers spend on site. There must be willingness on the part of the latter to be flexible enough to adopt local practices and ethics in avoiding friction between both groups. Munanura and Sabuhoro adopt the language of the UN Sustainable Development Goals with a focused approach on antecedents of livelihoods, livelihood resources, and sustainable wellbeing immersed within a model of constraints. While ecotourism is advanced as a manner with which to strengthen livelihoods, inattention to resilience leads to vulnerabilities that in the end have negative socio-ecological consequences.

The final chapter in this section by Pirnar, argues that female entrepreneurship is riding a wave of global acceptance in lesser developed and develop country contexts. This is especially true in conjunction with ecotourism for purposes of income generation, higher quality of life, higher employment, human capital accumulation, and sustainable and environmental awareness. Personal traits of female entrepreneurs, benefits, and motivations are discussed in providing a helpful baseline of knowledge to advance this under-represented area of ecotourism research.

Section 3 on the three "Cs" (change, conflict, and consumption) includes seven chapters that range from persons with disabilities as a neglected body of research in ecotourism, to ecotourism as a form of luxury consumption, again, showing diversity in how ecotourism is being investigated. Garrod convincingly argues that ecotourists with disabilities have vastly different expectations and motivations than general tourists and face tremendous challenges because of the nature-based settings in which ecotourism takes place. Furthermore, as providers attempt to make settings and attractions more accessible, they need to be mindful of the sustainability imperative that is inherent in ecotourism. As one of the hot buttons of global concern, Day and Noakes argue that climate change will have a considerable impact on the ecotourism industry from polar regions to tropical rainforests. The problem for ecotourism is that mitigation efforts are asymmetrical especially when it comes to managing greenhouse gases. There is a return to Binna Burra Lodge in Australia as a representative example of the types of lessons that can be learned from challenging, and changing, environmental circumstances.

Moving away from a focus on animals and environments stressed because of climate change, Wise investigates animals caught in the crossfire of war and conflict as another area of research that is untapped in ecotourism studies. Wise uses case studies on Mozambique (loss of giraffe and elephant herds from civil war), the Democratic Republic of the Congo (conflict efforts to protect endangered gorillas), Syria (the rescue of zoo animals), and how war either deters or catalyses poaching in Kashmir and Afghanistan. Education of local communities is said to be critical in changing values and attitudes. The conflict sub-theme is sustained through work by Pavelka on anti-ecotourism—a theme which could easily fit in other sections of this volume—not in a universal sense, but rather in reference to the community of Big Horn between Banff and Jasper National Parks in Alberta, Canada. Ecotourism is rejected on the basis of localism and the preservation of a traditional way of life by inhabitants that aims to protect historical cultural values, especially when progressive governments try to impose their will. Trust and cooperation are virtues to be nurtured, not imposed.

A volume of this nature would not be complete without a statement on social media. Indeed, technology of this nature has infiltrated society on so many different levels—from presidents governing by hand-held devices, to primary schoolchildren substituting sandlot football games for multiple hours watching TiKTok videos. Sarkar investigates the dynamics of socialisation and gratification through the use of social media as a major trip motivator. Furthermore, socialisation takes place post-trip as ecotourists share their experiences with others globally. Personal ecotourism experiences turn into globally oriented social experiences as a type of co-creative activity that leads to a collective consciousness, which can act as an agent to promote pro-environmental attitudes and behaviours.

The last two chapters of the section deal with non-traditional ecotourism markets (Do, Weaver, & Lawton) and ecotourism as a form of luxury consumption (Volo & D'Acunto). In the first of these, Do et al. found that Vietnamese ecotourists at Cat Tien National Park conformed to the Western characteristics of ecotourism: enthusiasm for nature-based destination, learning, and sustainability. However, the authors also found that the sample demonstrated soft-path ecotourism and anthropocentrism tendencies as dominant characteristics. Examples include touching plants and animals, visiting protected areas in larger numbers, and wariness of wilderness areas as dark, mysterious, and unsafe. Volo and D'Acunto suggest that although luxury consumption is vastly under-represented as a focus of study in ecotourism, there is clear indication that expectations of luxury have "radically evolved over time" and may be tied to the concept of sustainability in demonstrating legitimacy. Still, there is clear indication that ecotourist behaviours are determined by other values even though these are often immersed in greenwashing which confounds ecotourism from demand and supply perspectives.

The final section on learning and environment emphasises the important link that ecotourism has with not only using the natural world, but importantly, learning about it too, Learning is a core ecotourism criterion. González Herrera and Giralt-Escobar argue that learning about ecotourism can take place in formal or informal settings, and that teaching ecotourism ought to be about creativity along theoretical and methodological lines. Several different theories and models of learning are advanced in this chapter each has their own strengths and weaknesses. The second chapter in this section by Çakar critically analyses environmental governance for ecotourism through a review of 64 peer-reviewed publications. Findings suggest that the ideal governance model has yet to be developed because of gaps in key areas like overtourism and inclusion of the United Nations' Sustainable Development Goals. Even so, Çakar argues that adaptive co-management holds promise to overcome some of the inherent weaknesses in governance through learning and better integration of social, ecological and economic factors.

Burns' chapter on working animals in ecotourism—a topic of research that needs much more emphasis in ecotourism and tourism in general—stretches the interpretation of different venues and attractions that are sites for ecotourism. But such is necessary in articulating the sheer number of animals that continue to be used in the name of ecotourism—from birdwatching at one end of the spectrum, to captives, porters and labourers, and hunting and fishing. Practices that emphasise care and equity must be entertained if ecotourism is to live up to its promise of being more ethical and responsible. Hatipoglu, Denizci, and Imamoğlu demonstrate the tight connection between nature and culture in their case study on Ovacık,

Turkey, by underscoring the positive feedback loop that can take place between biodiversity conservation, community wellbeing, ecotourism, and sustainable agriculture. Aspects of co-ownership, co-management, gender inclusivity, learning, and awareness on the part of eco-tourists, are all part of the recipe for building community values from the preservation of heirloom seeds, growing a variety of different crops, and using of natural fertilizers, while at the same time improving livelihoods of people living in rural areas.

Departure into conservation and biodiversity continues through Elliot's deeply theoretical work on rewilding Europe, in the Netherlands, with the mission to restore altered wilderness areas to original states, reintroduce apex predators, and establish corridors between protected areas. Elliot enlists aspects of biopower, hybrid geographies and neoliberalist agendas through vastly different actors who enable or prevent rewilding based on institutional and instrumental priorities. Ecotourism is the preferred form of tourism to support such an initiative but may be subject to contradictions (e.g., greenwashing) that may hamper the aims of rewilding.

The need for more wild spaces is a theme discussed by Kopnina through the Half-Earth vision. This perspective proposes that stakeholders such as policymakers, industry, scientists, conservationists, and local people, work together to protect 50% of global landscapes and seascapes from intensive human economic activity (e.g., mining). Based on case studies from Mondulkiri, Cambodia, and Vlieland, the Netherlands, Kopnina argues that ecotourism must be a multidimensional process that integrates local economy with conservation, and the interests of nonhuman animals. The final chapter of the book by Stronza, Hunt and Fitzgerald is a summary of 30 years of research that investigates key successes and failures in ecotourism. Trends identified by the authors include misrepresentation of ecotourism as other forms of tourism or outdoor recreation, flawed research designs offering limited scope on ecotourism, and the need for comparative and longitudinal research to advance our thinking around the costs and benefits of ecotourism. Activating these studies now is especially important in view of the pressure that humans are placing on wildlife and ecosystems.

The journey is in no way complete. It never is when it comes to scholarship. But there is true representation of the essence of ecotourism in these pages. If strength comes from adversity and perseverance, the reader is in for a real treat.

References

Duffy, R. (2006). The politics of ecotourism and the developing world. *Journal of Ecotourism*, *5*(1–2), 1–6. doi:10.1080/14724040608668443.

Morrison-Saunders, A., Hughes, M., Pope, J., Douglas, A., & Wessels, J.-A. (2019). Understanding visitor expectations for responsible tourism in an iconic national park: Differences between local and international visitors. *Journal of Ecotourism*, *18*(3), 284–294.

Taylor & Francis. (2021). *Journal of Ecotourism Publishing Report*. London: Routledge.

Taylor, M., Hurst, C. E., Stinson, M. J., & Grimwood, B. S. R. (2020). *Becoming care-full*: Contextualizing moral development among captive elephant volunteer tourists to Thailand. *Journal of Ecotourism*, *19*(2), 113–131. doi:10.1080/14724049.2019.1657125.

THEME 1

Sustainability

1

ECOTOURISM AND THE SUSTAINABLE DEVELOPMENT GOALS

Anna Spenceley and Andrew Rylance[1]

Introduction

In 2015, governments adopted the 2030 Agenda for Sustainable Development and the Sustainable Development Goals (SDGs). The agenda established a global framework to end extreme poverty, fight inequality and injustice and remedy climate change. Building on the Millennium Development Goals (MDGs), 17 Sustainable Development Goals and 169 associated targets were agreed. Tourism has been included within the targets for Goal 8 on decent work and economic growth, Goal 12 on responsible consumption and production, and Goal 14 on life below water. However, tourism has the potential to contribute, directly or indirectly, to all of the goals (UNWTO, 2015).

During 2017, a review was undertaken of 64 countries' Voluntary National Reviews of the SDGs, and corporate social responsibility activities of 60 tourism companies (UNWTO, 2017). The findings included that SDGs 8, 12, and 17 have the strongest links with tourism, but that there are few linking SDGs 3, 4, 7, and 10 with the sector. Furthermore, challenges relate to irresponsible consumption and production, and poor management of resources related to SDGs 12, 14, and 11. Key findings included that policymakers should encourage and support the tourism private sector, and that active engagement and coherent dialogue are required to optimise progress. For the private sector, internalisation of the SDGs relates to their drive towards competitiveness and profitability, rather than philanthropy. Therefore, more inclusive and sustainable business models need to relate to core business activities.

Goal 1: No poverty: End poverty in all its forms everywhere

In 2013, it was estimated that 10.7% of the world's population lived on less than USD$1.90 per day (World Bank, 2016a,b). However, poverty is a multidimensional phenomenon, and it manifests where people have inadequate income, a lack of access to education, poor health, insecurity, low self-confidence, a sense of powerlessness, and where there is an absence of rights (Sen, 1999).

Bennett, Ashley, and Roe (1999) suggested that the tourism sector had the potential to contribute to poverty reduction in developing countries, because the market comes to the

DOI: 10.4324/9781003001768-1

producers, inter-sectoral linkages can be created, it is labour intensive (particularly for women, youth, and people will low-skills), can take place in marginal areas; and it has fewer barriers to entry than manufacturing or other export activities. There has been extensive research on 'pro-poor tourism' (e.g., Ashley, Roe, & Goodwin, 2001; Ashley, Boyd, & Goodwin, 2001; Ashley & Mitchell, 2007; Mitchell & Ashley 2009Mitchell 2009; Spenceley & Meyer, 2016), addressing the opportunity to harness markets for poverty reduction, and tools for doing so. In 2015, tourism generated an estimated USD$1.5 trillion in export earnings (UNWTO, 2017). Scheyvens (2009) estimated that approximately 40% of all international tourist arrivals accrue to developing countries, and so tourism can be a significant foreign exchange earner. Some of the poorest regions of the world are rich cultural and natural assets, which offer great potential as ecotourism attractions.

Ecotourism can provide a mechanism to re-distribute wealth from the rich to the poor, because as tourists travel they spend money on travel, accommodation, excursions, food, drinks and shopping (Spenceley & Meyer, 2015). For example, Pafuri Camp, a luxury ecotourism lodge in South Africa, employs around 52 permanent staff members, and 94% of them are from the local Makuleke community. Employees from the community receive approximately USD$298,000 in wages and related benefits annually, collectively, which makes a substantial impact in the local economy, and contributes to poverty reduction (Snyman & Spenceley, 2019).

Goal 2: Zero hunger: End hunger, achieve food security and improved nutrition, and promote sustainable agriculture

Tourism can catalyse sustainable agriculture by promoting the production and supplies to hotels and restaurants, and also through sales of local products to tourists. For example, agro-tourism can generate additional income for farmers while providing rich and educational tourism experiences (UNWTO & UNDP [United Nations World Tourism Organisation and United Nations Development Programme], 2017). Agriculture and the harvesting of natural resources continues to remain a predominant livelihood opportunity for poor communities working in rural areas, accounting for 55% of employment in developing countries and is the main source of income for the rural poor (Schiere & Kater, 2001). A review of 49 tropical-protected areas showed that they are becoming isolated as deforestation takes place around their boundaries and therefore effective management needs to address the wider local socio-economic developmental issues (Naughton-Treves, Holland, & Brandon, 2005). Diversification strategies are important for poor communities to reduce their dependence and associated risks on a single income stream, such as farming (Ashley, Mdoe, & Reynolds, 2002).

Marine-protected areas used by tourists and fishers provide different opportunities to reduce hunger. For example, small community-managed marine protected can provide (a) a refuge for breeding and nursing populations of fish to support the local subsistence fishing industry, and (b) provide areas for non-consumptive marine tourism (e.g., whale shark viewing and manta ray diving), which provide job opportunities for local people. Such a system is being developed by a luxury ecotourism company, andBeyond, in collaboration with local communities and authorities in Tanzania and Mozambique (Braack & Mearns, 2017). At this destination, efforts have been made to tackle overfishing and the killing of endangered marine species, protect reefs and endangered species, and also build capacity among local communities—regarding the management of local fishing stocks and responsible community fishing practices (Braack & Mearns, 2017).

Goal 3: Good health and wellbeing: Ensure healthy lives and promote wellbeing for all at all ages

There is increasing scientific evidence of the health benefits of protected areas, and 'Healthy Parks Healthy People' was one of the core themes of the IUCN 2014 World Parks Congress in Sydney, Australia (Spenceley, 2017). Visitation to areas of high biodiversity can be a tool in preventative medicine, and provide health benefits caused by certain lifestyle problems, such as obesity, cardiovascular disease, depression, and anxiety (Sparkes & Woods, 2009). In Australia, Parks Victoria formed a partnership with two major players in Australia's health care delivery system, Medibank Australia and the National Heart Foundation. This includes providing health care professionals the option to prescribe physical activity in protected areas as a proactive disease prevention approach, 2017). With the COVID-19 pandemic, the importance of visitation to natural areas has been heightened as a means to relieve the stress of lockdowns (Spenceley, 2021; Spenceley et al, 2021).

Promoting sport tourism is an increasing area of interest for protected areas as both a means of generating revenue to finance conservation efforts as well as demonstrating the wider social contribution of biodiversity to local communities. An excellent example comes from the cross-border tourism adventure products established in southern African transfrontier conservation areas, including the Tour de Tuli (a cross-border mountain biking event), Desert Knights (a cross-border canoe and mountain biking event), and Wildruns (cross-border trail runs) (Spenceley, 2018a).

Corporate social responsibility initiatives by ecotourism operators, working in areas of high biodiversity, and philanthropic efforts, also often contribute towards public health improvements. Reviews of philanthropic efforts demonstrate that tourism enterprises often use their corporate social responsibility funds to support the development of rural clinics, hospitals, or provide them with vital equipment. For example, the Africa Foundation is an organisation that channels donations from andBeyond's guests and donors into social initiatives in communities neighbouring their lodges, such as Phinda Game Reserve in South Africa. Over more than two decades, donations have supported health initiatives in communities neighboring Phinda such as construction of a clinic at Mduku, and refurbishment of another, a series of HIV/AIDS awareness workshops, improved water provision, and a school permaculture project (Snyman & Spenceley, 2019).

Goal 4: Quality education: Ensure inclusive and equitable quality education and promote lifelong learning opportunities for all

A skillful workforce is crucial in order to provide quality hospitality and experiences to tourists. The tourism sector provides professional development and training opportunities for direct and indirect jobs for youth, women, and those with special needs (UNWTO & UNDP, 2017). An example comes from !Xaus Lodge's, in the Kgalagadi Transfrontier Conservation Area. One of the lodge's cleaning staff, Melissa Mienies, joined the lodge soon after it opened and developed a special interest in guiding. Supported by the lodge managers, she taught herself key information (despite having failed her secondary school exams). !Xaus Lodge supported her with a distance-learning course, and she qualified as a guide. She was the first female nature guide in the Kgalagadi who was a member of the local community (Snyman & Spenceley, 2019).

Ecotourism also provides opportunities for environmental education of tourists, and of local community members and youth. For example, the ecotourism operator Wilderness Safaris has an NGO called Children in the Wilderness (CITW), which since 2005 has organised an annual international cycling event within the Mapungubwe Transfrontier Conservation Area called the Tour de Tuli. Between 2005 and 2017, donations from 3770 cyclists participating in the tour raised over USD$1.7 million. At the end of 2016, these funds had been used for over 5600 children to attend a CITW environmental awareness camp at a Wilderness Safaris lodge, for 6000 children to participate in an Eco-Club programme, as well as training over 200 Eco-mentors in six southern African countries (Spenceley, 2017).

Goal 5: Gender equality: Achieve gender equality and empower all women and girls

Research from the World Bank has found that women lag behind men in nearly all measures of economic opportunity in the world (World Bank, 2016a,b) and that the inequalities are most stark in low-income countries (Twining-Ward & Zhou, 2017). However, tourism can empower women, particularly through the provision of direct jobs and income-generation from small, medium, and micro enterprises (SMMEs) in tourism and hospitality-related enterprises (UNWTO & UNDP, 2017). Characteristics of the tourism sector that may explain the strong representation of women in tourism than other sectors include (1) a lower emphasis on formal education and training, greater emphasis on personal and hospitality skills, higher availability of part time and work-from-home options, and an option for entrepreneurship that does not require substantial start-up investment (Twining-Ward & Zhou, 2017).

In some countries, tourism has almost twice as many women employers as other sectors, and in a global study, the ILO has found that women make up between 60 and 70% of hotel labour force (ILO (International Labour Organisation), 2010). Despite this level of representation, they are generally paid 10–15% less than their male equivalents (UNWTO & UN Women, 2010). Women tend to dominate lower-paid jobs, such as clerical and cleaning roles, and are under-represented in higher-paid roles, such as tour guides, chefs, and particularly in management and decision-making positions (Twining-Ward & Zhou, 2017). The World Bank proposes for main strategic thrusts to improve the empowerment of women in tourism, namely (1) improving human development, (2) removing constraints for more and better jobs, (3) removing barriers to women's ownership of and control over assets, and (4) enhancing women's voice and agency (World Bank, 2016a,b).

Some examples of gender equality in ecotourism facilities include in Wolwedans collection of camps and lodges in the NamibRand Nature Reserve of Namibia. These lodges have a 50% female employment, and the same ratio of male-to-female employment is also illustrated at Simien Lodge in Ethopia (Snyman & Spenceley, 2019). There are other examples of women's groups supporting ecotourism facilities, such as the Umoja Women's Cooperative. The co-operative was given training by the Ruzizi Tented Lodge in Rwanda, including with training on tailoring and basket weaving in 2015. This helped them to generated sales of USD$1,700, including through the Akagera Park's curio shop (Snyman & Spenceley, 2019). However, some ecotourism facilities have encountered challenging in retaining female staff. For example, at Anvil Bay lodge in Mozambique, there are attrition rates following maternity leave, or migration to South Africa (Snyman & Spenceley, 2019).

Goal 6: Clean water and sanitation: Ensure availability and sustainable management of water and sanitation for all

Tourism investment requirement for providing utilities can play a critical role in achieving water access and security, as well as hygiene and sanitation for all. The efficient use of water in tourism, pollution control and technology efficiency can be key to safeguarding water (UNWTO & UNDP, 2017).

As an example of ecotourism investment improving water access and security, comes from a community-owned facility managed by a private operator, Covane Lodge in Mozambique. One of the main challenges that the Canhane community have is a lack of access to water. As part of a donor-funded infrastructure re-investment program from the Government of Mozambique's Mozbio project, a pipeline and pump system was installed from the Massingir dam to the community. To ensure sustainability, a community maintenance fund was established, so that people pay for use of the water (World Bank, 2014). Such initiatives can substantially improve the quality of life for the poor in dry areas, particularly for women, livestock owners, and farmers.

Goal 7: Affordable and clean energy: Ensure access to affordable, reliable, sustainable, and modern energy for all

Tourism can accelerate the shift towards increased use of renewable energy, and by promoting investments in clean energy sources, the sector can help to reduce greenhouse gases, mitigate climate change, and contribute to access of energy for all (UNWTO & UNDP, 2017).

While designing tourism infrastructure to integrate low energy use, and installing renewable energy technologies is easiest at the outset, some facilities decide to retrofit their operations to make them more energy efficient. For example, the ecotourism operation of Mombo camp in the Okavango Delta of Botswana switched from diesel generators to 100% solar energy in May 2012, following a capital investment of approximately USD$860,000. It was predicted that this change would lead to 93% reduction in carbon emissions from the camp, which previously relied on a diesel generate for energy. This meant that the lodge would emit only 22 tonnes of carbon dioxide equivalents during the year, compared to an estimated 287 tonnes during their 2012 financial year (Wilderness Holdings, 2014). In Ethiopia, Bale Mountain Lodge is 100% ecofriendly with power coming from its own 25 Kw micro-hydro power plant, biodegradable waste is processed through its bio-gas system to provide cooking gas, and firewood is sourced from sustainable plantations outside the national park (Snyman & Spenceley, 2019).

Goal 8: Decent work and economic growth: Promote sustained, inclusive, and sustainable economic growth; full and productive employment; and decent work for all

Tourism in protected areas represents one of the economic opportunities that can both help achieve the joint objectives of sustainable livelihood development of local communities and biodiversity conservation. Globally, tourism in 2015 accounted for 10% of global gross domestic product (GDP), 7% of global trade, and 1 in 10 jobs (UNWTO & UNDP, 2017). The flow of money from this sector provides opportunities for tourists to act as conduits to redistribute wealth from the rich to the poor. For example, when travellers visit developing countries they spend money on transport, accommodation, excursions, shopping, and on food and drink. Much of this money can be captured by local poor people if they are able to supply the products and services that tourists need, or by being employed in tourism businesses (Spenceley & Meyer, 2012).

At a national level, the U.S. Parks Service estimates that its protected area system contributed to creating 251,600 jobs, USD$9.34 billion in labour income, and USD$16.5 billion in value addition to the national economy in 2012 (Cui, Mahoney, & Herbowicz 2013).

At the local level, reports indicate that the benefits accrued by employees and host communities from tourism vary widely between enterprises and destinations, dependent on the institutional structures, partnership arrangements, and business viability of a particular venture (Dedeke, 2017; Spenceley, 2008). Emerton and Tessema (2001) argue that the livelihood benefits from tourism are not always sufficient to make up for the costs of living with wildlife, particularly when the costs and benefits are unequally distributed between people. Arguably, the greater livelihood benefits that are obtained, the more widely that wildlife in protected areas is appreciated and conserved by local people (Arntzen et al., 2003). It is often argued in the literature that tourism in conservation areas employs a large number of expatriates and limits employment of local people to menial jobs (e.g., Mbaiwa, 2003, 2005; Barber & Pittaway, 2000). However, Snyman and Spenceley's (2019) case studies of ecotourism facilities in Africa consistently illustrate more local people being employed than expatriate, and that they are empowered to fulfil management roles. Here the enterprises reviewed employed 1592 people, and at least 1216 (76%) of these came from local communities. Furthermore, collectively, the tourism businesses spent USD$8.5 million in 2017 on wages for local people.

Goal 9: Industry, innovation, and infrastructure: Build resilient infrastructure, promote inclusive and sustainable industrialisation, and foster innovation

Tourism development relies on good public and private infrastructure, and the sector can influence public policy for improvements that attract tourists, foreign investment (UNWTO & UNDP, 2017), while also supporting local communities.

Impact management approaches have been reviewed extensively by Buckley (2004, 2009, 2011, 2012), and can differ greatly in scale. Technologies for sewage and wastewater treatment, for example, may range from small-scale composting toilets for low-visitation infrastructure in warm, moist climates, to multi-stage industrial sewage treatment systems with artificial wetland and ponds, appropriate for infrastructure with high visitor volumes. Furthermore, there are numerous publications on sustainable design, and best practices in ecotourism infrastructure. For example, The International Finance Corporation's "Ecolodges: exploring opportunities for sustainable business" (IFC, 2005) provides background on the ecolodge marketplace (including what tourists are looking for), the business case and financial viability issues, and an overview of the potential positive and negative impacts on the environment and local communities.

Goal 10: Reduced inequalities: Reduce inequality within and among countries

Tourism can be a powerful tool for reducing inequalities if it engages local populations and all key stakeholders in its development, and can also contribute to urban renewal and rural development by giving people the opportunity to prosper in their place of origin (UNWTO & UNDP, 2017). In low- and middle-income countries, tourism can generate substantial foreign exchange. Estimates suggest that tourism can contribute up to 40% of GDP in less developed countries compared to 10% of GDP in more economically advanced countries (Sofield, De Lacy, Lipman, & Daughety, 2004).

Tourism has the potential to negatively impact on communities from social, environmental, and economic perspectives (Ashley et al., 2000; Diaz, 2001; Koea, 1977). As a result, tourism has often been promoted as an opportunity to both achieve livelihood diversification and poverty reduction, but also acknowledging that tourism can generate negative impacts for the poor, such as displacement, inflation of prices of local products, and increased competition with which local community businesses may not be able to compete (Roe & Urquhart, 2002). For example, a survey of 17 marine-protected areas (MPAs) in Thailand identified that local communities believed to receive negligible benefits for tourism livelihoods (Bennett and Dearden, 2014). It is therefore important to consider the negative influences of tourism (or lack of benefits) alongside the positive benefits.

Goal 11: Sustainable cities and communities: Make cities and human settlements inclusive, safe, resilient, and sustainable

Over recent decades, population sizes in urban areas have boomed and are expected to continue to grow by 61% by 2030. It is predicted that the volume of people living in cities will rise to 5 billion by 2030 (UNWTO, 2012). The growth of the tourism sector can provide employment opportunities for these bulging populations. Tourism can help to advance urban infrastructure and accessibility, promote regeneration, and preserve cultural and natural heritage, assets on which tourism depends. Furthermore, investment in green infrastructure (more efficient transport, reduced air pollution) should result in smarter and greener cities for not only residents but also tourists (UNWTO & UNDP, 2017).

In rural areas, ecotourism can contribute towards making rural communities safe, resilient, and sustainable. For example, philanthropic efforts by ecotourism facilities (e.g., andBeyond's Africa Foundation, Wilderness Safaris' Children in the Wilderness and Wilderness Wildlife Trust, and the Singita Community Development Trust) have generated substantial impacts in rural areas over long periods of time, particularly in the realms of education, health, and small business development (Snyman & Spenceley, 2019).

Goal 12: Responsible consumption and production: Ensure sustainable consumption and production patterns

The tourism sector can adopt sustainable consumption and production modes, and so accelerate the shift towards sustainability (UNWTO & UNDP, 2017). The UNWTO-led 10 Year Framework Program on Sustainable Consumption and Production's Sustainable Tourism Program aims to catalyse change in tourism, and promote transformation for sustainability (UNEP, n.d.).

For example, a program on the Great Barrier Reef in Australia encourages tourism companies (e.g., accommodation providers, tour operators) working in protected areas to operate sustainably. The Marine Park Authority preferentially promotes operators that are independently certified by recognised environmental certification schemes, such as EarthCheck and Ecotourism Australia (Great Barrier Reef Marine Park Authority [GBRMPA], 2018). Furthermore, protected area managers in Australia reward and encourage tour operators to become certified through longer licenses, exclusive access to sensitive sites, and promotional opportunities (R. Hillman, chief executive, Ecotourism Australia, pers. comm., 11 April 2016). However, the level of uptake of certification in the tourism sector is low. A study undertaken in Africa established that only around 3% of hotels had been independently established as operating sustainably (Spenceley, 2016). So, clearly, more needs to be done to mainstream monitoring and reporting of responsible consumption and production in the tourism sector.

Goal 13: Climate action: Take urgent action to combat climate change and its impacts

Tourism contributes to and is affected by climate change. By reducing its carbon footprint in the transport and accommodation sector, tourism can benefit from low carbon growth and help tackle one of the most pressing challenges of our time (UNWTO & UNDP, 2017).

The World Economic Forum suggests a series of options that the tourism industry could apply to mitigate GHG emissions that relate to land and air transport, water transport, and accommodation. For example, specific actions that tourism enterprises can undertake include hotel refurbishment to support the highest degree of energy-efficient heating, cooling, lighting, and building technology through incentives for energy-efficient investments or mandatory energy efficiency certificates (Chiesa & Gautam, 2009).

Energy saving options for ecotourism facilities include establishing environmental management systems, reducing energy use, using only renewable energy, reusing materials (e.g., packaging), recycling waste, using local food (with lower transport impacts), and constructing low-carbon buildings from recycled materials with high levels of insulation (Simpson, Gössling, Scott, Hall, & Gladin, 2008). The ecotourism company Wilderness Safaris reduced its carbon emissions per bednight by 11% between 2014 and 2018. By December 2017, 12 Wilderness camps that were totally powered by solar panels were producing 4740 kWh per day of usable energy. A further 24 camps had solar/battery-inverter hybrid systems. The company also had 861 solar geysers or solar thermodynamic geysers which reduce their reliance on diesel generators or electricity, and saving a total of 3444 kWh per day (Wilderness Holdings, 2018).

Goal 14: Life below water: Conserve and sustainably use the oceans, seas, and marine resources for sustainable development

Since the 1960s there has been a 10-fold increase in the number of Protected Areas (PAs) (Bishop et al., 2006). In 2016, there were 202,467 terrestrial and inland water PAs recorded in the World Database on Protected Areas (WDPA), covering 14.7% (19.8 million km^2) of the world's extent of ecosystems (excluding Antarctica) (UNEP–WCMC & IUCN [United Nations Environment Programme–World Conservation Monitoring Centre & International Union for Conservation of Nature], 2016).

Coastal and maritime tourism rely on healthy marine ecosystems, and tourism development must help conserve and preserve fragile marine ecosystems and serve as a vehicle to promote a blue economy, contributing to the sustainable use of marine resources (UNWTO & UNDP, 2017). For example, research in Guam found that asking diving tourists to watch their buoyancy, and avoid touching coral reefs led to a 75% reduction in accidental contacts with the reefs, so reducing damage to these highly sensitive systems (Williams & Raymundo, 2017). In coastal areas, ecotourism facilities must ensure safe solid and liquid waste disposal, and improve the collection, safe disposal, and recycling of waste.

As an example, Chumbe Island is an ecolodge in Tanzania, based within a marine park. Their work is characterised by a high regard for the natural, social, and cultural environment, and the lodge has received a series of prestigious awards (Olearnik & Barwicka, 2019). Since its establishment, the Reef Sanctuary has become one of the most pristine coral reefs in the region, with over 470 fish species and 200 species of hard coral, 90% of all recorded in East Africa. In 2005, researchers interviewed artisanal fishers, 94% confirmed the so-called spillover effect of the Reef Sanctuary by reporting increased yields in the vicinity (Snyman & Spenceley, 2019).

Goal 15: Life on land: Protect, restore and promote sustainable use of terrestrial ecosystems, sustainably manage forests

Rich biodiversity and natural heritage are often the main reasons why tourists visit a destination. Tourism can play a major role if sustainably managed in fragile zones, not only in conserving and preserving biodiversity, but also in generating revenue as an alternative livelihood to local communities (UNWTO & UNDP, 2017).

The ecotourism facility Nkwichi Lodge in Mozambique, created a 120,000-hectare new community conservation area that coordinated efforts of 16 communities to help regulate land use, and in particular, to stop hunting. At Phinda Private Reserve in South Africa, andBeyond restored cattle farming and exotic tree species farmlands, and re-stocked it with 15,000 head of game. In Rwanda, Bisate Lodge has established an extensive reforestation programme. Using seed and other material gathered only from outside the park, an indigenous tree nursery was established and more than 15,000 trees were planted between 2015 and 2017. A small permanent staff oversees the nursery and day-to-day planting, while up to 20 community members are also regularly employed on a casual basis to assist with reforestation during peak planting times. Trees have also been donated to reforest other areas around Volcanoes National Park (Snyman & Spenceley, 2019).

Goal 16: Peace, justice, and strong institutions: Promote peaceful and inclusive societies for sustainable development; provide access to justice for all; and build effective, accountable, and inclusive institutions at all levels

As tourism revolves around billions of encounters between people of diverse cultural backgrounds, the sector can foster multicultural and inter-faith tolerance and understanding, laying the foundation for more peaceful societies. Tourism, which benefits and engages local communities, can also consolidate peace in post-conflict societies (UNWTO & UNDP, 2017).

Transfrontier Conservation Areas (TFCAs) are described as relatively large areas encompassing one or more protected areas which straddle frontiers between one or more countries (World Bank, 1996), and are sometimes called 'peace parks'. By contrast to national parks, TFCAs have the potential to conserve a greater diversity of species within larger geographical areas and to promote cooperative wildlife management between countries (World Wildlife Fund, 1999). TFCAs may also improve opportunities for tourism, by allowing visitors to disperse over greater areas and obtain better quality experiences (Singh, 1999). In southern Africa, there have been extensive initiatives to establish transboundary tourism that support inclusive approaches to tourism across international borders. These include the development of guidelines for tourism concession in TFCAs (Spenceley, 2014) and also for cross-border tourism products (Spenceley, 2018a). These tools seek to establish transparent and well-governed processes for tourism investment in protected areas that benefit conservation and host communities.

Goal 17: Partnerships for the goals: Strengthen the means of implementation and revitalise the global partnership for sustainable development

Due to its cross-sectoral nature, tourism has the ability to strengthen private/public partnerships and engage multiple stakeholders—international, national, regional, and local—to work together to achieve the SDGs and other common goals (UNWTO & UNDP, 2017).

In protected areas that are attractive to tourists, and which present a commercially viable opportunity, authorities have been able to establish tourism concessions, or public private partnerships. These can be used by protected area authorities to spread the commercial risk associated with high value capital investment for infrastructure (e.g., luxury lodges, restaurants). In some instances, these partnerships can contribute meaningful revenue to protected area budgets, which in turn contribute towards funding for conservation management (Spenceley, Snyman, & Eagles, 2017). Funding for tourism partnerships, including joint ventures between private operators and rural communities, can be sourced from development banks, aid agencies, impact investment vehicles, NGOs, government grants, and private investment (Snyman & Spenceley, 2019).

In Australia, the Great Barrier Reef Authority has a partnership with tour operators to collect a marine conservation fee (an Environmental Management Charge). This fee was used to help finance conservation of the MPA. Although when introduced, the fee faced some controversy, following a decade of use and adaptation, the fee was well established and widely accepted (Skeat & Skeat, 2007).

Moving forward

This chapter has shown clearly that ecotourism can contribute to all of the SDGs, with individual ecolodges, and ecotourism companies, making substantial contributions at the local level. However, if there were a sufficient data set, it would be of value to systematically calculate the total contribution of ecotourism to each SDG. Compiling such a synthesis would be a massive undertaking though.

Recommendations from the UNWTO on improving achievement of the SDGs include to strengthen cooperation, and multi-stakeholder partnerships; design and implement incentives and smart subsidies; raise awareness of business opportunities created by the SDGs; and align development cooperation programs with the needs and priorities of developing countries, and also share experiences, good practices and lessons learned (UNWTO & UNDP, 2017). If such actions could be taken by all ecotourism operations globally, then this form of tourism could demonstrate its substantial contribution to sustainable development.

Note

1 This chapter is a substantially revised and reworked version of Spenceley, A., & Rylance, A. (2019). The contribution of tourism to achieving the United Nations Sustainable Development Goals. In S. F. McCool, & K. Bosak (Eds.), *A research agenda for sustainable tourism*. Cheltenham, UK and Northampton, USA: Edward Elgar Publishing.

References

Adiyia, B., Stoffelena, A., Jennesa, B., Vanneste, D., & Ahebwa, W. (2015). Analysing governance in tourism value chains to reshape the tourist bubble in developing countries: The case of cultural tourism in Uganda. *Journal of Ecotourism*, *14*(2–3), 111–129.

Ashley, C., Boyd, C., & Goodwin, H. (2000). *Pro-Poor Tourism: Putting Poverty at the Heart of the Tourism Agenda*. Natural Resource Perspectives No. 51. London: Overseas Development Institute.

Ashley, C. & Mitchell, J. (2007). *Assessing How Tourism Revenues Reach the Poor (ODI Briefing Paper No. 21)*. London: Overseas Development Institute.

Ashley, C., Roe, D., & Goodwin, H. (2001). Pro-poor tourism strategies. Making tourism work for the poor: a review of experience. Pro-poor tourism report No. 1, ODI/IIED/CRT, The Russell Press.

Ashley, C., Mdoe, N., & Reynolds, L. (2002). *Rethinking Wildlife for Livelihoods and Diversification in Rural Tanzania: A Case Study from Northern Selous.* Working Paper No. 15. London: Overseas Development Institute.

Ashley, C. (2009). *Harnessing Core Business for Development Impact: Evolving Ideas and Issues for Action.* London: Overseas Development Institute.

Arntzen, Jaap W., Molokomme, D. L., Terry, E. M., Moleele, N., Tshosa, O., & Mazambani, D. (2003). Main findings of the review of CBNRM in Botswana, CBNRM Support Programme Occasional Paper, No. 14, IUCN, https://portals.iucn.org/library/node/8629.

Barber, N., & Pittaway, L. (2000) Expatriate recruitment in South East Asia: Dilemma or opportunity? *International Journal of Contemporary Hospitality Management, 12*(6), 352–359. Retrieved from https://doi.org/10.1108/09596110010343530.

Bennett, N., & Dearden, P. (2014). Why local people do not support conservation: Community perceptions of marine protected area livelihood impacts, governance and management in Thailand. *Marine Policy, 44,* 107–116.

Bennett, O., Ashley, C., & Roe, D. (1999). *Sustainable Tourism and Poverty Elimination Study: A Report to the Department for International Development.* London: Deloitte & Touche.

Bishop, J., Emerton, L., & Lee, T. (2006). Sustainable financing of protected areas : a global review of challenges and options, IUCN Best Practice Guidelines series no. 13. https://portals.iucn.org/library/node/8800.

Braack, J., & Mearns, K. (2017). Oceans without borders. Presentation at the Conference on Sustainable Tourism in Small Island Developing States, 23–24 November, University of Seychelles, Mahe, The Seychelles.

Britton, S. (1991). Tourism, capital, and place: Towards a critical geography of tourism. *Environment and Planning D: Society & Space, 9*(4), 451–478.

Buckley, R. C. (ed.). (2004). *Environmental Impacts of Ecotourism.* Wallingford: CABI.

Buckley, R. C. (2009). *Ecotourism: Principles and Practices.* Wallingford: CABI.

Buckley, R. C. (2011). Tourism and environment. *Annual Review of Environment and Resources, 36,* 397–416.

Buckley, R. C. (2012). Sustainable tourism: Research and reality. *Annals of Tourism Research, 39*(2), 528–546. Retrieved from https://doi.org/10.1126/science.344.6182.358-b.

Buckley, R. C. (2013). Defining ecotourism: Consensus on core disagreement on detail. In R. Ballantyne, & J. Packer (Eds.), *International handbook on ecotourism.* Cheltenham:Edward Egar.

Chen, C., & Rothschild, R. (2010). An application of hedonic pricing analysis to the case of hotel rooms in Taipei. *Tourism Economics, 16*(3), 685–694.

Chiesa, T., & Gautam, A. (2009). *Towards a Low Carbon Travel and Tourism Sector.* Geneva: World Economic Forum.

Cui, Y., Mahoney, E., & Herbowicz, T. (2013). *Economic Benefits to Local Communities from National Park Visitation.* Natural Resource Report NPS/NRSS/EQD/NRTR—2013/631. Colorado: US National Park Service.

Dedeke, A. (2017). Creating sustainable tourism ventures in protected areas: An actor-network theory analysis. *Tourism Management, 61,* 161–172.

Department of National Parks. (2014). *Strategic Plan 2015–2019.* Australia: Queensland Government.

Diaz, D. (2001). *The Viability and Sustainability of International Tourism in Developing Countries.* Geneva: World Trade Organization.

Eagles, P. F. J., McCool, S. F., & Haynes, C. (2002). *Sustainable Tourism in Protected Areas: Guidelines for Planning and Management.* Best Practice Protected Area Guidelines Series No. 8. Gland. Gland, Switzerland: IUCN.

Emerton, L., & Tessema, Y. (2001). Marine protected areas : the case of Kisite Marine National Park and Mpunguti Marine National Reserve, Kenya, IUCN.

Great Barrier Reef Marine Park Authority (GBRMPA). (2018). Choosing a High Standard Tourism Operation. Retrieved from http://www.gbrmpa.gov.au/visit-the-reef/choose-a-high-standard-operator.

GSTC (Global Sustainable Tourism Council). (2017). What Is the GSTC? Retrieved from https://www.gstcouncil.org/about/about-us/ (accessed on 02.08.2018).

HPHP (Healthy Parks Healthy People). (2017). Healthy Parks Healthy People Central. Retrieved from http://www.hphpcentral.com (accessed on 15.02.2017).

IFC. (2005). Ecolodges: Exploring Opportunities for Sustainable Business. Retrieved from https://www.ifc.org/wps/wcm/connect/topics_ext_content/ifc_external_corporate_site/sustainability-at-ifc/publications/publications_report_ecolodges__wci__1319576869279.

ILO (International Labour Organization). (2010). *Developments and Challenges in the Hospitality and Tourism Sector – Issues Paper for Discussion at the Global Dialogue Forum, for Hotels, Catering, and Tourism.* Geneva: International Labour Organization.

Koea, A. (1977). Polynesian migration to New Zealand. In B. Finney, & A. Watson (Eds.), *A new kind of sugar: Tourism in the Pacific* (pp. 69–69). Santa Cruz: Centre for South Pacific Studies, University of California.

Leung, Y.-F., Spenceley, A., Hvenegaard, G., & Buckley, R. (2018). *Tourism and Visitor Management in Protected Areas: Guidelines for Sustainability.* Best Practice Protected Area Guideline Series No. 27. Geneva: IUCN.

Mbaiwa, J. E. (2003). The socio-economic and environmental impacts of tourism development on the Okavango Delta, north western Botswana. *Journal of Arid Environments, 54*(2), 447–467.

Mbaiwa, J. E. (2005). The socio-cultural impacts of tourism development in the Okavango Delta, Botswana. *Journal of Tourism and Cultural Change,* 2(3), 163–185.

Meyer, D. (2010). Pro-poor tourism: Can tourism contribute to poverty reduction in less economically developed countries? In S. Cole, & N. Morgan (Eds.), *Tourism and inequality: Problems and prospects* (pp. 164–182). London: CABI.

Mitchell, J., & Ashley, C. (2009). *Value Chain Analysis and Poverty Reduction at Scale (ODI Working Paper No. 49).* London: Overseas Development Institute.

Naughton-Treves, L., Holland, M., & Brandon, K. (2005). The role of protected areas in conserving biodiversity and sustaining local livelihoods. *Annual Review of Environment and Resources, 30,* 219–252.

Olearnik, J., & Barwicka, K. (2019). Chumbe Island Coral Park (Tanzania) as a model of an exemplary ecotourism enterprise. *Journal of Ecotourism, 19*(4), 373–387.

Rathnayake, R. (2016). Economic values for recreational planning at Horton Plains National Park, Sri Lanka. *Tourism Geographies, 18*(2), 213–232.

Roe, D., & Urquhart, P. (2002). *Pro-Poor Tourism: Harnessing the World's Largest Industry for the World's Poor.* London: IIED.

Rylance, A., & Spenceley, A. (2013a). *Having the Perseverance and Confidence In Yourself to Make the Right Business Decisions, Case Study: Mantovani Guest House.* Geneva: ILO.

Rylance, A., & Spenceley, A. (2013b). *Female-Owned Tourism Businesses in Lesotho, Case Study: Mokhotlong Hotel and Motlejoa Guest House.* Geneva: ILO.

Saarinen, J., & Rogerson, C. (2014). Tourism and the millennium development goals: Perspectives beyond 2015. *Tourism Geographies, 16*(1), 23–30.

Scheyvens, R. (2009). Pro-poor tourism: Is there value beyond the rhetoric? *Tourism Recreation Research, 34*(2), 191–196.

Schiere, H., & Kater, L. (2001). *Mixed Crop- Livestock Farming: A Review of Traditional Technologies on Literature and Field Experience.* Rome: FAO.

Sen, A. (1999). *Development as Freedom.* Oxford: Oxford University Press.

Simpson, M. C., Gössling, S., Scott, D., Hall, C. M., & Gladin, E. (2008). *Climate Change Adaptation and Mitigation in the Tourism Sector: Frameworks, Tools and Practices.* Paris: UNEP, University of Oxford, UNWTO, WMO.

Singh, J. (1999). *Study on the Development of Transboundary Natural Resource Management Areas in Southern Africa – Global Review: Lessons Learned.* Biodiversity Support Program, Reference No. 59. Washington, DC: World Wildlife Fund. Cited in Spenceley, A. (2007). Tourism in the Great Limpopo Transfrontier Park. *Development Southern Africa, 23*(5), 649–669.

Skeat, A., & Skeat, H. (2007). Tourism on the Great Barrier Reef: A partnership approach. In R. Bushell, & P. F. J. Eagles (Eds.), *Tourism and protected areas: Benefits beyond boundaries: The Vth IUCN World Parks Congress* (pp. 315–328). Wallingford: CABI.

Snyman, S., & Spenceley, A. (2019). *Private Sector Tourism in Conservation Areas in Africa.* Wallingford: CABI.

Sofield, T., De Lacy, T., Lipman, G., & Daughety, S. (2004). *Sustainable Tourism – Eliminating Poverty (ST–EP): An Overview.* Tasmania: Australian Cooperative Research Centre for Sustainable Tourism.

Sparkes, C., & Woods, C. (2009). *Linking People to Landscape: The Benefit of Sustainable Travel in Countryside Recreation and Tourism*. England: East of England Development Agency.

Spenceley, A. (2008). Impacts of wildlife tourism on rural livelihoods in Southern Africa. In A. Spenceley (Ed.), *Responsible tourism: Critical issues for conservation and development* (pp. 159–186). Sterling, VA: Earthscan.

Spenceley, A. (2014). *Tourism Concession Guidelines for Transfrontier Conservation Areas in SADC*. Botswana: GIZ/SADC.

Spenceley, A. (2016). *Green Certification in the Tourism Sector in Africa: Monitoring Water and Waste*. Abidjan: African Development Bank.

Spenceley, A. (2017). Tourism and protected areas: Comparing the 2003 and 2014 IUCN World Parks Congress. *Tourism and Hospitality Research*, *17*(1), 8–23.

Spenceley, A., Snyman, S., & Eagles, P. (2017). Guidelines for Tourism Partnerships and Concessions for Protected Areas: Generating Sustainable Revenues for Conservation and Development. Report to the CBD. Available at https://www.cbd.int/tourism/doc/tourism-partnerships-protected-areas-web.pdf.

Spenceley, A. (2018a). *SADC Guidelines for Cross-Border Tourism Products*. Report to GIZ, 26 March.

Spenceley, A. (2018b). Tourism in Africa and the Sustainable Development Goals. Presentation at the International Conference on Digitalisation and Sustainable Tourism, May, Le Meridien, Ponte au Piments, Mauritius.

Spenceley, A. (2018c). Sustainable tourism certification in the African hotel sector. *Tourism Review*, *74*(2).

Spenceley, A., Bashain, A., & Saini, A. (2011). *Design of Environmental Good Practice Guidelines for the Mauritius Hotel Industry*. Handbook and Implementation Plan, Project reference: X/MRT/027. Port Louis, Mauritius: Commonwealth Secretariat and AHRIM.

Spenceley, A., & Meyer, D. (2012). Tourism and poverty reduction: Theory and practice in less economically developed countries. *Journal of Sustainable Tourism*, *20*(3), 297–317.

Spenceley, A., & Meyer, D. (2016). *Tourism and Poverty Reduction: Principles and Impacts in Developing Countries*. London and New York: Routledge.

Tewes-Gradl, C., Van Gaalen, M., & Pirzer, C. (2014). *Destination Mutual Benefit: A Guide to Inclusive Business in Tourism*. Frankfurt: GIZ.

Twining-Ward, L., & Zhou, V. (2017). *Tourism for Development: Women and Tourism: Designing for Inclusion*. Washington, DC: World Bank.

UNEP. (n.d.). 10 YFP Sustainable Tourism Programme'. Retrieved from http://web.unep.org/10yfp/programmes/sustainable-tourism-programme.

UNEP–WCMC & IUCN (United Nations Environment Programme–World Conservation Monitoring Centre & International Union for Conservation of Nature). (2016). *Protected Planet Report 2016*. Cambridge, UK and Gland, Switzerland: UNEP-WCMC and IUCN.

UNWTO & UN Women (United Nations World Tourism Organization & UN Women). (2010). *Global Report on Women in Tourism*. Madrid: UNWTO.

UNWTO & UNDP (United Nations World Tourism Organization and United Nations Development Programme). (2017). *Tourism and the Sustainable Development Goals – Journey to 2030*. Madrid: UNWTO.

UNWTO (United Nations World Tourism Organization). (2002). The British Ecotourism Market. Retrieved from http://sdt.unwto.org/content/ecotourism-and-protected-areas.

UNWTO (United Nations World Tourism Organization). (2009). *From Davos to Copenhagen and Beyond: Advancing Tourism's Response to Climate Change*. UNWTO Background Paper. Madrid: UNWTO.

UNWTO (United Nations World Tourism Organization). (2012). Global Report on City Tourism—Cities 2012 Project'. AM Report No. 6. Madrid: UNWTO.

UNWTO (United Nations World Tourism Organization). (2015). *Tourism and the Sustainable Development Goals*. Madrid: UNWTO.

UNWTO (United Nations World Tourism Organization). (2017a). Tourism and the Sustainable Development Goals – Journey to 2030, Highlights. Retrieved from http://publications.unwto.org/publi cation/tourism-and-sustainable-development-goals-journey-2030 (accessed on 14.04.2018).

UNWTO (United Nations World Tourism Organization). (2017b). UNWTO Tourism Highlights, 2017 Edition. Retrieved from https://www.e-unwto.org/doi/book/10.18111/9789284419029 (accessed on 14.04.2018).

UNWTO (United Nations World Tourism Organization). (2017c). *UNWTO Annual Report 2016*. Madrid: UNWTO.

Wilderness Holdings. (2014). Integrated Annual Report for the Year Ended 28 February 2014. Wilderness Holdings, South Africa and Botswana.

Wilderness Holdings. (2018). Integrated Annual Report for the Year Ended 28 February 2018.https://africanfinancials.com/document/bw-wild-2018-ar-00/

Williams, A., & Raymundo, L. (2017). Ask and you shall receive: Reducing diver impacts on Guam's coral reefs with a coral-safe diving reminder. Presentation at the Sustainable Tourism in SIDS conference, 24–25 November, The Seychelles.

World Bank. (1996). *Mozambique, Transfrontier Conservation Areas Pilot and Institutional Strengthening Project.* Maputo: Global Environment Facility Project Document.

World Bank. (2014). Mozambique, Transfrontier Conservation Areas and Tourism Development Project (TFCATDP) Project Implementation Support Mission. 4–14 February, aide memoire

World Bank. (2016a). Taking on Inequality, Poverty and Shared Prosperity. Retrieved from http://www.worldbank.org/en/publication/poverty-and-shared-prosperity (accessed on 14.04.2018).

World Bank. (2016b). *Gender Strategy, 2016–2023: Gender Equality, Poverty Reduction and Inclusive Growth.* Washington, DC: World Bank Group. Retrieved from http://documents.worldbank.org/curated/en/820851467992505410/pdf/102114-REVISED-PUBLIC-WBG-Gender-Strategy.pdf.

World Wildlife Fund. (1999). *Study on the Development of Transboundary Natural Resource Management Areas in Southern Africa: Highlights and Findings.* Washington DC: World Wildlife Fund.

Wroughton, L. (2008). *More People Living Below Poverty Line.* Reuters. Retrieved from http://www.reuters.com/article/2008/08/26/idUSN26384266 (accessed on 28.12.2011).

WTO (World Trade Organization). (2002). International Ecolodge Guidelines. Retrieved from https://www.e-unwto.org/doi/book/10.18111/9789284405480.

WTTC (World Travel and Tourism Council). (2009). Leading the Challenge on Climate Change. Retrieved from http://www.wttc.org/eng/Tourism_Initiatives/Environment_Initiative/.

2

ECOTOURISM, REGENERATIVE TOURISM, AND THE CIRCULAR ECONOMY

Emerging trends and ecotourism

Jonathon Day, Sandra Sydnor, Maria Marshall, and Steve Noakes

Ecotourism organisations have often been at the cutting edge of new trends in sustainability. They were among the first tourism organisations adopting sustainability, they have developed techniques to operationalise key principles, and they have been quick to adopt new techniques to achieve their objectives. This chapter will examine three trends that seem likely to influence ecotourism in the coming years. These three interrelated trends are: the rise of the circular economy, the emerging focus on regenerative tourism, and the regenerative recovery. Each of these concepts is closely aligned with ecotourism principles. While some ecotourism organisations may already be applying these principles, each provides new opportunities for ecotourism products to improve their operational effectiveness.

Beyond sustainability

Sustainable tourism has been broadly adopted by policymakers and academics. In the years since the Brundtland report (WCED, 1987), the Rio Earth Summit and the development of the Agenda 21 framework (1992), and the development of the Agenda 21 for the Travel and Tourism Industry (UNWTO, 1997), sustainable tourism has been embraced by policymakers, practitioners, and academics. Moyle, Moyle, Ruhanen, Weaver, and Hadinejad (2020) note that sustainable tourism has become the "dominant discourse in academic, businesses, policy and governance" (p. 106). Despite the attention given to sustainable tourism, there is a growing concern that sustainable tourism is not delivering on the promise. Ruhanen, Moyle, and Moyle (2019) note that "both researchers and policymakers consistently question the effectiveness of sustainable tourism and its practices, applications and practical adaptation" (Ruhanen et al., 2019, p. 139). Indeed, the execution of sustainable tourism programs within destinations has proven to be challenging (Dodds & Butler, 2010; Maxim, 2014). There are several reasons proposed for the lack of progress in sustainable tourism. One factor is the "wicked nature" of sustainability (Day, 2020). It is a complex activity that incorporates a wide range of tasks to be undertaken over a long period of time. Progress can be uneven and difficult to measure.

DOI: 10.4324/9781003001768-2

Concurrent with the general growth of sustainable tourism discourse is the development of concepts built on a foundation of sustainable development for tourism principles. These new approaches, built on the foundations of sustainable development principles, emerge to focus attention on specific aspects of the complex task. For example, Responsible Tourism, popularised by Goodwin (2011), focuses on corporate and individual responsibility for implementing sustainable principles in tourism. Fair Tourism, emerging in South Korea, is a new derivation of sustainable tourism. Geotourism, popularised by National Geographic, places a focus on "all aspects of the geographical character" (of a destination), as well as environmental and cultural responsibility (National Geographic, 2020). Slow Tourism advocates the principles of the slow food movement to improve tourism sustainability. These new expressions of sustainability often provide focus across the wide range of activities required to fully implement sustainability programs in destinations and tourism organisations. At the same time, emerging research streams have begun to address issues associated with sustainability. Some of these topics are substantial; for example, there is a significant body of work addressing resilience in the tourism system both at a business and a destination level. Emerging areas of study, including the circular economy and the regenerative movement, influence—and are influenced by—sustainability.

To understand the relationship between ecotourism and these emerging trends, it is worthwhile to review ecotourism's relationship to sustainable tourism. At the turn of the century, a group of industry leaders decided on the "Mohonk Agreement—A Framework for the Certification of Sustainable and Ecotourism" from GEN (2000), a document that outlines the relationship between sustainable tourism and ecotourism. The agreement recognises that sustainable tourism principles can be applied to any destination or organisation and describes ecotourism as a subset of sustainable tourism. Indeed, ecotourism has been an important focus of sustainable development through tourism for over 40 years. Much of the research on sustainable tourism conducted in the 1990s focused on small-scale, nature-based tourism, or ecotourism (Ruhanen, Weiler, Moyle, & McLennan, 2015), and even today, ecotourism remains one of the key themes in sustainable tourism (Buckley, 2012).

Following a review of ecotourism definitions in academic articles, Fennell (2001) has described ecotourism as tourism that tends to take place in natural areas, is concerned with conservation, is respectful of local culture, and endeavors to ensure that the benefits of tourism accrue to local people. Consistent with those findings, ecotourism is defined by Ecotourism Australia (EA)—one of the oldest ecotourism organisations—as "ecologically sustainable tourism with a primary focus on experiencing natural areas that fosters environmental and cultural understanding, appreciation and conservation" (EA, 2020). The Global Ecotourism Network (GEN), a leading network of ecotourism organisations, states that ecotourism is "responsible travel to natural areas that conserves the environment, sustains the well-being of the local people, and creates knowledge and understanding through interpretation and education for all involved: Visitors, staff, and the visited" (GEN, 2016). GEN elaborates that the principles of ecotourism are to:

- Produce direct financial benefits for conservation
- Generate financial benefits for both local people and private industry
- Deliver memorable interpretative experiences to visitors that help raise sensitivity to host countries' political, environmental, and social climate
- Design, construct, and operate low-impact facilities
- Minimise physical, social, behavioural, and psychological impacts on fauna and flora
- Recognise the rights and spiritual beliefs of indigenous and local peoples and work in partnership to create empowerment (GEN, 2016)

Of course, while it is acknowledged that not every product that promotes itself as an "ecotourism" product meets these criteria, they are still a worthwhile foundation to explore regenerative tourism, regenerative recovery, and the circular economy.

A brief primer on the circular economy

The concept of the circular economy (CE) has been growing in influence in recent years. As an emerging movement, it has support from a variety of proponents including NGOs, most notably the Ellen MacArthur Foundation, policymakers, and academics. The Ellen McArthur Foundation has been influential in defining CE. It describes the circular economy as "looking beyond the current take-make-waste extractive industrial model, a circular economy aims to redefine growth, focusing on positive society-wide benefits. It entails gradually decoupling economic activity from the consumption of finite resources, designing waste out of the system." Underpinned by a transition to renewable energy sources, the circular model builds economic, natural, and social capital. It is based on three principles: (1) design out waste and pollution, (2) keep products and materials in use, and (3) regenerate natural systems (EllenMacArthurFoundation, 2017). Geissdoerfer, Savaget, Bocken, and Hultink (2017) define the circular economy as "a regenerative system in which resource input and waste, emission, and energy leakage are minimised by slowing, closing, narrowing material and energy loops. This is achieved through long-lasting design, maintenance, repair, reuse, remanufacturing, refurbishing, and recycling" (Geissdoerfer et al., 2017, p. 766). As such, there is a strong focus in circular economy research on waste management. Circular economy thinking assumes that, as is the case in nature, nothing is waste, but items traditionally considered waste should either be designed out of the production process or used as resources for other activities (Schumann, 2020). Ghisellini, Cialani, and Ulgiati (2016) note that CE emerges from the 3 Rs principles of Reduce, Reuse, and Recycle. Other researchers have expanded on these principles to 6 Rs—Reduce, Reuse, Recycle, Redesign, Recover, Remanufacture—with additional means of reducing waste (Jawahir & Bradley, 2016).

While the application of circular economy principles is gaining some traction in manufacturing and agriculture, it has been limited in tourism and hospitality (Jones & Wynn, 2019; Pattanaro & Gente, 2017). Nevertheless, there is a growing body of work on circular economy and tourism. Rodríguez, Florido, and Jacob (2020) conducted a review of CE and tourism articles that included 55 English-language articles. Jones and Wynn (2019) note that there are a growing number of Chinese-language articles on the topic as well. CE is also attracting conferences and seminars, including the United Nations Industrial Development Organisation (UNIDO) Conference on Circular Economy in Tourism in South East Europe (2020), "Towards a circular economy and sustainable tourism on islands," sponsored by the Association of cities and regions for sustainable resource Management (2018), and the "Advancing circularity solutions in tourism and construction" session presented by One Planet Sustainable Tourism Network (2018).

CE principles can be applied at various levels within the tourism system. From a meso perspective they can be applied to a destination community; from a micro perspective they can be applied at a business level. Rodríguez et al. (2020), in their review of circular economy contributions to the tourism industry, note several applications across several specific sectors including agritourism, cultural tourism, and maritime tourism, as well more generally in destination communities. Schumann (2020) identifies early steps undertaken by Guam in adopting CE principles, including buying local programs and CE education. Several authors identify the value of CE in ecotourism destinations (Zhang, 2014; Zhang & Tian, 2014). At a micro level, there is evidence of hotels recognising CE in their waste-management approaches (Pamfilie, Firoiu, Croitoru, & Ionescu, 2018; Rodríguez et al., 2020). Several authors have identified the need for greater policy support in implementing circular economy activities (Falcone, 2019). Jones and Wynn (2019) note that tourism and hospitality applications focus on environmental processes including water, waste, and energy management.

While much of the focus of CE principles in hospitality and tourism has been on environmental issues, particularly waste management, the CE "is a framework for an economy that is restorative and regenerative by design" (EllenMacArthurFoundation, 2017). As such, it is closely aligned with the regenerative movement. Nevertheless, the interest in regenerative processes is developing concurrently with the circular economy movement, and thus it is worthwhile to examine this concept separately.

The regenerative movement

The regenerative movement has been influenced by several emerging steams of thought. As noted, regeneration is a critical component of the circular economy. Other important streams of thinking about the role of regeneration in socioeconomic systems derive from agriculture, architecture, and design. At least part of the appeal of regenerative approaches is the frustration with the framing of sustainability. JWT Thompson's Innovation Group captured the sentiment by noting, "sustainability as we know it is dead. Doing less harm is no longer enough. The future of sustainability lies in regeneration: seeking to restore and replenish what we have lost, to build economies and communities that thrive, and that allow the planet to thrive too" (Stafford, Tilley, & Britton, 2018, p. 2). John Elkington, the sustainability thought leader who coined the term "triple-bottom-line," amplifies the sentiment that "doing less harm" in not enough and now advocates for adopting a regenerative approach to sustainability (Elkington, 2020).

Regenerative design is emergent, having evolved from sustainability science, where the above-mentioned focuses on minimising environmental damage and resource efficiency gave way to regenerative design and a whole system approach to development. Critical to regenerative design and development is the co-evolution of human and earth's living systems, including its politics. Systems thinking permits the inclusion of resilient systems embedded both in a community's needs and in nature's integrity. The American Heritage dictionary defines regenerate as "to revitalize, to give new life or energy to." Regenerative design and development are organised around nine premises, chief of which is the first premise: "Every living system has inherently within it the possibility to move to new levels of order, differentiation, and organisation" (Mang & Haggard, 2016).

Regenerative design (Mang & Reed, 2012b) is an approach whose origins harken back to general systems theory, an analytical lens that emphasises the interrelationships of a system's elements and parts embedded within the context of a larger system. As explained by Oliver, Thomas, and Thompson (2013, p. 2), "a regenerative system restores ecosystems, gives new life, and creates social and natural capital". Additionally, these systems adopt co-evolution within relationships that are less managerial in nature, but more of a partnership between sociocultural and ecological systems. It is also the case that regenerative approaches, by their very nature, inculcate current topics such as climate change and the loss of biodiversity in their planning horizons. Its accents on restoration, revitalisation, and renewal through innate sources of energy and materials ties together systems thinking, community engagement, and respect for place (core tenets). As a design approach involving both human and ecological systems, it also borrows from the elegance of biomimicry, the imitation of elements and systems in nature that naturally resolve complex issues.

Janine Benyus, a biologist and cofounder of the Biomimicry Institute, popularised the notion of biomimicry as a modelling of nature's graceful and inborn processes to create a healthier planet (Biomimicry Institute: https://biomimicry.org). There is abundant evidence (Mang & Reed, 2012a) of regenerative design thinking stemming from its biological subfield, ecology, and sustainability's ecological beginnings (as distinct from the engineering-based, minimal impact, greening—especially buildings—beginnings). Regenerative design and development's minimal requirement is a net-positive environmental benefit (Mang & Reed, 2012a). Furthermore, agriculture and forestry's influences on regenerative design and development can be tracked through its persistent references to permaculture, a combination of agriculture and forestry which, when combined, forms agroforestry. A design system based on permaculture holds as tenets that it is possible to discern, develop, and generate new patterns in both human and natural systems, that weave them together in a dynamic whole. Permaculture considers nature as the preeminent recycler where there is no waste, but rather, infinite resources (Mollison, Slay, & Girard, 1991). Mang and Reed (2012b) offer the example of "road systems that serve as water harvesting structures and erosion control features while supporting wind-break, wildlife habitat, and firebreak functions" (p. 31). Rhodes (2017) defines its primary goal as the improvement of soil health, to restore highly degraded soil in service of enhanced water quality, flora and foliage, and land productivity. Permaculture is embedded in regenerative agriculture, and it uses inherent qualities of animals and plants with landscapes and structures to yield an integrated design system modelled on nature.

Many regenerative design approaches take their cue from the built environment, architecture, and place traditions such as green building ratings systems. Landscape architect John Tillman Lyle proposed regenerative design as a dynamic process for cities in a co-evolving participation in environmental issues (Lyle, 1996). Regardless of the tradition, from LEED ratings (Leadership in Energy and Environmental Design) to frameworks such as LENSES (Living Environments in Natural, Social and Economic Systems) and REGEN (the REGENerative design framework), the co-evolution between human social systems, and their communities, and ecological ones is enunciated and weighted heavily. Regenerative design might be best considered by Aldo Leopold, who suggested conditions of ecological health as defined by "the capacity of the land for self-renewal" in response to the carelessness of our fossil-fueled growth economy (Mang & Haggard, 2016). Mang and Haggard suggest that a regenerative approach "shifts the focus [of sustainable design] from slowing down entropy to building the capability of living communities to evolve toward greater value" (p. 21).

There is regenerative design and there is regenerative development about which (Mang & Reed, 2012a) make a critically important distinction. While regenerative design builds the self-renewing capacities of designed and natural systems (the designed interventions), regenerative development creates the conditions necessary for its sustained, positive co-evolution. It might be helpful to use the metaphor of gardening and gardeners in further explicating designing and designers in regenerative processes. A gardener designs an ecosystem, layered within other ecosystems, creating the healthy growth of plants given their seasonality and environmental challenges. The garden's success stands on the shoulders of the gardener and her or his gardening acumen. So too, the designer in regenerative approaches incorporates natural, biological, and human systems to improve conditions for both.

Regenerative tourism

The application of regenerative design and principles in tourism is in its infancy. Even so, calls for regenerative approaches to tourism development have come from a variety of sources.

Several authors echo Elkington's observation that sustainability fails to provide inspiration and that regenerative practices contribute to tourism. Pollock (2015) calls for tourism leaders to transcend growth and greening and adopt regenerative principles. Day (2016) calls for tourism system actors to extend beyond sustainability and adopt regenerative processes. Howard, Hes, and Owen (2008) suggest that its development is a result of frustration with the pace and adequacy of approaches that safeguard a more sustainable future. These authors also note that sustainability lacks any motivating factors to connect environmentalism with sociopolitical dimensions and that its frame of reference on minimal impact suffers in comparison to regenerative design's focus on net-positive impacts. While the general approaches to adapting regenerative approaches are emerging, to date there are no broader frameworks to develop regenerative approaches to tourism systems. Pauline Sheldon, during her keynote address at the Travel and Tourism Research Association's International conference (Dredge, 2019), suggested adopting the eight principles proposed by Fullerton (2015). Fullerton's principles include:

1. In Right Relationship: Ensuring humanity's relationship with nature is appropriate.
2. Views Wealth Holistically: Advocates moving beyond just money to broader measures of wellbeing.
3. Innovative, Adaptive, Responsive: Able to respond to emerging challenges.
4. Empowered Participation: With each actor contributing to the health of the whole.
5. Honors Community and Place: Regenerative economy nurtures healthy and resilient communities.
6. Edge Effect Abundance: This principle notes that "creativity and abundance flourish at the 'edge' of systems where bonds holding the dominant pattern are weakest" (Fullerton, 2015, p. 9).
7. Robust Circulatory Flow: Advocates for a flow of money and information through the system and for the efficient use and reuse of materials.
8. Seeks Balance: Advocates harmonising multiple variables instead of optimising single variables.

While comprehensive approaches help to clarify the concepts that may be incorporated in regenerative tourism, some of the earliest examples of regenerative tourism come from the application of regenerative design to small-scale, nature-based, or ecotourism operations. Howard et al. (2008) examine the application of regenerative principles to community-based tourism on a small island in the Torres Strait (Australia). Playa Viva, an ecolodge in Mexico with a commitment to improving the local environment and community, is also presented as an example of regenerative tourism. Regenerative Travel, a travel supplier, curates a portfolio of more than 45 ecolodges they assert achieve the principles of regenerative travel.

Nevertheless, the concept of regenerative tourism, perhaps benefiting from the "do no harm is not enough" attitude and a desire of pandemic-weary citizens to improve the world from which they have been isolated, is gaining in profile with both consumers and industry. Examples of the rising profile of the topic include a *New York Times* article titled "Move Over, Sustainable Tourism. Regenerative Travel has Arrived" (Glusac, 2020); Canada's Impact Sustainability Travel and Tourism Conference, described as "impact 2020 explored how Canadian tourism can move beyond sustainability towards being a regenerative and restorative industry" (TourismVictoria, 2020); and Vermont's Farm to Plate program, offering a seminar of "Regenerative Tourism and Agriculture and the First International (virtual) Regenerative Travel Summit, which took place in October.

Regenerative resilience: Building back better

This chapter was written during the 2020 coronavirus pandemic, and discussions of recovery from system shocks seem more pertinent than ever. As has been well documented, the coronavirus had significant impacts on the tourism system, and since the earliest days of the pandemic, there have been calls for tourism to "build back better." Many academics and policymakers have seen the disruption in the tourism system as an opportunity to address some of the more significant negative impacts of tourism. At the same time, the tourism industry is reeling from the impacts of the virus on travel and hospitality.

Any discussion on how to build back better can be framed in terms of resilience and recovery. Resilience has been described as a "measure of the persistence of systems and their ability to absorb change and disturbance and still maintain the same relationships between populations or state variables" (Holling, 1973, p. 14). Sydnor-Bousso, Stafford, Tews, and Adler (2011) describe resilience as the ability to withstand shocks and rebuild. It is the ability of the hospitality industry to return after disasters to pre-disaster levels of functioning or better (Sydnor-Bousso, 2009). There is a significant body of research addressing the resilience of destination communities (Hamzah & Hampton, 2013; Ruiz-Ballesteros, 2011). Resilience is considered one of the critical capabilities of long-term sustainability of tourism organisations and destination communities.

While resilience can address the ability to respond to a variety of changes, including long-term gradual change, it is a response to and recovery from exogenous, non-normative shocks that is worthy of examination. The ability to recover from a disaster—whether a flood, fire, or storm—will only become more important as the impacts of climate change become more immediate.

Recovery is a complex task that must be considered from several perspectives. Three important considerations include: time frame, scale, and perception. Time frame represents various periods of time following the event, scale refers to the type of entity being examined, and perceptions refer to those of involved parties (Marshall & Schrank, 2014). Marshall and Schrank (2014) note the importance of assessing recovery over time. Assessment of recovery in the months immediately following a disaster may fail to appreciate the long-term trauma inflicted on businesses following an event. They also point out the importance of recognising that recovery can be considered against a continuum from mere survival to recovery to a state they describe as resilient in which owners perceive they are better than they were before the event. Although there is a growing body of research on business recovery from these types of events, there is little information on how businesses and destination communities recover to levels that exceed pre-disaster levels.

While the concept of "building back better" is often touted as an ideal, several authors have noted that tourism organisations lack even basic preparations for disasters (Drabek, 1995). If we are to achieve recovery that leads to better outcomes than before the disaster, then there is need for a greater understanding of the actions required to achieve these goals.

Binna Burra Lodge: Recovery and regeneration

Binna Burra Lodge (BBL) is an ecotourism property located in the subtropical rainforest of the Lamington National Park, one of the parks in the Gondwana World Heritage Site in Australia. BBL was first established in 1933 and has been a leader in nature-based tourism in Australia. On September 8, 2019, a bushfire destroyed the historic lodge and pioneering cabins. In Chapter 16, we examined Binna Burra Lodge in the context of climate change. In this chapter we'll look at BBL's recovery through the context of recovery and regeneration.

Be as prepared as possible

BBL has demonstrated a commitment to high-quality management. Their commitment to continuous performance improvement is evident in their over 20 years of certification as an "advanced ecotourism" operation, a recognition that requires achieving high standards across a range of business and operational criteria. BBL was prepared for the event of a disaster and had undertaken drills and training activities to be ready for disaster. BBL had considered business continuity issues and were able to quickly set up an off-site temporary office to manage immediate post-disaster issues. BBL also had insurance for different components of the operation, with the separate legal entity known as the Sky Lodges being fully covered by wildfire damage, but the central lodge and cabins could only be partially covered for wildfire damage. Despite these preparations, unanticipated challenges presented themselves in the months following the disaster. The impact of refunds and lost business were significant, and they created cashflow challenges. While BBL had insurance, there was a cap in wildfire-related funds. BBLs commitment to the wellbeing of their staff placed additional pressures on the business operations. Before the wildfire, BBL had 65 staff members of which 90% were retrenched within the first week of the disaster.

Bias toward action and confidence

Despite the devastation of the fires, Chairperson Steve Noakes recognised the importance of maintaining optimism on the future of BBL. Learning from other successful businesses that recovered from devastation, Noakes recognised the need to project an immediate positive message, even while those closely associated with the enterprise were still coming to terms with the shock and grief of the loss. BBL committed to a focused communication strategy with a wide range of stakeholders to maintain a positive orientation.

Throughout the process, BBL has been committed to sustainability principles that have informed their recovery response. BBL has a long commitment to ecotourism certification, has committed to continuing its Advanced Ecotourism certification in the post-wildfire era.

BBL has a clearly articulated set of principles that provide a foundation for all development. They include:

- Conservation: Identification, protection, conservation, presentation, and transmission to future generations of cultural and natural heritage.
- Action: "Walking the Talk" of sustainability being a daily activity, achieved through a process of engage, learn, do.
- World Heritage: Framework of World Heritage Values.
- Tourism: Sharing responsibility for conservation of our cultural and natural heritage through sustainable tourism management.
- Climate Change: Both a risk and an opportunity too big to ignore.

In addition to these, BBL is committed to the 10 principles of the United Nations Global Compact and the United Nations Sustainable Development Goals (SDGs).

Social capital, systems support, and regenerative recovery

As noted in Chapter 16, BBL developed strong relationships with key stakeholders including the tourism industry, supply chains, the government, consumers, and the community. In particular, the stakeholder support from the federal, state, and local government agencies in the days following the wildfire event laid the foundations for a solid recovery. Given the complex planning and regulatory issues BBL has to work with as a commercial enterprise in the World Heritage–listed national park, the leadership of BBL understood the need and benefits of rapid engagement with each level of government in Australia. Beyond the financial support—including a crowd-funding support program—and government support, goodwill encouraged the board and management to persevere.

BBL recognised the emotional dimensions of recovery. The Bushfire was traumatic not only for BBL management and staff, but for a wide range of stakeholders including guests and visitors from around the world. The Binna Burra Bush Fire Gallery was officially opened on the first anniversary of the wildfire and provides opportunities for stakeholders to come and share memories and testimonials of their connection to Binna Burra. These healing activities were supported by the most senior management at the BBL and included the chairperson allocating time to make "Aussie Billy-tea" on a campfire adjacent to the Bushfire Gallery and share anecdotes and remembrances with guests visiting the site. "Reflection Benches", with the message "Looking back so the view looking forward is clearer" were provided. These activities are an important "way to reflect on the loss caused by the Sarabah bushfire at Binna Burra and in the surrounding area and celebrate the strength of the community and the return of Binna Burra Lodge" (Schultz & Barnett, 2020, p. 40).

Physical recovery and development

In the year following the wildfire, the physical recovery of the site has been a priority, and despite challenges in gaining secure single-road access into the site and sources of capital during the year following the fire, plans are under way to rebuild. BBL's master plan, completed in 2008 after extensive stakeholder engagement, provided an important foundation on which to frame post-disaster planning. The fire created an opportunity to reimagine the BBL precinct. Binna Burra is being developed in a way that builds on their commitment to nature and has committed to ensuring that BBL will be a zero-waste, zero-emissions site. The design characteristics incorporate commitments to the following approaches:

- Climate responsive design principles.
- Incorporate salvaged materials from the property, both historically and recovered from the fires and through land management.
- Maximise use of recycled and natural materials.
- Colors to harmonise with the environment.
- Integrate water-sensitive urban design (WSUD) principles.
- Utilise renewable energy (e.g., solar, wind with battery) to power operations on the site.

The lighting plan minimises "obtrusive light" and "add[s] to the cultural values and heritage ambiance that makes our visitors feel warm and welcome" (Schultz & Barnett, 2020, p. 55). This plan is designed to use low amounts of energy, be heritage-listing compliant, and minimise any disturbance to nocturnal animals and birds.

Cultural heritage-based regeneration

BBL has maintained focus on long-term goals and principles while getting back to business. Several projects supporting the long-term goals, committed before the wildfire, ensured a continued focus on the future of BBL. BBL is committed not only to environmental sustainability, but also to sociocultural sustainability. Despite the twin challenges of the bushfires and the pandemic, BBL is committed to a Reconciliation Action Plan (RAP) designed to recognise and pay respect to the traditional custodians of the land on which BBL is located. BBL is working with Yugambeh Region Aboriginal Corporation Alliance (YRACA) in the development of reconciliation initiatives. The RAP has influenced the recovery process; for example, a traditional smoking and healing ceremony was undertaken by representatives of the Yugambeh Aboriginal language group (Schultz & Barnett, 2020) and was incorporated into the opening ceremony of the Binna Burra Fire Gallery.

Conclusions

Lessons from Binna Burra

Even without any onsite trading activities, BBL successfully survived the first year following the fires and seems well positioned to continue its recovery and "build back better." At this point in the recovery, there are several possible lessons from their process. Those lessons include:

- Be prepared for disaster. Regularly assess vulnerabilities and risks, and plan accordingly.
- Work to reopen as soon as practical and maintain communication with key stakeholders throughout the recovery process.
- Effective crisis and post-crisis communication are critical for recovery. Establish and maintain consistent messaging from the most senior person in the organisation.
- Expect the unexpected and build in capacity to meet contingencies.
- Invest in building relationships with key stakeholders. Social capital is an asset that endures through disaster.
- Take time to process the emotional toll. By establishing of the Bush Fire Gallery and creating a safe place for people to come and talk about the trauma of the fire, BBL enabled healing to take place.
- Stay true to your principles. Establishing a vision and a mission provides guidance through disruption.
- Focus on both the immediate and the long term. The highest priority for BBL was re-opening; nevertheless, while working on the immediate goal, they maintained a commitment to advancing strategic initiatives including updating the master plan and progressing the RAP.

As an ecotourism product, BBL embraces many of the principles of the circular economy and regenerative tourism. In both general operations and through the recovery, BBL places priority on the 6 Rs: Reuse, Repair, Reduce, Rethink, Recycle, Refuse, Reduce. As the rebuilding process continues, BBL will use salvaged materials from the property, both historically and recovered from the fires, and through land management. Additionally, new buildings and operations will be designed to ensure that BBL is a zero-waste site. BBL also undertakes many of the activities identified as regenerative. The conservation and perseveration of the surrounding environment is at the core of its operation. At the same time, their commitment to recognising the role of Indigenous peoples of the region and engaging them as key stakeholders in the future is consistent with the principles described as regenerative tourism.

Emerging insights

Early evidence suggests that the concepts of the circular economy and regenerative tourism are ideas whose times have come. There is no doubt that the positive proactive message of regenerative tourism benefits from a general frustration that the broad implementation of sustainable tourism practices is taking too long. Several authors have noted that despite the almost ubiquitous adoption of sustainable tourism rhetoric by policymakers and academics, the implementation of sustainable tourism practices has been slow. For some, "sustainability" is no longer an inspiring term. Indeed, sustainability has long been considered a jargon term (AdAge, 2010), and a poorly understood one. In addition to slow progress, it has been noted that sustainable tourism has been coopted as a means of justifying unsustainable practices (Collins, 1999). Elkington (2020), who first coined the term "triple-bottom-line," even tried to "recall" the concept through frustration with the way it was applied. Nevertheless, both the circular economy and the regenerative movement are closely tied to sustainability. Geissdoerfer et al. (2017) propose that circular economy is a condition of sustainability. Certainly, the focus on environmental management is key to the environmental dimension of sustainability. Similarly, while some criticise sustainability as merely "not doing harm," this may be more the fault of poorly implemented sustainability programs than a lack of concern for regenerative approaches in the core tenets of sustainability. At best, this is an exciting opportunity for these new approaches to refresh and revitalise sustainability efforts.

Regenerative recovery, or building back better, is a far less established concept than either the circular economy or regenerative tourism. This concept also has its roots in sustainability and resilience research, but it has been under research and is poorly understood. Nevertheless, as we persevere through the pandemic and prepare for emerging challenges presented by climate change, it is an important topic to address.

The general excitement of these new approaches, particularly regenerative tourism, should be tempered by some caution. The enthusiasm of the embrace of this concept is reminiscent of the Gartner Hype Cycle, where inflated expectations are preceded by a trough of disillusionment, before eventually reaching a plateau of productivity (Fenn & Blosch, 2018; Fenn & Raskino, 2008). It could be argued that sustainability faced just a cycle and is only now reaching a more productive stage. Certainly, these new concepts face many of the same challenges that sustainability faces. Each of these concepts addresses a wicked problem, and application of the concepts will vary in different locations. These concepts must be operationalised into specific actions that can be applied in a range of circumstances. Like sustainability, these approaches require systems thinking, and the full benefits of the approach will only be accrued when they are adopted throughout the system. Vargas-Sánchez (2018) notes that adoption of the circular economy principles will require change within the tourism system, and each of these processes requires the adoption of systems for "capture, processing, analysis, and reporting of data and information" (Jones & Wynn, 2019, p. 2547). Cave and Dredge (2020) suggest that regenerative tourism may require changes in the operating systems of society. Without a doubt, these are significant challenges to overcome to achieve a tourism system-wide adoption.

In the meantime, ecotourism remains a fertile ground for establishing "proof of concept" for these innovations. Just as ecotourism provided key insights for sustainable tourism studies, ecotourism operators have already adopted key techniques associated with the circular economy and regenerative tourism. These concepts align with the principles of ecotourism (GEN, 2016), and as research on the applications of the concepts continues, it is reasonable to expect that examples from ecotourism will remain central, as our understanding of both the circular economy and regenerative tourism grows.

References

AdAge. (2010). Book of Tens: Jargoniest Jargon We've Heard All Year. *Advertising Age*. Retrieved from http://adage.com/article/special-report-the-book-of-tens-2010/advertising-s-jargoniest-jargon/147583/.

Buckley, R. (2012). Sustainable tourism: Research and reality. *Annals of Tourism Research, 39*(2), 528–546. doi:https://doi.org/10.1016/j.annals.2012.02.003.

Cave, J., & Dredge, D. (2020). Regenerative tourism needs diverse economic practices. *Tourism Geographies, 22*(3), 503–513. doi:https://doi.org/10.1080/14616688.2020.1768434.

Collins, A. (1999). Tourism development and natural capital. *Annals of Tourism Research, 26*(1), 98–109. doi:https://doi.org/10.1016/s0160-7383(98)00059-0.

Day, J. (2016). *An Introduction to Sustainable Tourism and Responsible Travel* (Beta ed.). West Lafayette, IN: Placemark Solutions.

Day, J. (2020). Sustainable tourism in city is a wicked problem. In A. Morrison, & J. A. Coca-Stefaniak (Eds.), *Routledge handbook of tourism cities*. New York: Routledge.

Dodds, R., & Butler, R. (2010). Barriers to implementing sustainable tourism policy in mass tourism destinations. *Tourismos: An International Multidisciplinary Journal of Tourism, 5*(1), 35–53.

Drabek, T. (1995). Disaster planning and response by tourism business executives. *Cornell Hotel and Restaurant Adminsitration Quarterly*, June, 86–96.

Dredge, D. (2019). Shifting Sands, the Next Economy & Tourism. Retrieved from https://ttra.com/shifting-sands-the-next-economy-tourism/.

EA. (2020). Ecotourism Australia: Home. Retrieved from https://www.ecotourism.org.au/.

Elkington, J. (2020). *Green Swans: The Coming Boom in Regenerative Capitalism*. New York: Fast Company Press.

EllenMacArthurFoundation. (2017). What Is a Circular Economy. Retrieved from https://www.ellenmacarthurfoundation.org/circular-economy/concept.

Falcone, P. M. (2019). Tourism-based circular economy in Salento (South Italy): A SWOT-ANP analysis. *Social Sciences, 8*(7), 216. doi:https://doi.org/10.3390/socsci8070216.

Fenn, J., & Blosch, M. (2018). Understanding Gartner's Hype Cycles. Retrieved from https://www.gartner.com/en/documents/3887767.

Fenn, J., & Raskino, M. (2008). *Mastering the Hype Cycle: How to Choose the Right Innovation at the Right Time*. Boston, MA: Harvard Business Review Press.

Fennell, D. A. (2001). A content analysis of ecotourism definitions. *Current Issues in Tourism, 4*(5), 403–421. doi:10.1080/13683500108667896.

Fullerton, J. (2015). *Regenerative Capitalism: How Universal Principles and Patterns Will Shape Our New Economy*. Stonington,CT: Capital Institute. Retrieved from http://capitalinstitute.org/wp-content/uploads/2015/04/2015-Regenerative-Capitalism-4-20-15-final.pdf.

Geissdoerfer, M., Savaget, P., Bocken, N. M. P., & Hultink, E. J. (2017). The circular economy – A new sustainability paradigm? *Journal of Cleaner Production, 143*(1), 757–768. doi:https://doi.org/10.17863/cam.7193.

GEN. (2000). Mohonk Agreement: Proposal for an International Certification Program for Sustainable Tourism and Ecotourism. Retrieved from http://www.globalecotourismnetwork.org/wp-content/uploads/2019/02/mohonk.pdf.

GEN. (2016). Definations and Key Concepts. Retrieved from https://www.globalecotourismnetwork.org/definition-and-key-concepts/.

Ghisellini, P., Cialani, C., & Ulgiati, S. (2016). A review on circular economy: The expected transition to a balanced interplay of environmental and economic systems. *Journal of Cleaner Production, 114*, 11–32. doi:https://doi.org/10.1016/j.jclepro.2015.09.007.

Glusac, E. (2020, August 27). Move Over, Sustainable Travel. Regenerative Travel Has Arrived. *New York*

Times. Retrieved from https://www.nytimes.com/2020/08/27/travel/travel-future-coronavirus-sustainable.html.
Goodwin, H. (2011). *Taking Responsibility for Tourism: Responsible Tourism Management*. Woodeaton, Oxford: Goodfellow Publishers.
Hamzah, A., & Hampton, M. P. (2013). Resilience and non-linear change in island tourism. *Tourism Geographies, 15*(1), 43–67. doi:https://doi.org/10.1080/14616688.2012.675582.
Holling, C. S. (1973). Resilience and stability of ecological systems. *Annual Review of Ecology and Systematics, 4*(1), 1–23. doi:https://doi.org/10.1146/annurev.es.04.110173.000245.
Howard, P., Hes, D., & Owen, C. (2008). Exploring principles of regenerative tourism in a community driven ecotourism development in the Torres Strait Islands. Presentation at the 2008 World Sustainable Building Conference, Melbourne, Australia.
Jawahir, I. S., & Bradley, R. (2016). Technological elements of circular economy and the principles of 6R-based closed-loop material flow in sustainable manufacturing. *Procedia CIRP, 40*, 103–108. doi:https://doi.org/10.1016/j.procir.2016.01.067.
Jones, P., & Wynn, M. G. (2019). The circular economy, natural capital and resilience in tourism and hospitality. *International Journal of Contemporary Hospitality Management, 31*(6), 2544–2563. doi:https://doi.org/10.1108/IJCHM-05-2018-0370.
Lyle, J. T. (1996). *Regenerative Design for Sustainable Development.*United Kingdom: Wiley.
Mang, P., & Haggard, B. (2016). *Regenerative Development and Design: A Framework for Evolving Sustainability*. Hoboken, NJ: Wiley.
Mang, P., & Reed, B. (2012a). Designing from place: A regenerative framework and methodology. *Building Research & Information, 40*(1), 23–38. doi:https://doi.org/10.1080/09613218.2012.621341.
Mang, P., & Reed, B. (2012b). Regenerative development and design. In R. A. Meyers (Ed.), *Encyclopedia of sustainability science and technology* (pp. 8855–8879). New York, NY: Springer.
Marshall, M. I., & Schrank, H. L. (2014). Small business disaster recovery: A research framework. *Natural hazards (Dordrecht), 72*(2), 597–616. doi:https://doi.org/10.1007/s11069-013-1025-z.
Maxim, C. (2014). Drivers of success in implementing sustainable tourism policies in urban areas. *Tourism Planning & Development, 12*(1), 1–11. doi:https://doi.org/10.1080/21568316.2014.960599.
Mollison, B., Slay, R., & Girard, J. L. (1991). *Introduction to Permaculture*. Tyalgum, Australia: Tagari Publications.
Moyle, B., Moyle, C.-l., Ruhanen, L., Weaver, D., & Hadinejad, A. (2020). Are we really progressing sustainable tourism research? A bibliometric analysis. *Journal of Sustainable Tourism, 29*(1), 106–122. doi:https://doi.org/10.1080/09669582.2020.1817048.
National Geographic. (2020). About Geotourism. Retrieved from https://www.nationalgeographic.com/maps/geotourism/about/.
Oliver, A., Thomas, I., & Thompson, M. (2013). Resilient and Regenerative Design in New Orleans: The Case of the Make It Right Project. *S.A.P.I.EN.S, 6.1*. Retrieved from http://journals.openedition.org/sapiens/1610.
Pamfilie, R., Firoiu, D., Croitoru, A., & Ionescu, G. (2018). Circular economy - A new direction for the sustainability of the hotel industry in Romania. *Amfiteatru Economic, 28*(48), 388–404. doi:https://doi.org/10.24818/EA/2018/48/388.
Pattanaro, G., & Gente, V. (2017). Circular economy and new ways of doing business in the tourism sector. *European Journal of Service Management, 21*, 45–50. doi:https://doi.org/10.18276/ejsm.2017.21-06.
Pollock, A. (2015). Social Entreprenuership in Tourism: The Conscious Travel Approach. *TIPSE Tourism*.
Rhodes, C. J. (2017). The imperative for regenerative agriculture. *Science Progress, 100*(1), 80. doi:https://doi.org/10.3184/003685017X14876775256165.
Rodríguez, C., Florido, C., & Jacob, M. (2020). Circular economy contributions to the tourism sector: A critical literature review. *Sustainability (Basel, Switzerland), 12*(11), 4338. doi:https://doi.org/10.3390/su12114338.
Ruhanen, L., Moyle, C.-l., & Moyle, B. (2019). New directions in sustainable tourism research. *Tourism Review, 74*(2), 138–149. doi:https://doi.org/10.1108/tr-12-2017-0196.
Ruhanen, L., Weiler, B., Moyle, B. D., & McLennan, C.-L. J. (2015). Trends and patterns in sustainable tourism research: A 25-year bibliometric analysis. *Journal of Sustainable Tourism, 23*(4), 517–535. doi:https://doi.org/10.1080/09669582.2014.978790.

Ruiz-Ballesteros, E. (2011). Social-ecological resilience and community-based tourism: An approach from Agua Blanca, Ecuador. *Tourism Management, 32*(3), 655–666. doi:https://doi.org/10.1016/j.tourman.2010.05.021.

Schultz, T., & Barnett, B. (2020). *Binna Burra Strategic Visions, Masterplan, and RAP Background Summary Report.* Queensland, Australia: Binna Burra.

Schumann, F. R. (2020). Circular economy principles and small island tourism: Guam's initiatives to transform from linear tourism to circular tourism. *Journal of Global Tourism Research, 5*(1), 13–20. doi:https://doi.org/10.37020/jgtr.5.1_13.

Stafford, M., Tilley, S., & Britton, E. (2018). *The New Sustainability:Regeneration.* New York: Wunderman Thompson. Retrieved from https://intelligence.wundermanthompson.com/trend-reports/the-new-sustainability-regeneration/.

Sydnor-Bousso, S. (2009). *Assessing the Impact of Industry Resilience as a Function of Community Resilience: The Case of Natural Disasters* (Doctoral Dissertation, The Ohio State).

Sydnor-Bousso, S., Stafford, K., Tews, M., & Adler, H. (2011). Towards a resilience model for the hospitality and tourism industry. *Journal of Human Resources in Hospitality and Tourism, 10*(2), 195–217.

TourismVictoria. (2020). Impact 2020 Program. Retrieved from https://www.tourismvictoria.com/impact/program.

UNWTO. (1997). *Agenda 21 for the Travel and Tourism Industry: Towards Environmentally Sustainable Development.* Madrid, Spain: World Tourism Organization.

Vargas-Sánchez, A. (2018). The unavoidable disruption of the circular economy in tourism. *Worldwide Hospitality and Tourism Themes, 10*(6), 652–661. doi:10.1108/WHATT-08-2018-0056.

WCED. (1987). *Our Common Future.* Oxford: Oxford University Press.

Zhang, Y. (2014). Circular economy perspective research wanlu lake eco-tourism industry gathering area for innovation and development. *Advanced Materials Research, 962–965*, 2301–2309. doi: https://doi.org/10.4028/www.scientific.net/AMR.962-965.2301.

Zhang, Y., & Tian, L. (2014). The sustainable development of circular economy under the perspective of ecological tourism. *Advanced Materials Research, 1010*, 2090–2093. doi:https://doi.org/10.4028/www.scientific.net/AMR.962-965.2301.

3

ECOTOURISM AND THE TROUBLE WITH TRANSPORTATION

Stephen Schweinsberg and Simon Darcy

Introduction

In 1903, the 26th president of the United States, Theodore Roosevelt, travelled into the heart of Yosemite National Park with environmentalist John Muir with the intention of seeing for himself the grandeur and majesty of the American Wilderness. The resulting experience, which Wearing and Schweinsberg (2018) have used as the basis for an examination of a culturally framed ecotourism gaze was described in glowing terms by the then president. As Roosevelt observed in a post-trip letter to Muir:

> I trust I need not tell you (Muir), my dear sir, how happy were the days in the Yosemite I owed to you, and how greatly I appreciated them. I shall never forget our three camps; the first in the solemn temple of the great sequoias; the next in the snow storm among the silver firs near the brink of the cliff; and the third on the floor of the Yosemite, in the open valley fronting the stupendous rocky mass of El Capitan with the falls thundering in the distance on either hand. (Roosevelt, 1903)

Some readers might bristle with the idea of President Roosevelt being used as an exemplar of an ecotourism experience given his predilection for hunting and subsequent tension with Muir over the protection of the Hetch Hetchy Valley (Curry & Gordon, 2017; Richardson, 1959). While such concerns are valid, 'ecotourism' and 'ecotourist' are contested phenomena (see Fennell, 2015b) subject to the cultural framing of societies that participate in (Lorenzo-Romero, Alarcón-del-Amo, & Crespo-Jareño, 2019) and support (Schweinsberg, Darcy, & Wearing, 2018) nature-based travel. It is important to begin this discussion of ecotourism transportation with reference to history because "it is through the quoting of history with all of its inbuilt societal and cultural discourses that we are able to expose those contraditions and ruptures that brought science, art, capitalism, railroads and tourism together in the production of national park space" (Denzin, 2008, p. 453 in Wearing et al., 2019, p. 183).

The transport options available to ecotourists in environments like Yosemite National Park have evolved over time. Prior to the introduction of cars into America's national parks in 1916 the primary means of traversing the park landscape, whether as a private individual or the president of the United States was by wagon, stagecoach, or horseback (National Parks Service,

DOI: 10.4324/9781003001768-3

2016). The vivid experiences that were available to early travellers as they traversed the wide expanses of the American wilderness have progressively been juxtaposed with concern that 'improvements' in transportation options will result in overcrowding of many high visitation park regions (White, 2007). For President Roosevelt, his trip to Yosemite began routinely for an individual of his stature:

> Muir came from San Francisco on the train with the Presidential Party of eight, including Governor George C. Pardee of California, Benjamin Ide Wheeler, President of the University of California, and Roosevelt's personal secretary Mr Loeb. The group was placed in an eleven passenger coach ... Under Lieutenant Mays, thirty cavalrymen escorted this stage from Raymond directly to the Grizzly Giant in the Mariposa Grove. (Leidig, 1903/2003, p. 4 in Wearing & Schweinsberg, 2018, p. 138)

After dismissing most of his travelling entourage, Roosevelt transferred to horseback and began a journey that would help lay the policy groundwork for the protection of some of North America's most iconic ecotourism landscapes. Today, however, the negative effects of horse-based transport on conservation areas is widely recognised, both in terms of direct impacts e.g., trampling, as well as indirect impacts resulting from the effects of horse faeces and urine on local soil regimes (Newsome, Milewski, Phillips, & Annear, 2002; Newsome, Moore, & Dowling, 2012; Newsome, Smith, & Moore, 2008). Schiller (2018) has observed that the sustainability of a transport form will be determined on the basis of its resultant level of pollution and its ability to promote socially equitable access to sites that would otherwise be difficult to reach. This chapter will explore these and other sustainability issues, whilst also recognising that ecotourism transport occurs at different geographical scales—"... transport directly associated with the ecotourism experience, for example a boat trip around an ecotourism site; ... travel between various ecotourism sites or operations; and ... transport from the home location to the destination, where the ecotourism experience takes place" (Simmons & Becken, 2004, p. 15).

The ecotourism—Transport nexus

The local dimension

Fennell (2002) has identified that ecotourism and sustainable development discourse emerged simultaneously from the literature on ecodevelopment in 1970s and 1980s. It was in this period that people first started looking seriously for an alternative to traditional mass tourism, promoting an approach to development that sought to both protect the environment and maximise development opportunities for local people. With the rise of air transport being entwined with the development of mass tourism in the second half of the twentieth century (Lumsdon, 2000), alternative tourism forms like ecotourism have throughout history been forced to justify their reliance on transportation forms that are increasingly linked to anthropocentric climate change (Peeters, Higham, Cohen, Eijgelaar, & Gössling, 2019). While existing transport scholarship has considered mechanisms whereby airlines and other transport forms can be made more sustainable (Graham & Shaw, 2008; Lumsdon & Page, 2007; Sarker, Hossan, & Zaman, 2012), ecotourism scholarship has tended to focus on the management of the industry at a destination level (Hall, 2013). This includes research investigating the relevance of community based ecotourism (CBET) as a strategy for sustainable local development (Pookhao Sonjai, Bushell, Hawkins, & Staiff, 2018; Reggers, Grabowski, Wearing, Chatterton, & Schweinsberg, 2016;

Wearing, McDonald, Schweinsberg, Chatterton, & Bainbridge, 2019), as well as other co-management arrangements in different jurisdictional contexts (Abebe & Bekele, 2018; Reggers, Schweinsberg, & Wearing, 2013).

Transport represents an important management consideration for ecotourism at the local destination level on account of its ability to both facilitate experiences, as well as to facilitate unintended impacts from poor visitor behaviour and a lack of regulation. In the case of dive tourism for example, fins used for on-site personal transportation have been identified as a principle mechanism whereby tourists may inadvertently cause damage to reef ecosystems (Giglio, Luiz, Chadwick, & Ferreira, 2018; Hammerton, 2017). In Australia, award-winning ecotourism ventures such as the Penguin Parade on Summerland Peninsula (Phillip Island, Victoria—Head, 2000) have over the course of their history developed management responses to deal with the issue of fairy penguins killed by cars, as well as by feral dogs and cats (Anon, 2008). Internationally, the use of trail bikes and mountain bikes are associated with the development of informal trails and the often associated issues of erosion and damage to native vegetation (Newsome & Davies, 2009; Pickering, Rossi, & Barros, 2011; White, Waskey, Brodehl, & Foti, 2006). The use of elephants as local transport options in countries including Thailand and Malaysia has led to debates over the liberalisation of nature (Duffy & Moore, 2010) and the morality of methods used to domesticate and control wild animals as tourism transports and attractions (Kontogeorgopoulos, 2009; Taylor, Hurst, Stinson, & Grimwood, 2019). In the case of whale watching and whale diving (see Cunningham, Huijbens, & Wearing, 2012; Kessler, Harcourt, & Heller, 2013; Orams, 2001), local whale boats are often subject to critique on account of their ability to disrupt animal behaviour (Amerson & Parsons, 2018; Sullivan & Torres, 2018). While the potential for transport forms to impact on the sustainability of local ecosystems is therefore acknowledged in the literature, the challenge for ecotourism managers is how to balance a tourist's right to access a site in a manner acceptable to them, whilst simultaneously protecting the resources on which the experience is based. Non-automotive forms of transport including walking (Hall & Ram, 2019) and cycle tourism (Lumsdon, 2000) are often recognised for their sustainability credentials. However, as we will demonstrate later in this chapter many ecotourists are reliant on mechanised transport options to access otherwise inaccessible ecotourism destinations. The importance of balancing a right of access for all with the development of diverse sustainable transport forms will be an important issue for the sector going forwards.

The breadth of destinations that can broadly be considered part of the global ecotourism industry including the Monteverde Cloud Reserve, Kruger National Park, Patagonia, the New Zealand south island, and Australia's Great Barrier Reef are diverse with respect to the nature of their visiting clientele, as well as in terms of politics of land-use management and access. While organisations including the American Association of Travel Agents continue to espouse core principles of ecotourist behaviour through their *Ten Commandments of Ecotourism* (Edgell & Swanson, 2019); the operationalisation of a sustainable ecotourism industry is very much determined based on local conditions. The challenge for destination planners is how to integrate transport forms into a broader strategic development framework given the role that transport plays in facilitating ecotourism demand (Khadaroo & Seetanah, 2007). In established ecotourism locales like Australia, demand for ecotourism products and services was the impetus for the Queensland Shire of Douglas to commence an accreditation process, which saw it labelled the "world's first eco certified destination" by Ecotourism Australia (Ecotourism Australia, 2019b). The criteria for Eco Destination Certification includes a focus on sustainable mobility, which is concerned with the "impact from transportation to people, environment and climate" (Ecotourism Australia, 2019a, p. 11). With more and more localities e.g., the Commonwealth of Dominica seeking to market themselves as an ecotourism destination (see CS Global Partners, 2019) policymakers must increasingly explore the applicability of different environmental management frameworks including carrying capacity analysis and the Recreation Opportunity Spectrum (see, for example, Ferreira & Harmse, 2014; Smith, Tuffin, Taplin, Moore, & Tonge, 2014) to different ecotourism environments. At the heart of all these methods are a series of fundamental questions including: how much development is too much? And what level of alteration to a resource is a particular stakeholder group(s) willing to accept? Different stakeholder groups will respond to these questions on the basis of their own evolving value positions, as is evidenced in the continued inability for some travellers to reconcile the negative impacts from their use of air transport with their ability to have a sustainable holiday (see Hanna & Adams, 2019).

In the case of ecotourism, transport is therefore both the facilitator of experience and, if one is pessimistic, a necessary evil for ensuring the perpetuation of holiday forms that many stakeholders believe are more sustainable than mass packaged holiday travel. But is it really? As Horton (2009) has argued with respect to the Osa Peninsula in southwestern Costa Rica; in a few decades the area went from an "off the beaten track travel destination with very limited services for tourists and a way of life centred on traditional activities of agriculture, cattle ranching and gold panning … [to a thriving ecotourism destination characterised by] small planes transporting ecotourists … and local taxis and expatriate SUVs clogging the main streets of the peninsula's new ecotourism hub, Puerto Jiménez" (Horton, 2009, p. 93). Transport and ecotourism can in many respects be considered a zero-sum game. Yes, on the one hand it is indisputable that there are a range of transport related impacts on the destination. On the other hand, however, the development of ecotourism experiences carries with it the potential to protect societies and environments from potentially more damaging impacts from traditional mass tourism or other forms of resource extraction (Schweinsberg, Wearing, & Darcy, 2012). What a particular stakeholder or society considers sustainable will evolve over time according to the particular place based circumstances (see Schweinsberg, Wearing and Wearing, 2015). In the early twentieth century when the decision was made to allow automobiles into national parks for the first time, passionate objections were lodged by individuals including the superintendent of Yellowstone on account of the conflicts that would inevitably in their mind develop between car users and horse riders (National Parks Service, 2004). Writing a few decades after the introduction of cars, Edward Abbey observed that the introduction of cars had essentially shrunk the park, thus devaluing the experience that was on offer; "no more cars in national parks. Let the people walk. Or ride horses, bicycles, mules, wild pigs—anything—but keep the automobiles and the motor cycles and all their motorised relatives out" (Abbey, 1968, p. 160). Today, the use of automotive transport in national parks is commonplace. As Youngs, White, and Wodrich (2008) have observed with respect to Yosemite, automotive access has evolved from simply being an alternative to "uncomfortable and expensive" accommodation forms like train and stage coaches to become intimately connected to the cultural landscape of the region.

Ecotourism, transport, and diversity in a national park setting

Throughout the history of the national parks movement, managers have been forced to reconcile their dual mandate to provide experiences for visitors, whilst also protecting the ecology on which tourist experiences are based in its own right (see Wearing, McDonald, Ankor, & Schweinsberg, 2015; Wearing, Schweinsberg, & Tower, 2016). The potential for transport forms to be caught in the cross-hairs of this classic sustainable management trade-off between use and preservation can also be illustrated in the contemporary debate over helicopter national park tours. Helicopter and fixed-wing flyovers have been a part of national park management debates for many decades (see Wichelns, 2017). Environmentalists including Sigurd F. Olson have since the mid-twentieth century campaigned against the use of aircraft and other mechanised transportation forms in regions including the Boundary Waters Wilderness Area in the northern Minnesota (Harvey, 2002). More recently the National Parks Service and the Federal Aviation Administration have worked to develop the 2019 *Air Tourism Management Plans.* The primary goal of such regulation is to "protect park resources and visitor use without compromising aviation safety or the nation's air traffic control system" (National Parks Service, 2019). The importance of getting the balance right was illustrated in a fictitious case of two young adults who visited the Grand Canyon National Park in the edited volume *Cases in Sustainable Tourism: An Experiential Approach to Making Decisions* (see Herremans, 2006b). The two tourists, named Katie and Sam, had their hike constantly interrupted by the sounds of over flights:

> In the morning they discovered a whole new Canyon with interesting combinations of light and shadow providing a totally different view of the geology than they had enjoyed the night before. After a relaxing breakfast they broke camp and got their backpacks ready for another day's travel. They enjoyed an hour of 'natural quiet' before they were once again disturbed by the whirling sound of helicopter blades overhead. (Herremans, 2006a, p. 184)

However, in the course of watching people board charter flights at Grand Canyon Airport they were forced to re-assess the cost and benefits of different transport forms. Yes, it was true that tourists taking a half-hour scenic flight would:

> never hear the rushing water or the noise of the wildlife over the sounds of the aircraft ... [However, as Katie observed at the airport] did you notice the couple with the two small children? Do you think they could have enjoyed the same [hiking] adventure we did or should alternatives [like air tours] be available for those whose situations require a different experience? (Herremans, 2006a, p. 185)

If we are to answer this question, we must consider the specific requirements of visitor groups in national parks and the way they dovetail (only sometime successfully) with the availability of transport and mobility infrastructure. For example, Lovelock (2010) has identified that managers of national parks and other protected areas find themselves in a dilemma with respect to access provision for people with disabilities and seniors with access needs. While there will always be debate about site hardening of national park environments for reasons of transport access, amenity provision and other issues of carrying capacity; what has not been considered when decisions are made to provide access to these areas involving transport options is the social sustainability of the choices made. Darcy, Cameron, and Pegg (2010) argue that accessibility and sustainability are not mutually exclusive but should be considerations in sustainable destination management. While sustainability is quite often discussed without considering the three areas of environment, economic and social (+ governance), to do so with any infrastructure feasibility exercise is inappropriate, ineffective, and inefficient (Veal, 2010). People with disability and seniors with access needs make up 31% of tourists (Darcy & Dickson, 2009) and will account for 1.5 billion people worldwide by 2050 (World Health Organisation & World Bank, 2011). Amadeus (2017) and the GKF, University of

Surrey, Newman Consult, and Pro A Solutions (2015) identify that 2 billion people around the world have a disability or access needs and valued the market worldwide at US$500 billion annually.

To provide an example of the importance of considering the social sustainability of this group for the tourism industry we provide three examples of gondola design and operations in Skyrail Kuranda Queensland (Australia), Skyline Gondola Tours Queenstown (New Zealand), and the Ngong Ping gondola (Hong Kong). The two 'western' examples from Australasia chose a gondola design that prevents independent access by those who are wheelchair users, those who are frail aged, or are unsteady on their feet. From an operations perspective, those who are ambulant need to step up and across a vertical and horizontal gap from the platform to the carriage. For wheelchair users, the opening is only 57 cm (Queenstown, New Zealand) and 63 cm (Kuranda, Australia). This precludes all but the narrowest of manual wheelchairs with the exception of those that fold and can be lifted aboard the carriage (Queenstown, New Zealand) or tilted/lifted (Kuranda, Australia), as shown in Figure 3.1, with the person having to either walk themselves (most wheelchair users don't have this option) or be lifted or assisted on board, which can be a work, health, and safety issue if staff are not properly trained. To facilitate their experience requires that the gondola be stopped and restarted after access or egress. In the case of Skyrail (Kuranda) this is further complicated by having to change gondolas during the standard circuit.

In contrast, as shown in Figure 3.2, the case of Ngong Ping the cable car design is one based on a universal design principles where there are no vertical or horizontal gaps, a wide entry (120 cm), and circulation space for manual and power wheelchairs as well as other

Figure 3.1 Manual handling of wheelchair required for the Kuranda Skyrail

Source: Have Wheelchair Will Travel 2020 with permission

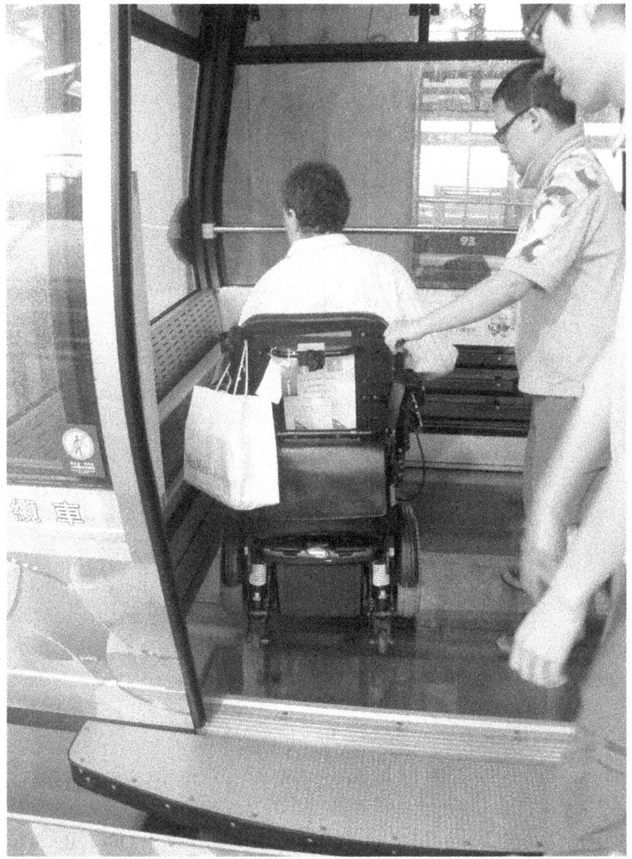

Figure 3.2 Ngong Ping cable car access showing universal design from platform to cable car without horizontal or vertical gap

Source: Fiona Darcy© 2010 with permission

mobility devices. The cable car only needs to be brought to a halt if requested by the individuals who are boarding. Given the 170 national signatories to the UN *Convention on the Rights of Persons with Disabilities* (United Nations, 2006), it would be hoped that all new tourism infrastructure in national parks would include universal design principles to provide transport access for all. Apart from the human rights arguments, this form of social sustainability built into the design and planning of parks infrastructure can provide a significant boost to a destination's and operator's economic bottom line as individuals with disabilities typically travel with families, friends, and business groups. Recent United Kingdom, United States, and Australian national visitor surveys have identified average group size of 2–5 people travelling with an individual with a disability on day trips or overnight travel (Amadeus, 2017; Darcy, McKercher, & Schweinsberg, 2020; GKF et al., 2015). As a footnote, the International Standards Organisation is currently preparing an international standard on accessible tourism due for completion in 2021 (see International Standards Organisation, 2020). This work, together with the UNWTO research, has established accessible tourism as part of mainstream considerations for destination competitiveness (Clara, Darcy, Garbero, & Almond, 2019; T. Vila, Darcy, & González, 2015).

A regional/international perspective

Sustainable transportation seeks to question "business as usual approaches to mobility ... It questions whether all trips are necessary and whether those ones most necessary could be done with minimal or no environmental burden" (Schiller, 2018, p. 234). As our understanding of what is or is not ecotourism evolves (see Fennell, 2015b), new forms of transport become essential for ensuring that desired ecotourism experiences can be met. Fennell (2015) has noted that to be sustainable, it is not enough for one part of the industry (e.g., attractions or transportation) to work in a sustainable way. Rather, the whole industry must be both working in concert, as well as working in a manner that acknowledges changing societal understanding of the value and role of nature (see Hay, 2002). For example, as humankind's attitude to the intrinsic value of the natural world has evolved, attitudes to the role of animals in the provision of ecotourism services have become more nuanced and sophisticated (Burns, Macbeth, & Moore, 2011; Fennell & Nowaczek, 2010; Lemelin, Fennell, & Smale, 2008; Stronza, Hunt, & Fitzgerald, 2019). Newsome and Rodger (2013) have argued that eco-tour operators tend to view wildlife tourism as sustainable on account of its educative nature and focus on minimal impacts on the animals concerned. As Wearing, Cunningham, Schweinsberg, and Jobberns (2014) have argued, however, whale-based ecotourism whilst non-extractive is not necessarily as sustainable as may be thought on first glance. Higham, Bejder, Allen, Corkeron, and Lusseau (2016) have argued that the sustainable management of whale watching operations requires that such activities be perceived in terms of their "sub-lethal anthropogenic stress and energetic impacts ... [and] for the industry to be managed within an architecture of strong national and international regulation" (p. 83). What this means for the present discussion is that as ecotourism resources dovetail with transport forms across international boundaries, the industry as a whole must not think of its sustainability solely in terms of its local impacts, but rather on a global scale with full acknowledgement of the impacts caused by transport and other parts of the tourism system that facilitate the movement of people to and from ecotourism destinations.

Quantifying such impacts and assigning responsibility for their management is often easier said than done. Gössling (2002) observed that some 90% of tourism's contribution to climate change comes from its transport dimension. The long-haul nature of many of the ecotourism destinations has meant that aviation policy has had a disproportionate impact on the sector's sustainable future relative to other parts of the global tourism industry. Gössling (1999) argues that fuel consumption, as an example of an environmental damage cost, needs to be factored into any robust measure of the 'total environmental impact' of a tourism operation. However, should we stop with fuel consumption? Weaver (2011) has asked, should we not also include the indirect impacts resulting from aircraft manufacture and maintenance? The partially industrialised nature of the global ecotourism industry makes such questions challenging to answer. For example, cruise ship tourism has been subject to critique over its sustainability in a variety of destinational contexts (Hritz & Cecil, 2008; Klein, 2011; Lamers, Eijgelaar, & Amelung, 2015; Larsen & Wolff, 2019). As Hopkins (2019) has observed, "maritime activities contribute to emissions by way of ship emissions, crane emissions, and the emissions associated with vehicles in the harbour" (p. 6). With cruise operators expanding into more remote localities as part of Last Chance Tourism (see Eijgelaar, Thaper, & Peeters, 2010), iconic localities including the North West Passage are now part of the cruise tourism itinerary (Sevunts, 2019; Snider, 2016). With Last Chance Tourism infrastructure such as ports being used for non-tourism purposes (see Schweinsberg, Wearing, & Lai, 2020), ecotourism's sustainability potential must increasingly be seen in relation to a multifaceted understanding of 'place'. At the same time, however, the development of new nature based tourism forms is not only precipitating the need for the development of new transport infrastructure (Lamars, Eijgelaar, & Amelung, 2012); it is also leading many commentators to question the ethics of travel to remote and threatened regions (M. Vila, Costa, Angulo-Preckler, Sarda, & Avila, 2016) and by implication to question the sustainability of different tourism related transport forms that may be involved in their facilitation. At a philosophical level the act of tourism related travel cannot be separated from transport; to have one without the other Gross and Grimm (2018) observe would be "inconceivable" (p. 402). Given that people also have an intrinsic right to travel, as per article seven of the World Tourism Organisation's *Code of Ethics for Tourism* (1999), the question becomes how one can make ecotourism transport options more sustainable within the framework local and global circumstance?

Pursuing a sustainable ecotourism transport future

Simmons and Becken (2004) have argued with respect to New Zealand that the ecotourism sector's contribution to greenhouse gas emissions is both a consequence of it being over 12,000 km from popular tourism-generating regions in Europe, as well as based on what they describe as an often over reliance on private cars for commutes to popular ecotourism sites. With respect to the first issue—travel to the destination—Wood (2017) has shown how an inefficient long-haul flight of a little over 12,000 km can emit more CO_2 than is recommended for an individual per year if we are to help keep global temperature increases to less than 2 degrees. Simmons and Becken (2004) have similarly demonstrated that the CO_2 emissions from a short-haul flight between New Zealand and Australia is roughly equivalent to the total CO_2 emissions of a hypothetical 20-day self-drive tour in New Zealand.

Any move to develop a more sustainable ecotourism transport future in these and other destinations must find a way to manage the interplay of two often inextricably opposing forces. On the one hand, there are technologies out there that can help to mitigate the negative environmental and other impacts from ecotourism transport options. These technologies include virtual reality experiences (Bristow, 1999) where tourists are offered the opportunity for remote experience of nature in lieu of being transported to the site in question through technologies like Google Earth (see Appnbrink, 2019). While the use of interactive multimedia is recognised as having a role to play in facilitating a mediated experience of ecologically vulnerable resources (Skibins & Sharp, 2019), it has also been observed that a range of independent variables will play role in a tourist's level of satisfaction with such experiences (Orru, Kask, & Nordlund, 2019). Similarly, the use of drones as a mechanism for conservation and destination marketing has been recognised by groups including Ecotourism Australia (Mills, 2016). While the potential for drones to impact on the behavioural patterns of local wildlife is acknowledged (Ditmer et al., 2015); to date, drones have received minimal attention in the ecotourism literature (exceptions include King, 2014). One of the key issues for tourism scholars around the use of new technologies will be to what degree does the average tourist care about the environmental consequences of their transport choices? Ecotourists are not a homogenous group with respect to their motivations to travel, their level of environmental awareness or behaviours (Weaver & Lawton, 2002). As such, there is the potential for some travellers to maintain a socially organised denial over the relationship between their transport choices and their impact (Hanna & Adams, 2019). In recent years, tourism scholars have begun to employ the concept of 'Akrasia' to explore some tourists' "deficient capacity to contain or restrain one's desires, broadly conceived; where the anticipation of pleasure overwhelms good judgement" (Fennell, 2015a, p. 95).

While much of the discussion around a sustainable transport future will focus on the use of different transport forms, what must also be considered is the potential for transport systems to render an ecotourist immobile for sustainability gains. Within the wider sustainable tourism literature the concept of de-growth has recently come into vogue (Andriotis, 2018; Higgins-Desbiolles, Carnicelli, Krolikowski, Wijesinghe, & Boluk, 2019). Referring to the need to decouple tourism's sustainable future from a growth-at-all-costs mantra, which had characterised neoliberal economies in the twentieth century (Fletcher, Murray Mas, Blanco-Romero, & Blázquez-Salom, 2019); de-growth is seen as offering a sustainable future for the industry, one where the "rights of local communities [are valued] above the rights of tourists for holidays or the rights of tourism corporations for profits" (Higgins-Desbiolles et al., 2019, p. 1926). Andriotis (2018) has argued that ecotourists who ascribe to a de-growth mantra will tend to make environmental protection the overriding factor when making decisions over transport choices. This will often mean avoiding long-haul air travel and automobiles in favor of slow tourism alternatives. Wearing, Wearing, and McDonald (2012) have argued that slowing down the pace of our ecotourism experiences carries with it the potential for the development of sustainable links between the tourism industry, nature, and communities. At the same time, however, any move to render a particular transport form accessible to some will equally often have the effect of rendering others immobile (Hopkins, 2019).

The development of urban ecotourism in areas along the urban/rural interface is illustrative of this phenomena. Urban ecotourism has been recognised for many years as being a means of cultivating tourism demand whilst also helping to green and make more sustainable what are often otherwise rundown urban precincts (Gibson, Dodds, Joppe, & Jamieson, 2003; Higham & Lück, 2002). While the literature exploring the transport dimensions of urban ecotourism is limited (see Aversa, Petrescu, Apicella, & Petrescu, 2017), there is recognition that urban ecotourism destinations serviced by effective public transport systems (Wu, Wang, & Ho, 2010) can have the effect of encouraging more sustainable destination futures (Sorupia, 2007). While such futures may increase the transport options available to some eco-travellers, others from emerging destinations may be precluded on the basis of their inability to navigate the complex array of regulatory restrictions governing the global tourism industry. Duffy (2006) has observed that at the national level "ecotourism requires a particular kind of national policy environment that favours the global market, accepts and facilitates global air travel and supports the development of local transport networks … to serve the ecotourism market" (p. 5). While many examples exist of best practice with respect to the facilitation of sustainable ecotourism experiences (Edgell, 2016; Queensland Government Department of Environment and Science, 2017–2019); what is still needed is comprehensive evaluation of ecotourism's transport needs and a recognition that not only does the industry need to "address the aviation issue, … [but also that they need to] give travellers the option of doing something to repair the damage they do" (Mader, 2019).

Conclusion

The chapter is entitled *Ecotourism and the trouble with transportation* to draw attention to the fact that whilst the aviation sector and other transport forms are vital for the enactment of the ecotourism experience (see Snow & Snow, 2003), the environmental damage from transport mechanisms is often disproportionate to other parts of the ecotourism system (Folke, Østrup, & Gössling, 2006). The scope of the problem was articulated by the World Tourism Organisation and the United Nations Environment Program who estimated that the "tourism sector contributed approximately 5% of all man-made CO_2 emissions in 2005, with transport representing the largest component, i.e., 75% of the overall emissions of the sector" (World Tourism Organisation & International Transport Forum, 2019, p. 12). How can we seek to move the industry out of the paradox that characterises a reliance on that very thing that poses the greatest risk to ecotourism's survival? Firstly, one must recognise that a business as usual approach to tourism mobility is inconsistent with global efforts to reduce levels of CO_2 emissions in line with the Paris Climate Agreements (see Peeters et al., 2019). Secondly, we must embrace the access—impact paradox. We must look to understand mobility needs, both in the destination and more regionally and develop coordinated transport plans that align transport offerings to both the carrying capacity of a locality and perceptions of visitors of the appropriateness of different tourist numbers (Scuttari, Orsi, & Bassani, 2019). Finally, we must recognise that the right to free movement and mobility for all is governed under both the International Covenant on Civil and Political Rights (see Perkumienė & Pranskūnienė, 2019). We should not try to halt ecotourism growth on account of transport, but rather recognise that as with sustainable development debates more broadly, the tourism industry has the potential to use its diverse systems-based characteristics to be a leader in the development and operationalisation of sustainable transport forms.

References

Abbey, E. (1968). *Desert Solitaire*. New York: McGraw Hill.

Abebe, F. B., & Bekele, S. E. (2018). Challenges to national park conservation and management in Ethiopia. *Journal of Agricultural Science, 10*(52), 10.5539.

Amadeus. (2017). Voyage of Discovery: Working Towards Inclusive and Accessible Travel for All. Retrieved from http://www.amadeus.com/blog/04/10/accessible-travel-report/.

Amerson, A., & Parsons, E. (2018). Evaluating the sustainability of the gray-whale-watching industry along the pacific coast of North America. *Journal of Sustainable Tourism, 26*(8), 1362–1380.

Andriotis, K. (2018). *Degrowth in Tourism: Conceptual, Theoretical and Philosphical Issues*. Oxfordshire: CABI.

Anon. (2008). Penguin Parade Has Tourists in Flap. Retrieved from https://www.heraldsun.com.au/travel/australia/penguins-have-tourists-in-flap/news-story/e5cb02e78f78aee57d1313c4cc2827b9.

Appnbrink, K. (2019). Visit the U.S. National Parks in Google Earth. Retrieved from https://www.blog.google/products/earth/visit-us-national-parks-google-earth/.

Aversa, R., Petrescu, R. V., Apicella, A., & Petrescu, F. I. (2017). Modern transportation and photovoltaic energy for urban ecotourism. *Transylvanian Review of Administrative Sciences, Special*(2017), 5–20.

Bristow, R. S. (1999). Commentary: Virtual tourism—the ultimate ecotourism? *Tourism Geographies, 1*(2), 219–225.

Burns, G. L., Macbeth, J., & Moore, S. (2011). Should dingoes die? Principles for engaging ecocentric ethics in wildlife tourism management. *Journal of Ecotourism, 10*(3), 179–196.

Clara, R., Darcy, S., Garbero, N., & Almond, B. (2019). Critical elements in accessible tourism for destination competitiveness and comparison: Principal component analysis from Oceania and South America. *Tourism Management, 75*, 169–185.

CS Global Partners. (2019). New Financial Times Documentary: Dominica Is Carving a Niche of Its Own in Ecotourism, Raising Foreign Capital via Citizenship by Investment. Retrieved from https://www.prnewswire.com/news-releases/new-financial-times-documentary-dominica-is-carving-a-niche-of-its-own-in-ecotourism-raising-foreign-capital-via-citizenship-by-investment-300957782.html.

Cunningham, P., Huijbens, E., & Wearing, S. (2012). From whaling to whale watching: Examining sustainability and cultural rhetoric. *Journal of Sustainable Tourism, 20*(1), 143–161.

Curry, T. J., & Gordon, K. O. (2017). Muir, Roosevelt, and Yosemite National Park as an emergent sacred symbol: An interaction ritual analysis of a camping trip. *Symbolic Interaction, 40*(2), 247–262.

Darcy, S., Cameron, B., & Pegg, S. (2010). Accessible tourism and sustainability: A discussion and case study. *Journal of Sustainable Tourism, 18*(4), 515–537.

Darcy, S., & Dickson, T. (2009). A whole-of-life approach to tourism: The case for accessible tourism experiences. *Journal of Hospitality and Tourism Management, 16*(1), 32–44.

Darcy, S., McKercher, B., & Schweinsberg, S. (2020). From tourism and disability to accessible tourism: a perspective article. *Tourism Review, 75*(1).

Ditmer, M. A., Vincent, J. B., Werden, L. K., Tanner, J. C., Laske, T. G., Iaizzo, P. A., ... Fieberg, J. R. (2015). Bears show a physiological but limited behavioral response to unmanned aerial vehicles. *Current Biology, 25*(17), 2278–2283.

Duffy, R. (2006). The politics of ecotourism and the developing world. *Journal of Ecotourism, 5*(1–2), 1–6.

Duffy, R., & Moore, L. (2010). Neoliberalising nature? Elephant-back tourism in Thailand and Botswana. *Antipode, 42*(3), 742–766.

Ecotourism Australia. (2019a). Eco Destination Certification: Essentials. Retrieved from https://www.ecotourism.org.au/assets/Destination-Certification/2019-Program-Essentials-ECO-Destination-Certification-compressed.pdf.

Ecotourism Australia. (2019b). MEDIA RELEASE: Douglas Shire Becomes World's First ECO Certified Destination. Retrieved from https://www.ecotourism.org.au/news/media-release-douglas-shire-becomes-worlds-first-eco-certified-destination/.

Edgell, D. (2016). *Managing Sustainable Tourism: A Legacy for the Future*. London: Routledge.

Edgell, D., & Swanson, J. (2019). *Tourism Policy and Planning: Yesterday, Today and Tomorrow*. London: Routledge.

Eijgelaar, E., Thaper, C., & Peeters, P. (2010). Antarctic cruise tourism: the paradoxes of ambassadorship,"last chance tourism" and greenhouse gas emissions. *Journal of Sustainable Tourism, 18*(3), 337–354.

Fennell, D. (2002). Ecotourism: Where we've been; where we're going. *Journal of Ecotourism, 1*(1), 1–16.

Fennell, D. (2015a). Akrasia and tourism: Why we sometimes act against our better judgement? *Tourism Recreation Research, 40*(1), 95–106.

Fennell, D. (2015b). *Ecotourism*. London: Routledge.

Fennell, D., & Nowaczek, A. (2010). Moral and empirical dimensions of human–animal interactions in ecotourism: Deepening an otherwise shallow pool of debate. *Journal of Ecotourism, 9*(3), 239–255.

Ferreira, S., & Harmse, A. (2014). Kruger National Park: Tourism development and issues around the management of large numbers of tourists. *Journal of Ecotourism, 13*(1), 16–34.

Fletcher, R., Murray Mas, I., Blanco-Romero, A., & Blázquez-Salom, M. (2019). Tourism and degrowth: An emerging agenda for research and praxis. *Journal of Sustainable Tourism, 27*(12), 1745–1763. doi:10.1080/09669582.2019.1679822.

Folke, J., Østrup, J., & Gössling, S. (2006). 14 Ecotourist choices of transport modes. In S. Gössling, & J. Hultman (Eds.), *Ecotourism in Scandinavia: Lessons in theory and practice* (Vol. 4, pp. 154–165). Wallingford, Oxfordshire, UK: CABI.

Gibson, A., Dodds, R., Joppe, M., & Jamieson, B. (2003). Ecotourism in the city? Toronto's green tourism association. *International Journal of Contemporary Hospitality Management, 15*(6), 324–327.

Giglio, V. J., Luiz, O. J., Chadwick, N. E., & Ferreira, C. E. (2018). Using an educational video-briefing to mitigate the ecological impacts of scuba diving. *Journal of Sustainable Tourism, 26*(5), 782–797.

GKF, University of Surrey, Newman Consult, & Pro A Solutions. (2015). Economic Impact and Travel Patterns of Accessible Tourism. Retrieved from https://www.google.com.au/url?sa=t&rct=j&q=&esrc=s&source=web&cd=1&cad=rja&uact=8&ved=0ahUKEwjRps-nwdTMAhVDlKYKHY1fBLEQFggcMAA&url=http%3A%2F%2Fec.europa.eu%2FDocsRoom%2Fdocuments%2F5566%2Fattachments%2F1%2Ftranslations%2Fen%2Frenditions%2Fnative&usg=AFQjCNEY6QmrblkXMcZ-ii0naFfkteR1dQ&sig2=nPjOfgzmcm75v6KgHT4LRQ.

Gössling, S. (1999). Ecotourism: A means to safeguard biodiversity and ecosystem functions? *Ecological Economics, 29*(2), 303–320.

Gössling, S. (2002). Global environmental consequences of tourism. *Global Environmental Change, 12*(4), 283–302.

Graham, B., & Shaw, J. (2008). Low-cost airlines in Europe: Reconciling liberalization and sustainability. *Geoforum, 39*(3), 1439–1451.

Gross, S., & Grimm, B. (2018). Sustainable mode of transport choices at the destination–public transport at German destinations. *Tourism Review, 73*(3), 401–420.

Hall, C. (2013). Ecotourism and global environmental change. In R. Ballantyne, & J. Packer (Eds.), *International handbook on ecotourism* (pp. 54–65).Cheltenham : Edward Elgar Pub. Ltd.

Hall, C., & Ram, Y. (2019). Measuring the relationship between tourism and walkability? Walk score and English tourist attractions. *Journal of Sustainable Tourism, 27*(2), 223–240.

Hammerton, Z. (2017). Low-impact diver training in management of SCUBA diver impacts. *Journal of Ecotourism, 16*(1), 69–94.

Hanna, P., & Adams, M. (2019). Positive self-representations, sustainability and socially organised denial in UK tourists: Discursive barriers to a sustainable transport future. *Journal of Sustainable Tourism, 27*(2), 189–206.

Harvey, M. (2002). Sound politics: Wilderness, recreation, and motors in the boundary waters, 1945-1964. *Minnesota History, 58*(3), 130–145.

Hay, P. (2002). *Main Currents in Western Environmental Thought*. Bloomington: Indiana University Press.

Head, L. M. (2000). Renovating the landscape and packaging the penguin: Culture and nature on Summerland Peninsula, Phillip Island, Victoria, Australia. *Australian Geographical Studies, 38*(1), 36–53.

Herremans, I. (2006a). Grand Canyon National Park: Tourists by land, tourists by air. In I. Herremans (Ed.), *Cases in sustainale tourism: Resource guides for an experiential learning environment* (pp. 175–186). New York: The Haworth Hospitality Press.

Herremans, I. (Ed.) (2006b). *Cases in Sustainable Tourism: An Experiential Approach to Descision Making*. London: Haworth Hospitality Press.

Higgins-Desbiolles, F., Carnicelli, S., Krolikowski, C., Wijesinghe, G., & Boluk, K. (2019). Degrowing tourism: rethinking tourism. *Journal of Sustainable Tourism*, 1–19.

Higham, J., & Lück, M. (2002). Urban ecotourism: A contradiction in terms? *Journal of Ecotourism, 1*(1), 36–51.

Higham, J. E., Bejder, L., Allen, S. J., Corkeron, P. J., & Lusseau, D. (2016). Managing whale-watching as a non-lethal consumptive activity. *Journal of Sustainable Tourism, 24*(1), 73–90.

Hopkins, D. (2019). Sustainable mobility at the interface of transport and tourism: Introduction to the special issue on 'Innovative approaches to the study and practice of sustainable transport, mobility and tourism'. *Journal of Sustainable Tourism*, 1–15.

Horton, L. R. (2009). Buying up nature: Economic and social impacts of Costa Rica's ecotourism boom. *Latin American Perspectives, 36*(3), 93–107.

Hritz, N., & Cecil, A. K. (2008). Investigating the sustainability of cruise tourism: A case study of Key West. *Journal of Sustainable Tourism, 16*(2), 168–181.

International Standards Organization. (2020). Iso/dis 21902 Tourism and Related Services — Accessible Tourism for All — Requirements and Recommendations. Retrieved from https://www.iso.org/standard/72126.html.

Kessler, M., Harcourt, R., & Heller, G. (2013). Swimming with whales in Tonga: Sustainable use or threatening process? *Marine Policy, 39*, 314–316.

Khadaroo, J., & Seetanah, B. (2007). Transport infrastructure and tourism development. *Annals of Tourism Research, 34*(4), 1021–1032.

King, L. M. (2014). Will drones revolutionise ecotourism? *Journal of Ecotourism, 13*(1), 85–92.

Klein, R. A. (2011). Responsible cruise tourism: Issues of cruise tourism and sustainability. *Journal of Hospitality and Tourism Management, 18*(1), 107–116.

Kontogeorgopoulos, N. (2009). Wildlife tourism in semi-captive settings: A case study of elephant camps in northern Thailand. *Current Issues in Tourism, 12*(5–6), 429–449.

Lamars, M., Eijgelaar, E., & Amelung, B. (2012). Last chance tourism in Antarctica: Cruising for change? In R. Lemelin, J. Dawson, & E. Stewart (Eds.), *Last chance tourism: Adapting tourism opportunities in a changing world* (pp. 25–41). London: Routledge.

Lamers, M., Eijgelaar, E., & Amelung, B. (2015). The environmental challenges of cruise tourism. In C. Hall, S. Gossling, & D. Scott (Eds.), *The Routledge handbook of tourism and sustainability* (pp. 430–439). London: Routledge.

Larsen, S., & Wolff, K. (2019). Aspects of the unsustainability of cruise tourism. Presentation at the CAUTHE 2019: Sustainability of Tourism, Hospitality & Events in a Disruptive Digital Age: Proceedings of the 29th Annual Conference.

Lemelin, R. H., Fennell, D., & Smale, B. (2008). Polar bear viewers as deep ecotourists: How specialised are they? *Journal of Sustainable Tourism, 16*(1), 42–62.

Lorenzo-Romero, C., Alarcón-del-Amo, M.-d.-C., & Crespo-Jareño, J.-A. (2019). Cross-cultural analysis of the ecological behavior of Chilean and Spanish ecotourists: A structural model. *Ecology and Society, 24*(4).

Lovelock, B. A. (2010). Planes, trains and wheelchairs in the bush: Attitudes of people with mobility-disabilities to enhanced motorised access in remote natural settings. *Tourism Management, 31*(3), 357–366.

Lumsdon, L. (2000). Transport and tourism: Cycle tourism–a model for sustainable development? *Journal of Sustainable Tourism, 8*(5), 361–377.

Lumsdon, L., & Page, S. (2007). *Tourism and Transport.* Amsterdam, Boston : Elsevier.

Mader, R. (2019). Sustainable Development of Ecotourism Web Conference Report (2002). Retrieved from https://planeta.com/iye-web-conference-english/.

Mills, E. (2016). How Drones Help Conservation Efforts. Retrieved from https://www.ecotourism.org.au/news/how-drones-help-conservation-efforts/.

National Parks Service. (2004). The Development of Park Roads - Autombiles: Yea or Nay. Retrieved from https://www.nps.gov/parkhistory/online_books/roads/shs2.htm.

National Parks Service. (2016). Early Visitors. Retrieved from https://www.nps.gov/yell/learn/historyculture/early-visitors.htm.

National Parks Service. (2019). Air Tours. Retrieved from https://www.nps.gov/subjects/sound/airtours.htm.

Newsome, D., & Davies, C. (2009). A case study in estimating the area of informal trail development and associated impacts caused by mountain bike activity in John Forrest National Park, Western Australia. *Journal of Ecotourism, 8*(3), 237–253.

Newsome, D., Milewski, A., Phillips, N., & Annear, R. (2002). Effects of horse riding on national parks and other natural ecosystems in Australia: Implications for management. *Journal of Ecotourism, 1*(1), 52–74.

Newsome, D., Moore, S., & Dowling, R. (2012). *Natural Area Tourism: Ecology, Impacts and Management* (Vol. 58). Clevedon: Channel View Publications.

Newsome, D., & Rodger, K. (2013). Wildlife tourism. In A. Holden, & D. Fennell (Eds.), *The Routledge handbook of tourism and the environment* (pp. 345–358). London and New York: Routledge.

Newsome, D., Smith, A., & Moore, S. (2008). Horse riding in protected areas: A critical review and implications for research and management. *Current Issues in Tourism, 11*(2), 144–166.

Orams, M. B. (2001). From whale hunting to whale watching in Tonga: A sustainable future? *Journal of Sustainable Tourism, 9*(2), 128–146.

Orru, K., Kask, S., & Nordlund, A. (2019). Satisfaction with virtual nature tour: The roles of the need for emotional arousal and pro-ecological motivations. *Journal of Ecotourism, 18*(3), 221–242.

Peeters, P., Higham, J., Cohen, S., Eijgelaar, E., & Gössling, S. (2019). Desirable tourism transport futures. *Journal of Sustainable Tourism, 27*(2), 173–188.

Perkumienė, D., & Pranskūnienė, R. (2019). Overtourism: Between the right to travel and residents' rights. *Sustainability, 11*(7), 2138.

Pickering, C. M., Rossi, S., & Barros, A. (2011). Assessing the impacts of mountain biking and hiking on subalpine grassland in Australia using an experimental protocol. *Journal of Environmental Management, 92*(12), 3049–3057.

Pookhao Sonjai, N., Bushell, R., Hawkins, M., & Staiff, R. (2018). Community-based ecotourism: Beyond authenticity and the commodification of local people. *Journal of Ecotourism, 17*(3), 252–267.

Queensland Government Department of Environment and Science. (2017–2019). Strategies and Plans: Queensland Ecotourism Plan 2016–2020. Retrieved from https://parks.des.qld.gov.au/tourism/ecotourism/strategies/.

Reggers, A., Grabowski, S., Wearing, S. L., Chatterton, P., & Schweinsberg, S. (2016). Exploring outcomes of community-based tourism on the Kokoda Track, Papua New Guinea: A longitudinal study of participatory rural appraisal techniques. *Journal of Sustainable Tourism, 24*(8–9), 1139–1155.

Reggers, A., Schweinsberg, S., & Wearing, S. (2013). Understanding stakeholder values in co-management arrangements for protected area establishment on the Kokoda Track, Papua New Guinea. *Journal of Park and Recreation Administration, 31*(3).

Richardson, E. R. (1959). The struggle for the valley: California's Hetch Hetchy controversy, 1905-1913. *California Historical Society Quarterly, 38*(3), 249–258.

Roosevelt, T. (1903). Letter from Theodore Roosevelt to John Muir. Retrieved from https://www.theodorerooseveltcenter.org/Research/Digital-Library/Record?libID=o184892.

Sarker, M. A. R., Hossan, C. G., & Zaman, L. (2012). Sustainability and growth of low cost airlines: An industry analysis in global perspective. *American Journal of Business and Management, 1*(3), 162–171.

Schiller, P. (2018). From feet to wheels to and feet machines and back. In J. Caradonna (Ed.), *Routledge handbook of the history of sustainability* (pp. 233–253). London: Routledge.

Schweinsberg, S., Darcy, S., & Wearing, S. (2018). Repertory grids and the measurement of levels of community support for rural ecotourism development. *Journal of Ecotourism, 17*(3), 239–251.

Schweinsberg, S., Wearing, S., & Darcy, S. (2012). Understanding communities' views of nature in rural industry renewal: The transition from forestry to nature-based tourism in Eden, Australia. *Journal of Sustainable Tourism, 22*(2), 195–213.
Schweinsberg, S., Wearing, S., & Wearing, M. (2015). Transforming nature's value – cultural change comes from below: Rural communities, the 'Othered' and host capacity building. In Y. Reisinger (Ed.), *Transforming Tourism: Host Perspectives* (pp. 102–113). Wallingford, UK: CABI.
Schweinsberg, S., Wearing, S., & Lai, P. (2020). Host communities and last chance tourism. *Tourism Geographies*. https://www.tandfonline.com/doi/abs/10.1080/14616688.2019.1708446
Scuttari, A., Orsi, F., & Bassani, R. (2019). Assessing the tourism-traffic paradox in mountain destinations. A stated preference survey on the Dolomites' passes (Italy). *Journal of Sustainable Tourism, 27*(2), 241–257.
Sevunts, L. (2019). 2019 Saw Increase in Commercial Shipping Through Northwest Passage. Retrieved from https://www.rcinet.ca/en/2019/12/11/2019-commercial-shipping-through-northwest-passage/.
Simmons, D., & Becken, S. (2004). The cost of getting there: Impacts of travel to ecotourism destinations. In R. Buckley (Ed.), *Environmental impacts of ecotourism*. E-Book: CABI.
Skibins, J. C., & Sharp, R. L. (2019). Binge watching bears: Efficacy of real vs. virtual flagship exposure. *Journal of Ecotourism, 18*(2), 152–164.
Smith, A., Tuffin, M., Taplin, R., Moore, S., & Tonge, J. (2014). Visitor segmentation for a park system using research and managerial judgement. *Journal of Ecotourism, 13*(2–3), 93–109.
Snider, D. (2016). Crystal Serenity–A New Chapter in Arctic Shipping or Just "Doing It Right"? *The Arctic Yearbook*. Retrieved from https://martechpolar.com/sites/default/files/AY%202016%20COM%20Snider%20Chrystal%20Serinity.pdf.
Snow, M. M., & Snow, R. K. (2003). Aviation opportunities in ecotourism. *Journal of Aviation/Aerospace Education & Research, 12*(2), 3.
Sorupia, E. (2007). Transport Networks and Ecotourism Destinations: The Aim for Sustainability. Retrieved from https://minerva-access.unimelb.edu.au/bitstream/handle/11343/39366/67632_00004015_01_ESorupia_Thesis.pdf?sequence=1.
Stronza, A. L., Hunt, C. A., & Fitzgerald, L. A. (2019). Ecotourism for conservation? *Annual Review of Environment and Resources, 44*, 229–253.
Sullivan, F. A., & Torres, L. G. (2018). Assessment of vessel disturbance to gray whales to inform sustainable ecotourism. *The Journal of Wildlife Management, 82*(5), 896–905.
Taylor, M., Hurst, C. E., Stinson, M. J., & Grimwood, B. S. (2019). Becoming care-full: Contextualizing moral development among captive elephant volunteer tourists to Thailand. *Journal of Ecotourism*, 1–19.
United Nations. (2006). Convention on the Rights of Persons with Disabilities (CRPD). Retrieved from https://www.un.org/development/desa/disabilities/convention-on-the-rights-of-persons-with-disabilities.html.
Veal, A. (2010). *Leisure, Sport and Tourism, Politics and Planning* (3rd ed.). Oxfordshire: Cabi Publishing.
Vila, M., Costa, G., Angulo-Preckler, C., Sarda, R., & Avila, C. (2016). Contrasting views on Antarctic tourism:'Last chance tourism'or 'ambassadorship' in the last of the wild. *Journal of Cleaner Production, 111*, 451–460.
Vila, T., Darcy, S., & González, E. (2015). Competing for the disability tourism market–a comparative exploration of the factors of accessible tourism competitiveness in Spain and Australia. *Tourism Management, 47*, 261–272.
Wearing, S., Cunningham, P., Schweinsberg, S., & Jobberns, C. (2014). Whale watching as ecotourism: How sustainable is it? *Cosmopolitan Civil Societies: An Interdisciplinary Journal, 6*(1), 38–55.
Wearing, S., McDonald, M., Ankor, J., & Schweinsberg, S. (2015). The nature of aesthetics: How consumer culture has changed our national parks. *Tourism Review International, 19*(4), 225–233.
Wearing, S., McDonald, M., Schweinsberg, S., Chatterton, P., & Bainbridge, T. (2019). Exploring tripartite praxis for the REDD+ forest climate change initiative through community based ecotourism. *Journal of Sustainable Tourism*, 1–17.
Wearing, S., Schweinsberg, S., & Darcy, S. (2019). Consuming our national parks: Cultural heritage in a consumer culture. In A. Campelo, L. Reynolds, A. Lindgreen, and M. Beverland, (Eds.), *Consumer culture* (pp. 183–194), London: Routledge.
Wearing, S., & Schweinsberg, S. (2018). *ECOTOURISM: Transitioning to the 22nd Century*. London: Routledge.
Wearing, S., Schweinsberg, S., & Tower, J. (2016). *The Marketing of National Parks for Sustainable Tourism*. Clevedon: Channel View.

Wearing, S., Wearing, M., & McDonald, M. (2012). *Slow'n Down the Town to Let Nature Grow: Ecotourism, Social Justice and Sustainability*. Bristol: Channel View.

Weaver, D. (2011). *Ecotourism* (2nd ed.). Milton, QLD: Wiley.

Weaver, D., & Lawton, L. (2002). Overnight ecotourist market segmentation in the Gold Coast hinterland of Australia. *Journal of Travel Research, 40*(3), 270–280.

White, D. D. (2007). An interpretive study of Yosemite National Park visitors' perspectives toward alternative transportation in Yosemite Valley. *Environmental Management, 39*(1), 50–62.

White, D. D., Waskey, M. T., Brodehl, G. P., & Foti, P. E. (2006). A comparative study of impacts to mountain bike trails in five common ecological regions of the Southwestern US. *Journal of Park & Recreation Administration, 24*(2).

Wichelns, R. (2017). For National Parks, Helicopter Tours Are a Noisy Problem. *Backpacker.*

Wood, M. E. (2017). *Sustainable Tourism on a Finite Planet: Environmental, Business and Policy Solutions.* London: Routledge.

World Health Organization, & World Bank. (2011). World Report on Disability.

World Tourism Organization, & International Transport Forum. (2019). Transport-Related CO2 Emissions of the Tourism Sector – Modelling Results. Retrieved from https://www.e-unwto.org/doi/pdf/10.18111/9789284416660.

Wu, Y.-Y., Wang, H.-L., & Ho, Y.-F. (2010). Urban ecotourism: Defining and assessing dimensions using fuzzy number construction. *Tourism Management, 31*(6), 739–743.

Youngs, Y. L., White, D. D., & Wodrich, J. A. (2008). Transportation systems as cultural landscapes in national parks: The case of Yosemite. *Society and Natural Resources, 21*(9), 797–811.

4

LINKING RESILIENCE THINKING AND SUSTAINABILITY PILLARS TO ECOTOURISM PRINCIPLES

Valerie A. Sheppard

Introduction

In 1998, the United Nations (UN) declared 2002 the International Year of Ecotourism (IYE) (United Nations Social and Economic Council, 1998). In so doing, the UN anticipated that governments, international and regional organisations, and non-governmental organisations (NGOs) would cooperate and collaborate in promoting ecotourism as a more sustainable form of tourism that would

> meet the needs of present tourists and host communities and regions while protecting and enhancing opportunities for the future, managing resources to fulfil economic, social, and aesthetic needs, and maintaining cultural integrity, essential ecological processes, biological diversity and life-support systems. (p. 2)

Certainly, these were lofty goals for the IYE, and particularly so for the ecotourism businesses that would be expected to deliver upon these goals. Indeed, many have observed the challenges associated with envisioning ecotourism as exemplary model for sustainable tourism development. For example, Butcher (2006, p. 155) contends it promotes the view that traditional knowledge is "as good, if not better, than modern technology," which then discourages modernisation. He adds that it empowers communities to modify, but not transform their relationship with the natural environment.

The ability of individuals, communities, organisations, and systems of governance to transform (see Chapter 23) is associated with the concept of resilience. Recently, a growing body of scholars have sought to understand how resilience thinking may supplement and complement sustainability approaches for encouraging better tourism business practices. Indeed, resilience is a concept and a development approach that is increasingly drawing the attention of scholars and development agencies. These agents and agencies contend that a resilience approach may assist humanity in better managing its responses to changes within socio-ecological systems (SESs) and that it may be effective in correcting the unsustainable trajectory of human societies (see Lebel et al., 2006). On the other hand, some contend there is a lack of scientific consensus that a resilience approach will lead to more sustainable forms of development (Simmie & Martin, 2010).

DOI: 10.4324/9781003001768-4

Despite Butcher's and others' caveats, ecotourism continues to be promoted as a form of tourism that reconciles the conflicting goals of tourism development and the conservation of nature (Baral, 2013). It also continues to be associated with the concepts of sustainability and, increasingly, resilience, as some consider ecotourism to be a business model that is better able to adapt and adjust to system disturbances. Given these competing perspectives there is value in conceptualising how a set of ideal ecotourism principles are linked with resilience thinking and the pillars of sustainability. This step is important for assisting in the development of practical business models to guide all tourism operations along more sustainable and resilient paths. Such a business model or models, are not a new focus. Indeed, scholars, governments, NGOs, businesses, and other interested organisations and individuals have worked collaboratively for decades to develop a variety of business-focused sustainability models and assessment criteria (see the Global Sustainable Tourism Council criteria at www.gstcouncil.org/gstc-criteria/). However, most of these frameworks have neglected to consider and apply a resilience perspective.

Consequently, the purpose of this chapter is to take a step back from sustainability models and assessment frameworks, to examine the nexus between the concepts of resilience and sustainability. This is an important step in order to conceptualise how both perspectives may enhance ecotourism practice, as well as inform other types of tourism practice. As volumes have been written about both concepts, the chapter begins by presenting a brief overview of tourism sustainability before moving on to explore the concept of resilience, generally, and then as it relates to tourism. From there the chapter briefly explores the links between resilience and sustainability. The intention here is to conceptualise how and where the three concepts, sustainability, resilience, and ecotourism, overlap and diverge, and to then present them as a conceptual model of ecotourism resilience and sustainability. The chapter concludes by suggesting areas for future research.

Sustainable tourism overview

McCool (2013) attributes the concept of sustainable tourism to two merging issues in the late 1980s. The first was related to the economic development of nations and regions characterised by high levels of poverty, a lack of access to health and education, and a limited ability to participate in the global economy. The second was related to the fact that society was paying increasing attention to the economic impacts of tourism. The economic impacts arose from the rapid growth in tourism travel, and the fact that tourists were penetrating destinations, many of which characterised by low income and high biological diversity. Natural and cultural heritage were considered means for economic growth, but at the same time there was increasing recognition of the associated negative social, environmental, and economic consequences of this focus.

The first documented use of the words *sustainable tourism* is attributed to J. J. Pigram, whose 1990 chapter entitled 'Sustainable Tourism – Policy Considerations', appeared in *The Journal of Tourism Studies* (McCool, 2013). Sustainable tourism was considered a "mental model" for guiding environmental, social, and economic issues through policy (McCool, Butler, Buckley, Weaver, & Wheeler, 2013). Specifically, it is often envisioned as a type of small-scale tourism venture, focused on social justice and minimising the negative environmental impacts (McCool, 2013). Ecotourism came to be seen as epitomising the concept of sustainable tourism (McCool et al., 2013).

However, many have observed the challenges with defining and achieving sustainable tourism development. Butler (2013, p. 224), for example, contends that the concept of sustainable tourism is now "distorted, politicized, and changed beyond general recognition from what may have been implied in the Brundtland Report" (see World Commission on Economic Development WCED, 1987). Others, such as McCool et al. (2013), consider sustainable tourism a utopian dream with little guidance to assess the actually sustainability of a venture. Similarly, Espiner, Orchiston, and Higham (2017, p. 4) contend that "steady-state sustainable tourism is an archaic form of thinking," that it is neither realistic nor relevant. Some researchers have observed the lack of focus on governance models and management strategies that will be more effective in shaping progress on sustainability (Gill & Williams, 2011; Rijke et al., 2012; Sheppard & Williams, 2016). Indeed, the lack of tourism sustainability progress is considered a policy problem within systems of governance that continue to promote tourism as a vehicle for economic growth, while ignoring its contribution to negative environmental and socio-cultural change (Hall, 2011; van Zeijl-Rozema, Corvers, Kemp, & Martens, 2008). Fennell and Sheppard (2021) observe that the concept of sustainable tourism is often spoken of in terms of stakeholders; yet, observe that non-human animals, despite the fact that they bear the brunt of many of the negative impacts of tourism development, are rarely considered stakeholders.

Interestingly, sustainability discourse has mostly ignored the ethical behaviour that is required to move the "utopian dream" to reality. Although Becker's (2012) research related to sustainability ethics lies outside the field of tourism, it is of relevance for the purposes of this chapter.[1] Specifically, Becker believes that many of the challenges associated with sustainable development, both in theory and in practice, arise from our failure to determine the most appropriate ethical approach. He argues that sustainability would more appropriately be defined around three key dimensions: continuance; orientation; and relational. He presents each of these dimensions as an ethical question, as follows: 1) Continuance: What systems, processes, or entities should be continued and for what purpose? 2) Orientation: How should one act and live? and 3) Relational: How should one live in relationship to contemporaries, future generations, and nature? It is this latter question that Becker suggests should function as the basic ethical question associated with a modern conceptualisation of sustainability. In other words, sustainability-focused decisions should be based upon a consideration of the moral relationship between humans and their contemporaries, humans and future generations, and humans and nature (Becker, 2012).

At its most fundamental level, sustainability ethics is centred around person identity and self-understanding as a "relational, interdependent, and virtuous person in the context of sustainability relations" (Becker, 2012, p. 67). Becker contends that the *sustainable person* possesses four core competencies. The first, a *relational identity*, incorporates the temporality, interdependency, and cultural contingency of human existence. Second is a set of relational virtues, which include *respect*, *care*, *responsibility*, and *tolerance*. Third is *relational competencies*, which include attentiveness and receptiveness. The fourth and final competency is a basic understanding of humans as *emotional*, *rational* (possessing *practical wisdom* and *reason*), *communicative*, and *creative* beings. Becker points out that a sustainable person is not perfect, nor are sustainability relations perfect. Rather, the sustainable person strives for ongoing self-development and perfection of relations, within a sustainability context.

Resilience overview

In response to these criticisms associated with sustainability, some researchers have turned their attention to the concept of resilience. They contend it may offer more hope for substantive progress on helping humanity better live within its means. The resilience literature can be traced back to the C. S. Holling, an ecological theorist, who first applied the concept of resilience within the ecology literature in the 1970s. He defined resilience as a measure of the ability of a system to persist in the presence of change and disturbance. Since that time the literature reveals an evolution in the concept and an understanding that it is not simply about persisting in the face of change and disturbance; rather, it is also about the ability of the system to continue to develop (Simonsen et al., n.d., p. 3). In other words, resilience is defined in terms of flexibility, or the capacity of the system to undergo some change (or adapt). Flexibility refers to the degree of manoeuvrability within the system or in activities (Smithers & Smit, 1997). However, Walker and Salt (2006, p. 32) warn that if the system crosses a threshold, it will enter into a different regime or state—or a system with a "different identity" (Walker & Salt, 2006, p. 32). This may not necessarily be a desired outcome. These two aspects of resilience, flexibility, and the ability of a system to undergo some degree of change, are considered important (McGlade, Murray, Baldwin, Ridgway, & Winder, 2006). As such, resilience is now defined as the ability to bounce back, renew, and reorganise (Folke, 2006). This occurs through the building of capacity within a system in order to successfully deal with unexpected ecological and or social change (Berkes, Colding, & Folke, 2003). Interestingly, a smaller body of resilience literature, within the developmental psychology field, is focused on individual resilience. It appears to run parallel to the ecological and social change literature. Within this body of literature, resilience is defined as the ability of individuals to adapt to adversity (see Kumpfer, 1999; Ungar, 2003, 2004, 2005).

As noted above, flexibility and the ability of a system to adapt to change are important concepts associated with resilience thinking. Adaptation is defined in terms of the capacity of humans to influence and manage potential damages or to take advantage of opportunities associated with changes in the environment (Janssen, Schoon, Ke, & Börner 2006; Walker, Holling, Carpenter, & Kinzig, 2004). The forces that affect the ability of a system to adapt also determine its adaptive capacity (see Adger, 2003; Walker et al., 2004; Smit & Wandel, 2006). Indeed, adaptive capacity is closely linked to adaptation and is defined as responses to risks associated with environmental hazards and human vulnerability (Adger & Vincent, 2005; Janssen & Ostrom, 2006). A variety of techniques exist to assess adaptive capacity (e.g., theory driven approaches, assessment of secondary data, self-assessment processes, and futures modelling) (see Lockwood, Raymond, Oczkowski, & Morrison, 2015). Indicators (e.g., social, educational, institutional) can also be assessed to determine adaptive capacity (Simpson, Gossling, Scott, Hall, & Gladin, 2008) at a variety of scales (i.e., individual to national level) (Adger & Vincent, 2005).

The concept of resilience has been a focus of tourism researchers since approximately the mid-1990s.[2] Much of this early tourism research was focused on linking economics with resilience, or a lack thereof (e.g., tourism market fluctuations) (see O'Hare & Barrett, 1994). Since the turn of the century, resilience has been applied to a broader range of tourism topics. For example, it is linked with the environmental impacts of tourism (see Nystrom, Folke, & Moberg, 2000), climate-environmental change and tourism sustainability (see Cheer & Lew, 2018; Holladay, 2018; Klint et al., 2012), disasters, and risks in association with tourism destinations (see Biggs, Hall, & Stoeckl 2011; Cochrane, 2010; Hall, 2011, Larsen, Calgaro, & Thomalla, 2011); communities as it relates to tourism development (see Biggs et al., 2011; Holladay & Powell, 2013); and ecotourism (see Baral, 2013; Jamaliah & Powell, 2017; Kumari, Behera, & Tewari, 2010).

Linking back to the general resilience literature, tourism resilience may be defined as the ability of tourism stakeholders (including destinations, tourism operators, employees, the natural environment, etc.) to bounce back, renew, and/or reorganise (adapt) after a significant shock or stressor. Shocks are defined as sudden events, that often precede a crisis (e.g., terrorism, natural disasters) while stressors are slow-moving events (e.g., climate change, the anti-flying movement). Tourism researchers contend that the tourism industry is increasingly confronted with a range of shocks and stressors (see Scott, de Freitas, & Matzarakis, 2008). Responses to them (successes and failures) can overlap and "compound over time" (Calgaro, Lloyd, & Dominey-Howes, 2014, p. 348). Shocks and stressors, while often associated with negative impacts, do present opportunities for renewal and innovative thinking (Moberg & Simonsen, n.d., p. 3).

There are conflicting perspectives on whether ecotourism is more resilient to shocks and stressors compared to other forms of tourism operations. An increasing number of tourism scholars are focusing their research around various aspects of this debate. For example, Jamaliah and Powell (2017) sought to understand whether the theoretical dimensions of resilience (social, environmental, governance, and economic) would enable local communities in the Dana Biosphere Reserve, located in Jordan, to persist and adapt to the effects of climate change. Their study suggested that the local communities had a "moderate" level of resilience as it relates to the environmental dimension, while the social, economic, and governance dimensions required further development. Another example is the research of Baral (2013) who assessed the level of resilience of ecotourism activities in the Annapurna Conservation Area, Nepal, in response to the Maoist insurgency from 1996 to 2006. His assessment involved four key areas: local capacity building, waste management, education, and infrastructure development. His findings suggest that local ecotourism entrepreneurs demonstrated resilience to the insurgency through self-organisation, local capacity building, and diversification of livelihoods. The activities of the entrepreneurs helped to maintain the stability of ecotourism businesses and assisted the operators in dealing with the uncertainly associated with the insurgency.

Indeed, a resilience approach, with its emphasis on flexibility, adaptation, and adaptive capacity may promote and enable successful responses to shocks and stressors, including the ability to recognise opportunities for positive change. Furthermore, it may be a solution for linking together society, the economy, and the biosphere (Moberg & Simonsen, p. 7). However, as with the concept of sustainability, some researchers observe the challenge of applying a resilience approach to tourism because of the social nature of tourism and the unpredictability of social interactions. For example, Calgaro et al. (2014) suggest that the notion of tipping points (or resilience thresholds) is not easily applied to social situations, in what Zahra and Ryan (2005) refer to as the unpredictability associated with human actions and outcomes. Further, McKercher (1999, p. 427) contends that tourism destinations, "shaken to the core by events" have "re-emerge[d] in an even more competitive manner." He adds that a focus on "orderly linear change", cannot account for "rogue" players who often influence the development of tourism destinations. In other words, tourism destinations often do not proceed through change in a linear fashion, due to the influences of outside and often unpredictable forces. Therefore, determinations and/or predictions of destination resilience, or otherwise, are difficult to make.

Linking resilience and sustainability

While it is true that the concepts resilience and sustainability are well linked in the literature, there is considerable debate swirling around these concepts. For example, some researchers contend that the concepts are synonymous (Holling & Walker, 2003; Maler, 2008; Perrings, 2006), others suggest they are divergent but complementary concepts (Espiner et al., 2017). Redman (2014) advocates for treating them as two separate concepts, sharing similar principles and objectives, but founded upon separate assumptions around the functioning of systems.

On the other hand, Rees (2014) links the two concepts, suggesting that sustainability infers the importance of maintaining socio-ecological systems and the avoidance of crossing critical thresholds. Similarly, Lebel et al. (2006, n.p.) suggest that the pursuit of more sustainable forms of development cannot occur without "strengthening the capacity of societies to manage resilience." Others suggest that resilience is a "theoretical construct for sustainability" (see Rees, 2014). In other words, it is a theoretical underpinning of sustainability, but that it lacks practicality as far as making progress on less impactful ways of human living. Adding to the confusion, some researchers contend that resilience is a precondition for sustainability, while others contend sustainability is a precondition for resilience (see Arrow et al., 1995; Lebel et al., 2006; Perrings, 2006).

In addition to the lack of consensus regarding the relationship between the two concepts, some literature questions whether a resilience approach will influence a more sustainable future. For example, some researchers have questioned how something can be judged resilient when the impact of future shocks and stressors is unknown (Christopherson, Michie, & Tyler, 2010). Carpenter et al. (2001) agree, stating that sustainability is always desirable; however, this is not the case with resilience. Resilience can be desirable if it promotes adaptation and regeneration (Hassink, 2010), or undesirable if promotes a return to existing structures that encourage the repetition of past mistakes.

Despite the confusion and points of debate and disagreement, the perfect system to push human behaviour toward a more sustainable form of living, that is likewise more resilient to system shocks and stressors has yet to be uncovered. Indeed, the concepts of sustainability and resilience may be complementary and together offer hope for changing the current trajectory of human behaviour and the societies in which they live. Researchers, governments, NGOs, and others have and continue to build upon these concepts in order to develop the best framework from which to develop a practical and operational set of guidelines, principles, and/or rules, for implementing sustainability and resilience, on a variety of levels (individual, community, regional, national, etc.). The following sections more fully explore the concept of resilience, particularly as it relates to community and individual resilience.

Community resilience-enhancing characteristics

Defining, characterising, and conceptualising community resilience is a more recent scholarly endeavour. The Resilience Alliance, the Stockholm Resilience Centre (2015), and the Centre for Community Enterprise (CCE) (n.d., p. 11) define a resilient community as being one "that takes intentional action to enhance the personal collective capacity of its citizens and institutions to respond to, and influence the course of social and economic change." They contend that a resilient community involves four aspects: people, organisations, resources, and processes. Patton and Johnson (2001, p. 273) add that a resilient community has the ability to "bounce back [after a shock or stressor] and recover using its own resources" (2001, p. 273). They define community resilience in terms of the physical infrastructure, the economic resources, and "ensuring that community members have the resources, capacities and capabilities necessary" to respond to adversity (p. 272). These researchers also contend that it is the ability of a community to conceptualise its "response to adversity," in advance of an event, that enables the community to draw upon its "internal resources and competencies to manage the demands, challenges and changes" associated with the event (p. 273). However, Tobin (1999, p. 23) warns that responses can perpetuate "the disaster-damage cycle, rather than addressing the root causes of the problem." He links the resilience of the community to the businesses that comprise the community, especially as it relates to large employers. As he observes, community resilience may be in the hands of "absentee" employers, rather than local businesspeople. Large employers "make decisions based upon shareholder profits rather than local concerns" (p. 16). He suggests sustainable and resilient communities must:

- Seek to lower the level of risk to all community members through reduced exposure to geophysical events. This can be achieved through structural and non-structural measures;
- Seek to reduce the level of vulnerability of all members of the community, especially those who are politically or economically marginalised;
- Plan for sustainability and resilience; Ensure commitments are long term and that sustainability goals stay at the forefront of all community planning efforts;
- Ensure high-level support and political will from agencies and political leaders; embrace partnerships and cooperation across the various levels of government and across organisations. This will ensure the involvement of leadership, skill and resource utilisation, and local knowledge to help develop mitigation projects and to ensure buy in for sustainability initiatives;
- Strengthen networks of independent and interdependent segments of society, and plan at the appropriate scale.

Identifying the factors or characteristics that build resilience at the local or community level is a critical undertaking in order to determine what resilience looks "on the ground" (Berkes & Seixas, 2005, p. 973). These factors can be assessed qualitatively and/or quantitatively, despite the difficulty associated with collecting the data (Berkes & Seixas, 2005). Ruiz-Ballesteros (2011) appears to be amongst the first tourism researchers to utilise a set of community-based socio-ecological resilience (SER) characteristics in a tourism context. His research involved an exploratory ethnographic case study, which took place in the tourism-focused community of Agua Blanca, Ecuador, a community of approximately 260 residents. The purpose of this research was to explore the practicality of examining SER in tourism-focused communities in order to develop a new methodology for analyzing tourism development. Ruiz-Ballesteros' (2011) research explored four groups of interrelated factors that other researchers suggest may nurture the development of SER at the local or community level. These factors are drawn mainly from the SES resilience research of Berkes and Seixas (2005) and Folke (2003); however, is important to note that the factors can be traced back to a body of resilience research that began in approximately 2000 (see Adger, 2000). The four groups of resilience enhancing factors are: 1) learning to live with change and uncertainty, 2) nurturing diversity for reorganisation and renewal, 3) combining different kinds of knowledge, and 4) creating opportunity for self-organisation. Table 4.1 provides detail associated with each of the four factors.

Individual resilience-enhancing characteristics

The lack of studies linking individual and community resilience is perplexing given the fact that SESs are comprised of individuals. Only a few recent studies have linked SES resilience at the community level with resilience at the individual level (see Sheppard & Williams, 2016; Sheppard, 2017a, 2017b). Until these more recent studies, individual resilience was mostly studied within the field of developmental psychology, which is focused on understanding the positive life adaptation factors in children and young adults. Here, resilience is defined as the behaviour and internal capacities of children and youth, as well as the structural conditions (socio-cultural, political) that enable adaptation in the face of adversity (Ungar, 2003, 2004, 2005). Of most relevance within this body of research is the work of Kumpfer (1999) who

Table 4.1 Resilience enhancing characteristics at the community level (from Ruiz-Ballesteros, 2011)

Community-Based Resilience Factors Nurturing SES Level	*Description*
Learning to live with change and uncertainty	• Creating learning environment, particularly as it relates to shocks and stressors • Building rapid feedback capacity to respond to environmental change • Managing disturbance • Building a portfolio of livelihood activities • Developing coping strategies
Nurturing diversity for reorganisation and renewal	• Nurturing ecological memory • Nurturing diversity in institutions to respond to change • Creating political space for experimentation • Building trust among users • Using social memory as a source for innovation and novelty
Combining different kinds of knowledge	• Incorporating systems of local knowledge into management and external decision-making authorities • Building capacity to monitor the environment • Building capacity for participatory management • Building institutions that frame learning, memory, and creativity • Building institutions that create cross-scale mechanisms to share knowledge
Creating opportunity for self-organisation	• Promoting participatory strategies that permit self-organisation of groups and communities • Promoting participatory strategies that consider the diversity and alteration inherent in resilience • Building capacity for user self-organisation • Building capacity for self-determined and self-organised fairness in resource access and allocation • Building capacity for self-organisation in response to external drivers • Building conflict management mechanisms • Matching scales of ecosystem governance • Creating multi-level governance

developed a resilience framework based upon a set of five internal resilience characteristics. These internal characteristics (spiritual or motivational, cognitive, behavioural/social, emotional stability/emotional management, and physical wellbeing/physical ability) enhance individual resilience, particularly in times of stress or challenge. They are positively or negatively affected in the presence of another set of factors (family, culture, community, school, and peers) that function as either risk or protective factors (see Table 4.2).

Ecotourism principles and resilience

As noted, there are conflicting perspectives on whether ecotourism is more sustainable and/or more resilient to shocks and stressors than other forms of tourism operations. While it is difficult to make comparisons to other types of tourism operations, the consensus of aforementioned studies is that ecotourism may be more resilient to shocks and stressors; however, the research

Table 4.2 Resilience enhancing characteristics at the individual level (Kumpfer, 1999)

Characteristics Nurturing Individual Resilience	*Description*
Spiritual or motivational	• Encompasses mostly cognitive capabilities or a belief system • Motivates individuals to proceed in a specific direction; success depends upon the direction taken
Cognitive	• Helps an individual achieve his or her dreams and goals • Includes intellectual, academic, and job skills • Includes moral reasoning, and ability to: 1) judge right from wrong; 2) internalise standards of the way things should be done or what is normative 3) value compassion, fairness, decency; and, 4) the desire to serve others • Connected to insight and reflective skills, high levels of self-esteem, creativity, and the ability to plan
Behavioural/social	• Similar to cognitive competencies, except that they require a behavioural action, as opposed to just thought • Includes problem solving skills, communication skills, street and peer resistance smarts, and ability to be empathetic to the needs of others
Emotional stability/ emotional management	• Includes happiness, the recognition of feelings, ability to control anger and depression • Includes ability to restore one's self-esteem, humour, and hopefulness
Physical wellbeing/physical ability	• Includes good health, health maintenance skills, physical talent development, and physical attractiveness • Good physical state is predicative of resilience

also reinforces the importance of governance systems in enabling and enhancing such resilience. If this is true, and ecotourism is more resilient than other forms of tourism, it is important to understand specifically what it is that makes it more resilient. Understanding how and where resilience factors overlap with the principles of ecotourism is important because it may enable other types of tourism businesses to develop and emulate these factors. Table 3.0 draws together the characteristics or principles of ecotourism business operations from the work of Stacey and Needham (1993), Fennell (2001), and Donohue and Needham (2006) with the themes of community resilience from Ruiz-Ballesteros (2011).

As evidenced in Table 4.3, all of the ecotourism principles (see Fennell, 2001; Stacey & Needham, 1993; Donohue & Needham, 2006) can be matched to the resilience themes that have been identified in the community resilience themes identified (see Ruiz-Ballesteros, 2011). For example, the ecotourism principles of *nature-based* and *focused on preservation and/or conservation* are both aligned to the resilience theme of *nurturing diversity for reorganisation and renewal.* All are focused on healthy ecosystems, which involves nurturing ecological memory and nurturing diversity, as well as collaboration between ecotourism providers and the community and building trust amongst among users. While ecotourism principles are focused on incorporating and/or implementing preservation/conservation into management plans, resilience thinking is focused on creating political space for exploration, innovation, and novelty.

Similarly, the *educational* component associated with ecotourism is aligned with the community resilience characteristic of *learning to live with change and uncertainty*. For example, both are focused at educating and creating awareness amongst stakeholders as well as empowering them. While the *educational* component (ecotourism) is focused at interactions and between guests and nature, *learning to live with change and uncertainty* (resilience) is focused on building the

Table 4.3 Comparison of ecotourism principles and community resilience themes (based upon Donohue & Needham, 2006; Fennell, 2001; Ruiz-Ballesteros, 2011; Stacey & Needham, 1993)

Key ecotourism principles	*Community-level resilience characteristics*
Nature-based: • Activities occurring mostly in nature, where there is minimal human interference • Focusing on healthy ecosystems • Providing opportunities to visit natural areas **Focused on preservation and/or conservation:** • Maintaining and/or enhancing ecosystems • Focusing awareness on ecosystem requirements • Collaboration between providers and community (protected area managers, local people, etc.) • Incorporating/implementing preservation/ conservation into management plans	**Nurturing diversity for reorganisation and renewal:** • Nurturing ecological memory • Nurturing diversity in institutions to respond to change • Creating political space for experimentation • Building trust among users • Using social memory as a source for innovation and novelty
Educational component: • Educating re biological-cultural aspects to all stakeholders (staff, guests, community) • Encouraging interaction between guests and nature (experiential/educational benefits) • Increasing stakeholder awareness and understanding area's natural heritage (including visitors) • Empowering stakeholders to become involved in issues affecting natural and cultural heritage (including visitors)	**Learning to live with change and uncertainty:** • Creating learning environment, particularly for shocks and stressors • Building rapid feedback capacity to respond to environmental change • Managing disturbance • Building a portfolio of livelihood activities • Developing coping strategies
Focused on sustainability: • Achieving equity and social justice • Maintaining ecological integrity • Satisfying human needs • Achieving social self-determination • Integrating conservation and development **Ethics, responsibility, and awareness:** • Taking an ethics-based environmentally, socially and culturally responsible approach • Making decisions based upon ecological principles • Considering impacts and consequences of travel to and within natural areas • Leading by example/increasing awareness of values-based ethics action and business approach	**Combining different kinds of knowledge:** • Incorporating systems of local knowledge into management and external decision-making authorities • Building capacity to monitor the environment • Building capacity for participatory management • Building institutions that frame learning, memory, and creativity • Building institutions that create cross-scale mechanisms to share knowledge
Benefits distributed amongst a variety of stakeholders: • Ensuring equitable access to resources, costs, and benefits • Benefits complement rather than replace traditional local practices, activities	**Creating opportunity for self-organisation:** • Promoting participatory strategies that permit self-organisation of groups and communities • Promoting participatory strategies that consider diversity and alteration inherent in resilience • Building capacity for user self-organisation

(*Continued*)

Table 4.3 (Continued)

Key ecotourism principles	*Community-level resilience characteristics*
• Maximising short- and long-term benefits for stakeholders, including visitors • Improving quality of life for local people • Complementing existing tourism infrastructure	• Building capacity for self-determined and self-organised fairness in resource access and allocation • Building capacity for self-organisation in response to external drivers • Building conflict management mechanisms • Matching scales of ecosystem governance • Creating multi-level governance

ability to respond to and manage disturbance, as well as an emphasis on the development of coping strategies.

The ecotourism principles of *sustainability and ethics, responsibility, and awareness* can be linked with the community resilience characteristic of *combining different kinds of knowledge*. In this regard, ecotourism seeks to combine equity, social justice, self-determination (social pillar), with satisfying human needs and development (economic pillar), and maintaining ecological integrity and integrating conservation (environmental pillar). Likewise, *combining different kinds of knowledge* (resilience) is characterised by a focus on local knowledge, sharing of knowledge, and institution building (social pillar), with monitoring the environment (environmental pillar), and participatory management and decision making (economic pillar). Importantly, *ethics, responsibility, and awareness* (ecotourism principle) can also be linked with *sustainability* and with *combining different kinds of knowledge*. For example, an ethics approach involves environmental and socio-cultural approaches, as well as a values-based business approach (sustainability pillar). *Responsibility* and *awareness*, aspects associated with ethics, are linked with the sharing of knowledge, building capacity for participatory management, incorporating local knowledge, and also monitoring the environment (resilience).

Finally, the ecotourism principle of *benefits distributed amongst a variety of stakeholders* can be linked with the community resilience characteristic of *creating opportunity for self-organisation*. For example, ensuring equitable access to resources, costs and benefits and improving the quality of life for local people (ecotourism) are both linked with building capacity for self-determination and self-organised fairness in resource access and allocation (resilience). Similarly, benefits of ecotourism complement rather than replace traditional local practices and activities are linked with building conflict management mechanisms and a focus on creating multi-level social and ecosystem governance (resilience).

The comparison, as depicted in Table 4.3, linking the principles of ecotourism with the characteristics of resilience, could be reorganised in a number of different ways. Indeed, the various principles and characteristics overlap, merge, and diverge on many other levels than described above. What is clear is that the principles of ecotourism and the characteristics of resilience bear many similarities. However, what is missing in this analysis is a consideration of the role individuals play in the success of ecotourism. In other words, what are the characteristics that individual ecotourism operators and other stakeholders should or must have in order to be resilient on an individual level? A resilient ecotourism operation requires a resilient owner-manager, a resilient community, a resilient governance system, and a resilient ecosystem. The following section conceptualises the addition of the individual resilience characteristics from Table 4.2 into the sustainability-resilience of ecotourism.

Reconceptualising the sustainability and resilience of ecotourism operations

Figure 4.1 conceptualises the centrality of individual resilience in the sustainability and overall resilience of ecotourism operations. Specifically, if individuals are resilient and focused on their personal sustainably, the community is more likely to also be resilient. Likewise, if both the community and the individuals who comprise the community are resilient, then it is likely that their businesses, tourism focused, and otherwise, will be more resilient and also focused on sustainability. Central to the resilience of the community and the tourism businesses within the community, is the degree to which individuals, their businesses, and the community are all focused on decision-making with a consideration of the pillars of sustainability. As demonstrated, Figure 4.1 includes a fourth pillar, governance, as sustainability and resilience cannot occur within governance systems that are not modelled in such a manner as to enable sustainability approaches and resilience thinking. Consequently, the ecotourism business model that adheres to the principles of ecotourism and pillars of sustainability, and which is supported by a community of resilient individuals and leaders, may serve as a model for other tourism operations, particularly those that function within and/or are dependent upon the natural environment.

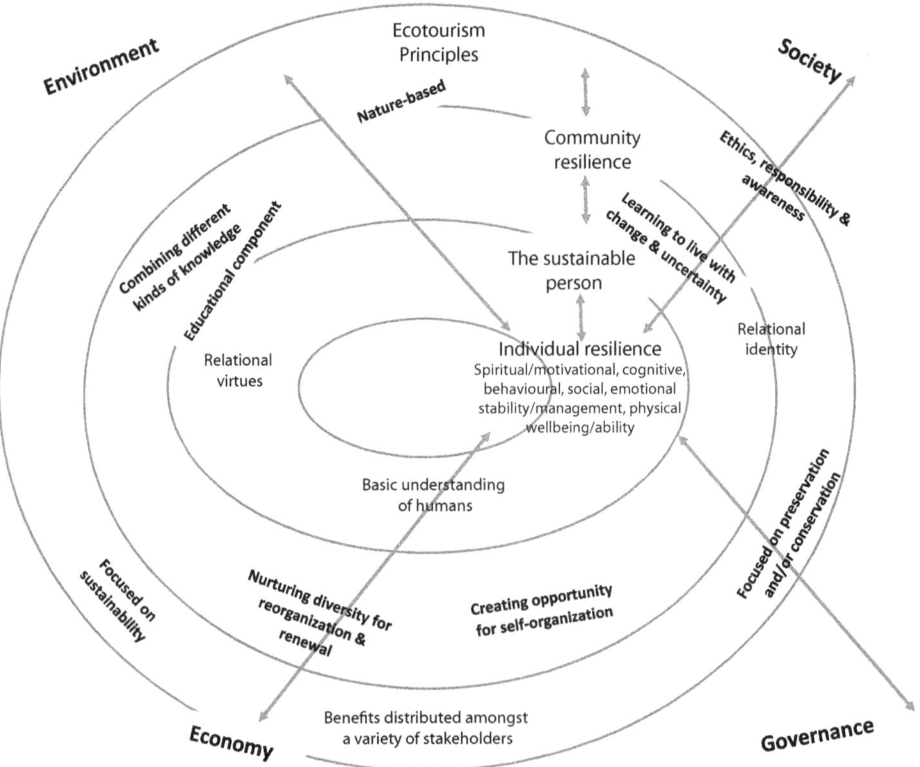

Figure 4.1 Diagram showing the centrality of individual resilience in the sustainability and overall resilience of ecotourism operations

Conclusion

The purpose of this chapter was to examine the nexus between the concepts of resilience and sustainability, in order to conceptualise how both perspectives may enhance ecotourism practice, as well as inform other types of tourism business practice. It has been suggested here that the lack of focus on the individual, both in terms of resilience and sustainability, may be hindering more timely and effective sustainability progress. Indeed, there appears to be an urgent need for a sustainable tourism business model that encompasses aspects of ecotourism principles and resilience characteristics, which are then applied at a variety of levels (personal, business, community). Within such a model, business owners and managers would take personal responsibility for minimising the environmental and socio-cultural impacts of their operations, in addition to striving for profitability. Most importantly, their behaviour would emulate ethics-based values, responsibility, and awareness.

This chapter has also attempted to move the ecotourism agenda forward by suggesting a framework for more resilient and sustainable tourism business operations. Admittedly, it has not taken the next step, which is to develop and test a prototype tourism business model for not only ecotourism operations, but also other forms of tourism operations. Therefore, a recommendation is made that future research could explore the practicality of such business models. Specifically, researchers would test the ability of tourism businesses operating under a resilience and sustainability focused business model to respond and adapt to system shocks and stressors (e.g., economic, political, socio-cultural, environmental, communicable disease and illness, etc.) as well as recover, and successfully move on.

The years 2018 and 2019 were challenging years for the tourism industry as evidenced by the collapse of Thomas Cook, a major travel group, and a number of small airlines around the world (i.e., Jet Airways, Fly Jamaica Airways, WOW Air, Asian Express Airline). The 2020 outbreak of the coronavirus in China and its rapid spread across the globe is further evidence of the vulnerability of tourism operations to system shocks and stressors and the need for actions that enhance their resilience and overall sustainability (see www.ctvnews.ca/canada/chinese-tourists-cancel-trips-to-canadian-hotspots-such-as-banff-yellowknife-1.4791629). While the business model, as described, may seem like just another utopian dream, its need is undeniable and urgent. Determining which model will be most effective, in terms of individual, business and community resilience and sustainability, is the next logical step for making timely and substantive sustainability progress.

Notes

1 See Fennell (2019) for a tourism-focused application of Becker's sustainability ethics model.
2 There is a body of research dating back to the 1970s that discusses resilience in terms of the management and conservation of parks and protected areas, particularly as it relates to area degradation (see Western & Henry, 1979).

References

Adger, W. N. (2000). Social and ecological resilience: Are they related? *Progress in Human Geography*, *24*(3), 347–364.

Adger, W. N. (2003). Social capital, collective action, and adaptation to climate change. *Economic Geography*, *79*(4), 387–404. Retrieved from https://ezproxy.royalroads.ca/login?url=http://search.ebscohost.com.ezproxy.royalroads.ca/login.aspx?direct=true&db=aph&AN=11617060&site=ehost-live.

Adger, W. N., & Vincent, K. (2005). Uncertainty in adaptive capacity. *Comptes Rendus Geoscience, 337*(4), 399–410. doi: 10.1016/j.crte.2004.11.004.

Arrow, K., Bolin, B., Costanza, R., Dasgupta, P., Folke, C., Holling, C. S., ... Pimentel, D. (1995). Economic growth, carrying capacity, and the environment. *Science, 268*(5210), 520–521. Retrieved from http://www.jstor.org.ezproxy.royalroads.ca/stable/2886637.

Baral, N. (2013). Evaluation and resilience of ecotourism in Anapurna Conservation Area, Nepal. *Environmental Conservation, 41*(1), 84–92. doi:10.1017/S0376892913000350.

Becker, C. U. (2012). *Sustainability Ethics and Sustainability Research*. New York: Springer.

Berkes, F., Colding, J., & Folke, C. (Eds.). (2003). *Navigating Social-Ecological Systems Building Resilience for Complexity and Change*. Cambridge and New York: Cambridge University Press. Retrieved from http://voyager.royalroads.ca.ezproxy.royalroads.ca/vwebv/holdingsInfo?bibId=278714.

Berkes, F., & Seixas, C. S. (2005). Building resilience in lagoon social-ecological systems: A local-level perspective. *Ecosystems*, 8, 967–974.

Biggs, D., Hall, C. M., & Stoeckl, N. (2011). The resilience of formal and informal tourism enterprises to disasters: Reef tourism in Phuket, Thailand. *Journal of Sustainable Tourism, 20*(5), 645–665. doi:10.1080/09669582.2011.630080.

Butcher, J. (2006). The United Nations International Year of Ecotourism: A critical analysis of development implications. *Progress in Development Studies, 6*(2), 146–156. doi:10.1191/1464993406ps133oa.

Butler, R. (2013). Sustainable tourism – The undefinable and unachievable pursed by the unrealisitic? *Tourism Recreation Research, 38*(2), 221–226.

Calgaro, E., & Lloyd, K., & Dominey-Howes, D. (2014). From vulnerability to transformation: A framework for assessing the vulnerability and resilience of tourism destinations. *Journal of Sustainable Tourism, 22*(3), 341–360. doi:10.1080/09669582.2013.826229.

Carpenter, S., Walker, B., Anderies, J. M., & Abel, N. (2001). From metaphor to measurement: Resilience of what to what? *Ecosystems, 4*(8), 765–781. Retrieved from http://www.jstor.org.ezproxy.royalroads.ca/stable/3659056.

Centre for Community Enterprise. (2020). Tools & techniques: Community recovery & renewal. Retrieved from www.communityrenewal.ca/tools-and-techniques

Cheer, J. M., & Lew, A. A. (Eds.) (2018). *Tourism, Resilience and Sustainability: Adaping to Social, Political and Economic Change*. New York: Routledge.

Christopherson, S., Michie, J., & Tyler, P. (2010). Regional resilience: Theoretical and empirical perspectives. *Cambridge Journal of Regions, Economy and Society, 3*(1), 3–10. doi:10.1093/cjres/rsq004.

Cochrane, J. (2010). The sphere of tourism resilience. *Tourism Recreation Research, 35*(2), 173–185.

Donohue, H. M., & Needham, R. D. (2006). Ecotourism: The evolving contemporary definition. *Journal of Ecotourism, 5*(3), 192–210. doi:10.2167/joe152.0.

Espiner, S., Orchiston, C., & Higham, J. E. S. (2017). Resilience and sustainability: a complementary relationship? Towards a practical conceptual model for the sustainability *resilience* nexus in tourism. *Journal of Sustainable Tourism, 25*(10), 1385–1400. http://dx.doi.org/10.1080/09669582.2017.1281929.

Fennell, D. A. (2001). A content analysis of ecotourism definitions. *Current Issues in Tourism, 4*(5), 403–421.

Fennell, D. A. (2019). Sustainability ethics in tourism: The imperative next imperative. *Tourism Recreation Research, 44*(1), 117–130. doi:10.1080/02508281.2018.1548090.

Fennell, D. A., & Sheppard, V. A. (2021). Tourism, animals, and the scales of justice. *Journal of Sustainable Tourism*, 29(2-3), 314–335. doi: https://doi.org/10.1080/09669582.2020.1768263.

Folke, C. (2003). Socio-ecological resilience and behavioural responses. In A. Biel, B. Hansson, & M. Martensson (Eds.), *Individual and structural determinants of environmental practice* (pp. 437–440). London: Ashgate Publishers.

Folke, C. (2006). Resilience: The emergence of a perspective for social–ecological systems analyses. *Global Environmental Change, 16*(3), 253–267. doi:10.1016/j.gloenvcha.2006.04.002.

Gill. A., & Williams, P. (2011). Mindful deviation in creating a governance path towards sustainability in resort destinations. *Tourism Geographies, 16*(4), 546–562. doi:10.1080/14616688.2014.925964.

Hall, C. M. (2011). A typology of governance and its implications for tourism policy analysis. *Journal of Sustainable Tourism, 19*(4–5), 437–457. doi:10.1080/09669582.2011.570346.

Hassink, R. (2010). Regional resilience: A promising concept to explain the differences in regional economic adaptability? *Cambridge Journal of Regions, Economy* and *Society, 3*, 45–58. doi:10.1093/cjres/rsp033.

Holladay, P. (2018). Destination resilience and sustainable tourism development. *Tourism Review International, 22*, 251–261. doi:https://doi.org/10.3727/154427218X15369305779029.

Holling, C. S. (1973). Resilience and stability of ecological systems. *Annual Review of Ecology and Systematics, 4*, 1–23. Retrieved from http://www.jstor.org.ezproxy.royalroads.ca/stable/2096802.

Holling, C. S., & Walker, B. H. (2003). Resilience defined. In International Society of Ecological Economics (Ed.), *Internet encyclopedia of ecological economics* (Online ed., pp. 1–2). The International Society for Ecological Economics. Retrieved from isecoeco.org/pdf/resilience.pdf

Holladay, P. J., & Powell, R. B. (2013). Resident perceptions of socio-ecological resilience and the sustainability of community-based tourism development in the Commonwealth of Dominica. *Journal of Sustainable Tourism, 21*(8), 1188–1211. doi:http://dx.doi.org/10.1080/09669582.2013.776059.

Jamaliah, M. M., & Powell, R. B. (2017). Ecotourism resilience to climate change in Dana Biosphere Reserve, Jordan. *Journal of Sustainable Tourism*, August. doi:10.1080/09669582.2017.1360893.

Janssen, M. A., & Ostrom, E. (2006). Resilience, vulnerability, and adaptation: A cross-cutting theme of the international human dimensions programme on global environmental change. *Global Environmental Change, 16*(3), 237–239. doi:10.1016/j.gloenvcha.2006.04.003.

Janssen, M. A., Schoon, M. L., Ke, W., & Börner, K. (2006). Scholarly networks on resilience, vulnerability and adaptation within the human dimensions of global environmental change. *Global Environmental Change, 16*(3), 240–252. doi:10.1016/j.gloenvcha.2006.04.001.

Klint, L. M., Wong, E., Jiang, M., Delacy, T., Harrison, D., & Dominey-Howes, D. (2012). Climate change adaptation in the pacific island tourism sector: Analysing the policy environment in Vanuatu. *Current Issues in Tourism, 15*(3). doi:10.1080/13683500.2011.608841.

Kumari, S., Behera, M. D., & Tewari, H. R. (2010). Identification of potential ecotourism sites in West District, Sikkim using geospatial tools. *Tropical Ecology, 51*(1), 75–85.

Kumpfer, K. L. (1999). Factors & processes contributing to resilience: The resilience framework. In M. D. Glantz, & J. L. Johnson (Eds.), *Resilience & development: Positive life adaptations* (pp. 179–224). New York: Kluwer Academic/Plenum Publishers.

Larsen, R. K., Calgaro, E., & Thomalla, F. (2011). Governing resilience building in Thailand's tourism-dependent coastal communities: Conceptualising stakeholder agency in social–ecological systems. *Global Environmental Change, 21*(2), 481–491. doi:10.1016/j.gloenvcha.2010.12.009

Lebel, L., Anderies, J. M., Campbell, B., Folke, C., Hatfield-Dodds, S., Hughes, T., & Wilson, J. (2006). Governance and the capacity to manage resilience in regional social-ecological systems. *Ecology and Society, 11*(1).

Lockwood, M., Raymond, C. M., Oczkowski, E., & Morrison, M. (2015). Measuring the dimensions of adaptive capacity: A psychometric approach. *Ecology and Society, 1*(20), 37. doi:http://dx.doi.org/10.5751/ES-07203-200137.

Maler, K. (2008). Sustainable development and resilience in ecosystems. *Environmental & Resource Economics, 39*(1), 17–24. doi:10.1007/s10640-007-9175-7.

McCool, S. (2013). Sustainable tourism: Guiding fiction, social trap or path to resilience? *Tourism Recreation Research, 38*(1), 214–221.

McCool, S., Butler, R., Buckley, R., & Weaver, B., & Wheeler, B. (2013). Is concept of sustainability utopian: Ideally perfect but impractical? *Tourism Recreation Research, 38*(1), 213.

McGlade, J., Murray, R., Baldwin, J., Ridgway, K., & Winder, B. (2006). Industrial resilience and decline: A co-evolutionary approach. In E. Garnsey, & J. McGlade (Eds.), *Complexity and co-evolution: Continuity and change in socio-economic systems* (pp. 147–176). Cheltenham: Edward Elgar.

McKercher, B. (1999). A chaos approach to tourism. *Tourism Management, 20*, 425–435.

Moberg, F., & Simonsen, S. H. (n.d.). *What Is Resilience? An Introduction to Socio-Ecological Research*. Stockholm University. Stockholm Resilience Centre. Retrieved from http://www.stockholmresilience.org/download/18.10119fc11455d3c557d6d21/1398172490555/SU_SRC_whatisresilience_sidaApril2014.pdf.

Nystrom, M., Folke, C., & Moberg, F. (2000). Coral reef disturbance and resilience in a human-dominated environment. *Trends in Ecology & Evolution, 15*(10), 413–417. doi:10.1016/S0169-5347(00)01948-0.

O'Hare, G., & Barrett, H. (1994). Effects of market fluctuations on the Sri Lankan tourism industry: Resilience and change, 1981–1991. *Tijdschrift Voor Economische En Sociale Geografie, 85*(1), 39–52.

Patton, D., & Johnson, D. (2001). Disasters and communities: vulnerability, resilience and preparedness. *Prevention and Management, 10*(4), 270–277.

Perrings, C. (2006). Resilience and sustainable development. *Environment and Development Economics, 11*(4), 417–427. doi:10.1017/S1355770X06003020.

Pigram, J. J. (1990). Sustainable tourism - Policy considerations. *The Journal of Tourism Studies, 1*(2), 2–9.

Redman, C. L. (2014). Should sustainability and resilience be combined or remain distinct pursuits. *Ecology and Society, 19*(2), n.p. doi:http://dx.doi.org/10.5751/ES- 06390-190237.

Rees, W. (2014). Sustainability vs. Resilience. Retrieved from http://www.resilience.org/stories/2014-07-16/sustainability-vs-resilience# (accessed on 01.10.2015).

Rijke, J., Brown, R., Zevenbergen, C., Ashley, R., Farrelly, M., Morison, P., & van Herk, S. (2012). Fit-for-purpose governance: A framework to make adaptive governance operational. *Environmental Science & Policy, 22*(0), 73–84. doi:10.1016/j.envsci.2012.06.010.

Ruiz-Ballesteros, E. (2011). Social-ecological resilience and community-based tourism: An approach from Agua Blanca, Ecuador. *Tourism Management, 32*(3), 655–666. doi:10.1016/j.tourman.2010.05.021.

Scott, D., de Freitas, C., & Matzarakis, A. (2008). Adaptation in the tourism and recreation sector. In K. L. Ebi, & I. Burton (Eds.), *Biometeorology for adaptation to climate variability and change* (pp. 171–194). Dordrecht: Kluwer Academic Publishing.

Sheppard, V. A. (2017a). Resilience & destination governance: methodology and application in Whistler, British Columbia. In R. Butler (Ed.), *Tourism & Resilience* (pp. 53–68). UK: CABI.

Sheppard, V. A. (2017b). Destination governance: resident perceptions of resilience factors, Whistler, British Columbia: In R. Butler (Ed.), *Tourism & resilience* (pp. 69–80). UK: CABI.

Sheppard, V. A., & Fennell, D. A. (2019). Progress in tourism public sector policy: Toward an ethic for non-human animals. *Tourism Management, 73*(August), 134–142. doi:https://doi.org/10.1016/j.tourman.2018.11.017.

Sheppard, V. A., & Williams, P. W. (2016). Factors strengthening tourism resort resilience. *Journal of Hospitality and Tourism Management, 28*(September), 20–30. doi:10.1016/j.jhtm.2016.04.006.

Simmie, J., & Martin, R. (2010). The economic resilience of regions: Towards an evolutionary approach. *Cambridge Journal of Regions, Economy and Society, 3*(1), 27–43. doi:10.1093/cjres/rsp029.

Simonsen, S. H., Biggs, R., Schluter, M. Schoon, M., Bohensky, E., Cundill, G., … Mobeg, F. (n.d.). *Applying Resilience Thinking: Seven Principles for Building Resilience in Social-Ecological Systems.* Stockholm, Sweden: Stockholm Resilience Centre. Retrieved from: http://www.stockholmresilience.org/download/18.10119fc11455d3c557d6928/13981 50799790/SRC+Applying+Resilience+final.pdf.

Simpson, M., Gossling, S., Scott, D., Hall, M. C., & Gladin, E. (2008). *Climate Change Adaption and Mitigation in the Tourism Sector: Frameworks, Tools and Practices.* Paris, France: UNEP, University of Oxford, UNWTO, WMO.

Smit, B., & Wandel, J. (2006). Adaptation, adaptive capacity and vulnerability. *Global Environmental Change, 16*(3), 282–292. doi:10.1016/j.gloenvcha.2006.03.008.

Smithers, J., & Smit, B. (1997). Human adaptation to climatic variability and change. *Global Environmental Change*, 7(2), 129–146. doi:10.1016/S0959-3780(97)00003-4.

Stacey, C. L. & Needham, R. D. (1993). Heritage: A catalyst for innovative community development. In D. Bruce, & M. Whitla (Eds.), *Community-based approaches to rural development: Principles and practices* (pp. 21–44). Sackville, New Brunswick: Mount Allison University.

Stockholm Resilience Centre. (2015). What Is Resilience. Retrieved from http://www.stockholmresilience.org/21/research/research-news/2-19-2015-what-is-resilience.html.

Tobin, G. A. (1999). Sustainability and community resilience: the holy grail of hazards planning? *Environmental Hazards, 1*(1), 13–25. doi:10.3763/ehaz.1999.0103.

Ungar M. (2003). Resilience, resources and relationships: Making integrated services more family-like. *Relational Child and Youth Care Practices, 16*(3), 45–57.

Ungar, M. (2004). *Nurturing Hidden Resilience in Troubled Youth.* Toronto: University of Toronto Press.

Ungar, M. (2005). Resilience among children in child welfare corrections, mental health and educational settings: Recommendations for service. *Child & Youth Care Forum, 34*(6), 445–464. doi:10.1007/s105 66-005-7756-6.

United Nations Social and Economic Council. (1998). Economic and Environmental Questions: Sustainable Development. Declaring the Year 2020 as the International Year of Ecotourism (draft resolution). Retrieved from https://documents-dds-ny.un.org/doc/UNDOC/LTD/N98/217/26/pdf/N9821726.pdf?OpenElement.

van Zeijl-Rozema, A., Corvers, R., Kemp, R., & Martens, P. (2008). Governance for sustainable development: A framework. *Sustainable Development, 16*(6). doi:10.1002/sd.367.

Walker, B., Holling, C. S., Carpenter, S. R., & Kinzig, A. (2004). Resilience adaptability and transformability in social-ecological systems. *Ecology and Society, 9*(2), 5. Retrieved from http://www.ecologyandsociety.org/vol9/iss2/art5/print.pdf.

Walker, B. H., & Salt, D. (2006). *Resilience Thinking. Sustaining Ecosystems and People in a Changing World.* Washington, DC: Island Press.
Western, D., & Henry, W. (1979). Economics and conservation in third world national parks. *Biospheres*, 29(7), 414-418. https://doi.org/10.2307/130764.
World Commission on Economic Development (WCED). (1987). Our Common Future (Brundtland Report). Retrieved from http://un-documents.net/wced-ocf.htm.
Zahra, A., & Ryan, C. (2005). Complexity in tourism structures - the imbedded system of New Zealand's regional tourism organisation. *International Journal of Tourism Sciences*, *5*(1), 1–17.

5

OVERTOURISM IN PETRA PROTECTED AREA

Tour guides' perspectives

Areej Shabib Aloudat

Introduction

Tourism has become an attractive planning and management tool for the economy of countries. Its direct benefits, mainly, employment, the generation of revenues and foreign exchange, are viewed as an alternative for the economic development of tourism destinations with many countries trying to tap into it—Jordan being one of them (Shdeifat, Mohsen, Mustafa, Al-Ali, & Al-Mhaisen, 2006).

Ecotourism is one of the fastest-growing forms of tourism (Bayrama et al., 2017) and its success, as with all kinds of tourism, is based around its growth, intuitively, with tourism destination agencies exerting many efforts to increase tourist flows for the purpose of elevating the economic benefits. Accordingly, overtourism (OT) is a result of the success of these agencies. However, tourism benefits have the potential to cause negative impacts. The resultant overcrowding may cause direct environmental, infrastructural, and cultural damage to a number of destinations, and indirect impacts on local residents' lives based on pollution, increased prices, traffic congestion, and economic leakages. Overtourism, or overcrowding, occurs from high-use levels of visitors leading to effects on residents, tourists, and resources. These causes and effects all bring about concerns with OT that is fast becoming one of the most hotly debated issues, especially in ecotourism which relies on protected areas (PAs). It is important to address the issue of OT in PAs and highlight any inappropriate or poorly management tourism that may cause negative impacts on PAs biodiversity, landscapes, and resources base (Leung, Spenceley, Hvenegaard, & Buckley, 2018).

This chapter explores the phenomena of OT in PAs from the perspectives of tour guides. The focus on tour guides as respondents in this study is because this group is an important part of the communication and interpretation of values and cultural [and natural] heritage resources in PAs (Eagles, McCool, & Haynes, 2002). Undoubtedly, tour guides are the main providers and mediators of such interpretation. Their interpretation covers the environment with its physical location (the surroundings), and its related aspects including emotional, sensory, mental, and cultural aspects (Schaller, 2016).

Aim and scope of the research

The research focuses on Jordanian tour guides and seeks to explore how they perceive OT in Petra Archeological Park (PAP), which is a UNESCO world heritage site, the best-known and

 DOI: 10.4324/9781003001768

most-visited attraction in the Hashemite Kingdom of Jordan. The aim of this research, therefore, is to increase the understanding of the concept of OT in PAs from the perceptive of tour guides who are regarded as a main and maybe sole direct, day-to-day service providers, yet, are often considered a marginalised group within the tourism industry (Aloudat, 2017). The study also explores the strategies and techniques that tour guides employ to safeguard PAs and control visitor behaviour. Multiple journeys were taken to the Petra site to listen to tour guides talk about their field experience on OT.

Petra Archeological Park (PAP) has been a UNESCO World Heritage site since 1985, and was selected as one of the New Seven Wonders of the world in 2007. Its landscape and biodiversity combined with a very rich culture heritage makes the site the most frequented by foreign visitors to Jordan. The rock-cut façades are the iconic monuments of Petra. Of these, the most famous is the so-called Treasury (or Khazneh). Petra is characterised by its entrance, which is the outer Siq (path) with a length of 1200 m, which is one the most severely impacted parts of the site from random climbing by tourists, erosion of the sandstone cliffs, in addition to the dusts caused by the horses and chariots used by tourists (Moustafa & Balaawi, 2013).

It was only in 2017 that Petra joined The National Network for Natural Reserves in Jordan and declared as a natural PA in parallel to its status as an archeological PA. The ecological importance of Petra is based on the four different biogeographical zones that it spans according to the distribution of plants: Mediterranean and Irano-Turanian at higher elevations, and Saharo-Arabian and the Afro-tropical penetration zone at medium to low altitudes (The Royal Society for the Conservation of Nature, 2018).

The distinctive biodiversity that Petra abounds in includes unique plants, shrubs and animals, in addition to its geological and natural characteristics. At least 25 flora species of Petra are considered endemic to Jordan and to the Eastern Mediterranean region and more than 23 species are considered endangered at both national and regional levels, including tree species (The Royal Society for the Conservation of Nature, 2018).

Being a natural PA might preserve the city's biodiversity, much of which has been endangered due to decades of focusing on the historical and archaeological features of the city, at the expense of its ecosystem (Ministry of Environment, 2017). Different approaches to preserve Petra Protected Area (PPA) have included enhancing interpretation methods such as the use of signs and brochures to spread the environmental awareness, in addition to restricting duration and group size (Moustafa & Balaawi, 2013). Different leading initiatives are being taken to sustain environmental practices, especially in the tourism sector. This is manifested by the establishment of the Tourism Green Unit (TGU) with the aim of encouraging more eco-friendly actions, behaviours, and work practices that are in sync with environment and natural assets of Jordan (United Nations Development Programme, 2020).

Literature review

Ecotourism has been defined as traveling to relatively undisturbed or uncontaminated natural areas with the specific object of studying, admiring, and enjoying the scenery and its wild plants and animals, as well as any existing cultural aspects (Weaver, 2005, p. 19). Another definitions focus on the benefits that ecotourism strives to achieve including economic development and political empowerment of local community, and fostering respect for different cultures (Honey, 2008, p. 33). Ecotourism typically occurs in natural areas, and should contribute to the conservation of such areas (Fennell, 2007, p. 2). The International Ecotourism Society, TIES (2015) defined it as responsible travel to natural areas that conserves the environment, sustains the wellbeing of the local people, and involves interpretation and education. It involves both the natural environment and the social and natural aspects of local people. Managed ecotourism ensures the balance between the needs of economy, society, and the needs of conserving the ecological system of a destination. Other definitions of ecotourism highlight how management approaches should integrate natural, biological, social, and cultural components of an environment for the purpose of sustainability (Keitumetse, 2008).

Overtourism

Overtourism is a new term in tourism studies. In 2018, the Oxford English Dictionary made "overtourism," one of its words of the year. It is defined as an excessive number of visitors heading to famous locations, damaging the environment and having a detrimental impact on resident's lives (CNN Travel, 2019). OT is fast becoming one of the most hotly debated issues in the modern age of travel and tourism industry (Koens et al., 2018). The sharp rise in international tourists has resulted from a number of factors, including cheap air fares offered by budget airlines, rising incomes, social media's marketing, and the widespread popularity of rental platforms such as Airbnb (Koens et al., 2018). Thus, more travellers are descending on places that can no longer cope with their own popularity, contributing to the OT phenomenon. The term 'Trexit' (tourist and exit) expresses the actions that destination agencies introduce to face OT by reducing or even stopping tourists from visiting popular destinations and attractions (Seraphin, Sheeran, & Pilato, 2018). Accordingly, some countries put severe measures in place to regulate tourism in specific 'over-touristed' sites such as India, Netherlands, and Iceland (CNN Travel, 2019). For example, the Netherlands Tourist Board is engaged in destination management rather than destination promotion in its Perspectives 2030 Report. In Iceland, entry fees to national parks are being charged, while in India visits to the Taj Mahal are limited to three hours as a method for regulating numbers.

However, many destination agencies are powerless to control tourist flows as air travel and seaports are operated by air companies and cruise companies (Dodds & Butler, 2019). Accordingly, factors responsible for the growth of OT may increase rather than decrease unless serious action is taken (ibid). Optimistically speaking, one might suggest that COVID-19's dramatic results could be seen as a natural force approach allowing nature to recover in PAs.

Areas set aside to protect nature have a long history (Eagles et al., 2002). In India, for example, protected areas were established to safeguard natural resources over two millennia ago (Holdgate, 1999), while in Europe, hunting grounds were protected for rich and powerful people 1000 years ago (Eagles et al., 2002). These sites became open for public use and it was the first nucleus of community involvement in tourism (ibid). Nowadays, there are over 236,200 PAs according to the 12th and final update of the (International Union for the Conservation of Nature [IUCN], 2020). However, PAs are established and managed for a variety of purposes, which makes it difficult to identify one definition that encompasses all diverged managerial objectives. One definition by IUCN is: 'A PA is a clearly defined geographical space, recognized, dedicated and managed, through legal or other effective means, to achieve the long term conservation of nature with associated ecosystem services and cultural values' (IUCN, 2008).

The IUCN has developed a six-category system based on the purpose of the management of PA as: 1) Scientific Reserve/Strict Nature Reserve; 2) National Park; 3) Natural Monument/ Natural Landmark; 4) Nature Conservation Reserve/Managed Nature Reserve/Wildlife Sanctuary; 5) Protected Landscape or Seascape; 6) Managed Resource Protected Area (International Union for the Conservation of Nature, 1985), with many PAs fitting under categories five and six. Petra is a protected landscape that contains a resource with high historical, architectural, and archeological significance, describing the core testimony of human settlement in Jordan. Well-managed tourism in Petra can assist, therefore, not only in preserving nature but also in protecting and restoring cultural resource as well.

However, PAs are increasingly under pressure from several factors: demands for 'multiple use' parks allowing extractive industries; demand of lobby groups seeking access for a range of recreational activities; four-wheel driving, horse riding, hunting, fishing; and the aspirations of indigenous groups for title and management of parks (Neil and Wearing, 1999, p. 39). Tourism activities result in a variety of negative impacts on critical environments of PAs such as: trail creation, crowding, overdevelopment, impacts on vegetation, soil compactions or erosion, taking souvenirs, user conflicts, and damage of archeological sites (Cole et al., 1987; McNeely & Thorsell, 1989; Dowling, 1993; Buckley & Pannell, 1990; Wight, 1996) (cited in Eagles et al., 2002). Yet, the literature on PAs also identifies successes in protecting biodiversity from external threats (Holland, 2012); yet, this success, as Gaston, Jackson, Aantu-Satuzar, and Cruz Pinon (2008) argue, varies and depends on the types and levels of management regimes applied in different PAs.

The role of the tour guides

The literature on tour guides indicates that they are performers who carry out significant roles in tourism (Weaver, 2001; Weiler & Ham, 2001; Weiler & Black, 2015; Aloudat, 2017). In fact, tour guides accomplish multiple, infinite, and overlapped roles. Among the first to study the roles of tour guides were Holloway (1981), Cohen (1985), and Pond (1993), all of whom identified tour guides as information givers, ambassadors, public relation representatives, and mediators. In a seminal study that categorised the roles of tour guides but from their perceptions and found how they see their worldview, the tour guides perceived their roles to fit under four main roles namely: ambassadorial, managerial, promotional, and mediating (Aloudat, 2017). These inseparable roles often cause pressure and worry for tour guides and 'tour guides may be the most maligned people in the world of travel' (Prakash, Chowdhary, & Sunayana, 2011, p. 66). The complexity in tour guides' roles and features such as tour cancellations, extended tours, irregular work, delays, long working hours, and irregularity of employment and seasonality make this job a stressful and demanded career job (Aloudat, 2017). Another new study has examined the role of tour guides as facilitators of spiritual tourism (Parsons, Houge Mackenzie, & Filep, 2019). The study revealed that tour guides act as brokers to spiritual tourist through facilitating physical access to sites and facilitating encounters, and understanding, empathy, and self-development. In terms of examining the occupational commitments of tour guides, a study found a strong positive relationship between burnout and economic anxiety levels (Yetgina & Benligiray, 2019).

Tour guiding in ecotourism has been researched from different scholars (Weiler & Ham, 2001; Weiler & Ham, 2002; Christie & Mason, 2003; Skanavis, Sakellari, & Petreniti, 2005; Tripathi, 2016). Tour guiding is identified as an effective tool to achieve the purpose of ecotourism which is to engage tourists in low impact, non-consumptive, and locally oriented environments in order to maintain species and habitats (Tripathi, 2016, p. 27). Thus, tour guides are the supporters and, most importantly, the implementers of ecotourism goals in the field.

Interpretation is a main role for tour guides and it has been argued that the role of tour guides as interpreters is the most predominate role that the tour guides play (Black & Weiler, 2005; Black & Ham, 2005). It is a crucial factor in protecting tourism assets as it reveals the significance of the visited places and to stimulate sustainable behaviour (Black & Ham, 2005). Interpretation has been identified as the process of transforming heritage into language

(Howard, 2003), as the heart of ecotourism (Weiler & Ham, 2002), and a direct way of enhancing the experience and the understanding of the site (Kong, 2014). Accordingly, tour guides have been regarded as stimulators of positive behaviour toward the environment. Through interpretation, tour guides can elevate the understanding and appreciation of resources (Eagles et al., 2002). As interpreters, tour guides are often the main awareness and educational source for visitors to natural and cultural PAs (Skanavis & Giannoulis, 2010). They are the motivator of environmentally responsible behaviour and conservation values, and specialist information givers (Black & Weiler, 2005; Black & Ham, 2005). At a deeper level, tour guides are seen as main factors that direct the tourist to or away from sustainable practices, thus, they contribute to the success or failure of an ecotourism venture (Pond, 1993 as cited in Christie & Mason, 2003; Weiler & Ham, 2002; Christie & Mason, 2003; Black & Ham, 2005; Imon, 2013). As such, they act as conservators of the natural resources of the destination. Interpretation is an approach to communicate environmental messages in PAs where natural and cultural resources are predominating such as national parks, national forests, museums, zoos, aquariums, and botanical gardens (Ham, 1992).

Well-trained guides, using high-quality performances, maximise the value of the ecotourists' experience and, therefore, help in achieving the conservation goals of ecotourism (Steward et al., 1998, cited in Christie & Mason, 2003; Black, Ham, & Weiler, 2001; Weiler & Ham, 2002). Advancing the education and training of tour guides is an important tool to improve the quality of ecotourism experience. Skanavis and Giannoulis (2010) have developed a model for environmental interpreter guides in PAs in Greece. They urge local people to train themselves as interpretive tour guides to achieve not only ecological sustainability but also economic sustainability (Skanavis & Giannoulis, 2010).

Good quality guiding stimulates the visitor to be committed to the visited place (Aloudat, 2017) and to behave in a responsible and sustainable way (Weiler & Kim, 2011; Poudel & Nyaupane, 2013) as they enhance the sustainable behaviour of visitors (Alazaizeh, Jamaliah, Mgonja, & Ababneh, 2019).

The literature on tour guides has also focused on revealing the roles and importance of tour guides in the tourism industry. Most of the literature has taken an instrumental approach to investigate tour guiding with a professional and functioning focus (Holloway, 1981; Cohen, 1985; Pond, 1993; Weiler & Ham, 2002; Christie & Mason, 2003; Salazar, 2005; Macdonald, 2006; Jennings & Weiler, 2006; Scherle & Nonnenmann, 2008) and very few studies have considered the situational features of tour guiding particularly from the tour guides' perspective (Salazar, 2005; Aloudat, 2017). More specifically, few studies have referred to them as key informants on tourism issues despite their close, deep, and long contacts with tourists. Tour guides' daily contact with tourists make them a valuable source of information related to different tourism issues including the performance of tourism destination, and tourist behaviour (Aloudat, 2013). Accordingly, their perceptions may be regarded as an informative source on the performance of a destination and their knowledge may be utilised by tourism agencies to gain valuable and up-to-date feedback on the tourism performance of a given destination (Aloudat, 2017). Thus, their input on OT is worthy of exploration especially in sensitive areas such as PAs where the behaviour of tourists is important to monitor and control.

Methodology

The research enquiry used a qualitative approach. The data was collected from tour guides joining tourists in PPA, and semi-structured interviews took place with 12 tour guides. The participants were all Jordanian tour guides and members of the Jordanian Tour Guides

Association (JTGA), which is the professional association representing and registering all tour guides in Jordan. There are around 200 on-site guides in addition to 1180 national guides.

The interviews took three months from November 2019 to January 2020. The time of interviews varied from 25 minutes to 70 minutes. The participants shared their experiences and perceptions of aspects of OT in PPA. The interviews provided information about the views on and the experience of tour guides in OT, and operational strategies they follow to deal with OT. The sample strategy started as a purposive. Two interviews were with two tour guides known to the researcher thereafter the snowballing technique was used. The later strategy was suitable in finding informants from tour guides to participate in the research given the fact that the data collection was during a peak season before the interruption by COVID-19. All interviews were recorded after getting the permission of the interviewees, and all participants remained anonymous in this study. After recording, the interviews were transcribed and then translated to English. Quotes used in the text are retrieved from these transcripts.

The themes and main topics of the interviews consisted of first, the interviewee's sociodemographic profile (age, gender, years of work experience, foreign language spoken), and second, emergent themes relating to tour guides' perceptions on OT in PPA. The analysis was based on open and axial coding. The findings of the interviews were categorised into three themes namely tour guides' perspectives on OT; tourists' behaviour and attitude changes; and tour guides' strategies, techniques, and future solutions. The following table represents the main characteristics of the respondents of this study (Table 5.1).

It came as no surprise that the majority of participants were male guides and only one female guide was interviewed. In Jordan, there is a big disparity in the numbers of tour guides in terms of gender. Currently, out of the total number of government-approved tour guides listed in the JTGA, there are 65 female guides in the list and 1113 males (Ministry of Tourism and Antiquities, 2019).

Findings

The data provided rich insights into how the tour guides in this study perceive overtourism in PAs and how they perceive their role in this phenomena. Overall, three key thematic areas emerged that are discussed in this section of the chapter with quotations from the tour guides illustrating the key issues. The three key themes are as below:

Table 5.1 Demographic variables of the tour guides

Acronym	*Gender*	*Age*	*Years of work experience*	*Foreign language spoken*
Guide 1	M	46	25	English/French/Indonesian
Guide 2	M	35	13	French
Guide 3	F	42	15	English
Guide 4	M	43	10	English
Guide 5	M	53	15	Germany
Guide 6	M	33	6	English
Guide 7	M	40	6	English
Guide 8	M	32	3	Spanish
Guide 9	M	36	15	French/English
Guide 10	M	38	19	English/Dutch
Guide 11	M	36	15	English
Guide 12	M	37	13	German/English

Tour guides' perspectives on overtourism

The phenomenon of OT has existed for a long time in tourism destinations. Scholars have identified OT through discussing the excessive numbers of visitors and the changes brought to tourism destinations including negative reactions by residents and changes and damages to cultural and natural resources. The tour guides found that OT is observed in PPA in the peak season mainly from April to October. They feel uncomfortable with high tourist numbers and, therefore, their activity as they escort tourists through the main entrance of PPA (The Siq)

> … Petra is suffering especially in the peak … when I find a lot of tourists I feel unconformable and this affects my performance … (Guide 1)

Several participants expressed the aspects of OT in PPA as: crowding, stepping on the fragile rocks and plants, raising the voices, and pollution.

> … negative activities like stepping on the plants and herbs, raising the voices, the pollution including water bottles, the pressure on the sewage and drainage services, the dung of horses and its smell, and even the tough way the carriage drivers treat the horses … (Guide 4)

Some participants went even further to reveal that PPA is visited as a cultural and archeological site as a main motivation of tourists rather than a natural site. Thus, tourists are less respectful to the ecosystem of PPA. However, the participants asserted their role in making the tourists aware of the significance of Petra as a PA by interpreting the value of its ecosystem.

> … unless the tour guides revealed the significance of the ecosystem of Petra because when visiting PPA they focus on it as a cultural and archeological place rather than a natural site … (Guide 2)

While early PAs were established predominantly to protect and preserve the wildlife, the International Union for the Conservation of Nature (IUCN; 2008) has more recently described PAs as 'clearly defined geographical space, recognized, dedicated and managed, through legal or other effective means, to achieve the long term conservation of nature with associated ecosystem services and cultural values' (Dudley, 2008, p. 8). This definition incorporated and emphasised the cultural value and the resource use of the PA making the concept of PA wide to include many aspects that may relate to the culture of the place, the local community and any resources regardless of its type. A guide expressed this idea as: 'I always make the tourists aware of the local community, because the tourists may don't know how they feel … this is very sensitive …' (Guide 5).

Additionally, the tour guides identify their efforts in sustaining the benefits of the local community in the PA by explaining about tradition and customs of the community and by promoting their local products. ', … I explain about customs and traditions … I promote the products of the local' (Guide 1).

The Siq and the Treasury are the main places to expose the OT effects in PPA. Of interest is how tour guides describe themselves as responsible about any negative behaviour of tourists. The same guide (Guide 1) with experience of 25 years continues as: 'I try to find and manage the tour …, I feel guilty when I see any tourists do something wrong' (Guide 1).

The tour guides noted a number of visible problems that traced to the lack of management effectiveness of PPA. Issues appears in the comments of most of the participants is related to the management approach in PPA. 'well coordination should be between the different parts of the tourism … there are a lot of problems in managing PPA, this includes the number of visitors, time of visit, the horses used to transfer the tourists' (Guide 7). Another guide said: 'The management is ad hoc … no control on those who ride the horse carriages …' (Guide 12).

According to the time of visits, almost all of the participants expressed that the managers of PPA should organise the visiting time slots to not go above the carrying capacity of the site: 'a coordination between PPA and the tour operators in order to know the number of visitors estimated in a day … they may find a strategy that ensures not exceeding the carrying capacity of Petra' (Guide 5).

Another guide found that the problem of OT is manifested in an observed way when the tourists who come in cruise ships enter Petra. 'The problem is when the tourists of cruise ships visit the site …' (Guide 4).

Tourists' behaviour and changes in attitudes

Although tour guides see their main role as information providers they also acknowledge that their role is more than disseminating information or mediating between the place and the tourists. When asked about their role in affecting the tourist's behaviour they asserted that the role of the tour guide comes with responsibility toward the PA. One guide said: 'the tour guide is responsible about everything, you understand the value of every assets plants, herbs, birds, etc.' (Guide 3).

The tour guides perceived their role as safe guardians of the PPA. They transfer different educational messages through their work not only for the tourist but also to other partners in the tourism process. 'I am always keen to educate the tourists and the drivers' (Guide 8). Another guide said: 'I behave responsible in front of my clients...' (Guide 1).

It is interesting to see how tour guides perceived their role toward other partners of the tourism process including the tourists and the drivers whom they influence their behaviour and attitudes. Several guides described their attitude as ideal.

Tour guides' strategies, techniques, and future solutions

Tour guides in this study talked about their experience in minimising the effects of OT through various strategies and techniques. The most significant one is timing of the visits. The majority of tour guides found that OT may be solved by informing managers in PPA about the schedules of cruise ships as this will offer site management the opportunity to organise and control the numbers of visitors and the time of visits. Another suggestion is to manage the flow of groups in a stratified manner. '… to make a distance between each group, this will insure no crowding, no damages to the site ...' (Guide 9).

Another guide told:

> you can't tell the tourists that there is OT in Petra we can't go there … they pay money and the tour operators can't refuse any demand … so the only way to minimize the negative effects is to make more effective management. (Guide 7)

Another technique proposed by some of the participants to overcome OT is to manage the entrance and exit of the site. Some guides suggested to make two different ways and separate the entrance from the exit. The tour guides even went further to mention that this approach will not only control the OT effects but will ensure the safety of the tourists. The majority of the participants talked about floods that happened in Petra last winter when they faced a safety problem. It was mentioned that flash floods hit in the ancient city of Petra in 2018 which killed at least seven people and nearly 4000 tourists were forced to evacuate to safe areas and this was through the main entrance.

Another guide asserted the importance of some strategies that may help in minimising any anticipated negative impacts of OT. ' ... more information should be given to the tourists, the guides, and the drivers, about the site, it's capacity, and the climate' (Guide 10).

Safety was a focus to manage the PA, as the topography of Petra is considered as difficult terrain, thus, the tour guides proposed to establish an area for first aid uses. When any of the tourists get tired because of illness or any accident inside the site the way is too long to send him back to the entrance... (Guide 6).

The development of more infrastructure for world heritage site is challenging, as there are restrictions on what managers can add to the site without affecting its identity and authenticity. This development and managerial problem is revealed by some of the participants. However, of interest, other participants expressed that the problems also come from people working in the industry themselves like carriage drivers and souvenir buyers. A female tour guide expressed her occupational feelings toward the site as: 'the way the carriage drivers treat the horses is so tough and harsh ... I feel sorry when I see them like this ...' (Guide 3).

She continues:

> The souvenir buyers also have negative impacts on the site .... They put their stuff on the rocks or under the trees ... We as tour guides can't do that because they are from the local community and they feel that the site is for them ... (Guide 3)

The previous extracts revealed how the effects of OT are traced back not only to the tourists who are always the blamed users of tourism attractions, but also to tourism service providers. The awareness on the part of industry people and the local community is a strategy that the management bodies in PAs should focus on and formulate regulations regarding OT effects.

Discussion and conclusion

The findings of this study revealed the importance of tour guides in promoting and achieving the goals of ecotourism in PAs. The tour guides regarded themselves as protectors and conservatives of the environment and its surroundings. These findings confirm previous literature on tour guides' role as interpreters of the environment, motivators of environmentally responsible behaviour and conservation values, and as special information givers (Eagles et al., 2002; Black & Ham, 2005; Yamada, 2011). Thus, more concern should be made to the performance of tour guides and the quality of interpretation they provided, which reflected on the positive experience for the tourists aligning with the principles of ecotourism. This may be achieved through the continuous education and training of tour guides in PAs. Appropriate guide training provides tour guides with necessary skills and knowledge that enables them to ensure the 'safety of visitors, to provide accurate and compelling interpretation of sites and modeling appropriate environmental and cultural behaviours' (Black et al., 2001, p. 147).

Even though the phenomenon of OT is a term that is manifested by changes brought by the increasing numbers of tourists including the negative impacts on the resources and the local community, the tour guides analysed in this study found that OT's negative impacts, including environmental degradation, on PPA are also brought by local people as well as industry. Tour guides provide an essential role in raising the environmental awareness of tourists, other partners in the industry, and the local community. As safe-guardians of PPA, tour guides emphasised their capability to influence the behaviour of, not only the tourists, but also other tourism partners such as drivers and souvenir buyers in addition to the local community. Their views support the literature in the field of ecotourism that viewed them as major contributors to the success or failure of an ecotourism trip, and promoters of positive behaviours and practices toward the environment (Pond, 1993; Stewart et al., 1998 as cited in Christie & Mason, 2000, 2003; Weiler & Ham, 2002; Alazaizeh et al., 2019). It is thus possible that tourism managers and planners concerned with PAs may refer to the tour guides to give feedback on behaviour of tourists on PAs. The tour guides, therefore, are capable in achieving the goal of environmental education in ecotourism.

The guides in this study expressed their concern about OT in PPA because of the many impacts. As direct observers, day-to-day field workers, tour guides mentioned that the management in PPA is lacking updated strategies and regulations to overcome the impacts of OT. They conveyed that OT is harming PPA, and putting infrastructure under enormous strain. The tour guides also indicated that OT in PPA is an issue that is oversimplified and not managed well, thus, confirming the findings of previous studies that showed a lack of organisational cohesion and effectiveness among organisations and lack of authority for on-site managers in PPA (Comer, 2012; Moustafa & Balaawi, 2013). One might ask if the management in PPA is able to cope with OT impacts and maintain the balance between keeping the site for future tourism demand, in one hand, and avoiding more degradation of the ecosystem of PPA in the other hand.

The findings raise some interesting managerial implications. The tour guides offered several suggestions that may enhance the effectiveness of the management of PAs. They felt that because they are not decision makers or, at least, a part of the decision-making process that their voices would not be heard. Accordingly, the study advocates tour guides as direct observers of and key informants on the tourism performance of PAs. The informative skills of tour guides, which are cumulatively acquired from their daily work in the field and their closeness to the tourists and the tourism attractions, including PAs, make them capable of offering significant managerial insights. As such, they stand as a considerable source of data for national tourism management bodies. Tour guides' capability as in-depth information givers on the management of PAs confirmed other studies that advocated the tour guides as important for the management of tourism sites in general and for the PAs in specific (Moustafa & Balaawi, 2013; Schaller, 2016; Aloudat, 2017). The findings also confirm studies that position them as key informants of national tourism performance (Aloudat, 2013). The implications of this study are considerable, since the data suggest their contact with tourists may offer significant insight into the performance of PAs. Precisely because of their location at the "sharp end" of tourism as the sole agents who mediate tourism itineraries "on the ground" to visitors, they appear to be in a position to provide "ground level" insights on current and evolving issues in the delivery of the tourism experience that may be less apparent or available to managers more remote from the operational levels of tourism (Aloudat, 2013).

References

Alazaizeh, M., Jamaliah, M., Mgonja, J., & Ababneh, A. (2019). Tour guide performance and sustainable visitor behavior at cultural heritage sites. *Journal of Sustainable Tourism*, *27*(11), 1708–1724.

Aloudat, A. Sh. (2013). Tour guides as source of tourism performance data. In M. Koerts, & Smith (Eds.), *Conference proceedings of 3rd International Research Forum on Guided Tours*. Breda:NHTV-NRIT.

Aloudat, A. Sh. (2017). *The Worldview of Tour Guides: A Grounded Theory Study*. Germany: Noor Publishing.

Al-Shorman, A., Rawashdeh, A., Makhadmih, A., Oudat, A., & Darabsih, A. (2016). Middle Eastern political instability and Jordan's tourism. *Journal of Tourism Research and Hospitality*, *5*(1), 1–5.

Balmford, A., Rodrigues, A. S. L., Walpole, L., Ten Brink, P., Bratt, L., & Groot, R. D. (2008). *The Economics of Ecosystems and Biodiversity: Scoping the Science*. University of Cambridge: European Commission.

Bayrama, G., Karaçarb, E., & Turan, A. (2017). Karabük Üniversitesi Sosyal Bilimler Enstitüsü Dergisi. *Özel Sayı*, 3, 40–50.

Black, R., Ham, S., & Weiler, B. (2001). Ecotour guide training in less developed countries: Some preliminary research findings. *Journal of Sustainable Tourism*, *9*(2), 147–156.

Black, R., & Ham, S. (2005). Improving the quality of tour guiding: Towards a model for tour guide certification. *Journal of Ecotourism*, 4, 178–195. 10.1080/14724040608668442.

Black, R., & Weiler, B. (2005). Quality assurance and regulatory mechanisms in the tour guiding industry: A systematic review. *The Journal of Tourism Studies*, *16*(1), 24–35.

Buckley, R. & Pannell, J. (1990). Environmental impacts of tourism and recreation in national parks and conservation reserves. *Journal of Tourism Studies*, 1(1), 24–32.

Christie, M., & Mason, P. (2003). Transformative tour guiding: Training tour guides to be critically reflective practitioners. *Journal of Ecotourism*, *2*(1), 1–16.

CNN Travel. (2019). Destination Trouble: Can Overtourism Be Stopped in Its Tracks? Retrieved from https://edition.cnn.com/travel/article/how-to-stop-overtourism/index.html (accessed on 25.03.2020).

Cohen, E. (1985). The tourist guide: The origins, structure, and dynamics of a role. *Annals of Tourism Research*, *12*, 5–29.

Cole, D. N., Petersen, M. E., & Lucas, R. C. (1987). Managing wilderness recreation use: common problems and potential solutions. Gen. Tech. Rep. INT-GTR-230.USDA Forest Service, Intermountain Research Station, Ogden, UT, USA.

Comer, D. (2012). *Tourism and Archaeological Heritage Management at Petra*. New York: Springer.

Daher, R. (2007). *Tourism in the Middle East, Continuity, Change and Transformation*. Clevedon, Buffalo: Channel View Publications.

Dodds, R., & Butler, R. (2019). The phenomena of overtourism: A review. *International Journal of Tourism Cities*, *5*(4), 519–528.

Dowling, R. K. (1993). Tourism planning, people and the environment in Western Australia. *Journal of Travel Research*, 31, 52–58. 10.1177/004728759303100408.

Dudley, N. (2008). Guidelines for Applying Protected Area Categories, IUCN, Gland.

Eagles, P., McCool, S., & Haynes, C. (2002). *Sustainable Tourism in Protected Areas*. Gland: IUCN.

Ekanayake, E., & Long, A. (2012). Tourism development and economic growth in developing countries. *The International Journal of Business and Finance Research*, *6*(1), 61–63.

Fennell, D. (2007). *Ecotourism* (3rd ed.). New York: Routledge.

Gaston, K. J., Jackson, S. F., Aantu-Satuzar, L., Cruz Pinon, G. (2008). The ecological performance of protected areas. *Annual Review of Ecology, Evolution, and Systematics*, *39*, 93–113.

Ham, S. H. (1992). *Environmental Interpretations: A Practical Guide for People with Big Ideas and Small Budgets*. Golden, CO: North American Press.

Holdgate, M. (1999). *The Green Web: A Union for World Conversation*. London, UK: Earthscan.

Holland, M. (2012). The role of protected areas for conserving biodiversity and reducing poverty. In Jane Carter Ingram, Fabrice DeClerck, & Cristina Rumbaitis del Rio (Eds.), *The application of ecology in development solution: Integrating ecology and poverty reduction* (pp. 253–270). New York: Springer Science & Business Media.

Holloway, J. (1981). The guided tour, a sociological approach. *Annals of Tourism Research*, *8*(3), 377–402.

Honey, M. (2008). *Ecotourism and Sustainable Development: Who Owns Paradise?* (2nd ed). Washington, DC: Island Press.

Howard, P. (2003). *Heritage: Management, Interpretation, Identity*. New York: Continuum.

Imon, S. S. (2013). 13 Issues of sustainable tourism at heritage sites in Asia. *Asian Heritage Management: Contexts, Concerns, and Prospect*, *39*, 253.

International Union for the Conservation of Nature. (1985). *United Nations List of National Parks and Protected Areas*. Gland, Gstaad: IUCN.

International Union for the Conservation of Nature. (2020). World Database on Protected Area. Retrieved from https://www.iucn.org/theme/protected-areas/our-work/quality-and-effectiveness/world-database-protected-areas-wdpa (accessed on 28.03.2020).

International Ecotourism Society. (2016). What Is Ecotourism? Retrieved from https://www. https://ecotourism.org/what-is-ecotourism/ (accessed on 08.04.2020).

IUCN. (2008). Defining protected areas: an international conference in Almeria, Spain. Gland, Switzerland: IUCN, Nigel Dudley and Sue Stolton (eds.), p. 220.

Jennings, G., & Weiler, B. (2006). Mediating meaning: Perspectives on brokering quality tourist experiences. In G. Jennings, & N. Nickerson (Eds.), *Quality tourism experiences* (pp. 57–78). USA, UK: Butterworth-Heinemann.

Keitumetse, S. O. (2008). The Eco-tourism of Cultural Heritage Management (ECT-CHM): Linking heritage and 'environment' in the Okavango Delta regions of Botswana. *International Journal of Heritage Studies, 15*(2–3), 223–244.

Koens, K., Postma, A., & Papp, B. (2018). Is overtourism overused? Understanding the impact of tourism,in a city context. *Sustainability, 10*(12), 4384.

Kong, H. (2014). Are tour guides in China ready for ecotourism? An importance-performance analysis of perceptions and performances. *Asian Pacific Journal of Tourism Research, 19*(1), 3741.

Leung, Y., Spenceley, A., Hvenegaard, G., & Buckley, R. (2018). *Tourism and Visitor Management in Protected Areas: Guidelines for Sustainability*. Gland, Switzerland: International Union for the Conservation of Nature.

Macdonald, S. (2006). Mediating heritage: Tour guides at the Former Nazi Party Rally Grounds, Nuremberg. *Tourist Studies, 6*(2), 119–138.

McNeely, J. A. & Thorsell, J. W. (1989). Jungles, mountains, and islands: How tourism can help conserve the natural heritage. *World Leisure & Recreation*, 31, 29–39. 10.1080/10261133.1989.10559089.

Ministry of Environment. (2017). Petra. Jordan: Ministry of Environment. Retrieved from http://moenv.gov.jo/En/OurNews/Pages/enew12720171.aspx (accessed on 28.03.2020).

Ministry of Tourism and Antiquities. (2019). Statistics. Retrieved from https://www.mota.gov.jo/Contents/Statistics.aspx (accessed on 20.03.2020).

Moustafa, M., & Balaawi, F. (2013). Evaluation visitor management at the archeological site of Petra. *Mediterranean Archeology and Archaeometry, 13*(1), 77–87.

Neil, J., & Wearing, S. (1999). *Ecotourism: Impacts, Potentials, and Possibilities*. Oxford: Butterworth-Heinemann.

Parsons, H., Houge Mackenzie, S., & Filep, S. (2019). Facilitating self-development: How tour guides broker spiritual tourist experiences. *Tourism Recreation Research, 44*(2), 141–152.

Pond, K. (1993). *The Professional Guide*. New York: VNR.

Poudel, S., & Nyaupane, G. P. (2013). The role of interpretive tour guiding in sustainable destination management: A comparison between guided and non-guided tourists. *Journal of Travel Research, 52*(5), 659–672.

Prakash, M., Chowdhary, N., & Kumar, S. (2011). Tour guiding: Interpreting the challenges. *Tourismos: An International Multidisciplinary Journal of Tourism, 6*(2), 65–81.

Reid, M., & Schwab, W. (2006). Barriers to sustainable development: Jordan's sustainable tourism strategy. *Journal of Asian and African Studies, 41*(5/4), 439–457.

The Royal Society for the Conservation of Nature. (2018). Biodiversity Information Management System (BIMS). Retrieved from https://www.rscn.org.jo/biodiversity-information-management-system-bims-website (accessed on 08.04.2020).

Salazar, N. (2005). Tourism and glocalization "local" tour guiding. *Annals of Tourism Research, 32*(3), 628–646.

Schaller, H. (2016). Tour guides in nature-based tourism: Perceptions of nature and governance of protected areas, the case of Skaftafell at the Vatnajökull National Park. In D. Rancew-Sikora, & U. D. Skaptadóttir (Eds.), *Mobility on the edges of Europe: The case of Iceland and Poland* (pp. 187–214). Warsaw: SCHOLAR Publishing House.

Scherle, N., & Nonnenmann, A. (2008). Swimming in cultural flows: Conseptualisating tour guide as intercultural mediators and cosmopolitans. *Journal of Tourism and Cultural Change, 6*(2), 120–137.

Seraphin, H., Sheeran, P., & Pilato, M. (2018). Over-tourism and the fall of Venice as a destination. *Journal of Marketing and Management, 9*, 374.

Shdeifat, O., Mohsen, M., Mustafa, M., Al-Ali, Y., & Al-Mhaisen, B. (2006). Tourism in Jordan. LIFE Third Countries. *The Hashemite University*, (1), 1–68.

Skanavis, C., & Giannoulis, C. (2010). Improving quality of ecotourism through advancing education and training for eco-tourism guides: The role of training in environmental interpretation. *TURISMOS: An International Multidisciplinary Journal of Tourism, 5*(2), 49–68.

Skanavis, C., Sakellari, M., & Petreniti, V. (2005). The potential of free choice learning of environmental participation in Greece. *Environmental Education Research, 11*(3), 321–333.

Tripathi, K. (2016). Tourism to ecotourism: A tour. *International Journal of Humanities and Social Science, 3*(6), 27–30.

United Nations Development Programme. (2020). UNDP Jordan Supports the Ministry of Tourism & Antiquities to Take the First Step Towards Tourism Transition into Green Economy. Retrieved from https://www.jo.undp.org/content/jordan/en/home/ourwork/environmentandenergy/successstories/undp-jordan-supports-the-ministry-of-tourism---antiquities-to-ta.html (accessed on 22.03.2020).

United Nations Educational, Scientific and Cultural Organization. (2011). *Wadi Rum Protected Area.* Paris: United Nations Educational, Scientific and Cultural Organization. Retrieved from https://whc.unesco.org/en/list/1377 (accessed on 25.03.2020).

United Nations of World Tourism Organization. (2020). *Why Tourism?* Madrid, Spain: United Nations of World Tourism Organization. Retrieved from https://www.unwto.org/why-tourism (accessed on 27.03.2020).

Weaver, D. (2001). *The Encyclopedia of Ecotourism.* UK: CABI.

Weiler, B., & Black, R. (2015). *Tour Guiding Research: Insights, Issues, and Applications: Aspects of Tourism.* UK: Channel View Publication.

Weiler, B., & Ham, S. (2001). Tour guides and interpretation. In D. Weaver (Ed.), *Encyclopedia of ecotourism* (pp. 549–563). Oxford: CABI.

Weiler, B., & Ham, S. (2002). Tour guide training: A model for sustainable capacity building in developing countries. *Journal of Sustainable Tourism, 10*(1), 52–69.

Weiler, B., & Kim, A. K. (2011). Tour guides as agents of sustainability: Rhetoric, reality, and implications for research. *Tourism Recreation Research, 36*(2), 113–125.

Weaver, D. B. (2005). Mass and urban ecotourism, new manifestations of an old concept. *Tourism Recreation Research, 30*, 19–26.

Wight, P. A. (1996). Planning for Success in Sustainable Tourism. Invited paper presented to "Plan for Success". Canadian Institute of Planners National Conference, Saskatoon, Saskatchewan, June 2–5.

Yamada, N. (2011).Why tour guiding is important for ecotourism: Enhancing guiding quality with the ecotourism promotion policy in Japan. *Asia Pacific Journal of Tourism Research*, 16, 139–152.

Yetgina, D., & Benligiray, S. (2019). The effect of economic anxiety and occupational burnout levels of tour guides on their occupational commitment. *Asia Pacific Journal of Tourism Research, 24*(4), 333–347.

6

TECHNOLOGY AND THE SUSTAINABLE TOURIST IN THE NEW AGE OF DISRUPTION

David A. Fennell

Introduction

The global impact of the recent COVID-19 virus has had a disruptive impact on the planet in unprecedented ways. How we negotiate the new realities around this virus, other global health and security issues, chaos in financial markets, and environmental crises in the near future will require a range of adaptive strategies that are inherently resilient and sustainable.

The tourism industry was especially hard hit by COVID-19 as mobility restrictions in all world regions presented the tourism industry "with a major and evolving challenge" (UNWTO, 2020). The tourism scholarly community (TRINET) initiated perhaps the lengthiest discussion in recent memory on the consequences of post-viral tourism. Topics included safe airlines, safe hotels, proper hygiene, codes of ethics, certification, and safety assurance (Maingi, 2020); the financial realities for small businesses, many of which will never recover (Ruane, 2020); reduction of taxes for governments and far fewer jobs (Buhalis, 2020); and the importance of sustainable communities (Macbeth, 2020). Cvelbar (2020) wrote that, "If the pandemic crisis lasts up to six months, we will probably go back to the pre-virus tourism development scenario. Yet if the situation prolongs, then we will never go back to tourism as we know it today." There have also been calls for tourists to simply stay home and practice localism in appreciating what we have in our own backyards. Such would ensure that local businesses would stay viable with the added benefit of enhancing sustainability through the reduction of long-haul travel.

Scholars argue that tourism is premised on a traditional five-stage model based on anticipation and preparation, travel to the site, the on-site experience, travel home, and recollection (Clawson & Knetsch, 1966). "Being there" is of obvious significance for tourists in their search for novelty and escape (Iso-Ahola, 1982), even in times of crisis. We see this presently with the COVID-19 virus, where tourists appear undaunted by the existential threat by maintaining a business-as-usual stance (Brito, 2020; Osumi, 2020). This may simply be a personal cost-benefit calculation in following through with travel that has already been booked and paid for. But as governments enforce new rules around travel, tourism in the new age of disruption must be open to transformation.

Technology has already emerged as a key driver of change in the tourism industry (Neuhofer, Buhalis, & Ladkin, 2014). In their state-of-the-art review of information and

DOI: 10.4324/9781003001768-6

communication technologies (ICTs) and the future of tourism management, Hughes and Moscardo (2019) observe that:

- Tourism scholars have traditionally focused on the implications of ICTs for supply rather than tourists (see Pearce, 2016).
- ICTs will change the way we manage tourist experiences, including how tourists connect with other people immediately in time and space (see Werthner et al., 2015).
- Technology is stimulating a move away from intermediaries in directly connecting tourists with destinations (see Atzori, Iera, & Morabito, 2014).
- Tourists are tapping into specialised tourism providers for tailor-made and co-created experiences.
- New opportunities are emerging for ICT infrastructure and IT literacy to facilitate new opportunities for tourists at the destination (see Manyika et al., 2013).
- Greater concern over war, conflict, and security threats among the Gen Y and Z cohorts in developed economies is forcing tourism providers to develop new approaches to serving the needs of these cohorts (see Coca-Stefaniak & Morrison, 2018).

What Hughes and Moscardo (2019) did not touch on in their account, however, is how ICTs may contribute to a more sustainable future in tourism. This is the point of departure for the present chapter. I explore how personalised, interactive, real-time tours, or PIRTs (5G streaming in real-time using 360-degree view cameras, webcams, drones, and with appropriate hardware, software and infrastructure) will change how we consume touristic experiences in the new age of disruption. New technologies will enable "tourists" to stay at home whilst using the benefits of a tour guide at the destination as a realistic alternative to conventional travel, with further implications on the ways in which we theorise and practice sustainability.

There are companies that presently use different forms of ICTs to promote their attractions and destinations. For example, EarthCam promotes their products as a "Visual doorway to a tourist's attention" by using live-streaming video of an attraction (e.g., conditions at a ski resort) (www.earthcam.com/company/join_network.php). The Live Royal Albatross Cam, a 24-hour live-stream at Dunedin, New Zealand, is a case in point (www.doc.govt.nz/nature/native-animals/birds/birds-a-z/albatrosses/royal-albatross-toroa/royal-cam/). Several destinations use virtual reality and 360-degree content technology for a variety of purposes, including the reduction of overtourism. Examples include Table Rock Mountain in South Carolina, Egypt's Tomb of Nefertari, and Ha'ena State Park in Hawaii (Haugen, 2020). Other examples include aquarium tours www.youtube.com/watch?v=1PpOrGbqCBA&list=PLb0l37SaBAgWvECrYnBfliN776Dm5d582&index=14&t=0s, walking tours www.youtube.com/watch?v=m65NkIAIMSw. www.youtube.com/channel/UCPur06mx78RtwgHJzxpu2ew/videos, and as this research note was going to press, live Instagram safari tours that can be scheduled two times per day (www.youtube.com/watch?v=hQ_5EQqwKD0).

The model proposed in this research note takes a step forward in the use of technology, to better connect people and settings in a much different way.

Theoretical and methodologial frameworks

The theoretical framework for this research note is assembled around three perspectives: alternative hedonism, the sustainable citizen, and ecological footprint analysis. Soper (2008) argued that in an era of looming environmental crisis (e.g., climate change), consumerism, instant gratification, capitalism, and neo-liberalism, alternative hedonism represents a radical change. In the search for the moral good, consumers gain pleasure in purchasing products that are good for the environment and other people, or by not purchasing products at all in their search of the moral good. Alternative hedonism resonates with research on the sustainable citizen and the environmental subject. The former includes the willingness to pay for sustainable practices in an effort to protect the planet through their purchasing power (Miller, 2003). The environmental subject discourse places emphasis on how exposure to nature, environmental matters, and sustainability at a young age may stimulate environmentalist attitudes, values, and behaviours later on in life (Wells & Lekies, 2006). Ecological footprint (EF) analysis is an accounting system that measures the level of human demand on the natural world (Rees, 1992). EF has spawned several parallel schemes that pertain to specific resources and their utilisation. These include carbon, water, and energy footprints, all of which are regularly cited in the literature, and increasingly in tourism (Hunter, 2002).

The methodology stems from recreation program planning, and in particular, a technique by which to implement the best program in the most efficient and comprehensive way possible. Russell (1982) argued that in order to select such a program, different alternatives would need to be built from several key criteria (see Fennell 2002 for an application of this methodology to ecotourism). The criteria selected for this study are identified in Table 6.1 and correspond to four main categories: environmental impacts, economic impacts, socio-cultural impacts, and existential factors.

In order to remove subjectivity in the development of program criteria and alternatives, weighting and scoring techniques are often employed. Criteria are typically assigned weights to reflect the importance of one criterion relative to others (Russell, 1982). Furthermore, alternatives are scored along a continuum to gauge the level of impact or value that a criterion has within a given alternative. Criteria were weighted in this study based on a scale of "1 = least amount of importance," to "5 = most amount of importance." The two alternatives in Table 6.1 were scored according to "Very high negative impact/value (-3)," to "Very high positive impact/value (+3)." (See the bottom of Table 6.1 for more details on these methods.) The first (column B in Table 6.1) is a traditional ecotour based on a hypothetical scenario. The second (column C in Table 6.1) is a surrogate ecotour based on the use of PIRT technologies, as described previously, in which the tourist does not leave the generating region to consume the experience. The summed scores for each alternative provide an indication of the importance of one alternative against others.

Results

Table 6.1 shows the results of the comparison of both alternatives. Not surprisingly, the largest positive impact for a conventional ecotour is in the area of economics (+83). Clearly being at the destination provides a much higher degree of integration of the many needed elements of service provision that are called upon to facilitate tourism. There is an associated high negative impact for economies in Alternative 2 (–75) because ecotourists are staying home. The recent example of COVID-19 illustrates how a shutdown of the tourism industry can have a debilitating impact on economies, especially those almost completely reliant upon tourism.

The environmental impacts domain of Table 6.1 also shows considerable differences between both alternatives. The most emphasis was placed on international transportation (weighted as 5) because the tourism industry, and especially ecotourism, is heavily criticised over the movement of

Table 6.1 Sustainability dimensions of contrasting ecotours

A. Sustainability Criteria (Weighted in importance as 1–5)[1]	*B. Alternative 1 Traditional Ecotour (Scored –3 to +3)*[2] *(value) in column A × (value) in column B, then summed*	*C. Alternative 2 PIRT (Scored –3 to +3) (value) in column A × (value) in column C, then summed*
Environmental impacts		
Generating region transport to airport (1)	Some negative impact in fuel used (-1)	Some positive impact in fuel saved (+1)
Airport services like food & beverage (1)	Some negative impact in resources used (-1)	Some positive impact in resources saved (+1)
International transportation (5)	Very high negative impact in fuel, etc. used (-3)	Very high positive impact in fuel, etc. saved (+3)
Destination region ground transportation (1)	High negative impact in fuel used (-2)	High positive impact in fuel saved (+2)
Daily tour transportation (2)	Some negative impact in fuel used (-1)	Some positive impact in fuel saved (+1)
Direct impacts on wildlife (2)	High negative impact from tour group (-2)	Some impact from guide alone (-1)
Indirect impacts on wildlife (2)	Some negative impact based on tour activities (-1)	Some impact from guide alone (-1)
Food production and consumption (3)	High negative impact in resources used (-2)	High positive impact in resources saved (+2)
Waste production and consumption (3)	High negative impact in waste used (-2)	High positive impact in waste saved (+2)
Energy production and consumption (3)	High negative impact in energy used (-2)	High positive impact in energy saved (+2)
Water consumption (3)	High negative impact in water used (-2)	High positive impact in water saved (+2)
SUB-TOTAL	-51	+41
Economic impacts		
Generating region ground transport to airport (2)	Some positive impact in gained revenue (+1)	Some negative impact in lost revenue (-1)
International transportation (5)	Very high positive impact in gained revenue (+3)	Very low negative impact in lost revenue (-3)
Destination region ground transportation (2)	Some positive impact in gained revenue (+1)	Some negative impact in lost revenue (-1)
Daily tour transportation (2)	High positive impact in gained revenue (+2)	High negative impact in lost revenue (-2)
Economic impacts on local food and beverage businesses (4)	Very high positive impact in gained revenue (+3)	Very high negative impact in lost revenue (-3)
Economic impact on accommodation (4)	Very high positive impact in gained revenue (+3)	Very high negative impact in lost revenue (-3)
Other attractions/amenities (4)	Very high positive impact in gained revenue (+3)	Very high negative impact in lost revenue (-3)
Impacts on tour operator (4)	Very high positive impact in gained revenue (+3)	High negative impact in lost revenue (-2)
Creation of new jobs (4)	High positive impact in gained salary (+2)	High negative impact in lost salary (-2)
SUB-TOTAL	+83	-75

(Continued)

Table 6.1 (Continued)

A. Sustainability Criteria (Weighted in importance as 1–5)[1]	*B. Alternative 1 Traditional Ecotour (Scored –3 to +3)*[2] *(value) in column A × (value) in column B, then summed*	*C. Alternative 2 PIRT (Scored –3 to +3) (value) in column A × (value) in column C, then summed*
Socio-cultural impacts		
Crowding and density (3)	Some negative impact on local people (-1)	Some positive impact (+1) by absence
Cultural erosion like food and dress (3)	High negative impact (-2)	High positive impact (+2)
Positive intercultural interaction (2)	Some positive impact by presence (+1)	Some negative impact by absence (-1)
Demonstration effect (3)	Some negative impact given wilderness setting (-1)	Some positive impact based on absence (+1)
Issues around authenticity and commodification (4)	Some negative impact due to presence (-1)	Some positive impact based on absence (+1)
SUB-TOTAL	-14	+14
Existential factors		
Being at the destination (5)	Very high positive value by being there (+3)	Being there via technology only (-2)
Being a sustainable citizen (5)	Limited opportunity because of impacts (-2)	Very high positive value by staying home (+3)
Safety/security/risk (4)	High negative value based on possible exposure (-2)	Very high positive value based on no exposure (+3)
SUB-TOTAL	-3	+17
GRAND TOTAL	(-51, +83, -20, -3) = **+9**	(+41, -75, +14, +17) = **-3**

Notes

1 Criteria weighted: 1 = least amount of importance; 5 = most amount of importance

2 Alternatives scored: Very high negative impact/value (-3); high negative impact/value (-2); some negative impact/value (-1); negligible impact/value (0); some positive impact/value (+1); high positive impact/value (+2); very high positive impact/value (+3)

tourists from generating to destination countries under the guise of moral superiority (Butcher, 2003). Overall there is a difference of 96 points between Alternative 1 (–51) and Alternative 2 (+41) showing clearly how being there is more environmentally damaging than not being there.

The figures in the socio-cultural impact domain are not as high given that fewer criteria were used to compare both alternatives. Issues around crowding and density, cultural erosion, demonstration effect, and authenticity and commodification are more serious because of the appearance of ecotourists at the destination. These may be tempered by the fact that intercultural communication is important for both tourists and local people. Additionally, the existential domain also has lower values for the same reason as the socio-cultural domain. Being at the destination is weighted high and scored high because of novelty and escape dimensions discussed in the Introduction. Similarly, high value is placed on being an environmental citizen in Alternative 2, and so it was also weighted and scored high. The feeling of being a sustainable citizen for the conventional tour is tempered by the use of so many different resources and sectors to facilitate the ecotourism event. Considerable differences exist around safety/security and risk given the nature of both alternatives. Overall, given the criteria selected, weighting and scores, Alternative 1 is the best scenario over Alternative 2 by a small margin of 12 points.

Discussion

The comparative analysis of the two contrasting ecotours suggests a number of important issues and trade-offs. The most significant drawback of Alternative 2 is not being at the destination to enjoy a more holistic experience with nature, local people and other elements of the tour. Ecotourism has been defined as a participatory event, where actually being in nature is a component part of the experience itself (Fennell, 2015). However, other key defining characteristics of ecotourism include learning, sustainability, and ethics. I argue that tourists can learn just as much from PIRTs as they can from conventional tours, and sustainability (at least from an ecological standpoint) and ethics dimensions are superior in Alternative 2 because of the elimination of key impact dimensions like transportation (Wang, Hu, He, & Wang, 2017).

The reduction or elimination of tourists at the destination, however, will have several negative consequences for the economic pillar of sustainability (e.g., service provision). In the aftermath of 9/11, President Bush very quickly commented that New York City was "open for business", indicating how important the tourism industry is to the fabric of communities (CNN.com./U.S., 2001). Part of the promise of ecotourism, and indeed all forms of tourism, is the economic impact that the industry has in time and space. For ecotourism, these benefits provide incentive for local people to conserve biodiversity in the face of other competing land uses such as forestry and mining (Osano et al., 2013). As commerce would be compromised in the absence of an ecotourist presence at the destination, part of the price of PIRTs should include extra fees designed for other purposes at the destination, like infrastructure development, biodiversity conservation, or social programs. These extra fees would be not unlike the higher gate fees at many protected areas for ongoing park management purposes including employment.

While the elimination of ecotours at a destination would appear to be rather draconian, indeed, PIRTs may be viewed as poor imitations of the "real thing", at the very least PIRTs may be used as a resiliency buffer for ecotour operators if the flow of tourists stops because of global events such as COVID-19. Furthermore, the model may stand as an alternative for those who wish to decrease their ecological footprint on the planet as the most ethical and sustainable method of "travel" (Hunter, 2002). Other cohorts such as persons with disabilities and elderly people who are faced with accessibility constraints may see this as an attractive option, and still others may view it as cheaper alternative to conventional travel.

In terms of PIRT demand, the most popular might be a personalised tour that an individual, family or small group is attracted to. This would allow for a more intimate experience with guides providing freedom and flexibility in time and space, and where participants could ask questions, and get answers, in real-time. As popularity of this "travel" option improves, the potential to employ many more guides in the same location becomes a distinct possibility increasing the number of jobs and economic impact for local communities. Other tour options could possibly be built around larger tours that include participants from around the world, but with an associated reduction in flexibility and personal interaction with the tour guide.

A major concern with Alternative 2 is the technological infrastructure needed to run these tours in certain areas. ICT infrastructure is likely in place in most urban environments; however, there will be more significant challenges in wilderness settings that do not have the infrastructure to allow transmission at all. Furthermore, success of such a model would necessitate the need for more ICT expertise in remote regions. Training local people would therefore be a priority in efforts to secure decent work and economic growth (UNEP, 2018).

While employing expertise externally to facilitate these sorts of experiences may be essential at initial stages, a move away from a traditional model of tourism development that excludes local people from meaningful involvement is paramount (Roe & Urquhart, 2002). Another type of concern is that this type of experience might have the effect of making "PIRTers" crave the real thing even more. Coupled with a strong rebound in travel volume and frequency post-COVID-19, there may in fact be more travel to sensitive cultures and destinations, when an important theme for the development of these tours is protection of the destination.

Conclusion

As crises continue to present challenges in the new age of disruption, radically different ways of providing services will be required in the tourism industry. In the past, tourism has proven itself to be a resilient industry, but recent events are now testing our ability to recover. This research note has presented a different model of service provision that reflects a vision of transformation and adaptation during rapidly changing times. I recommend more intensive investigation of the proposed model, both conceptually and methodologically, in an effort to refine what is at present a rough sketch. Delphi panels, workshops, and interviews with tourism industry stakeholders would be useful in better identifying criteria, as well as how these criteria ought to be weighted and scored in a variety of situational contexts.

References

Atzori, L., Iera, A., & Morabito, G. (2014). 'From' smart objects 'to' social objects. *IEEE Communications Magazine, 52*(1), 97–105.

Brito, C. (2020). Spring Breakers Say Coronavirus Pandemic Won't Stop Them From Partying. Retrieved from https://www.cbsnews.com/news/spring-break-party-coronavirus-pandemic-miami-beaches/.

Buhalis, D. (2020). Post-viral Tourism. *TRINET Communication.*

Butcher, J. (2003). *The Moralisation of Tourism: Sun, Sand...and Saving the world?* London: Routledge.

Clawson, M., & Knetsch, J. L. (1966). *Economics of Outdoor Recreation.* Baltimore, MD: Johns Hopkins University Press.

CNN.com./U.S. (2001). Text of Bush's Address. Retrieved from http://edition.cnn.com/2001/US/09/11/bush.speech.text/.

Coca-Stefaniak, A., & Morrison, A. M. (2018). City tourism destinations and terrorism – a worrying trend for now, but could it get worse? *International Journal of Tourism Cities, 4*(4), 409–412.

Cvelbar, L. (2020). Post-viral Tourism. *TRINET Communication.*

Fennell, D. A. (2002). *Ecotourism Programme Planning.* Wallingford, UK: CABI.

Fennell, D. A. (2015). *Ecotourism* (4th ed.). London: Routledge.

Haugen, J. (2020). How Tech Could Help Tourist Destinations Struggling in the Wake of COVID-19. Retrieved from https://www.fastcompany.com/90478303/how-tech-could-help-tourist-destinations-struggling-in-the-wake-of-covid-19.

Hunter, C. (2002). Sustainable tourism and the touristic ecological footprint. *Environment, Development and Sustainability, 4,* 7–20.

Hughes, K., & Moscardo, G. (2019). ICT and the future of tourist management.*Journal of Tourism Futures,* 5(3), 228–240.

Iso-Ahola, S. (1982). Toward a social psychological theory of tourism motivation: A rejoinder. *Annals of Tourism Research, 9*(2), 256–262.

Macbeth, J. (2020). Post-viral Tourism. *TRINET Communication.*

Maingi, S. (2020). Post-viral Tourism. *TRINET Communication.*

Manyika, J., Chiu, M., Bughin, J., Dobbs, R., Bisson, P., & Marrs, A. (2013). *Disruptive Technologies.* San Francisco, CA: McKinsey Global Institute.

Miller, G. (2003) Consumerism in sustainable tourism: A survey of UK consumers. *Journal of Sustainable Tourism, 11*(1), 17–39.
Neuhofer, B., Buhalis, D., & Ladkin, A. (2014). A typology of technology-enhanced tourism experiences. *International Journal of Tourism Research, 16*, 340–350.
Osano, P. M., Said, M. Y., de Leeuw, J., Ndiwa, N., Kaelo, D., Schomers, S., ... Ogutu, J. O. (2013). Why keep lions instead of livestock? Assessing wildlife tourism-based payment for ecosystem services involving herders in the Maasai Mara, Kenya. *Natural Resources Forum, 37*, 242–256.
Osumi, M. (2020). Hotels in Japan Rollout Unusual Bids to Woo Tourists in COVID-19 Crisis. Retrieved from https://www.japantimes.co.jp/news/2020/03/19/business/hotels-japan-unconventional-offers-coronavirus/#.XnTlZi0ZNcA.
Pearce, D. (2016). Interdependent destination management functions. *Tourism Recreation Research, 41*(1), 37–48.
Rees, W. (1992). Ecological footprints and appropriated carrying capacity: What urban economics leaves out. *Environment and Urbanization, 4*(2), 121–130.
Roe, D., & Urquhart, P. (2002) *Pro-Poor Tourism: Harnessing the World's Largest Industry for the World's Poor*. London: International Institute for Environment and Development. Retrieved from www.propoortourism.org.uk/Dilys%20IIED%20paper.pdf.
Russell, R. V. (1982). *Planning Programs in Recreation*. St. Louis, Missouri: C. V. Mosby.
Ruane, S. (2020). Post-viral Tourism. *TRINET Communication*.
Soper, K. (2008). Alternative hedonism, cultural theory and the role of aesthetic revisioning. *Cultural Studies, 22*(5), 567–587.
UNEP. (2018). Goal No. 8: Decent Work and Economic Growth. Retrieved from https://www.undp.org/content/undp/en/home/sustainable-development-goals/goal-8-decent-work-and-economic-growth.html.
UNWTO. (2020). COVID-19: Putting People First. Retrieved from https://www.unwto.org/tourism-covid-19-coronavirus.
Wang, S., Hu, Y., He, H., & Wang, G. (2017). Progress and prospects for tourism footprint research. *Sustainability, 9*, 1847.
Wells, N. M., & Lekies, K. S. (2006). Nature and the life course: Pathways from childhood nature experiences to adult environmentalism. *Children, Youth and Environments, 16*(1), 1–24.
Werthner, H., Alzua-Sorzabal, A., Cantoni, L., Dickinger, A., Gretzel, U., Jannach, D., ... Zanker, M. (2015). Future research issues in IT and tourism. *Information Technology & Tourism, 15*(1), 1–15.

THEME 2

Ethics and identities

7

ENCHANTMENT

Feeding care within the cracks of ecotourism

Kellee Caton, Chris E. Hurst, and Bryan S.R. Grimwood

Introduction

The foundation of ecotourism—what holds it together as an experience, concept, and ethical ideal for tourism encounters and enterprises—is fractured. We are referring of course to the significance of "nature" as the life blood of ecotourism from its earliest articulations in the 1980s by Ceballos-Lascura´in (1987) to the more contemporary and regularly referenced definition advanced by The International Ecotourism Society (2020). While community development, learning, and responsible travel have been recurrent dimensions of ecotourism, the common core idea of nature is what steers the moral values and experiential desires of ecotourists and the conservation orientation of ecotourism operators and enterprises (Donohoe & Needham, 2006). Nature is what we are told and sold to care about in ecotourism. Nature is, however, no stable, homogenous, or apolitical thing, as scholars from across the social and natural sciences have demonstrated for some time (e.g., Barad, 2007; Cronon, 1996; Haraway, 1991). Indeed, nature is (and historically has been) tethered to discursive and material relations that constitute colonialist and capitalist systems, and is also representative of worldly phenomena (from flora and fauna to weather patterns and geological dynamics) that actively shape the social, cultural, and ecological systems that enable, and give meaning to, life as we know it. If the ontology of nature is shaky and suspicious and shifting, what else might we latch on to as an ethical anchoring for ecotourism? How might we enact an ethics of ecotourism without reifying some sort of transcendental or static vision like nature? How do we forge care within a broken and transforming system? In this chapter, we explore the potential for ecotourism encounters to serve as spaces of enchantment and care for more than the human world. We explore how enchantment might energise us in our capacities for compassion and care, even as we are emplaced in the mix and midst of things—including those things that are ugly, damaged, and destructive.

Ecotourism and the limits of nature

Imaginaries of nature as an external, non-social environment with intrinsic or universal qualities have been troubled for some time within tourism studies and allied disciplines and fields (e.g., Castree & Braun, 2001; Glacken, 1967; Grimwood, Caton, & Cooke, 2018). Yet, as Reis and

DOI: 10.4324/9781003001768-7

Shelton (2011) observe, much of the tourism industry accepts nature simply, and rather uncritically, as one environment in which tourism occurs. This seems especially the case in the production and consumption of ecotourism, where representations and perceptions of low-impact visitation to undisturbed natural areas continue to circulate and structure a nature conservation-oriented ethic. Adhering to "taken-for-granted" or "common sense" understandings of nature gives rise to several difficult assumptions that amplify a nature/culture divide: nature is an inherent, biophysical aspect of the world; natural environments are the sole inspiration for care and personal and social transformation (i.e., underlying social-cultural influences are ignored); nature requires improved technological management; and touring nature destinations is distinct from everyday contexts (Donohoe, 2011; Grimwood et al., 2018; Holden, 2003; Vespestad & Lindberg, 2010).

Cater (2006) suggested that the nature/culture divide lies close to the heart of ecotourism, and that this privileges a dominant, Eurocentric environmental imagination. In addition to constructing certain places as destinations to desire and consume, the dangers of this ethnocentric worldview include its ability to be superimposed upon the lives, knowledge, and values of people inhabiting those locations (Cater, 2006). In other words, caring for nature can eclipse the care extended to our human kin. Many Indigenous communities have taken the brunt of this burden through colonial and neocolonial touristic encounters within ancestral lands. On one hand, imaginations of an ahistorical and ecologically "noble savage" have translated into mistaken assumptions that ecotourism, environmentalism, and Indigeneity have common ecological visions (Higgins-Desbiolles, 2009). On the other hand, patterns of traditional subsistence and self-reliant resource use can be undermined by the non-consumptive ideals of many ecotourists. In both cases, the "nature of ecotourism" is a source of potential conflict (Hinch, 1998; Grimwood, 2015). If, as Donohoe (2011) suggested, a single privileged conception of nature remains unchallenged, potential exists for community level tourism benefits to be replaced with insecurity, resentment, degradation, and economic loss.

Nature, as objectified and held in contrast with culture, is also commodified, and the ecotourism industry benefits from the nature/culture binary in the sale of experiences with "pure" wilderness, often offering the moral argument to consumers that taking part in such tourism is key to protecting these landscapes. As such, ecotourism exemplifies neoliberalism *par excellence*, as public goods are shifted to the private commodity system for stewardship (Fletcher & Neves, 2012). With the destruction of ecosystems, however, the pristine product is slipping away. But capitalism is undeterred. Enter "last-chance tourism" (Lemelin, Dawson, Stewart, Maher, & Lueck, 2010) wherein the industry capitalises not only on nature, but specifically on its destruction, as the Anthropocene "progresses" to pass it by. Fletcher (2019) keenly identifies last chance tourism as a form of what Klein (2007) calls "disaster capitalism," wherein capitalism manages to capture crises it itself has created and then turn them into new opportunities for accumulation. In Klein's (2007, p. 6) words, disaster capitalism involves "orchestrated raids on the public sphere in the wake of catastrophic events, combined with the treatment of disasters as exciting marketing opportunities." In the case of ecotourism, this process allows capital to continue to wring profit from raw materials that are fast disappearing by instead selling their very disappearance as the product. The body is also not exempt from capitalism's disciplining power, as ecotourists purchase expensive gear, both materially and symbolically useful, to adequately equip themselves for their endeavor. Once properly attired, if they find themselves on a package tour, they are then choreographed—"told where and when to sit, stand, look, walk, move, stay still, and so on. They are also told which senses to use, how, and when. They are often told what to feel and when" (Fletcher & Neves, 2012, p. 67).

What to do in such a dispiriting predicament? Are there other values that can be brought to ecotourism that can bring support for human and other animal communities who receive "nature" guests? Are there possibilities for resistance that might strike out rhizomatically, surprisingly, even within a practice that is so much a product of contemporary colonial and capitalist exploitation?

Ecotourism and care ethics

Care ethics is an unusual position in Western moral philosophy. Rather than taking abstract principles as its point of departure, it is instead rooted in understanding the lived moral experiences of actual people, particularly those of women, as care ethics derives from feminist philosophy (Gilligan, 1982/1993; Noddings, 2012).

> Care theorists have observed that the historical experiences of women, despite being as "important, relevant, and philosophically interesting" as those of men, have been intellectually ignored, with women's interactions in the private sphere being imagined as somehow simply natural and instinctive, rooted in the biological realm of reproduction. This is opposed to the experiences of men in the public sphere, which have been imagined as cultural, creative, and complex, and therefore worthy of philosophical exploration. Central to those supposedly complexity- and creativity-bereft experiences in the private sphere, however, is what is actually the quite intense and thorny work of care-giving and relationship cultivation. The wisdom gained from navigating that domain of work offers a window of critique onto moral philosophies that consider only the complexities of negotiating rights and responsibilities between strangers in the public sphere. (Hales & Caton, 2017, p. 97, quoting Held, 2006, p. 23)

Care theorists problematise the notion of liberal individualism so central to many dominant moral theories (e.g., Kantianism, utilitarianism), arguing instead that we are born into a web of relations that precedes any ontological notion of individual personhood, and they emphasise the moral obligations of responsiveness to need that inhere in concrete relationships with proximal others (Held, 2006). As a relational moral ethic, care is conceptualised as a practice rather than a disposition, involving relations that are characterised by attentiveness to context, responsibility, and responsiveness (Eger, Scarles, & Miller, 2019; Held, 2004). Traditionally framed within human care interactions (particularly in private settings), care ethics has more recently been applied across a variety of other contexts—including the public sphere—and has been extended to humans and nonhumans, things, and places (Jamal, 2019; Kheel, 2008).

Previous literature on care and ecotourism has helpfully sought to demonstrate the value of this perspective for guiding better ecotourism practices, as well as to problematise particular interpretations of care in the ecotourism context. Dangi (2018) and Eger et al. (2019), for example, examined how care ethics could be adopted to provide mutual benefit in host–tourist interactions and nature tourism business stakeholder relationships. Their studies examined how co-constructed caring relationships between stakeholders (i.e., tourists, host communities, tourism businesses) can contribute to more equitable and beneficial outcomes for all parties. Similarly, in their study of care ethics and justice in the development and management of a Mayan tourism destination, Jamal and Camargo (2014) sought to attend to the ecocultural relationships, cultural heritage, and wellbeing of Mayan peoples and their territories. For Jamal and Camargo (2014), an ethic of care complements an ethic of justice, contributing to the development of just tourism products and destinations characterised by care towards representations of Mayan cultural heritage, places, and local peoples. Their study also engaged with politics of care, wherein care ethics challenged the institutional and historically embedded racism and cultural discrimination experienced by Mayan peoples (Jamal & Camargo, 2014).

For Walker and Moscardo (2016), care ethics involves fostering care of place. Their study examined a sustainable tourism destination on an island off the coast of Australia, and illustrated how care of place can contribute to transformative change in the attitudes and behaviours of tourists towards local Indigenous peoples and the terrestrial and marine ecosystem on which they rely. Care ethics on an ecosystem level is a topic also explored by Waitt and Cook (2007), in their study of a kayak ecotour destination. In contrast with the previous authors, who focused on the benefits of care ethics in ecotourism, Waitt and Cook (2007, p. 542) problematise the way the ecotourism industry promotes a duty of care by the ecotourist that is translated into 'no touch' policies, which frame human presence and disturbance as "unnatural in ecotour spaces."

Care ethics is also commonly employed in studies concerning animals in ecotourism, with varied implications for the tourism industry and policymakers. Yudina and Grimwood (2016), for example, explored how care ethics could inform the reconceptualisation of sustainable tourism by examining representations of polar bear tourism in Churchill, Manitoba (Canada). Their study challenges the anthropocentric and instrumentalist discourses that marginalise non-human animal others (i.e., spectacle bears), and proposes a conceptualisation of sustainability that prioritises care, compassion, respect, and the inclusion of all tourism stakeholders (including animals); attends to the communications and desires of animal others in ethical decision-making; and embraces social justice and animal welfare concerns. Contextualised within the widespread bushfires of Australia in 2019 and 2020, Markwell (2020) examined how an ethics of care, specifically based on compassion, should inform industry responses to the destructive impacts of the fires on koalas and their habitat. For Markwell, care ethics would recognise that animals are not a tourism resource and that there is an obligation of care for animals, and ecosystems that support them, by the tourism industry. Taylor, Hurst, Stinson, and Grimwood (2019) and Fennell and Sheppard (2020) also offer recent examples of how care ethics can help us navigate our responsibilities to animals in tourism, in both cases considering the lives of elephants in touristic contexts. All of these authors also engage with a politics of care, with implications for animal rights, tourism regulation, and industry standards.

These examples illustrate how care ethics in tourism relationships and with nature—including animals (both human and otherwise), ecosystems, and places—can contribute to more equitable, beneficial, and just engagements with the peoples, critters, places, and things of tourism. Building on this growing body of literature on care and ecotourism, we are interested in a deeper layer of questioning: What feeds care? Can ecotourism play a role in providing it?

Enchantment

Fletcher's identification of ecotourism as an expression of neoliberalism, a disciplining of the body, and even a type of disaster capitalism is prescient but bleak. In many ways, it is a Frankfurt School style argument, where commodification is a zero-sum game (Bennett, 2001). Commodity culture is so total, and so totally alienating to the humans operating within it, that there is no "possibility of an affective response to commodities able to challenge the socioeconomic system" that created them. There is no "aesthetic sphere" independent of capital, which could potentially platform resistance (Bennett, 2001, pp. 121–122). Fletcher and Neves (2012, p. 67) suggest this directly, arguing that "what ecotourism sells most centrally is a particular affective state—excitement, satisfaction, peace, contentment, pleasure, and so forth—attached to the outdoor, generally 'wilderness' experience it offers." Our very affect is coopted for capitalist gain. It's an utterly normative and existential-level grip.

In contrast, together with Bennett, and also with Deleuze and Guattari, we believe that there is still potential for resistance even within a commodity culture that can feel hopelessly total. Experience never creates perfect repetition. There is always room for a new sensibility, a new recognition, an "Aha!" moment, growth. As Deleuze and Guattari (1987, p. 216) put it, "There is always something that flows or flees, that escapes … the overcoding machine" (quoted in Bennett, 2001, pp. 115–116).

Capitalism unchained it may be, but to emphasise only this aspect of ecotourism is to overcode it. It can also be a force for care, for healing and reconciling humans with our generally poorly acknowledged more-than-human kin. More specifically, it can be a force for care because it is a source of that which feeds care: *enchantment*. As Bennett (2001) argues, ethics requires more than a moral code—it also requires an affective dimension that fills us up, opens our generosity, and moves us to act in good ways. Experiences of enchantment, she contends, are important pathways to our better selves.

People have long seen the modern world as disenchanted. Weber was famous for this sentiment, lamenting in 1917 that "The fate of our times is characterized by rationalization and intellectualization and, above all, by the 'disenchantment of the world'" (quoted in Landy & Saler, 2009). Long before, of course, Nietzsche had observed that God was dead, meaning that religion had lost its grip as an overall organiser of life meaning and guarantor of ontological security (i.e., what it is and means to be human). In the aftermath, many humans have been left to seek in alternative directions the many functions which, as Nietzsche observed, religion used to provide: mystery, wonder, order, and purpose; a significance of objects and events; the possibility of redemption; a sense of the infinite, the sacred, and miracles; and epiphanies, "moments of being in which, for a brief instant, the center appears to hold, and the promise is held out … of union with something larger than oneself" (Landy & Saler, 2009, p. 2). Whether we have done that more or less successfully is a matter of debate. Both Bennett and Landy and Saler make a strong case that re-enchantment is not only possible but happening all around us in small ways. They even assert the danger of denying this pattern, suggesting that viewing the world as disenchanted is at best a misreading of modernity and at worst a form of micropolitics that militates against hope.

While we don't doubt moments of enchantment in the contemporary human experience, we tend to be less optimistic. It's hard to ignore the onto-epistemic unfolding of modernity. Berman (1984) and Kohák (1984) each offer a take on this process. Berman (1984, p. 351) emphasises the transition away from "participating consciousness," or the idea that "everything in the universe is alive and interrelated, and that we know the world through direct identification with it, or immersion in its phenomena (subject/object merger)." While this was a long transition, dating back even to early Jewish theology and some branches of Greek philosophy, it picked up steam in Europe with the eclipsing of alchemy (Berman's last great example of participating consciousness) by the epistemology of the Scientific Revolution. Through this transition, the Western world entered a state of non-participating consciousness, still dominant today: "a state of mind in which the knower, or subject 'in here,' sees himself as radically disparate from the objects he confronts, which he sees as being 'out there.' In this view, the phenomena of the world remain the same whether or not we are present to observe them, and knowledge is acquired by recognizing a distance between ourselves and nature" (Berman, 1984, p. 355). It is easy to see how this shift in thinking feeds a nature/culture binary. Like Berman, Kohák (1984) emphasises the importance of dichotomous thinking to modern onto-epistemics but concentrates more on the consequences for our ability to experience the world in a rich, direct, and embodied way. Separation of the knower from the object to be known sets us up to reduce the object to its parts, to abstract it into concepts that move increasingly further from our lived experience with it. In his explanation,

> Western scientific thought—and popular thought in its wake—gradually substituted a theoretical nature-construct for the nature of lived experience in the role of "reality." Far more than we ourselves usually realize, when we make seemingly obvious assertions about "nature," we are no longer speaking about the natural environment of our lived experience, the living, purposive *physis* which humans can recognize as kin and in which they can feel at home. Our statements are far more likely to refer to a highly sophisticated construct, say, matter in motion, ordered by efficient causality, which is the counterpart of the method and purpose of the natural sciences rather than an object of lived experience. Within such a construct, to be sure, there is no place for a moral subject, simply because that construct was not designed to deal with him. (Kohák, 1984, p. 12)

We are the poorer for this situation.

And so unlike Bennett (2001) and Landy and Saler (2009), who focus on the ways enchantment is still alive and well among humanity, when we look at the world, we instead see a largely disenchanted place. We see a place where people struggle with connection, meaning, and purpose (Stroh, 2019; Boddy, 2017), and a sense of being at home in the world (Kohák, 1984; Haraway, 2016). We see a place where our disconnection from this home is so intense that we are destroying its ability to sustain life for many species, including ourselves, and where we are destroying beloved land-, sky-, and seascapes (Wernick, 2016; Bogard, 2013; Joyce, 2018) that give human life meaning and possibility. This is in no way to argue that moments of enchantment don't occur or that there is no hope for a widespread re-enchantment; indeed, it is the opposite—it is to argue that widespread re-enchantment *can* and *must* occur. But we're not there yet. We have to work for it.

Like seemingly all scholars who chose to work with the concept of enchantment, we are not advocating simply turning the clock back to pre-Renaissance ways of thinking about the world, and we are certainly not arguing for a wholesale rejection of scientific epistemology. Rather, we are curious about what participating consciousness might look like *here*, on the other side of modernity, in the Anthropocene, the Capitalocene. How might a re-enchantment of the world, twenty-first-century style, ignite our desire and give us the strength, courage, and inspiration to live relationally, in care-full and response-able (Haraway, 2016) ways, in our earth home?

All this talk of enchantment: but what do lived experiences of enchantment actually look like? Several key features of this experience have been offered. The most often highlighted are deep sensory engagement, quiet, and an alternative experience of time. As noted above, Kohák (1984) has argued that our ways of knowing have, since at least the seventeenth century, moved away from direct lived experience with nature and toward abstract understandings of it; conceptual ways of knowing have come to dominate visceral ones. This departure from the sensory in our understanding of nature is unhelpful in living a relational ethics. Sensory experiences cut through the abstraction. They allow us to be in the thick of things, in the mode Haraway (2016) calls *being-with.* They allow us to cultivate a kind of perception, a "meticulous attentiveness to the singular specificity of things" (Bennett, 2001, p. 37). In being with earth others in concrete and sensorily attentive ways, we put ourselves in a position to be struck with awe at their specificity—their irreplaceability.

Enchantment often emerges in quiet. As Kohák (1984, p. 31) expresses it:

> The stars do not insist: even the glare of a white gas lantern or the reflected glow of neon will drown them out. Only where humans respect the night can they see the wonder of the starry heaven as the Psalmist saw it.

Contemporary life is full of noise. In his environmental memoir *The End of Night*, Paul Bogard (2013) tells us that two-thirds of Europeans and North Americans no longer experience real night. Writing from the United States, he notes that 8 out of 10 of his compatriots born today will never see the Milky Way. But moments of quiet are so important for giving us our bearings, helping us to locate ourselves as very small pieces of something potentially infinitely larger. Alain de Botton (2003) describes the quiet experience of being small in sublime places. His lesson, as Thomas (2020, p. 137) paraphrases it, "is that the universe is mightier than we are, that we are frail and temporary and must accept limitations on our will, bow to necessities greater than ourselves." Paradoxically, this enchantment with the sublime leaves us not discouraged but inspired by that which goes beyond ourselves (de Botton, 2003).

Enchantment opens a different sense of time. We are used to time either as highly personal or highly abstract. We live by the clock, an abstraction of the diurnal cycle that strains credibility, particularly for folks like us, living and working at fairly high latitudes, who routinely experience sunset at anywhere from 4 to 10 p.m., depending on the month. And yet we live in time subjectively. Kellee's last visit to see a friend in the Mediterranean went by in a heartbeat. She's been writing this paragraph forever.

But there is a third sense of time, neither abstract nor subjective. As Kohák (1984, p. 16) explains:

> It has all the hardness of the real, a logic of its own—the rhythm of vigor and fatigue, of day and night, the cycle of the seasons in the life of nature and humans alike. Its stages, though personal, are not in the least arbitrary. Primordially, human experience simply is not a sequence of discrete events which need to be ordered by a clock and a calendar or by free association within a stream of consciousness à la Proust or Joyce. It is, rather, set within the matrix of nature's rhythm which establishes personal yet nonarbitrary reference points: when I have rested, when I grow weary, when the shadows lengthen, when life draws to a close. Though we may speak that way, it is simply not the case that at "six of the clock" certain events will occur—the shadows will lengthen, my axe will grow heavy in my hands. Stopping the clock does not stop the event. The primordial time reference is the opposite: it is the experience of the evening, lodged in the shadows about me and in the weariness of my arms, which is the primordial given. Only secondarily do we designate it by a clock reference or acknowledge it in an internal time consciousness.

Leaning into one's place in this rhythm facilitates presence. And being present is fundamental for living well with earth others—as Donna Haraway (2016) calls it, for staying with the trouble. She calls us to take our place, "not as a vanishing pivot between awful or edenic pasts and apocalyptic or salvific futures," but as fully present, with all those beings and earth processes with which we are already entangled but so easily forget to attend to (Haraway, 2016, p. 1).

And there is also a fourth sense of time, where, as Kohák (1984) describes it, eternity intersects with earthly time. Kohák describes exploring the cellar hole of an old farmhouse near the cabin where he has retreated to live in a forest clearing. A wall buckles under him to reveal a file that had long ago fallen through a crack in the wall. His thoughts turn immediately to the owner of the file, and he imagines how the owner must have been searching for it, must have been frustrated that he had just had it in his hands only a moment ago. Where had it gone? He found himself desperately hoping that the file's owner had not taken out his anger in losing the file on those around him, as we so often do to those closest to us. Some piece of us lives on in our artifacts, and in the way others can wonder about us. Then Kohák (1984, p. 122) imagines the man saying to his wife, "I will love you forever"—an "incredibly audacious statement," Kohák acknowledges, "for a being who dwells in ever-flowing time. Yet he said it." There is truth in eternal time, in legacy, and in the infinite expanse of a present moment. This is what Haraway (2016) means, when she talks of mourning as part of staying with the trouble. Lost species are remembered in the imprints they have left on others: for example, orchids that formed their flowers to match the anatomy of bees, now long gone. Enchanted by the specificity and irreplaceability of what has been lost, we too can keep its memory alive.

Synthesis

Moments of enchantment slip the overcoding machine. They turn us back to direct experience, to the expansive present, to a state of attentiveness wherein we can marvel at the specificity and integrity of all things and find our place among them. Moments of embodied sensory encounter break down abstract dualisms, bring mind back into body, bring body back into the earth as home.

The nature/culture binary that is so often reinforced through ecotourism suggests that humans are independent of others, even made of a different stuff, civilisational stuff, not the wild stuff of nature. Much like liberal theories of morality, the assumption is that we are free agents, negotiating our rights and responsibilities with others from the standpoint of individualism. But there is no self-made being, no such thing as autopoiesis, wherein beings organise themselves and progress through life on their own work and merit. There is only sympoiesis—making-with—where "critters interpenetrate one another, get indigestion, and partially digest and partially assimilate one another, and thereby establish sympoietic arrangements that are otherwise known as cells, organisms, and ecological assemblages" (Haraway, 2016, p. 58). We are all made of the same stuff. And our relations precede us.

Care ethics understands this fundamental relationality. But we need energy to exist in entanglement in caring ways, to practice care within the relationships that move us and invite us in. Living and dying together isn't easy. Care is an action, not only a sentiment, and we need resources to support us as we work to live care-fully. It is possible for ecotourism to be one such resource; it must be, lest we accept commodifying and colonising forces as totalising and impervious. Perhaps, rather than learning about or conserving nature, the value of ecotourism lies in its capacity to arouse our sensory engagements within the world. Perhaps, rather than fostering the economic development needs of particular communities, the value of ecotourism lies in its capacity to bring us all—tourists, operators, destinations, as well as our more-than-human kin—into sympoietic relationships that may invigorate more compassionate, vibrant worlds. Enchantment can provide the spark that moves our desire to be good to one another. Insofar as ecotourism can provide such moments of enchantment, there is still hope.

References

Barad, K. (2007). *Meeting the Universe Halfway: Quantum Physics and the Entanglement of Matter and Meaning*. Durham, NC: Duke University Press.

Bennett, J. (2001). *The Enchantment of Modern Life: Attachments, Crossings, and Ethics*. Princeton, NJ: Princeton University Press.

Berman, M. (1984). *The Reenchantment of the World*. Toronto, ON: Bantam.

Boddy, J. (2017). The Forces Driving Middle-Aged White People's 'Deaths Of Despair.' *National Public Radio*, March 23.

Bogard, P. (2013). *The End of Night: Searching for Natural Darkness in an Age of Artificial Light*. New York, NY: Back Bay.

Castree, N., & Braun, B. (Eds.). (2001). *Social Nature: Theory, Practice, and Politics*. Malden, MA: Blackwell Publishing Ltd.

Cater, E. (2006). Ecotourism as a western construct. *Journal of Ecotourism*, *5*(1&2), 23–39.

Ceballos-Lascura´in, H. (1987). The future of ecotourism. *Mexico Journal*. January, 13–14.

Cronon, W. (Ed.). (1996). *Uncommon Ground: Toward Reinventing Nature*. New York, NY: W.W. Norton.

Dangi, T. B. (2018). Exploring the intersections of emotional solidarity and ethic of care: An analysis of their synergistic contributions to sustainable community development. *Sustainability*, *10*(8), 2713–2733.

de Botton, A. (2003). *The Art of Travel*. London, UK: Penguin.

Deleuze, G., & Guattari, F. (1987). *A Thousand Plateaus*. Minneapolis, MN: University of Minnesota Press.

Donohoe, H. M. (2011). Defining culturally sensitive ecotourism: A Delphi consensus. *Current Issues in Tourism*, *14*(1), 27–45.

Donohoe, H. M., & Needham, R. D. (2006). Ecotourism: The evolving contemporary definition. *Journal of Ecotourism*, *5*(3), 192–210.

Eger, C., Scarles, C., & Miller, G. (2019). Caring at a distance: A model of business care, trust and displaced responsibility. *Journal of Sustainable Tourism*, *1*(1), 34–51.

Fennell, D. A., & Sheppard, V. (2020). Tourism, animals and the scales of justice. *Journal of Sustainable Tourism*, *29*(2&3), 314–335. doi:https://doi.org/10.1080/09669582.2020.1768263.

Fletcher, R. (2019). Ecotourism after nature: Anthropocene tourism as a new capitalist "fix." *Journal of Sustainable Tourism*, *27*(4), 522–535.

Fletcher, R., & Neves, K. (2012). Contradictions in tourism: The promise and pitfalls of ecotourism as a manifold capitalist fix. *Environment and Society*, *3*(1), 60–77.

Gilligan, C. (1982/1993). *In a Different Voice: Psychological Theory and Women's Development*. Cambridge, MA: Harvard University Press.

Glacken, C. (1967). *Traces on a Rhodian Shore: Nature and Culture in Western Thought From Ancient Times to the End of the Eighteenth Century*. Berkeley, CA: University of California Press.

Grimwood, B. S. R. (2015). Advancing tourism's moral morphology: Relational metaphors for just and sustainable Arctic tourism. *Tourist Studies*, *15*(1), 3–26.

Grimwood, B. S. R., Caton, K., & Cooke, L. (Eds.) (2018). *New Moral Natures in Tourism*. New York, NY: Routledge.

Hales, R., & Caton, K. (2017). Proximity ethics, climate change and the flyer's dilemma: Ethical negotiations of the hypermobile traveller. *Tourist Studies*, *17*(1), 94–113.

Haraway, D. J. (1991). *Simians, Cyborgs, and Women: The Reinvention of Nature*. New York, NY: Routledge.

Haraway, D. J. (2016). *Staying with the Trouble: Making Kin in the Chthulucene*. Durham, NC: Duke University Press.

Held, V. (2004). Care and Justice in the Global Context. *Ratio Juris*, *17*(2), 141–155.

Held, V. (2006). *The Ethics of Care: Personal, Political, and Global*. Oxford, UK: Oxford University Press.

Higgins-Desbiolles, F. (2009). Indigenous ecotourism's role in transforming ecological consciousness. *Journal of Ecotourism*, *8*(2), 144–160.

Hinch, T. (1998). Ecotourists and indigenous hosts: Diverging views on their relationship with nature. *Current Issues in Tourism*, *1*(1), 120–124.

Holden, A. (2003). In need of new environmental ethics for tourism? *Annals of Tourism Research*, *30*(1), 94–108.

The International Ecotourism Society. (2020). What is Ecotourism. Retrieved from https://ecotourism.org/what-is-ecotourism/ (accessed on 29.09.2020).

Jamal, T., & Camargo, B. A. (2014). Sustainable tourism, justice and an ethic of care: Toward the just destination. *Journal of Sustainable Tourism, 22*(1), 11–30.
Jamal, T. (2019). *Justice and Ethics in Tourism*. New York, NY:Routledge.
Joyce, C. (2018). We're Drowning in Plastic Trash. *National Public Radio*, July 24.
Kheel, M. (2008). *Nature Ethics: An Ecofeminist Perspective*. Lanham, MD:Rowman & Littlefield Publishers, Inc.
Klein, N. (2007). *The Shock Doctrine: The Rise of Disaster Capitalism*. New York, NY: Metropolitan.
Kohák, E. (1984). *The Embers and the Stars: A Philosophical Inquiry into the Moral Sense of Nature*. Chicago, IL: University of Chicago Press.
Landy, J., & Saler, M. (Eds.). (2009). *The Re-enchantment of the World: Secular Magic in a Rational Age*. Stanford, CA: Stanford University Press.
Lemelin, R. H., Dawson, J., Stewart, E. J., Maher, P., & Lueck, M. (2010). Last-chance tourism: The boom, doom, and gloom of visiting vanishing destinations. *Current Issues in Tourism, 13*, 477–493.
Markwell, K. (2020). Koalas, bushfires and climate change: Towards an ethic of care. *Annals of Tourism Research, 84*. doi:https://doi.org/10.1016/j.annals.2020.103003.
Noddings, N. (2012). The language of care ethics. *Knowledge Quest, 40*(5), 52–56.
Reis, A. C., & Shelton, E. (2011). The nature of tourism studies. *Tourism Analysis, 16*, 375–384.
Stroh, P. (2019). Feeling Lonely? You're Not Alone—and It Could Be Affecting Your Physical Health. *Canadian Broadcasting Corporation*, January 19.
Taylor, M., Hurst, C. E., Stinson, M. J., & Grimwood, B. S. R. (2019). Becoming care-full: Contextualizing moral development among captive elephant volunteer tourists to Thailand. *Journal of Ecotourism, 19*(2), 113–131.
Thomas, E. (2020). *The Meaning of Travel: Philosophers Abroad*. Oxford, UK: Oxford University Press.
Vespestad, M. K., & Lindberg, F. (2010). Understanding nature-based tourist experience: An ontological analysis. *Current Issues in Tourism, 14*(6), 563–580.
Waitt, G., & Cook, L. (2007). Leaving nothing but ripples on the water: Performing ecotourism natures. *Social & Cultural Geography, 8*(4), 535–550.
Walker, K., & Moscardo, G. (2016). Moving beyond sense of place to place of care: The role of Indigenous values and interpretation in promoting transformative change in tourists' place images and personal values. *Journal of Sustainable Tourism, 24*(8–9), 1243–1261.
Wernick, A. (2016). Fighting the Haze at the Grand Canyon. *The World*, April 16.
Yudina, O., & Grimwood, B. S. R. (2016). Situating the wildlife spectacle: Ecofeminism, representation and polar bear tourism. *Journal of Sustainable Tourism, 24*(5), 715–734.

8

ECOTOURISM DEVELOPMENT THROUGH CULTURALLY SENSITIVE UNIVERSALISM

John B. Read IV and Bryan S. R. Grimwood

Introduction

Ecotourism is often held up as the archetype for ethical and responsible tourism. As the negative social and environmental impacts associated with tourism became increasingly identified and understood (Jafari, 2001; Macbeth, 2005), ecotourism's profile as an ethically ideal mode of tourism became entrenched within tourism discourse (Fennell, 2003). There has, of course, been much debate over the years about the extent to which contemporary practices of ecotourism are actually able to fulfil their ethical promise and potential. Nowaczek, Moran-Cahusac, and Fennell (2007), for instance, suggest that ecotourism remains largely driven by profit-making ideologies and underpinned by Western philosophical concepts and meanings (e.g., "nature"; see also Grimwood, Caton, & Cooke, 2018). Bianchi and de Man's (2020) Marxian-inspired critique of UN Sustainable Development Goals and Moore's (2019) ethnography of small-scale tourism developments in the Bahamas show that these concerns remain central to the provision and management of ecotourism. On the demand or consumption side of ecotourism, critics have asserted that ecotourists are not any more or less ethical than their mass tourism counterparts, but are instead driven by similar desires to satisfy ego interests (Wheeller, 1993).

One explanation for why ecotourism, and other related modes of alternative tourism, consistently fail to realise any ethical ideal is the dearth of moral theory underpinning its practice. Fennell (2006, 2008, 2018) has frequently argued along these lines, suggesting for instance that greater emphasis be placed on contemplating the obligations, consequences, and deeper values and meanings associated with tourism behaviours and development decisions. Similarly, Caton (2012) advocates for enhanced philosophical and theoretical engagement within tourism—a *moral turn*—as "matters of morality and ethics [are] among *the most important* [topics to be addressed] in our field" (p. 1912). In recent years, this *moral turn* has been marked by several philosophically and theoretically informed investigations being advanced within the subfields of ecotourism, and nature-based tourism more broadly. For example, scholars have engaged various ecofeminist (Yudina & Grimwood, 2016), animal rights (Fennell, 2014; Sheppard & Fennell, 2019), ecocentric (Fennell, 2003; Holden, 2003, 2015, 2018), and relational (Grimwood & Doubleday, 2013b; Mullins, 2009) perspectives in ecotourism contexts.

DOI: 10.4324/9781003001768-8

To supplement literatures on ecotourism and ethics, the purpose of this chapter is to introduce the concept of culturally sensitive universalism and examine its utility in relation to tourism codes of conduct. Culturally sensitive universalism—a concept we borrow from Dower (2019)—is advanced as a bridge linking the emerging fields of global ethics and development ethics. In more specific terms, culturally sensitive universalism represents an approach to ethics that "allows for different forms of expression in different societies and at the same time retains a framework of trans-boundary responsibility towards humans anywhere" (Dower, 2019, p. 19). Here, ethics is understood as the rules or norms that determine the basis and justification for actions that are deemed right or wrong, good or bad, fair or unfair, or other related value-based distinctions (Fennell, 2018). A key issue we grapple with in this chapter is how culturally sensitive universalism provides a framework for negotiating, if not reconciling, classic tensions in ethical discourse (including the study of tourism ethics) between theoretical and applied approaches and concerns. Indeed, as it will become apparent later, global ethics and development ethics as fields of study blur the conventional distinctions between theoretical ethics or applied ethics. According to Fennell (2018), the former coalesces around two normative principles: either deontological, where moral duties align with principles and rules that embody cultural values of what is morally right or wrong—for example the golden rule—or teleological, where actions are ends-based; that is, maximising pleasure and minimising suffering for the greatest number of people. In contrast, applied ethics emphasises a "[fundamental devotion] to clarifying the ethical meanings of the situations on which they focus" (Lurie, 2018, p. 475); for example, situations within business, tourism, or health care among others. In this regard, applied ethics involves the critical analysis and application of moral philosophical reasoning in real-world contexts (Beauchamp, 2007); or, as Lurie (2018) states, "an interpretive and critical discourse of both ethics and life situation" (p. 475).

In the following sections, we introduce global ethics and development ethics[1] as emerging ethical discourses that bridge theoretical and applied ethics. This chapter—in a manner analogous to Fennell's (2019) negotiation of ethical pluralisms towards a "comprehensive framework for ethics in tourism" (p. 173)—aims to provide an ideal proxy for scholars, practitioners, and governments to consider how global ethics and development ethics contribute to ethical ecotourism development. Our intent is not to provide a comprehensive or nuanced overview of these literatures, but rather to convey the significance and utility of drawing these fields together in and through the concept of culturally sensitive universalism (Dower, 2019). Our discussion then considers how culturally sensitive universalism might be useful for developing ecotourism codes of conduct. More specifically, by considering a specific tourism geography (i.e., the Sir John A. Franklin shipwrecks in the Northwest Passage of Arctic Canada) and drawing connections to various declarations and codes associated with the self-determination of Indigenous peoples, we illustrate the utility of culturally sensitive universalism as a framework for code of conduct development. Finally, we outline a series of recommendations for the potential use of culturally sensitive universalism towards creating *with* the Uqsuqtuuq (Gjoa Haven) Inuit a code of conduct for tourism to the Franklin shipwrecks.

Global ethics

According to Chadwick and O'Conner (2015), global ethics represents "a distinct field of study, going beyond applied ethics to include insight from political philosophy" (p. 25). First gaining traction in the 1970s through the scholarly works of ethicists like Peter Singer and Onora O'Neill, global ethics' prominence increased as worldwide connections and mobilities intensified and universal moral issues—like famine, Indigenous rights, and climate change—demanded global scale deliberation and response. Among the early proponents of global ethics, it was common to apply different ethical theories (e.g., Singer favoured utilitarianism, O'Neill often adopted a Kantian approach) to address the global scale issues at hand. Such an approach effectively situated global ethics as an off-shoot of applied ethics. In the decades of literature since, global ethics has evolved to incorporate both normative and metaethical questions and concerns. Evidence that global ethics has distinguished itself from its applied roots is captured in Widdows' (2011) succinct observation that "the globe [is seen] as the proper sphere of all ethics" (p. 519).

As global ethics has evolved, scholars have been challenged with articulating both the philosophical basis and substantive concerns that unite the field. Dower (2011b, 2019), a leading voice on these matters, notes that global ethics as a philosophical field of study is principally concerned with:

- "The relationships between nation-states and
- The relations between all human beings in the world and, for some environmentalists, relations with nonhumans anywhere and the planet" (p. 505)

In this regard, global ethics interrogates the ethical underpinnings of human, non-human, and even planetary relationships at a transnational scale, including those relationships that occur across borders and in relation to global bodies such as the United Nations. Dower (2019) clarifies that "Global ethics … is often focused on the nature, extent and justification of ethical values and norms that make up what may be called a 'global ethic'", that is assumed to be universal (p. 21). A 'global ethic' according to Dower (2011b) embodies:

- the assertion that there are globally, universal values, and
- that actors anywhere have a duty to act towards those values.

In exploring and providing justifications within areas such as conflict and violence, poverty and development, economic justice, bioethics and health justice, as well as environmental and climate ethics (Dower, 2007; Moellendorf & Widdows, 2015), global ethics has advanced arguments towards values and norms that warrant universal acceptance.

A salient example of universal values that are embodied within a 'global ethic' can be seen in the United Nations Declaration on the Rights of Indigenous Peoples (UNDRIP). The UNDRIP is grounded upon multiple universal values that exist concerning the rights of Indigenous peoples, and which assert that nations-states—and subsequently their private citizens—are morally implicated as members of the global collective in upholding those values. Indigenous self-determination is one such universal value that is codified within the UNDRIP. The Declaration explains Indigenous self-determination in the following way:

> by virtue of that right [Indigenous peoples] freely determine their political status and freely pursue their economic, social and cultural development…[and] have the right to autonomy or self-government in matters relating to their internal and local affairs, as well as ways and means for financing their autonomous functions.
>
> *(United Nations, 2007, p. 8)*

Indigenous self-determination is further affirmed as salient to Indigenous tourism practices through the Larrakia Declaration (United Nations World Tourism Organization [UNWTO], 2012). Centred upon six principles including respect, protection, empowerment, consultation, business, and community, the Larrakia Declaration emphasises respectful, collaborative, sustainable, and equitable tourism practices that align with a 'global ethic' of Indigenous self-determination.

Within ecotourism, Indigenous self-determination has been discussed by Johnston (2000) as the understanding that "[I]ndigenous peoples can set the terms for visitation to their traditional territories, as well as other third party uses of their collective cultural property (p. 91). However, Higgins-Desbiolles (2007) identified that in tourism, Indigenous self-determination has often been side-lined, with Indigenous peoples being seen as one of many stakeholders and not central to the tourism development processes. More recently, scholars have argued for the central role of Indigenous self-determination in directing ecotourism development within Indigenous communities (Dowsley, 2009; Hitchner, Apu, Tarawe, Aran, & Yesaya, 2009). While not specifically highlighting global ethics in this research, these articles demonstrate the infiltration that has already occurred within the ecotourism and tourism literature of Indigenous self-determination as a global value and norm.

While global ethics has much to offer, it is important to note that the field has not avoided critique. As scholars of global ethics attempt to tackle a plethora of issues at a global scale—one example being global environmental problems such as climate change—commentators have questioned the field's grounding within European values and norms that have arisen from traditional academic perspectives and are complicit in maintaining historic disenfranchisement and devaluation of other knowledge systems (Dunford, 2017; Hutchings, 2019). This is salient, for instance, in regard to how Indigenous self-determination is framed within the UNDRIP such that it reproduces an "ethic of self" that is situated within governance structures constructed upon the laws of colonial powers (Franke, 2007, p. 372). Importantly, however, a path forward has been presented within certain critiques that call for the inclusion of pluralist ways of developing global ethics that articulate and implement other (non-European) ways of knowing and being (Dunford, 2017; Mignolo, 2011).

Development ethics

Development ethics is another emerging field of study that equally warrants consideration within ecotourism contexts. According to Marangos, Astroulakis, and Triarchi (2019), development ethics represents an "ethical reflection of the ends and means for any purposeful socio-economic activity towards development and the achievement of a "good society at a local, national and global scale" (p. 257). This has been more simply characterised as pursuing "justifications for judgments about the right and wrong ways of conducting development" (Drydyk & Keleher, 2019b, p. 2) as well as their associated goods and benefits (Dower, 2011a). Initially couched within the ethical discourses of Louis-Joseph Lebret and Denis Goulet, development ethics focused on the scientific, industrial, and socioeconomic growth of underdeveloped areas between the end of the second world war and the 1970s (Culp, 2015; Dower, 2011a; Hutchings, 2010). Development ethics has emphasised attaining "the good life/good living/well-being" or "Eudaimonia" (Marangos et al., 2019, p. 523) within established fields such as Aristotelian, Kantian, utilitarian, and Rawlsian ethics (Dower, 2011a). These philosophical understandings were subsequently built upon through virtue ethics, which "draws attention to the importance of social and cultural practices and institutions for establishing the conditions for human flourishing (Hutchings, 2010, p. 95). The inclusion of the capability approach furthered development ethics' normative focus on quality of life (Garza-Vázquez & Deneulin, 2019) emphasising the provision of the "actual freedom of choice a person has over alternative lives that [they] can lead" (Sen, 1990, p. 114). In the decades of literature since, scholars within development ethics have continued to wrestle with the divide between relativist and universalist thought (Dower, 2011a).

Similar to global ethics, the evolution of development ethics has challenged scholars to convey the philosophical and substantive nature of the field. Dower (2011a, 2019) continues to provide significant direction on these matters, noting that development ethics as a philosophical field of study is principally grounded upon the ideas that:

- "Development is Ethical in character;
- Development ethics is multidisciplinary;
- Reduction of poverty is central to authentic development;
- Development is ultimately human development;
- What is appropriate is context-sensitive;
- Conventional development *qua* economic growth [is] part of the problem" (p. 781)

In this regard, development is understood to be locally situated, emphasising relationships that imbue moral and pluralist engagements. This situated approach to development ethics "look[s] at the values and norms involved in development, often comparing different approaches and seeking a justification for what seems the right approach" (Dower, 2019, p. 23). Within development ethics, Dower (2011a) advances that there have been three normative approaches that have been taken up by researchers:

- Through "economic growth as central", emphasising free markets, equitable distributions, and poverty reduction;
- Through economic growth as conditional, which requires other values and norms in addition to economic growth;
- Through alternatives to economic growth, which calls for "redefinitions of authentic development or rejection of development altogether" (p. 781)

These approaches have each been uniquely embodied in addressing issues such as wellbeing, social and global justice, empowerment and agency, environmental sustainability, human rights, cultural freedom, and responsibilities (Drydyk & Keleher, 2019b).

Distinct from the philosophical manner in which development ethics is discussed, a 'development ethic' is a situated set of values and norms that are accepted/embodied by an individual or community (Dower, 2019). Since the turn of the century, these discussions of community values and norms within the ecotourism literature frequently occur as economically oriented. For example, these discussions often appear as *for* alleviating poverty or *for* increasing fairness and equity—through Pro-Poor or Fair-Trade Tourism (Boluk, 2011b, 2011a). Alternatively, in the case of Holmes, Grimwood, King, and the Lutsel K'e Dene First Nation (2016) and this chapter, we consider the potential of tourism development as community driven, specifically *for* Indigenous self-determination. Holmes et al.'s (2016) research was "guided by community-based, participatory, and narrative methodologies" (p. 1190) to enable Indigenous self-determination through the articulation of developmental values and norms of the Lutsel K'e Dene First Nation Holmes the Lutsel K'e Dene First Nation 2016. As previously discussed, Indigenous self-determination can be conceptualised as a 'global ethic'. Similarly, Indigenous self-determination can also be conceptualised as a 'development ethic' arising from within a community (e.g., the Lutsel K'e Dene First Nation). This conceptualisation at developmental and global scales engages Indigenous self-determination as both a process and outcome that "requires both the transformation of governance and law, as well as space to enable [I]ndigenous peoples to articulate, pursue, and realize lives they value" (Watene & Merino, 2019, p. 143). As such, Indigenous self-determination as a process and outcome of ecotourism development must be situated within specific communities or peoples. This emphasises Indigenous expectations of how to conduct authentic ecotourism that generate community specific understandings of wellbeing.

While there is much that development ethics can offer, it has borne critique of its historical development practices, especially in countries where colonial practices are attributed with the rise of poverty (Hutchings, 2010) (e.g., the United Kingdom influenced the development of many countries from agrarian to capitalist societies which changed existing social structures and increased poverty). Development ethics also experiences critique for its contribution to the proliferation of top-down policies that attempt to address development in a utilitarian manner (i.e., as a tool to achieve development). These critiques appear to arise from development that was centred around or favoured economics as a means to generate well-being for communities where development occurred. Dower's (2011a) identification of development that seeks alternative means to generate wellbeing outside of economic systems, appears to demonstrate that global values and norms are additionally salient to development ethics.

Towards culturally sensitive universalism

Until recently, the fields of global ethics and development ethics were considered to be distinct. However, as global and development ethics have continued to evolve, their interconnectedness has been explored in multiple ways. Recently, Dower (2019) has advanced culturally sensitive universalism where global ethics must be informed by the philosophical underpinnings of development ethics and, likewise, development ethics must align with universal values and norms that are embodied within a 'global ethic'. As such, global ethics has been argued as a field that acknowledges "culturally sensitive universal values and norms, which both allow for different forms of expression in different societies and at the same time retains a framework of trans-boundary responsibility towards humans anywhere" (Dower, 2019, p. 19).

Culturally sensitive universalism intertwines global and development ethics within the understanding that individuals and groups have unique views resulting in their acceptance or rejection of universal values and norms (e.g., tourism practitioners and Indigenous communities may have differing views on the values of development). Highlighting Parekh's analysis of global ethics, Dower (2019) discusses how both global and development ethics are intertwined within the understanding that individuals and groups will either consent or assent based upon their foundational ethical understanding of how development values or global norms might be enacted. This understanding of consent and assent allows for multiple global ethics and development ethics to be taken up by different individuals or groups (e.g., tourism scholars, non-governmental organisations, tourism associations, etc.) at different times (Dower, 2019). In this regard, global ethics' transnational quest to identify and validate a set of universal norms and values cannot be successful unless it acknowledges the validity of all worldviews to allow for cultural sensitivity at local/community levels. As such, in a manner that invites pluralistic collaborations at local/community levels (Drydyk & Keleher, 2019a; Dunford, 2017; Hutchings, 2019), culturally sensitive universalism requires "that a consensus of shared values derived from various sources is accepted, [and] the importance of flexibility is underlined" (Dower, 2019, p. 26).

Dower's (2019) introduction of culturally sensitive universalism provides a philosophically informed argument linking development ethics as the practical application of values and norms within global ethics. This is similar to Fennell's (2019) "pluralistic, integrated model of tourism ethics" (p. 164) that also provides an alternative to relativism (i.e., "there are no universal moral codes, but, rather unique practices" (Fennell, 2018, p. 86). While Dower (2019) advances culturally sensitive universalism to critique the development practices of countries and companies globally, Fennell (2019) negotiates the coming together of micro and macro ethical spaces towards moral governance by employing social contract theory. Rather than critiquing ecotourism development practices, we follow Fennell's (2019) work towards ethical frameworks by illuminating how culturally sensitive universalism might be useful in informing ecotourism development. Using the Northwest Passage within the Canadian Arctic as a proxy, we underscore how a culturally sensitive universalism that embraces Indigenous self-determination embodies a salient universal value for ecotourism development. We further explore how Indigenous self-determination occurs as both process and outcome in creating a code of conduct for the Franklin Shipwrecks.

The Franklin shipwrecks

Within the Northwest Passage, there lays a confluence where the distinctions between water, land, and ice (the Alexandra and Simpson Straights, the Adelaide Peninsula and Qikiqtaq (King William Island)), the Arctic ecosystem, and more specifically the ancestral and sovereign territory of Inuit from Uqsuqtuuq (Gjoa Haven, Nunavut) are blurred (ITK & NRI, 2006; Têtu, Lasserre, Pelletier, & Dawson, 2019). This confluence was further blurred in 2018 when Parks Canada and the Inuit Heritage Trust obtained joint management and responsibility of the Franklin Shipwreck (Parks Canada, 2018a). The HMS Erebus and HMS Terror, unseen since 1845, were 'rediscovered' in 2014 and 2016, respectively, though the locations of the shipwrecks were never truly hidden. Parks Canada (2018a) attributes Inuit Qaujimajatuqangit (the tradition of oral history and knowledge) as playing a pivotal role in supplying the knowledge required for the 'rediscovery' of the Shipwrecks. This role in discovery and joint management has resulted in a public commitment where "the Government of Canada will continue to collaborate with Inuit to share the story of the Franklin Expedition, and the important role of Inuit in the discovery and on-going protection of the Franklin wrecks" (Parks Canada, 2018a, n.p.). This commitment was further reiterated by Fred Pederson, the Chair of the Franklin Interim Advisory Committee who stated that "Canada and Inuit will be joint owners of the artifacts going forward which provides a great opportunity for Inuit to be involved with, and guide, how the rest of the story unfolds" (Parks Canada, 2018a, n.p.).

An unfolding plotline in the story of the Franklin Shipwrecks relates to their promotion as an ecotourism destination within the Northwest Passage. There is precedent for this notion as the current *Inuit Guardians Program* is intended to "play a key role in hosting visitors to the wreck sites—sharing knowledge and Inuit culture and presenting the Franklin story as well as monitoring the two wreck sites" (Parks Canada, 2018b, n.p.). While this potential growth of tourism to sites within the Northwest Passage that "respects both nature and culture" is openly desired by some Inuit (Stewart, Draper, & Dawson, 2011, p. 47), tourism development can contribute to increased vulnerability and decreased resilience, "potentially altering [Arctic] community structure and cohesion" (Sisneros-Kidd, Monz, Hausner, Schmidt, & Clark, 2019, pp. 1270–1271). As such, active local community involvement and coordination must occur in planning for the tourism futures of the Northwest Passage (Stewart et al., 2011). For the Franklin Shipwrecks, this has already begun to take shape as the *Inuit Guardians Program* further "supports Indigenous land management and oversight in their territories based on a cultural responsibility for the land" (Parks Canada, 2018b, n.p.). Understanding that an intention for tourism to the Franklin Shipwrecks exists and that the Inuit Guardians Program is comprised of community members from Uqsuqtuuq (Gjoa Haven) who possess inherent rights to the management, oversight, and cultural responsibility in and on their traditional territory, we ask:

how might culturally sensitive universalism usefully inform potential ecotourism development in the context of the Franklin Shipwrecks? To answer this question, we explore an opportunity for a code of conduct for tourists visiting the Franklin Shipwrecks.

Opportunities for a code of conduct

As a result of the historic economic, environmental, and cultural damage created by tourism, scholars have recommended that governments/practitioners imbed a code of conduct during the tourism development phase to educate, manage, and raise tourist awareness in regard to acceptable/unacceptable tourist behaviours (Fennell & Malloy, 2007; Mason & Mowforth, 1995). Frequently, this is reactionary as tourists have already begun to flock to outdoor destinations before management is able to clearly set out parameters for the development of tourism. In the case of the Franklin Shipwrecks, the public is currently prohibited from visiting the site (Parks Canada, 2018c) and this provides the opportunity to carefully think through the development and implementation of a code of conduct prior to the arrival of tourism. One strategy for implementing a code of conduct for the Franklin Shipwrecks is to adapt an existing code used in a similar context.

In the Canadian Arctic, we highlight three different codes of conduct that have been used to promote desired behaviours from tourists that visit these areas. First, we look to a code of conduct that is used at co-managed national parks and national historic sites in the Canadian Arctic. This code of conduct—the principles of Leave No Trace (LNT)—is highly promoted by Parks Canada across all park units including those located within the Arctic (Parks Canada, 2019). LNT was produced by non-governmental organisations in an attempt to generate responsible behaviours among those that participate in nature based activities (Leave No Trace, 2019). Designed as an overarching set of guidelines for wilderness etiquette, LNT is general and applicable for environmental protection in outdoor areas around the world. Second, we look to Arctic specific codes of conduct for tourism like the 10 principles for Arctic Tourism (WWF International Arctic Program [WWF], 2001; see original in Mason & Mowforth, 1995) which were produced by academia and non-governmental organisations like the World Wildlife Foundation (WWF) in an attempt to create overarching moral responsibilities for tourism practices in this transnational region. This code of conduct provides salient concepts that are general in nature and applicable across the Arctic especially in regard to environmental protection. Third, we turn to other codes of conduct within the Arctic that are community-based, appearing in the form of co-created documents between academia and communities such as Holmes et al.'s (2016) an *Indigenized Visitor Code of Conduct* which provides expectations of tourist behaviours within a localised region of the Arctic. While not the first code of conduct that implores tourists to be sensitive to Indigenous cultures [see e.g., Colvin (1994), as well as the guidelines produced by the Association of Arctic Cruise Expedition Operators], Holmes et al.'s (2016) *Indigenized Visitor Code of Conduct* was developed through processes that adhered to Indigenous self-determination as a value within specific community context. This process for developing a code of conduct provides a relevant and localised understanding of what tourism could look like when centring the narratives of specific Indigenous communities within the Arctic.

In reflecting on these three codes of conduct, we find that LNT and the WWF codes while useful in their overarching contexts would not be appropriate when considering tourism development to the Franklin Shipwrecks. This is due to the LNT and WWF codes of conduct not centring the site management, oversite, and cultural responsibility that is inherent to the Inuit of Uqsuqtuuq (Gjoa Haven). Likewise, we find that applying the Holmes et al. (2016) code of conduct to the Franklin Shipwrecks would be equally inappropriate as this would infringeon the rights of the Uqsuqtuuq Inuit and be tantamount to cultural amalgamation. Given these aforementioned limitations, we consider Dower's (2019) culturally sensitive universalism as a philosophically grounded and theoretically informed approach to illuminate future ecotourism development of Franklin's Shipwrecks. To embark on this inquiry we acknowledge Rigney's (1999) words: "therefore, the Indigenous context of knowledge production and research methodologies is about countering racism and including Indigenous knowledges and experiences for Indigenous emancipation" (p. 119; see also Seale, 1992). As such, we centre this engagement as one that is situated within processes that enact collaboration with the Uqsuqtuuq (Gjoa Haven) Inuit or as Holmes et al. (2016) state: an "'Indigenist paradigm', in which the priorities, expertise, control, and benefits associated with tourism research are centred on Indigenous peoples and communities" (p. 1181).

Processes and outcomes

To embrace an "Indigenist paradigm" and centre culturally sensitive universalism regarding the Franklin Shipwrecks and the blurred confluence within which they exist, we must engage with the 'global ethic' of Indigenous self-determination. As a 'global ethic', Indigenous self-determination intertwines the understanding that the Inuit Guardians' Program for the Franklin Shipwrecks was always already comprised of community members from Uqsuqtuuq (Gjoa Haven) who possess inherent rights to management, oversight, and cultural responsibility. These inherent rights built into the core principles of the Inuit Guardians Program further re-inscribe the values within Indigenous self-determination through the UNDRIP and Larrakia Declaration as salient to the Uqsuqtuuq (Gjoa Haven) Inuit. In this way, the 'global ethic' of Indigenous self-determination becomes both a process and an outcome. The process of Indigenous self-determination occurs though ecotourism development practices that incorporate research methodologies that evoke "a consensus of shared values" (Dower, 2019, p. 26) through morally and culturally significant narratives. This engagement in consensus seeking collaborations fosters space for the elucidation of Indigenous communities understanding of authentic development (Holmes et al., 2016). Indigenous self-determination within these methods are premised upon "prompt respectful dialogue among various actors and avenues toward collectively realising sustainable, responsible, and culturally appropriate visitor behaviours on Indigenous homelands" (Holmes et al., 2016, p. 1190). However, this work must not, as Higgins-Desbiolles (2007) identified, enact Indigenous self-determination as a stakeholder issue but rather Indigenous self-determination as autonomy in the who, what, when, where, and why ecotourism occurs on and within their traditional territory. In this way, culturally sensitive ecotourism development must intentionally combat what Grimwood, Muldoon, and Stevens (2019) identify as harmful narratives and practices within tourism that continue to "obscure, historicize, and essentialize Indigenous cultures" (p. 244) and their self-determination.

We previously outlined that the Holmes et al. (2016) code of conduct would not be applicable within the context of our case study. However, we can glean insights from the narrative and consensus building processes used by Holmes et al. (2016) that would be relevant when engaging Indigenous self-determination through culturally sensitive universalism. Holmes et al.'s (2016) process operationalised collaborative community-based research (CCBR) which as Glass et al. (2018) identify:

> Seek[s] to hear and respond respectfully and with critical care to the voices, truths, and visions of communities long marginalized in the dominant world ... [Being]

> committed to working with [local] communities to conduct research that mobilizes knowledge so that it can speak with ethical, epistemic, and political force.
>
> *(Glass et al., 2018, p. 525)*

Holmes et al. (2016) go beyond purely *working with* an Indigenous community and instead engage in Indigenous-driven research specifically as it "affords opportunities for enhancing [Indigenous] self-determination" (p. 1188). Through a series of consultation workshops, Holmes et al. (2016) demonstrate a process of: first, seeking initial agreement and mutual benefits for both the researchers and community; and second, pinpointing "expectations, priorities, and timelines for the research project, including the various roles and responsibilities of community and university researchers" (Holmes et al., 2016, p. 1182) prior to data collection. This engagement with Indigenous voices prior to the development process enacted by Holmes et al. (2016) did not only contribute to the development of a code of conduct. Rather, the communities self-determined values create the "justifications for judgments about the right and wrong ways of conducting [both research and ecotourism]" (Drydyk & Keleher, 2019b, p. 2) within their context.

For us as authors to speculate *exactly* on a code of conduct for ecotourism to the Franklin Shipwrecks would, in this case study, seem incongruent with culturally sensitive universalism. Rather—we acknowledge that for the Franklin Shipwrecks—the substantive principles or practices of the code of conduct require the direct involvement and capacities of the Uqsuqtuuq (Gjoa Haven) Inuit. As such, we offer a recommended process for creating a code of conduct that embraces culturally sensitive universalism and Indigenous self-determination. We previously highlighted three normative approaches that have been researched within development. These are: development as either economically essential or economically conditional to growth, or growth that is unrelated to economics and development altogether (Dower, 2011a). Further, engaging cultural sensitivity in ecotourism development could also materialise as no development whatsoever. In this manner, integrating culturally sensitive universalism as a foundation for ecotourism development with Indigenous communities engages the process of Indigenous self-determination in the co-creation of (non)development. This additionally forefronts Indigenous self-determination as an outcome generated by the community and enables the potential for other processes and outcomes (e.g., the development of a code of conduct).

Recommendations and conclusions

Abiding the commitment of the Canadian government to empower Uqsuqtuuq (Gjoa Haven) Inuit in the management of the Franklin Shipwrecks and presuming that future tourism is desired to these sites, we recommend that Parks Canada or their assigned agent recall Dower's (2019) note that to arrive at a consensus of values, flexibility is required. In a practical sense, this is similar to what Grimwood and Doubleday (2013a) identify within adaptive co-management as a space where "differences can be shared for common aims … related to and affecting place, meaning, and management" (p. 13). As such, the Franklin Shipwrecks as a co-managed site requires respect for and commitment to Indigenous paradigms when developing a code of conduct. Similar to Holmes et al.'s (2016) engagement with the Lutsel K'e Dene First Nation, Parks Canada could enact culturally sensitive universalism through the following processes and outcomes:

1. Enact Indigenous self-determination as an inherent 'global ethic'.
   a. Other universal values and norms must appertain to this 'global ethic' (e.g., the other values and norms within the UNDRIP).

2. Non-Inuit stakeholders of the Franklin Shipwrecks must embrace Dower's (2019) culturally sensitive universalism, understanding that their values must be flexible such that a consensus of Uqsuqtuuq (Gjoa Haven) Inuit community values can be elucidated.
3. Conduct consultation workshops where:
   a. Codes of conduct are determined to be desired and are mutually beneficial.
   b. The research process and outcomes identified by the Uqsuqtuuq (Gjoa Haven) Inuit are agreed to by all external parties.
      i. Understand that the process and outcomes will be unique to the Uqsuqtuuq (Gjoa Haven) Inuit.
4. Follow the agreed upon processes while researching and developing the code of conduct.
   a. Conduct Indigenous-driven research that engages CCBR and the Indigenist Paradigm of the Uqsuqtuuq (Gjoa Haven) Inuit.
5. Verify that the produced code of conduct is a culturally sensitive representation of the desires of Uqsuqtuuq (Gjoa Haven) Inuit.

In this chapter, we've aimed to supplement the tourism literature with ethical, moral, and theoretical insights by introducing Global and Development Ethics as salient concepts for ecotourism practitioners, especially those working in ecotourism development. Through articulating Dower's (2019) culturally sensitive universalism we demonstrate the bridging that can be achieved when the universal values and norms of global ethics connect with local priorities of development ethics to justify development outcomes. In framing our discussion around the Franklin Shipwrecks and the Uqsuqtuuq (Gjoa Haven) Inuit, we identify a 'global ethic'—Indigenous self-determination—as a salient universal concept for tourism practices that engage with Indigenous communities like Uqsuqtuuq (Gjoa Haven).

While we emphasised Indigenous self-determination as a salient 'global ethic' within this context, the UNDRIP encompasses 45 other articles that are affirmed in the Larrakia Declaration as equally salient to any ecotourism development that might occur in the context of the Franklin Shipwrecks or the Uqsuqtuuq (Gjoa Haven) Inuit (United Nations, 2007; United Nations World Tourism Organization (UNWTO), 2012). In this way, Dower's (2019) culturally sensitive universalism enables tourism development researchers, practitioners and governments to embrace "values that are universal and responsibilities that are global in scope" (Dower, 2011b, p. 505) while enacting "authentic development … [that] is ultimately human development; [where] what is appropriate is context specific" (Dower, 2011a, p. 781).

This chapter supplements current ethical and moral discourses (e.g., Fennell, 2019) within ecotourism and enables scholars, practitioners, and governments to think through culturally sensitive universalism. As such, we suggest that engagement with ecotourism and more specifically Indigenous ecotourism should affirm the values and norms of contextually appropriate 'global ethic' while maintaining cultural sensitivity that is situated within community realised understandings of development outcomes. In the end, a code of conduct for the Franklin Shipwrecks is not provided, rather, we provide recommendations for how culturally sensitive universalism might be enacted when developing a code of conduct. This work continues the *moral turn* in tourism, by demonstrating the utility of using global ethics and development ethics to think through culturally sensitive universalism as process and outcome that enables engagement with Indigenous-led approaches.

Note

1 We use global ethics and development ethics to discuss the fields of study. We use 'global ethic' and 'development ethic' to discuss codes of ethics or other accepted ethical principles.

References

Beauchamp, T. L. (2007). History and theory in "applied ethics." *Kennedy Institute of Ethics Journal; Baltimore, 17*(1), 55–64.

Boluk, K. (2011a). Fair trade tourism South Africa: Consumer virtue or moral selving? *Journal of Ecotourism, 10*(3), 235–249. doi:https://doi.org/10.1080/14724049.2011.617451.

Boluk, K. (2011b). In consideration of a new approach to tourism: A critical review of fair trade tourism. *The Journal of Tourism and Peace Research, 2*(1), 27–37.

Bianchi, R. V.& de Man, F. (2020). Tourism, inclusive growth and decent work: a political economy critique. *Journal of Sustainable Tourism*, 29, 353–371. 10.1080/09669582.2020.1730862.

Caton, K. (2012). Taking the moral turn in tourism studies. *Annals of Tourism Research, 39*(4), 1906–1928. doi:https://doi.org/10.1016/j.annals.2012.05.021.

Chadwick, R., & O'Connor, A. (2015). Ethical theory and global challenges. In D. Moellendorf, & H. Widdows (Eds.), *The Routledge Handbook of Global Ethics* (pp. 24–34). Routledge.

Colvin, J. G. (1994). Capirona: A model of indigenous ecotourism. *Journal of Sustainable Tourism*, 2, 174–17710.1080/09669589409510693.

Culp, J. (2015). Development. In D. Moellendorf, & H. Widdows (Eds.), *The Routledge handbook of global ethics* (pp. 170–181). Abingdon: Routledge. doi:https://doi.org/10.4324/9781315744520-16.

Dower, N. (2007). *World Ethics: The New Agenda*. Edinburgh: Edinburgh University Press.

Dower, N. (2011a). Development ethics. In R. Chadwick, D. Callahan, & P. Singer (Eds.), *Encyclopaedia of applied ethics* (2nd ed., Vol. 2, pp. 504–513). San Diego, United States: Elsevier Science & Technology. Retrieved from http://ebookcentral.proquest.com/lib/waterloo/detail.action?docID=858617.

Dower, N. (2011b). Global ethics, approaches. In R. Chadwick, D. Callahan, & P. Singer (Eds.), *Encyclopaedia of applied ethics* (2nd ed., Vol. 2, pp. 504–513). San Diego, United States: Elsevier Science & Technology. Retrieved from http://ebookcentral.proquest.com/lib/waterloo/detail.action?docID=858617.

Dower, N. (2019). Global ethics: Development ethics as global ethics. In J. Drydyk, & L. Keleher (Eds.), *Routledge handbook of development ethics* (pp. 17–28). London: Routledge.

Dowsley, M. (2009). Inuit-organised polar bear sport hunting in Nunavut territory, Canada. *Journal of Ecotourism, 8*(2), 161–175. doi:https://doi.org/10.1080/14724040802696049.

Drydyk, J., & Keleher, L. (Eds.). (2019a). Introduction: What is development ethics? In *Routledge handbook of development ethics* (pp. 1–14). London: Routledge.

Drydyk, J., & Keleher, L. (Eds.). (2019b). *Routledge Handbook of Development Ethics*. London: Routledge.

Dunford, R. (2017). Toward a decolonial global ethics. *Journal of Global Ethics, 13*(3), 380–397. doi:https://doi.org/10.1080/17449626.2017.1373140.

Fennell, D. A. (2003). *Ecotourism: An Introduction* (2nd ed.). London; Routledge.

Fennell, D. A. (2006). *Tourism Ethics*. Bristol: Channel View Publications.

Fennell, D. A. (2008). Responsible tourism: A Kierkegaardian interpretation. *Tourism Recreation Research, 33*(1), 3–12. doi:https://doi.org/10.1080/02508281.2008.11081285.

Fennell, D. A. (2014). Exploring the boundaries of a new moral order for tourism's global code of ethics: An opinion piece on the position of animals in the tourism industry. *Journal of Sustainable Tourism, 22*(7), 983–996.

Fennell, D. A. (2018). *Tourism Ethics* (2nd ed.). Bristol: Channel View Publications.

Fennell, D. A. (2019). The future of ethics in tourism. In E. Fayos-Solà, & C. Cooper (Eds.), *The future of tourism: Innovation and sustainability* (pp. 155–177). Cham: Springer International Publishing.

Fennell, D. A., & Malloy, D. C. (2007). *Codes of Ethics in Tourism: Practice, Theory, Synthesis*. Clevedon: Channel View Publications.

Franke, M. F. N. (2007). Self-determination versus the determination of self: A critical reading of the colonial ethics inherent to the United Nations Declaration on the Rights of Indigenous Peoples. *Journal of Global Ethics, 3*(3), 359–379.

Garza-Vázquez, O., & Deneulin, S. (2019). The capability approach. In J. Drydyk, & L. Keleher (Eds.), *Routledge handbook of development ethics* (pp. 68–83). London: Routledge.

Glass, R. D., Morton, J. M., King, J. E., Krueger-Henney, P., Moses, M. S., Sabati, S., & Richardson, T. (2018). The ethical stakes of collaborative community-based social science research. *Urban Education, 53*(4), 503–531. doi:https://doi.org/10.1177/0042085918762522.

Grimwood, B. S. R., Caton, K., & Cooke, L. (2018). *New Moral Natures in Tourism.* London: Routledge.

Grimwood, B. S. R., & Doubleday, N. C. (2013a). From river trails to adaptive co-management: Learning and relating with inuit inhabitants of the Thelon River, Canada. *Indigenous Policy Journal, 23*(4).

Grimwood, B. S. R., & Doubleday, N. C. (2013b). Illuminating traces: Enactments of responsibility in practices of Arctic river tourists and inhabitants. *Journal of Ecotourism, 12*(2), 53–74. doi:https://doi.org/10.1080/14724049.2013.797427.

Grimwood, B. S. R., Muldoon, M. L., & Stevens, Z. M. (2019). Settler colonialism, Indigenous cultures, and the promotional landscape of tourism in Ontario, Canada's 'near North.' *Journal of Heritage Tourism, 14*(3), 233–248. doi:https://doi.org/10.1080/1743873X.2018.1527845.

Higgins-Desbiolles, F. (2007). Taming tourism: Indigenous rights as a check to unbridled tourism. In P. Burns, & M. Novelli (Eds.), *Tourism and politics: Global frameworks and local realities* (pp. 83–107). Amsterdam: Elsevier.

Higgins-Desbiolles, F. (2009). Indigenous ecotourism's role in transforming ecological consciousness. *Journal of Ecotourism, 8*(2), 144–160. Doi:https://doi.org/10.1080/14724040802696031.

Hitchner, S. L., Apu, F. L., Tarawe, L., Aran, S. G. N., & Yesaya, E. (2009). Community-based transboundary ecotourism in the Heart of Borneo: A case study of the Kelabit Highlands of Malaysia and the Kerayan Highlands of Indonesia. *Journal of Ecotourism, 8*(2), 193–213. doi:https://doi.org/10.1080/14724040802696064.

Holden, A. (2003). In need of new environmental ethics for tourism? *Annals of Tourism Research, 30*(1), 94–108. doi:https://doi.org/10.1016/S0160-7383(02)00030-0.

Holden, A. (2015). Evolving perspectives on tourism's interaction with nature during the last 40 years. *Tourism Recreation Research, 40*(2), 133–143. doi:https://doi.org/10.1080/02508281.2015.1039332.

Holden, A. (2018). Environmental ethics for tourism- the state of the art. *Tourism Review, 74*(3). doi:https://doi.org/10.1108/TR-03-2017-0066.

Holmes, A. P., Grimwood, B. S. R., King, L. J., & the Lutsel K'e Dene First Nation. (2016). Creating an indigenized visitor code of conduct: The development of Denesoline self-determination for sustainable tourism. *Journal of Sustainable Tourism, 24*(8–9), 1177–1193. doi:https://doi.org/10.1080/09669582.2016.1158828.

Hutchings, K. (2010). *Global Ethics: An Introduction.* Cambridge: Polity Press.

Hutchings, K. (2019). Decolonizing global ethics: Thinking with the pluriverse. *Ethics & International Affairs, 33*(02), 115–125. doi:https://doi.org/10.1017/S0892679419000169.

ITK, & NRI. (2006). *Negotiating Research Relationships with Inuit Communities: A Guide for Researchers.* Ottawa and Iqaluit: Inuit Tapiriit Kanatami and Nunavut Research Institute. Retrieved from http://epub.sub.uni-hamburg.de/epub/volltexte/2011/6959/pdf/06_068_ITK_NRR_booklet.pdf.

Jafari, J. (2001). The scientification of tourism. *Hosts and Guests Revisited: Tourism Issues of the 21st Century*, 28–41.

Johnston, A. (2000). Indigenous peoples and ecotourism: Bringing indigenous knowledge and rights into the sustainability equation. *Tourism Recreation Research, 25*(2), 89–96. doi:https://doi.org/10.1080/02508281.2000.11014914.

Leave No Trace. (2019). The Leave No Trace Story. Retrieved from https://lnt.org/why/ (accessed on 20.10.2019).

Lurie, Y. (2018). Thick and thin methodology in applied ethics. *Metaphilosophy, 49*(4), 474–488. doi:https://doi.org/10.1111/meta.12311.

Macbeth, J. (2005). Towards an ethics platform for tourism. *Annals of Tourism Research, 32*(4), 962–984. doi:https://doi.org/10.1016/j.annals.2004.11.005.

Marangos, J., Astroulakis, N., & Triarchi, E. (2019). The philosophical roots of development ethics. *International Journal of Social Economics, 46*(4), 523–531. doi:https://doi.org/10.1108/IJSE-05-2018-0279.

Mason, P., & Mowforth, M. (1995). *Codes of Conduct in Tourism.* Plymouth, England: University of Plymouth, Department of Geographical Sciences.

Mignolo, W. D. (2011). *The Darker Side of Western Modernity: Global Futures, Decolonial Options.* Durham: Duke University Press. doi:https://doi.org/10.1215/9780822394501.

Moellendorf, D., & Widdows, H. (Eds.) (2015). Introduction. In *The Routledge Handbook of Global Ethics* (pp. 1–3). Abingdon: Routledge. doi:https://doi.org/10.4324/9781315744520-5.

Moore, A. (2019). Selling Anthropocene space: situated adventures in sustainable tourism. *Journal of Sustainable Tourism*, 27(4), 436–451, 10.1080/09669582.2018.1477783.

Mullins, P. M. (2009). Living stories of the landscape: Perception of place through Canoeing in Canada's north. *Tourism Geographies, 11*(2), 233–255. doi:https://doi.org/10.1080/14616680902827191.

Nowaczek, A. M., Moran-Cahusac, C., & Fennell, D. A. (2007). Against the current: Striving for ethical ecotourism. In J. Higham (Ed.), *Critical issues in ecotourism: Understanding a complex tourism phenomenon* (pp. 136–157). Oxford: Elsevier.

Parks Canada. (2018a, April 26). Government of Canada Receives Historic Gift of Franklin Shipwrecks from United Kingdom [News releases]. Retrieved from https://www.canada.ca/en/parks-canada/news/2018/04/government-of-canada-receives-historic-gift-of-franklin-shipwrecks-from-united-kingdom.html (accessed on 01.10.2019).

Parks Canada. (2018b, October 30). Inuit Guardians Program—Wrecks of HMS Erebus and HMS Terror National Historic Site. Retrieved from https://www.pc.gc.ca/en/lhn-nhs/nu/epaveswrecks/culture/inuit/gardiens-guardians (accessed on 18.06.2020).

Parks Canada. (2018c, November 1). Superintendent's Order—Wrecks of HMS Erebus and HMS Terror National Historic Site. Retrieved from https://www.pc.gc.ca/en/lhn-nhs/nu/epaveswrecks/info/plan (accessed on 19.06.2020).

Parks Canada. (2019, October 10). Visitor Guidelines—Plan Your Visit. Retrieved from https://www.pc.gc.ca/en/voyage-travel/regles-rules (accessed 18.06.2020).

Rigney, L.-I. (1999). Internationalization of an indigenous anticolonial cultural critique of research methodologies: A guide to indigenist research methodology and its principles. *Wicazo Sa Review*, 14, 109. 10.2307/1409555.

Seale, R. G. (1992). Aboriginal societies, tourism and conservation: The case of Canada's Northwest Territories. Fourth World Congress on Parks and protected Areas, Caracas.

Sen, A. (1990). Justice: Means versus freedoms. *Philosophy & Public Affairs, 19*(2), 111–121.

Sheppard, V. A., & Fennell, D. A. (2019). Progress in tourism public sector policy: Toward an ethic for non-human animals. *Tourism Management, 73*(Complete), 134–142. doi:https://doi.org/10.1016/j.tourman.2018.11.017.

Sisneros-Kidd, A. M., Monz, C., Hausner, V., Schmidt, J., & Clark, D. (2019). Nature-based tourism, resource dependence, and resilience of Arctic communities: Framing complex issues in a changing environment. *Journal of Sustainable Tourism, 27*(8), 1259–1276. doi:https://doi.org/10.1080/09669582.2019.1612905.

Stewart, E. J., Draper, D., & Dawson, J. (2011). Coping with change and vulnerability: A case study of resident attitudes toward tourism in Cambridge Bay and Pond Inlet, Nunavut, Canada. In P. T. Maher, E. J. Stewart, & M. Lück (Eds.), *Arctic tourism: Human, environmental and governance dimensions* (pp. 33–53). New York: Cognizant Communication Corporation.

Têtu, P.-L., Lasserre, F., Pelletier, S., & Dawson, J. (2019). 'Sovereignty' over submerged cultural heritage in the Canadian Arctic waters: Case study from the Franklin expedition wrecks (1845-48). *Polar Geography, 42*(2), 71–88. doi:https://doi.org/10.1080/1088937X.2019.1578288.

United Nations. (2007). United Nations Declaration on the Rights of Indigenous Peoples., Pub. L. No. 295, 61 Resolution 32.

United Nations World Tourism Organization [UNWTO]. (2012). *Addendum: Larrakia Declaration on the Development of Indigenous Tourism* (No. CE/94/5(a) Add.1; p. 3). Campeche: UNWTO.

Watene, K., & Merino, R. (2019). Indigenous peoples: Self-determination, decolonization, and indigenous philosophies. In J. Drydyk, & L. Keleher (Eds.), *Routledge handbook of development ethics* (pp. 134–147). London: Routledge.

Wheeller, B. (1993). Sustaining the Ego. *Journal of Sustainable Tourism, 1*(2), 121–129. doi:https://doi.org/10.1080/09669589309450710.

Widdows, H. (2011). Global ethics, Overview. In R. Chadwick, D.Callahan, P.Singer (Eds.), *Encyclopedia of Applied Ethics*, 2nd ed. (Vol. 2, pp. 514–522. Elsevier Science & Technology. http://ebookcentral.proquest.com/lib/waterloo/detail.action?docID=858617.

WWF International Arctic Program. (2001). Ten Principles for Arctic Tourism. Retrieved from https://arcticwwf.org/newsroom/publications/code-of-conduct-for-arctic-tourists/.

Yudina, O., & Grimwood, B. S. R. (2016). Situating the wildlife spectacle: Ecofeminism, representation, and polar bear tourism. *Journal of Sustainable Tourism, 24*(5), 715–734. doi:https://doi.org/10.1080/09669582.2015.1083996.

9
WOLF ECOTOURISM
A posthumanist approach to wildlife ecotourism

Bastian Thomsen

Introduction

Humans frequently exercise unilateral power over nonhuman animals at wildlife tourist attractions (WTAs) in a myriad of ways. We dictate nearly every aspect of nonhumans' quotidian life and milieu: when and what to eat, when to perform or socialise, and where to bed down or play—the list goes on. Our fascination and interest in nonhumans are perhaps fundamental to our identity and understanding of the world (Clayton, 2003; van der Werff, Steg, & Keizer, 2013), evidenced by more than 100,000,000 nonhumans working in entertainment annually (Fennell, 2013a). Some scholars estimate that 20–40% of all tourist attractions involve animals in some capacity (Moorhouse, D'Cruze, & Macdonald, 2017), and for many humans, WTAs may be one of the few opportunities to encounter other species as the current ecological crisis has devastated wildlife populations, and billions of 'wildlife' now live in captivity (Mason, 2010; Baker & Winkler, 2020). More than 60% of the world's wildlife has died off in the past 50 years (Grooten & Almond, 2018), urban sprawl and habitat destruction show no signs of slowing (Matović, 2020), and the human-caused ecological crisis has led to Earth's sixth mass-extinction event (Steffen et al., 2011). These conditions demand that we, as global citizens, problematise our relationship with 'other' species at the abstract and individual levels to question: how can we, as humans, equitably speak for nonhumans and foreground their agency, welfare, rights, and interests in any decision that affects them? This question engenders arduous ethical and ontological issues as it is inherently an exercise of humans speaking for and about nonhumans without their consent.

Nonetheless, wildlife ecotourism provides an optimal lens to explore this ethical question for its focus on human-nonhuman encounters, and its ubiquitous spectrum of wildlife welfare and justice across different types of WTAs (Fennell & Sheppard, 2021). In this chapter, I draw on a multispecies, multi-sited ethnographic study that investigated wolf-human conflict and coexistence in the western United States during the Trump administration. The gray wolf (*Canis lupus*) was chosen for its exceptional ability to provoke human-human and wildlife-human conflict as a symbol for other sociocultural and socio-environmental issues entrenched in Euro-American culture, politics, and history (Jürgens & Hackett, 2017; Lappalainen, 2019). During the study, the only opportunity to directly interact with wolves at the individual level occurred at five WTAs. Four of the WTAs could be described as 'wolf sanctuaries' that promoted tourists

DOI: 10.4324/9781003001768-9

to interact with wolves at different levels ranging from petting wolves and taking 'selfies', to 'hands-off' observations through chain-link fences. The fifth was Yellowstone National Park (YNP), which is a protected area where park rangers strictly enforce laws aimed to safeguard wolves. Preconceived notions of what ethical wildlife ecotourism should look like may propagate initial conceptions as to which of these models optimally foregrounds wolf's interests. However, through a posthumanist analysis I unpack the ethical entanglements of each WTA model to proffer an alternative perspective that considers the complexity and context in which they operate. I will first present a brief history of wolf's plight in the United States, and then engage relevant theoretical perspectives before analysing each of the WTAs' models.

Short history of the wolf in the United States

Except for a small population in Minnesota that historically persisted, U.S. and state-level governments sanctioned bounty programs to systematically extirpate wolf throughout the lower-48 states from the 1860s to 1940s (Mech, 2012). Following World War II, attitudes towards wolf, and wildlife in general, positively shifted, culminating in the passing of the Endangered Species Act (ESA) in 1973 and the wolf's listing on it in 1978 (Ripple et al., 2014; Carroll, Rohlf, vonHoldt, Treves, & Hendricks, 2021). Sixty-six gray wolves were re-introduced to YNP and central Idaho in 1995–1996 (Wilson, 1997; Foreyt, Drew, Atkinson, & McCauley, 2009), and the species has since dispersed to surrounding states. Wolf was delisted in the western United States of Montana, Wyoming, Idaho, Utah, and the eastern third of Washington and Oregon in 2011, even though the wolf has only recovered to an estimated 6000 individuals nationwide (U.S. Fish and Wildlife Services, 2019).

In October 2020, the Trump administration announced the wolf's delisting from the ESA, set to come into effect 4 January 2021, pending legal challenges (Rott, 2020). The decision was highly controversial, as critics charge that the decision was socio-politically motivated and not driven by science. If the delisting survives the lawsuits, wolf 'management' would be turned over to individual states, where anti-wolf sentiments proliferated in rural, conservative-leaning areas in the western United States For example, in 2019 the Idaho Fish and Game Commission awarded a $23,065 grant to a northern Idaho nonprofit, Foundation for Wildlife Management, to compensate hunters and trappers up to $1,000 per wolf 'harvested' (Peacher, 2019). Between 1 July 2019 and 30 June 2020, a record 570 wolves were killed in Idaho, mostly by hunters and trappers, up from an average of 400 annual deaths (Plank, 2020). In this study, wildlife sanctuary operators universally condemned these practices and cited these data as evidence to support their missions of educating humans about wolves.

Nonhuman animal ethics in wildlife ecotourism

The emergence of animal ethics in the tourism literature (see Fennell, 2008; Shani & Pizam, 2008) offers insight into how humans can exercise their power to promote equality with subaltern nonhumans inside and out of WTAs. Subaltern studies developed out of postcolonial theory to signify how indigenous, racial and ethnic minorities, women, and nonhuman groups, among others, have been marginalised and made subordinate to more dominant patriarchal interests on the basis of power (Spivak, 1988; Mitchell, 2002; Thomsen et al., 2021a). Subaltern studies, coupled with posthumanism, provide a complementary lens to reflexively analyse our power over nonhumans and to deconstruct human exceptionalism in our collective treatment of 'other' species. Posthumanism is a postmodern philosophical line of research "rooted in late twentieth-century feminist and anti-racist critiques of modern Western social and political institutions … [that] seeks to erase the human–animal divide, thereby rejecting the basic premise of human exceptionalism" (Cohen & Fennell, 2019, p. 416). Cohen recognises the inherent paradoxes of posthumanism where humanists criticise posthumanists' reliance on human exceptionalism to develop their scholarly arguments (Chagani, 2014). He cites Soper's (2012) defense of human exceptionalism in that "we need […] to defend human exceptionalism, and resist blurring the human-animal divide", as exceptionalism provides the needed 'footing' for binding moral obligations (p. 423). Cohen declares:

> Posthumanist thinkers thus failed to recognize sufficiently the importance of a critical point: that there is no reciprocity from animals in the sphere of morals; they do not share our ethical precepts nor respond to them. Posthumanist ethics cannot serve as the basis of a covenant between humans and animals. (p. 423)

However, posthumanists would counter to argue that humanists' insistence on contouring arbitrary boundaries between species is counterproductive in this ecological crisis, especially when their core argument is rooted in a 'we were here first' mentality (Badley, 2017). Posthumanists would also refute Cohen's assertation for being anthropocentric and falling into the same logical 'trap' lodged by humanists, by stating that animals 'do not share our ethical precepts nor respond to them'. Under humanistic logic, how can he know this?

Cohen does acknowledge that animal ethology may help us to untangle our human limitations and (mis)understandings of interspecies communication, intelligence, and emotion (De Waal, 2016; Turnbull & Bär, 2020). Cohen has advocated for nonhuman welfare and justice on multiple occasions and his critique of posthumanism is a scholarly exercise rather than a criticism of nonhuman-welfare (Cohen, 2009; Cohen, 2019b; Cohen & Fennell, 2019). He states that posthumanism is essentially absent from tourism scholarship, and suggests how it could contribute to deconstructing the human-nonhuman divide:

> Insistence on animal personhood and critique of anthropocentrism, could reinforce contemporary efforts to reduce animal abuse in tourism, and help to balance the presently often one-sided approach in contemporary tourism studies to the relationship between tourists and animals. Posthumanist attitudes could certainly sensitize tourist practitioners and researchers to issues overlooked in current touristic practices, especially in the field of embodied human-animal interaction. (p. 424)

The following analysis considers how posthumanism applies in wildlife ecotourism, in an effort to respond to Cohen's call to balance nonhumans' welfare in theory and practice. It also cogitates about the inherent paradoxes and complexities in determining whether a specific WTA is truly ecotourism and, if so, to what degree.

Applying posthumanism to WTAs

At their best, WTAs provide the potential to observe nonhumans exercising their own agency under "natural conditions that are completely unframed, according to Cohen (2009)" (Fennell & Sheppard, 2021, p. 330). At their worst, WTAs perpetuate the depravity of humanity, condemning nonhumans to slavery, torture, mental and physical abuse, and even death (Idfwru, Wkh, Wdeoh, & Xvlqj, 2013; Moorhouse, Dahlsjö, Baker, D'Cruze, & Macdonald, 2015; D'Cruze et al., 2017; Fennell & Sheppard, 2021). Most WTAs operate somewhere between these extremes, and research on nonhuman welfare and rights in (eco) tourism has materialised over the past two decades (see Cohen, 2009; Hayward et al., 2012; Fennell, 2013b; Cohen & Fennell, 2019; von Essen, Lindsjö, & Berg, 2020; Thomsen & Thomsen, 2020). Wildlife ecotourism transcends wildlife tourism by embracing non-consumptive activities such as wildlife sightseeing, compared to consumptive practices such as hunting or fishing (Burns, 2017), to equally stress ecological sustainability and human (economic) livelihoods (Duffy & Moore, 2010; Karanth, DeFries, Srivathsa, & Sankaraman, 2012; Sheppard & Fennell, 2019; Thomsen, Thomsen, Cipollone, & Coose, 2021b).

Moorhouse et al. (2015) suggest that tourists are not typically educated on how to identify poor animal-welfare conditions that may reinforce horrific practices such as inadvertently funding the illegal wildlife trade. Newsome (2017) contends that in addition to demographics such as age, and sex, culture was a key influence concerning who engages in wildlife-based tourism and how animals are treated. Accordingly, von Essen et al. (2020) caution against a cultural relativism approach for its anthropocentricity, as it can normalise and reinforce utilitarian views of human dominance over nature (Peterson & Nelson, 2017). von Essen et al. cite Juvan and Dolnicar's (2014) attitude-behaviour gap, as well as Kline (2018) to argue that it is common for some people to leave their pro-environmental values 'at home' when travelling and engage in WTA activities that may perpetuate negative animal welfare.

Fennell and Sheppard (2021) coalesce two common animal ethics approaches in the tourism literature, normative ethics (i.e., what we are told to do or believe based on rights, welfare, utilitarianism, ecocentrism, and contractarianism perspectives) against virtue ethics, in conceptualising a 'scales of justice' approach to animal welfare in ecotourism. Utilitarianism, which is most applicable to the present study, is "informed by the doctrine of "the greatest good for the greatest number" (p. 318), whereas virtue ethics "is focused not on what we are told to do, but rather on 'what sort of person I should be'. Virtues are positive traits of character such as altruism, compassion and loyalty, which if practiced regularly in the proper shaping of our desires through reason and will, allow individuals to flourish" (p. 320). Based on their framework, WTAs can be assessed for their ethical treatment of nonhumans, moving from 'No Justice' at the low-end, through 'Shallow Justice' and 'Intermediate Justice', towards 'Deep Justice' at the high end. The deeper the justice, the greater the emphasis for nonhuman animals in moving from welfare to rights in the recognition of animal agency.

Baker and Winkler (2020) argue that the ability for many species to exist outside of captivity is severely limited by human power dynamics, social carrying capacities, and habitat destruction. In the case of the wolf, natural habitat is ecologically abundant in the western United States as wolves are 'habitat generalists' able to exist in most landscapes, but are hindered by livestock ranching, natural resources extraction, and hunting activities (Bangs et al., 2005). Socio-political policy and discourse pose a 'real' threat of state-legitimised hunting, trapping, and culling programs, and illegal poaching endures with little or no negative consequences (Brasch, 2020). Coupled with the practical complexities of releasing captive, human-dependent wolves into 'wilderness', captive wolf ecotourism WTAs present a morally ambiguous case study. A posthumanist conceptual framework should not argue for maintaining the status quo in wildlife ecotourism (e.g., commodification), but rather aim to improve the quality of an animal's life, and protect them from existential harm in our efforts to be ethical and just.

Conceptualising a posthumanist approach to wolf ecotourism

There are well-established animal welfare models that evaluate an individual's welfare based on a combination of 'physical or functional domains', and 'affective experience domains' such as Mellor and Beausoleil's (2015) 'Five Domains model for animal welfare assessment'. Similarly,

Fraser's 'Practical Ethics for Animals' (2012) examines four principle criteria of human-animal welfare relations: keeping animals, causing intentional harm to animals, affecting animals in direct but unintended ways, and affecting animals indirectly by distributing life-sustaining processes and balances of nature (Fraser & MacRae, 2011). However, a posthumanist perspective extends existing models by considering the opportunity cost of what an individual's existence could be under specific temporal and spatial contexts. Wolf presents an even more nuanced analysis for the aforementioned sociocultural, socio-political, and socio-environmental conditions they face in the United States during the Trump administration. Like elephants (*Loxodonta*), the wolf is a charismatic species that polarises humans' perceptions. Fennell's (2013b) chapter, 'Contesting the zoo as a setting for ecotourism, and the design of a first principle', provides a convincing outlook as to "what might constitute the ethical use of animals in an ecotourism setting". He departs from utilitarianism to present his 'first principle of ecotourism' that "corresponds to the deontological school of ethics, which reasons that what is morally right is that which abides by rules, guidelines, duties or principles independent of the consequences of actions or inactions" (p. 10). Fennell posits that we should:

> Reject as ecotourism all practices that are based on or support animal capture and confinement, or other forms of animal use that cause suffering, for human pleasure and entertainment. Embrace as ecotourism interactions that place the interests of animals over the interests of humans. This would include encounters with free-living animals that would have the liberty to engage or terminate interactions independent of human influence. (p. 10)

On this account, Fennell declares that zoos are not ecotourism, because they severely limit the ability of animals to participate in normal behaviour, and in so doing, deny nonhuman animal rights and agency. Some of the enclosures for wolves in this study may appear similar to zoos in terms of size and setting, but small differences between the WTAs either reinforce or violate Fennell's *first principle of ecotourism*. Four key criteria should be considered to develop a posthumanist conceptual framework for wildlife ecotourism (see Figure 9.1). These include: 1) Quality of life of the animal prior to living in the WTA; 2) animal welfare conditions at the WTA that includes the physical, mental, and emotional health and wellbeing; 3) agency of animal to determine what to do and when based on rights granted; and 4.) right to exist free of exogenous threats. These criteria are not presented in a linear progression, but rather as equally weighted elements that must be considered to evaluate the ethics of a specific WTA's operations. This posthumanist conceptual approach departs from Fennell's *first principle of ecotourism* to consider whether or not a WTA's model is ecotourism by focusing not on what the species' rights and welfare should be in a non-captive setting, but by comparing the quality of life that the individual animal experienced prior to, and at the WTA. Following the figure, I ethnographically describe each wolf WTA's model and then assess it compared to the posthumanist conceptual framework.

1. Quality of life of the animal prior to living in the WTA	2. Animal welfare conditions at the WTA that includes the physical, mental, and emotional health and well-being	3. Agency of animal to determine what to do and when, based on rights granted	4. Right to exist free of exogenous threats

Figure 9.1 Posthumanist conceptual framework for wildlife ecotourism

Posthumanist analysist of wolf WTAs

Wolf 'Sanctuary' #1: No contact, wolves paired in separate enclosures

The rain finally broke as we pulled into the sanctuary's parking lot. As my research assistant and partner, Jenn, and I stepped out of the truck, the hair on the back of my neck stood up and a grin spread across her face. About a half-a-kilometer away wolves were howling. It had been three years since we saw a wolf in person, and I was eager to observe wolves as part of an actual academic study. Five minutes later, the executive director (ED) was giving us a private tour through the sanctuary, providing a mix of standard tour rhetoric with her own insights as a biologist regarding the welfare of the wolves in her care, as well as the conflicts that plagued their 'wild cousins', as she phrased it. Tourists are not guaranteed a wolf sighting, humans do not interact with wolves, and even the sanctuary workers have extremely limited contact. About 50 individual wolves are in residence, and once an individual enters the sanctuary its physical, emotional, and mental wellbeing, such as 'enrichment' activities (mental and physical games and stimuli), are provided for the rest of their life. The sanctuary changed its original model from breeding gray wolves decades ago to only taking in 'at risk' individuals in need of care. The sanctuary is one of 170 sites nationwide that engages in captive-breeding programs of either red (*Canis rufus*) or Mexican wolves (*Canis lupus baileyi*) (U.S. Fish and Wildlife Services, 2020). These 'species survival programs' (SSP) are partnerships between the United States Fish and Wildlife Service (USFWS), the Association of Zoos and Aquariums (AZA), and various sanctuaries and zoos nationwide. In these SSPs, wolf pups are selected for cross-fostering re-introduction at the discretion of USFWS to ensure genetic variability. This is explained in great detail on tours at the sanctuary, as wolf education is emphasised, and visitors learn about current wolf affairs in the United States (Figure 9.2).

After the ED opened the gate, we approached the first enclosure cautiously as the 13-year-old male had recently lost his partner. The ED's eyes teared-up as she described how he was still in mourning, and that he regularly 'set off the choir' by initiating constant howls of sorrow. As we slowly circled the sanctuary, we learned innate details about each wolf, their background, and why they were paired with each other. Every wolf or wolf-dog hybrid was born in captivity, and wolves were typically brought to the sanctuary once they reached full maturity, and when humans realised that 'the cute pups aren't domesticated dogs and demand different needs than a companion dog would'. In one instance, the wolf had lived for over a year-and-a-half in a two-bedroom apartment without even going outside—ever.

Enclosures, though modest compared to non-captive environments, range between 1/3 and 3 acres, and are a massive quality of life improvement for the wolves in their care. The ED stressed that if mature, adult wolves were released into the wild they would not survive long due to a lack of survival skills and dependence on humans. The ED described that with increased monetary or in-kind donations they could further improve the size and quality of the enclosures. The ED strongly objected to any type of wolf-human physical contact, and the focus of the sanctuary was to stay current on academic literature, avoid wolf-human contact, and use the best available science to foster wolf welfare. In this case, wolves' rights, agency, and welfare were foregrounded, and since they are guaranteed lifetime care once they enter the sanctuary, their quality of life seemed to be drastically improved. Each individual always has the ability to ignore humans, and though conditions could always improve, most posthumanists might agree that these conditions were beyond adequate given the previous abuse these individual wolves endured. By applying the posthumanist conceptual framework to Fennell and

Figure 9.2 Wolf in enclosure at wolf sanctuary #1

Photo credit: Jennifer Thomsen

Sheppard's (2021) scales of justice framework, wolves at this WTA experience 'deep justice' for the degree to which their quality of life improved.

Wolf 'Sanctuary' #2: Wolves paired in separate enclosures, wolf-human contact encouraged

Nine months later, we headed to a Colorado-based sanctuary right before the Proposition #114 vote to learn how this ballot initiative to reintroduce the gray wolf to the state may influence discourse on-site (Blevins, 2020). Before Jenn and I conducted interviews with four volunteers and employees we paid to take the standard hour-long tour. Like the first sanctuary, none of the wolves were purposely born in the sanctuary except for the Mexican wolves as part of the SSP. What stood out in this regard was the extra level of security surrounding the Mexican wolves, as all wolves in the SSPs are 'property' of the USFWS. At wolf sanctuary #1, there were few obvious differences between the enclosures, whereas at wolf sanctuary #2 the Mexican wolves had extra fencing and were not allowed to engage with tourists. The first two models were similar in many regards, including the enhanced quality of life at the sanctuary compared to the individual's previous captive conditions (Figure 9.3).

Wolves generally had the right to ignore tourists, as guides attempted to attract them with treats but quickly moved to the next enclosure if the wolf ignored them. However, this sanctuary diverted from the first for its overt commercialisation of wolves. The educational components of the tour were scientifically accurate, but were tempered with guides trying to

Figure 9.3 Fencing differences at wolf sanctuary: gray wolf (left), Mexican wolf (right)
Photo Credit: Jennifer Thomsen

'up-sell' the ability to pet a wolf and take a selfie for a charge 10–20 times the price of admission, depending on the package purchased. In this way, wolves were commodified beyond their educational contribution (Burns, 2017; Belicia & Islam, 2018; Cohen & Fennell, 2019), redolent of any other zoo experience that violates Fennell's 'first principle of ecotourism' (2013b). From a posthumanist lens, this sanctuary rates highly for wolf's improved captivity conditions and guaranteed lifetime care. However, the treatment of wolves as capitalistic prop is counterproductive and problematises the idea of what a sanctuary should be. Wolf's commodification was anthropocentric, and the economic focus promulgated sentiments of deception and increased opacity concerning the individual wolves' rights, agency, and welfare. For these reasons, this sanctuary rates rather low at 'shallow justice'.

Wolf 'Sanctuary' #3: No wolf-human contact, wolves in packs

Before embarking on the wolf tour, we gathered with other visitors in an educational room that was filled with posters, books, pictures, and skulls. As I watched the kids in the room, I felt a sense of nostalgia-like comradery with them, as if we had come here together on a primary school class field trip. The entire experience centred on education, and we spent 20 minutes learning about wolf as a species including their diet, mating, and behavioural norms. As we transitioned outside, six-metre-tall fences and an unkindness of ravens (*Corvus corax*) dominated the scenery. We learned that this was natural behaviour for ravens, as they not only tried to steal pieces of flesh from wolves' meals at the sanctuary, but this behaviour was commonplace in non-captive settings as well (Stahler, Heinrich, & Smith, 2002). Only four gray wolves resided here, and were rescued from other captive environments.

The primary focus of this organisation was the Mexican wolf and it was also a member of the SSP. There were two packs of Mexican wolves, and no human interaction was permitted except for occasional contact for staff veterinarians. The wolves were allowed to live together as they were members of a pack, though packs were split from each other. The tour guide informed us that aside from the wolf pups involved in the SSP's cross-fostering program, no wolves would ever purposely be reintroduced to non-captive settings. They were strict in limiting human contact so that if something catastrophic occurred such as wildfire, which was a constant threat, then wolves would be better equipped to survive on their own. The sanctuary also had metal fire-proof boxes that would protect the wolves in case of wildfire, and on a couple of occasions, they had to place the wolves in them as wildfires came over the ridge before being fought off by local fire crews.

This sanctuary had a small gift shop and engaged in traditional nonprofit fundraising activities to support operational costs including fundraisers and sponsorships. However, unlike wolf sanctuary #2, they did not commodify the wolves for financial gain, and all calls for funding were aimed at humans. Though captive breeding occurs for the SSPs, the wolves were able to exercise some agency and live with family members. Those who came from previous captive situations had their quality of life improved, and wolves were guaranteed life-long care and protections from natural disasters such as wildfire to the best of the sanctuary's ability. Short of a fully non-captive existence, this sanctuary WTA model ideally promotes a posthumanist wolf ecotourism paradigm and qualifies as 'deep justice'.

Wolf 'Sanctuary' #4: Wolf pack, wolf-human contact encouraged for education

Out of the five wolf WTAs we visited, this organisation produced the most challenging ethical dilemma from a posthumanist perspective. The two breeding wolves were rescued from similar captive situations as the wolves in sanctuaries #1 and #2. However, the wolves were purposely bred to create a pack. Initial criticisms of captive breeding could be made, but counter-arguments of a nonhuman's right to reproduce also muddy the context (Wickins-Dražilová, 2006). The wolves were empowered to fully practise their agency in the captive context. The co-founders of the sanctuary co-existed in a quasi-domesticated setting where the wolves lived like pets inside the home, and each wolf was provided its own physical space in the form of a custom-made den-like crate. The wolves were able to go outside into a 10-acre fully fenced area at any time, and regularly practised natural behaviours such as digging dens. This is possible because all of the wolves are from the same pack, and their welfare, rights, and agency were always foregrounded.

Both directors worked externally to offset the costs of the wolves' care, and it would be difficult for anyone to question their dedication and compassion to the wolves. They were politically active fighting for the species' rights and welfare in the United States, and regularly promoted wolf and wildlife education in local schools. The directors were hard-pressed to take even a week's vacation a year, and spend nearly $20,000 per month for their food, medical, and insurance requirements. Jenn, a doctor of physical therapy and certified canine rehabilitation practitioner, performed physical exams on two of the wolves during our most recent visit. We visited this pack on multiple occasions before and during the study. As the directors give private educational seminars during the 'wolf experience', tourists could engage in wolf-selfies. The wolves can come and go as they please, but only inside the captive environment. Though wildlife-selfies are generally condemned for good reason (Moorhouse et al., 2017), and the captive breeding of gray wolves is questionable and should not be encouraged, this organisation seems to be one of the rare exceptions where a posthumanist may condone these acts considering how well the wolves are treated 'behind the scenes'. This ethical context was by far the most complex from a posthumanist perspective. It elucidated the challenges of trying to develop an overarching theory concerning wolves' rights, agency, and welfare across WTAs, and further research into the ethical dilemmas of captive wildlife is needed from a posthumanist lens. The complexities of this specific case are rife in ambiguity, resulting in a rating of high 'shallow justice' to low 'intermediate justice'.

Wolf 'Sanctuary' #5: Wolves in protected areas (Yellowstone National Park)

Yellowstone National Park (YNP) is arguably the most famous protected area for wolves in the world. It is the epicenter of wolf reintroduction in the western United States, and has contributed to billions in increased revenue over the past 25 years (Middleton et al., 2020). "Park visitors spent more per person to the see wolves (around US$160 per day), compared to elk hunters who are a primary voice of opposition to the wolves' reintroduction (at about $39 per day)" (Thomsen, Muurlink, & Best, 2018, p. 203). In December, we spent two 8-hour days with a private ecotourism operator as it is one of the best times of year to observe wildlife. We stayed on the Montana side of the North Park Entrance, and had to leave the hotel at 6:45 a.m. to be able to spend as much time in the park before sunrise when wolves are most active. Our guide had been operating his business for seven years and had developed strong connections with 'regulars' at the park that included the winter wolf research team and nearby locals who come to 'glass' wolves through telescopes as a hobby. Wolf sightings weren't guaranteed, and it is extremely rare to see one with the 'naked eye', though other species such as bison (*Bison bison*) and elk (*Cervus canadensis*) are commonly found roadside. What transpired over the two days was awe-inspiring as we glassed wolves chasing herds of elk, bison, and big horned sheep (*Ovis canadensis*), bedding (lying) down and howling, yearlings playing, and consuming an elk carcass as ravens and magpies (*Pica pica*) competed for scraps (Figure 9.4).

The YNP experience was a microcosm of the omnipresent wildlife-human conflict and coexistence issues in the western United States. On the one hand, wolves were living a natural life free from captivity—in most regards. For scientific purposes, wolves are routinely tranquilised via helicopters where biologists take blood samples and collar wolves with bulky GPS monitors around their necks. From a posthumanist perspective, this is ethically ambiguous as scientific data is used to inform policy, but simultaneously infringes on an individual wolf's agency and welfare. Wolves enjoy strong protections inside the park's boundaries including laws that limit humans coming closer than 100 yards or engaging in any activity that may alter their natural behaviour, though this is arguably hypocritical given the air-induced tranquilisations that occur. This percolates questions concerning how environmental policy is socially constructed and special-interest human groups legitimise what is and isn't acceptable human behaviour toward wildlife. We witnessed two park rangers investigate and then fine an amateur wildlife photographer who hiked up the side of the mountain to photograph wolves consuming the elk carcass, but was clearly visible through scopes to more than 20 people watching the activity.

Human–wolf conflict in the surrounding states of Montana, Wyoming, and Idaho is prevalent. In Wyoming, humans can legally kill a wolf without a permit in 85% of the state as long as they report the 'take' within ten days. As these borders are socially constructed by humans and invisible to wolves, exogenous threats to wolf safety is a real threat and demands us to

Figure 9.4 Glassing wolves at Yellowstone National Park

Photo Credit: Bastian Thomsen

question how ethical protected areas are if the surrounding areas are insecure with limited ecological corridors. Posthumanists would argue that protected area WTAs are just as ethically complex as captive WTAs due to the compounding threats and lack of wildlife laws and policies that equitably represent nonhumans' welfare, rights, and agency outside of them, at least in the U.S. context. The coupling of helicopter-based tranquilising and exogenous threats reduce the ranking of this WTA from 'deep justice' down to 'intermediate justice' at best.

Pathways toward a posthumanist future in wildlife ecotourism

Cohen's (2019) 'Posthumanism in Tourism' article was influential as it was one the first to apply posthumanism to the tourism literature. Though Cohen theoretically supported humanists' objections to posthumanism as a valid theory divorced from humanism, he supported its application to tourism in practice. However, humanists' can launch these claims because they represent utilitarian human-dominance perspectives who have maintained power since the Enlightenment. Posthumanism blurs these lines and in the context of subaltern wildlife, a posthumanist conceptual framework provides wildlife ecotourism theorists and practitioners an approach to shed binary humanist arguments and consider wildlife rights, agency, and welfare at the individual level. Chrulew (2021) argues that "wildlife conservation should [...] be seen as an art and a science of ontological ethopolitics: an anthropogenic apparatus tasked with maintaining animals in their very being, a cosmo-political and -ecological experiment in coexistence".

This chapter responded to Fennell and Sheppard's (2021) call for further research into the ethics of 'animals-in-tourism research', and Cohen's (2019) call to balance nonhumans' welfare in theory and practice, in an attempt to untangle complex ethical situations at five wolf WTAs in the United States. Since all captive wolves had never experienced a non-captive existence, a post-humanist approach demanded that the WTAs be evaluated for their treatment of wolves at the individual level, rather than what the species should experience in a non-captive environment. Fennell and Sheppard's *scales of justice* framework provided an optimal model to compare the posthumanist conceptual framework for its focus on emphasising nonhuman welfare, rights, and agency. Due to the captive focus of the first four WTAs, and the omnipresent exogenous threats that linger outside of YNP, not a single model in this study could be championed as a 'utopian model' for wildlife ecotourism. However, all five WTAs could be considered ecotourism for at least providing shallow justice, as the YNP wolves experienced non-captive lives and the wolf sanctuaries improved the quality of life for individual captive wolves.

Captive wildlife sanctuaries should be considered a 'stop-gap' effort to remedy poor welfare conditions of individual nonhumans. The magnitude of the ecological crisis demands that we no longer tolerate the status quo of treating nonhumans as subaltern. Wildlife eco-tourism researchers and practitioners should advocate for three key goals to promote equitable wildlife-human coexistence: 1) facilitate deep justice conditions for all nonhumans at all times; 2) promote non-captive conditions whenever possible for nonhumans where their welfare, rights, and agency are equitably considered in any decision that affects them; and 3) lobby for nonhumans to be granted personhood and even citizenship. Gombay (2015) describes the meaning of personhood where, "as a person, one is a holder of rights and responsibilities that form the basis of citizenship" (p. 13). Posthumanists must continue to deconstruct postcolonial power structures at WTAs and in wildlife policy to overcome the horrific practices that reinforce abhorrent welfare conditions of subaltern nonhumans, if they are to ever consistently experience deep justice, let alone personhood or citizenship.

References

Badley, G. F. (2017). Manifold creatures: A response to the posthumanist challenge. *Qualitative Inquiry*, *24*(6), 421–432.

Baker, L., & Winkler, R. (2020). Asian elephant rescue, rehabilitation and rewilding. *Animal Sentience*, *5*(28), 1.

Bangs, E. E., Fontaine, J. A., Jimenez, M. D., Meier, T. J., Bradley, E. H., Niemeyer, C. C.,... & Oakleaf, J. K. (2005). Managing wolf-human conflict in the northwestern United States. *Conservation Biology Series-Cambridge*, *9*, 340.

Belicia, T. X. Y., & Islam, M. S. (2018). Towards a decommodified wildlife tourism: Why market environmentalism is not enough for conservation. *Societies*, *8*(3), 59.

Blevins, J. (2020). Colorado Wildlife Officials Are Reluctant to OK Gray Wolf Reintroduction. So Advocates Want Voters to Do it. *Colorado Sun*. Retrieved from https://coloradosun.com/2019/04/25/colorado-gray-wolf-reintroduction-ballot-proposal/ (accessed on 01.11.2020).

Brasch, S. (2020). Three of Colorado's 'Pioneer Wolves' May Have Been Killed in Wyoming. *Colorado Public Radio*. Retrieved from https://www.cpr.org/2020/09/09/colorado-wolves-may-have-been-killed-in-wyoming/ (accessed on 02.11.2020).

Burns, G. L. (2017). Ethics and responsibility in wildlife tourism: lessons from compassionate conservation in the anthropocene. InIsmar Borges de Lima and R. Green (Eds.) *Wildlife tourism, environmental learning and ethical encounters* (pp. 213–220). Cham: Springer.

Carroll, C., Rohlf, D. J., vonHoldt, B. M., Treves, A., & Hendricks, S. A. (2021). Wolf delisting challenges demonstrate need for an improved framework for conserving intraspecific variation under the Endangered Species Act. *BioScience*, *71*(1), 73–84.

Chagani, F. (2014), Critical political ecology and the seductions of posthumanism. *Journal of Political Ecology*, *21*, 424–436.

Chrulew, M. (2021). The Ontological Ethopolitics of Conservation. *Theorizing the Contemporary, Fieldsights*. Retrieved from https://culanth.org/fieldsights/the-ontological-ethopolitics-of-conservation.

Clayton, L. W. (2003). *Identity and the Natural Environment: The Psychological Significance of Nature*. Cambridge, MA: MIT Press.

Cohen, E. (2009). The wild and the humanized: Animals in Thai tourism. *Anatolia*, *20*(1), 100–118. doi:10.1080/13032917.2009.10518898.

Cohen, E. (2019a). Posthumanism and tourism. *Tourism Review*, *74*(3), 416–427. doi:https://doi.org/10.1108/TR-06-2018-0089.

Cohen, E. (2019b). Crocodile tourism: The emasculation of ferocity. *Tourism Culture & Communication*, *19*(2), 83–102.

Cohen, E., & Fennell, D. (2019). Plants and tourism: Not seeing the forest [n] or the trees. *Tourist Studies*, *19*(4), 585–606.

D'Cruze, N., Machado, F. C., Matthews, N., Balaskas, M., Carder, G., Richardson, V., & Vieto, R. (2017). A review of wildlife ecotourism in Manaus, Brazil. *Nature Conservation*, *22*, 1.

De Waal, F. (2016). *Are We Smart Enough to Know How Smart Animals Are?* New York, NY: WW Norton & Company.

Duffy, R., & Moore, L. (2010). Neoliberalising nature? Elephant-back tourism in Thailand and Botswana. *Antipode*, *42*(3), 742–766. doi:https://doi.org/10.1111/j.1467-8330.2010.00771.x.

Fennell, D. A. (2008). Tourism ethics needs more than a surface approach. *Tourism Recreation Research*, *33*(2), 223–224.

Fennell, D. A. (2013a). Tourism and animal welfare. *Tourism Recreation Research*, *38*(3), 325–340.

Fennell, D. A. (2013b). Contesting the zoo as a setting for ecotourism, and the design of a first principle. *Journal of Ecotourism*, *12*(1), 1–14.

Fennell, D. A., & Sheppard, V. (2021). Tourism, animals and the scales of justice. *Journal of Sustainable Tourism*, *29*(2–3), 314–335.

Foreyt, William J., Drew, M. L., Atkinson, M., & McCauley, D. (2009). Echinococcus granulosus in gray wolves and ungulates in Idaho and Montana, USA. *Journal of Wildlife Diseases*, *45*(4), 1208–1212.

Fraser, D. (2012). A "practical" ethic for animals. *Journal of Agricultural and Environmental Ethics*, *25*(5), 721–746.

Fraser, D., & MacRae, A. M. (2011). Four types of activities that affect animals: Implications for animal welfare science and animal ethics philosophy. *Animal Welfare*, 20(4), 581–590.

Gombay, N. (2015). There are mentalities that need changing: Constructing personhood, formulating citizenship, and performing subjectivities on a settler colonial frontier. *Political Geography*, *48*, 11–23.

Grooten, M., & Almond, R. E. A. (2018). Living Planet Report-2018: Aiming Higher.https://www.wwf.org.uk/sites/default/files/2018-10/wwfintl_livingplanet_full.pdf

Hayward, M. W., Somers, M. J., Kerley, G. I., Perrin, M. R., Bester, M. N., Dalerum, F.,... & Owen-Smith, N. (2012). Animal ethics and ecotourism. *African Journal of Wildlife Research*, *42*(2).

Idfwru, F., Wkh, W. R., Wdeoh, S. U. R., & Xvlqj, W. (2013). Slow lorises as photo props in Thailand. *TRAFFIC Bulletin*, *27*(1).

Jürgens, U. M., & Hackett, P. M. (2017). The big bad wolf: The formation of a tereotype. *Ecopsychology*, *9*(1), 33–43.

Juvan, E., & Dolnicar, S. (2014). The attitude—Behavior gap in sustainable tourism. *Annals of Tourism Research*, *48*, 76–95. doi:10.1016/j.annals.2014.05.012.

Karanth, K. K., DeFries, R., Srivathsa, A., & Sankaraman, V. (2012). Wildlife tourists in India's emerging economy: Potential for a conservation constituency? *Oryx*, *46*(3), 382–390.

Kline, C. (2018). Abstracting animals through tourism. In C. Kline (Ed.), *Tourism experiences and animal consumption: Contested values, morality and ethics* (pp. 209–217). London, UK: Routledge.

Lappalainen, K. (2019). Recall of the fairy-tale wolf: "Little Red Riding Hood" in the dialogic ension of contemporary wolf politics in the US West. *ISLE: Interdisciplinary Studies in Literature and Environment*, *26*(3), 744–767.

Mason, G. J. (2010). Species differences in responses to captivity: Stress, welfare, and the comparative method. *Trends in Ecology and Evolution*, *25*, 713–721.

Matović, S. (2020). Habitat loss. *Climate Action*, 565–573.

Mech, L. D. (2012). *Wolf*. Garden City, NY: Doubleday.

Mellor, D. J., & Beausoleil, N. J. (2015). Extending the 'Five Domains' model for animal welfare assessment to incorporate positive welfare states. *Animal Welfare, 24*(3), 241.
Middleton, A. D., Stoellinger, T., Karandikar, H., Leonard, B., Doremus, H., & Kremen, C. (2020). Harnessing visitors' enthusiasm for national parks to fund cooperative large-landscape conservation. *Conservation Science and Practice*, e335.
Mitchell, T. (2002). *Rule of Experts: Egypt, Techno-politics, Modernity*. CA: University of California Press.
Moorhouse, T. P., Dahlsjö, C. A., Baker, S. E., D'Cruze, N. C., & Macdonald, D. W. (2015). The customer isn't always right—conservation and animal welfare implications of the increasing demand for wildlife tourism. *PloS one, 10*(10), e0138939.
Moorhouse, T., D'Cruze, N. C., & Macdonald, D. W. (2017). Unethical use of wildlife in tourism: What's the problem, who is responsible, and what can be done? *Journal of Sustainable Tourism, 25*(4), 505–516. doi:https://doi.org/10.1080/09669582.2016.1223087.
Newsome, D. (2017). A brief consideration of the nature of wildlife tourism. In J. K. Fatima (Ed.) *Wilderness of Wildlife Tourism* (pp. 1–5). Oakville, ON, Canada: Apple Academic Press.
Peacher, A. (2019). State of IDAHO FUNDS Controversial Wolf Bounty Program. Retrieved from https://www.boisestatepublicradio.org/post/state-idaho-funds-controversial-wolf-bounty-program#stream/0 (accessed on 03.02.2021).
Peterson, M. N., & Nelson, M. P. (2017). Why the North American model of wildlife conservation is problematic for modern wildlife management. *Human Dimensions of Wildlife, 22*(1), 43–54.
Plank, T. (2020, September 18). Idaho Wolf Killings Up to 570 Over Past Year. Retrieved from https://www.idahopress.com/news/local/idaho-wolf-killings-up-to-570-over-past-year/article_889f36f4-d883-5a14-b4d7-c7c4d0b70c58.html (accessed on 02.02.2021).
Ripple, W. J., Estes, J. A., Beschta, R. L., Wilmers, C. C., Ritchie, E. G., Hebblewhite, M.,… & Wirsing, A. J. (2014). Status and ecological effects of the world's largest carnivores. *Science, 343*(6167).
Rott, N. (2020). Gray Wolves To Be Removed From Endangered Species List. *National Public Radio*. Retrieved from https://www.npr.org/2020/10/29/929095979/gray-wolves-to-be-removed-from-endangered-species-list (accessed on 02.11.2020).
Shani, A., & Pizam, A. (2008). Towards an ethical framework for animal-based attractions. *International Journal of Contemporary Hospitality Management, 20*(6).
Sheppard, V. A., & Fennell, D. A. (2019). Progress in tourism public sector policy: Toward an ethic for non-human animals. *Tourism Management, 73*, 134–142.
Spivak, G. C. (1988). Can the subaltern speak? In C. Nelson, & L. Grossberg (Eds.), *Marxism and the interpretation of culture* (pp. 271–313). Urbana: University of Illinois Press.
Stahler, D., Heinrich, B., & Smith, D. (2002). Common ravens, Corvus corax, preferentially associate with grey wolves, Canis lupus, as a foraging strategy in winter. *Animal Behaviour, 64*(2), 283–290.
Steffen, W., Persson, A., Deutsch, L., Zalasiewicz, J., Williams, M., Richardson, K.,… & Svedin, U. (2011). The anthropocene: From global change to planetary stewardship. *Ambio, 40*(7), 739–761. doi:10.1007/s13280-011-0185-x.
Soper, K. (2012). The humanism in posthumanism. *Comparative Critical Studies, 9*(3), 365–375.
Thomsen, B., Muurlink, O., & Best, T. (2018). The political ecology of university-based social entrepreneurship ecosystems. *Journal of Enterprising Communities: People and Places in the Global Economy, 12*(2).
Thomsen, B., & Thomsen, J. (2020) Multispecies livelihoods: Partnering for sustainable development and biodiversity conservation. In: W. Leal Filho, A. M. Azul, L. Brandli, A. Lange Salvia, & T. Wall (Eds.), *Partnerships for the goals. Encyclopedia of the UN Sustainable Development Goals*. Cham: Springer. https://doi.org/10.1007/978-3-319-71067-9_99-1.
Thomsen, B., Thomsen, J., Copeland, K., Coose, S., Arnold, E., Bryan, H. Prokop,… & Chalich, G. (2021a). *Multispecies Livelihoods: A Posthumanist Approach to Wildlife Ecotourism That Promotes Animal Welfare*. Working paper, Oxford, UK.
Thomsen, B., Thomsen, J., Cipollone, M., & Coose, S. (2021b). Let's save the bear: A multispecies livelihoods approach to wildlife conservation and achieving the SDGs. *Journal of the International Council for Small Business, 2*(2).
Turnbull, O. H., & Bär, A. (2020). Animal minds: The case for emotion, based on neuroscience. *Neuropsychoanalysis*, 1–20.
U.S. Fish and Wildlife Service. (2019). History of Decline, Protection and Recovery. Retrieved from https://www.fws.gov/midwest/wolf/history/2011FinalDelisting/PostDelistFWSRole.html (accessed on 07.05.2019).

U.S. Fish and Wildlife Services. (2020, June). Southwest Region. Retrieved from https://www.fws.gov/southwest/es/mexicanwolf/captivemanage.html (accessed on 02.02.2021).

van der Werff, E., Steg, L., & Keizer, K. (2013). The value of environmental self-identity: The relationship between biospheric values, environmental self-identity and environmental preferences, intentions and behaviour. *Journal of Environmental Psychology*, *34*, 55–63.

von Essen, E., Lindsjö, J., & Berg, C. (2020). Instagranimal: Animal welfare and animal ethics challenges of animal-based tourism. *Animals*, *10*(10), 1830.

Wickins-Dražilová, D. (2006). Zoo animal welfare. *Journal of Agricultural and Environmental Ethics*, *19*(1), 27–36.

Wilson, M. A. (1997). The wolf in Yellowstone: Science, symbol, or politics? Deconstructing the conflict between environmentalism and wise use. *Society & Natural Resources*, *10*(5): 453–468.

10
INDIGENOUS ECOTOURISM IN CANADA

Sonya Graci

Introduction

Ecotourism is a subcomponent of sustainable tourism that places emphasis on environmental, social, and economic sustainability. Ceballos-Lascurain (1996) articulately coined the term "ecotourism" as: "traveling to relatively undisturbed or uncontaminated natural areas with the specific objectives of studying, admiring, and enjoying the scenery and its wild plants and animals, as well as any existing cultural manifestations (both past and present) found in these areas (Ceballos-Lascurain, 1996, pp. 4–5). Sirakaya, Sasidharan, and Sönmez (1999) reviewed the literature on ecotourism by surveying 282 U.S.-based eco-tour operators. The plethora of perspectives provided in this study converge to state that ecotourism is constructed around the sustained conservation of resources in an almost non-consumptive manner involving non-intrusive exploitation of resources through the management of cultural and environmental resources (Sirakaya et al., 1999). Ecotourism is also considered "responsible travel to natural areas that conserves the environment, sustains the well-being of the local people and creates knowledge and understanding through interpretation and education of all involved (visitors, staff and the visited)" (Global Ecotourism Network, 2016, p. 1). Ecotourism has undergone exponential growth and is being considered as one of the fastest-growing sectors in the tourism industry (Bostick, 2020).

Ecotourism fits well with Indigenous values. Indigenous lands are rich in biodiversity and have existing cultural manifestations which makes these landscapes a natural setting for ecotourism activities (Johnston, 2000). The general elements that are integral to Indigenous tourism include a commitment to environmental sustainability, environmental education, and the sharing and promotion of the Indigenous host community's culture (Graci, 2010). Therefore, Indigenous values that focus on honouring nature and culture, are in line with the concept of ecotourism.

Several arguments suggest that Indigenous knowledge and biodiversity conservation are complementary phenomena leading to ecotourism development. Maintaining and strengthening a distinctive spiritual relationship with their traditional land by utilising their 'traditional ecological knowledge', is a foundational element for Indigenous communities (Davis, 2008). Ecological knowledge is an accumulation of traditional knowledge, belief and practices, evolving by adaptive processes, culturally transmitted to next generations, about the human

 DOI: 10.4324/9781003001768-10

relationship with the environment (Menzies, 2006). Second, Indigenous knowledge reflects generations of experience and problem-solving, providing an important opportunity for systematic in-situ maintenance of genetic resources in a destination, which is well aligned with the expected outcomes of ecotourism (Warren, 1996). Finally, Indigenous tourism is a drawcard for ecotourism as it's mainly concentrated in the peripheral regions with a high degree of Indigenous biodiversity (Hall, 2007). The limited accessibility, number of tourists, and the distance from physical human structures often create a higher perceived value of naturalness and preserves the ecological integrity of the destination (Hall, 2007). All of these factors reiterate the commonalities between the concept of ecotourism and Indigenous cultural and natural attractions.

Definition of Indigenous ecotourism

Indigenous tourism in Canada is defined as an accumulation of all tourism businesses "majority owned, operated and/or controlled by First Nations, Metis or Inuit peoples that can demonstrate a connection and responsibility to the local Indigenous community and traditional territory where the operation resides" (ITAC, 2017, p. 4). Butler and Hinch (2007) have characterised the 'Indigenous tourism product' by examining the role of Indigenous societies in tourism and how they interact within the tourism framework. This 'Indigenous tourism product' includes activities in which "Indigenous peoples directly own or operate the tourism business or are indirectly involved by having their culture serve as the essence of the tourist attraction" (Butler & Hinch, 2007, p. 5). The Indigenous tourism product encapsulates a wide range of "special events (dances, festivals, powwows), experiential tourism (guided hikes, cultural-interpretation programs, wildlife tourism, applied activities), arts and crafts, museums, historical recreations, restaurants, and accommodations, lodges, and resorts that celebrate Aboriginal culture and are offered by or located in indigenous communities" (Gets & Jamieson, 1997 as cited in Lemelin, Koster, & Youroukos, 2015, p. 318).

In the context of Indigenous ecotourism, the tourism products include ethically managed non-consumptive, low-impact and locally oriented experiences that focus on natural and cultural attractions within Indigenous territories; managed and developed by Indigenous peoples (Fennell, 1999). In 2006, Zeppel defined Indigenous ecotourism as "Tourism which cares for the environment and which involves (Indigenous) people in decision making and management" (ANTA, 2001 as cited in Zeppel, 2006, p. 11). It includes nature-based tourism products or accommodation owned by Indigenous groups and Indigenous cultural tours or attractions in a natural setting (Zeppel, 2006, p. 11). This Indigenous ecotourism product usually refers to a variety of experiences including cultural performances such as singing and storytelling, nature-based activities such as canoeing, snowshoeing, kayaking, walking or sailing-based local tours, and incorporating Indigenous tourism experiences in the existing tourism businesses, "encouraging sustainable hunting, fishing, and harvesting; wildlife viewing; and lodge-based overnight stays" (Tides Canada, 2018, p. 9). According to Tides Canada (2018), Indigenous-led ecotourism can be defined as an initiative that is owned and operated by Indigenous peoples. The main purpose of Indigenous-led ecotourism is to educate the public about Indigenous culture, values, and ways of life while maximising community benefits. Considering such benefits, the definition of Indigenous ecotourism may be broadened to incorporate Indigenous cultural tourism (Tides Canada, 2018). Indigenous cultural tourism meets the Indigenous tourism criteria and, in addition, a significant portion of the experience incorporates Indigenous culture in a manner that is appropriate, respectful, and true to the

Indigenous culture being portrayed. Authenticity is ensured through the active involvement of Indigenous people in the development and delivery of the experience (ITAC, 2017).

Benefits and barriers of Indigenous ecotourism

There are several potential benefits as well as challenges of Indigenous ecotourism as observed in the Tides Canada (2018) report, noted previously. The report recognises 'flexibility' and 'attainability' as the two main characteristics of ecotourism. Based on these characteristics it concludes that there is a scope for building on existing infrastructures and capacity, as well as piloting initiatives that have proven to be successful elsewhere. This report highlights Indigenous-led ecotourism potential to create lasting and rewarding community impacts. According to the report, Indigenous communities can benefit from this type of tourism as it supports auxiliary businesses and employs youth. Community solidarity and pride strengthen when people share their culture, heritage, languages, and arts with others. As a result, ecotourism allows for communities to connect with their land, thereby reinforcing their commitment to environmental stewardship. Community leaders can benefit by paving an alternative path for economic development, steering away from natural resource extraction industries. Indigenous communities can also retain their youth and keep them engaged in ecotourism, who might otherwise leave in search of employment. Finally, when Indigenous communities integrate a sustainable relationship with their hereditary land into their personal lives, ecotourism has the capacity to provide healing, health, and well-being (Tides Canada, 2018).

In practice, however, there are many barriers hindering the implementation of ecotourism. Much of the literature analyzing the link between biodiversity conservation and community development assumes that ecotourism managed by Indigenous peoples will not only result in environmental conservation but also increased development. However, a study by Coria & Calfucura (2012) indicated that in practice, ecotourism has failed to deliver the anticipated benefits to Indigenous communities. Based on this study, the major factors limiting Indigenous communities from reaping the benefits of Indigenous ecotourism development include shortages of human endowments, lack of financial and social capital within the community, land insecurity and the absence of a proper mechanism for the fair distribution of economic benefits of ecotourism (Coria & Calfucura, 2012). A study conducted on the Indigenous community of Long Lamai in Malaysia established that the marketing of destination areas, formulation of tourism policies, and development of tourist attractions are mainly dictated by the needs of the visitors. Several scholars have argued that this process fails to take into account the local perspective, and often results in a lack of interest of the local Indigenous communities, dissatisfied residents and tourists, and a loss of the destination's culture (Liu, 2006; Falak, Chiun, & Wee, 2016).

The first Sustainable Indigenous Tourism Symposium held in Nanaimo, Canada, in 2017 facilitated talking circles with stakeholders and determined the three main barriers to Indigenous ecotourism development. The first barrier pertains to the concept of authenticity and the process of creating an authentic Indigenous tourism product. The delegates were concerned about their positive intentions being misinterpreted and feared portraying an attitude of appropriation and/or suggesting homogeneity between cultures (Graci, Maher, Peterson, Hardy, & Vaugeois, 2019). The second barrier emerged from discussions on reciprocity. The delegates believed that due to the history of Indigenous people in Canada, there is a sense of distrust between the respective parties, which limits engagement opportunities of the Indigenous communities in the decision-making process. Finally, the third barrier concerned the shortage of, or failure to allocate, tangible resources for Indigenous ecotourism development (Graci et al., 2019).

According to a study conducted by Graci (2010) on the potential for Indigenous ecotourism in Ontario, a major barrier identified was the competition with other tour operators within and between other provinces in Canada (Graci, 2010). Other barriers included insufficient funding and a lack of education and training within Indigenous communities for developing ecotourism products. In addition, the variations in seasonal travel lead to fluctuations in the utilisation capacity of the facilities as well as issues relating to staffing and employment. Lastly, the lack of proper marketing and support from the provincial and federal government can also make it difficult to attract tourists to Indigenous destinations (Graci, 2010). The barriers and benefits to Indigenous ecotourism development affect both the prevalence and success of this type of tourism.

The Larrakia Declaration and the N'autsamawt Declaration

The Declaration on the Rights of Indigenous peoples in 2007 raised several themes related to Indigenous tourism (United Nations, 2007). However, it was not until 2012 that the Pacific Asia Travel Association organised a global Indigenous tourism conference to formulate the guiding principles for Indigenous tourism development. These Indigenous tourism development principles are now being referred to as the Larrakia Declaration, named after the meeting's Australian Aboriginal host community (PATA/WINTA, 2014).

The following principles were listed in the Larrakia Declaration to guide all culturally respectful Indigenous tourism business development (World Indigenous Tourism Alliance, 2012, pp. 1–2):

- Respect for customary law and lore, land and water, traditional knowledge, traditional cultural expressions, cultural heritage that will underpin all tourism decisions.
- Indigenous culture, the land and waters on which it is based, will be protected and promoted through well-managed tourism practices and appropriate interpretation.
- Indigenous peoples will determine the extent, nature, and organisational arrangements for their participation in tourism and that governments and multilateral agencies will support the empowerment of Indigenous people.
- That governments have a duty to consult and accommodate Indigenous peoples before undertaking decisions on public policy and programs designed to foster the development of Indigenous tourism.
- The tourism industry will respect Indigenous intellectual property rights, cultures and traditional practices, the need for sustainable and equitable business partnerships and the proper care of the environment and communities that support them.
- That equitable partnerships between the tourism industry and Indigenous people will include the sharing of cultural awareness and skills development which support the well-being of communities and enable enhancement of individual livelihoods. (PATA/WINTA, 2014, p. 13)

A think tank at the Sustainable Indigenous Tourism Symposium in Nanaimo, Canada, in 2017 facilitated knowledge sharing for supporting "community empowerment, cultural expression and economic prosperity" (Graci et al., 2019, p. 1). The think tank formed the Naut'sa mawt Declaration, which is built upon the Larrakia declaration and presents the following principles to guide Indigenous tourism development:

- Recognising that often Indigenous people are marginalised, disadvantaged, and remote from the opportunity for social, economic and political advancement.
- Recognising that whilst tourism provides the strongest driver to restore, protect, and promote Indigenous cultures, it has the potential to diminish and destroy those cultures when improperly developed.
- Recognising that as the world becomes increasingly homogenous Indigenous cultures will become increasingly important to provide differentiation, authenticity, and the enrichment of visitor experiences.
- Recognising that for Indigenous tourism to be successful and sustainable, Indigenous tourism needs to be based on traditional knowledge, cultures, and practices and it must contribute to the wellbeing of Indigenous communities and the environment.
- Recognising that Indigenous tourism provides a strong vehicle for cultural understanding, social interaction, and peace.
- Recognising that universal Indigenous values underpin intergenerational stewardship of cultural resources and understanding, social interaction, and peace. (Graci et al., 2019, p. 4)

Additionally, with an aim to promote environmental and cultural tourism practice, the representatives at the Sustainable Indigenous Tourism Symposium (2017), formulated the following action principles:

- Recognising the need to acknowledge and understand both the community and ecological feasibility prior to embarking on tourism development.
- Recognising the need for cultural preservation in Indigenous tourism development and to work in collaboration to ensure the preservation of traditional knowledge and the environment.
- Recognising that Indigenous tourism development should be based on intrinsic authenticities and should protect what is sacred to the community.
- Recognising that Indigenous tourism development should be focused on reciprocity and that open communication and collaboration between Indigenous communities, non-indigenous communities, and tourism industry stakeholders is essential.
- Recognising the need to mentor and support Indigenous youth and future leaders to ensure that there is capacity, self-determination, and continuance in the values of the community in Indigenous tourism development.
- Recognising the important role of education in Indigenous tourism development including experiential hands-on learning opportunities, place-based learning, and exposure to best practices.
- Recognising that financial resources need to be identified, secured, and accessible in order to enable tourism development priorities of Indigenous communities and businesses. (Graci et al., 2019, p. 4)

These two declarations identify the importance of developing Indigenous ecotourism in line with these principles. Indigenous ecotourism should hold true to the principles identified in the Declarations and provide benefits to the local community while protecting the natural and cultural environment.

Elements of Indigenous ecotourism

In order for Indigenous ecotourism to be successful it must incorporate the following elements: be community focused, authentic, nature based, focus on partnerships, collaboration, and sustainability.

Community focused

Given the close ties between Indigenous peoples and nature, ecotourism widely depends on the 'sustainability' aspect in regard to tourism development (Nepal, 2004). Wachtel (1989) described sustainability as the building block for community development, that makes the people more aware of what they have and the long-term ripple effect of the short-term choices they make. Indigenous ecotourism explores the ways of putting sustainable concepts into practice, providing the best micro-solutions (Wheeler, 1991, p. 93).

Scheyvens (1999) argues that ecotourism ventures should only be deemed successful if local communities equitably share the benefits of the emerging ecotourism activities. The United Nations (2007) has also asserted the importance of community empowerment and participation, as well as the need for strong political leadership to ensure consensus building for sustainable Indigenous tourism development. Carr, Ruhanen, and Whitford (2016) explored an array of issues pertaining to sustainable Indigenous tourism and reinstated the capacity of tourism as a powerful tool for realising the potential of Indigenous community development. Carr et al. (2016) highlights the positive impacts of capacity building and the negative realities of commodification on Indigenous tourism development. The authors believe that this empowerment could successively help Indigenous communities gain global leadership within the tourism sector (Carr et al., 2016).

Authenticity

Indigenous ecotourism relies on delivering culturally authentic experiences as important contributors to Indigenous heritage. According to Paulauskaite, Powell, Coca-Stefaniak, and Morrison (2017) 'authenticity' is a core feature for experiencing and sharing economies through psychological and spiritual philosophies. Arguably, it is a prominent trend that has caused a 'tourist-to-traveller shift' by providing more opportunities to have meaningful interactions with locals (Paulauskaite et al., 2017). Indigenous cultural tourism ensures authenticity through active involvement of Indigenous peoples in the formulation and delivery of their travel experience (ITAC, 2016). In addition to this, Indigenous cultural tourism also incorporates an experience that is appropriate, respectful, and true to the Indigenous culture being portrayed (ITAC, 2016).

Nature based

Nature-based tourism is a core component of ecotourism that integrates many trends relative to a variety of outdoor recreational activities and adventure travel. Nature-based tourism can be defined as any form of tourism that mainly motivates travellers to observe and appreciate nature as well as the cultural manifestations in the area (UNWTO, 2002). According to the UNWTO (2002), the attributes of nature-based tourism can be broken down as follows:

- Interpretational and educational features;
- Most commonly but not exclusively, organised by specialised tour operators for smaller groups of travelers;
- Mitigating the negative implications of tourism on the socio-cultural and natural environment;
- Support the in-situ maintenance of the natural areas utilised as ecotourism attractions by:

- "Generating economic benefits for host communities, organizations and authorities managing natural areas with conservation purposes;
- Providing alternative employment and income opportunities for local communities;
- Increasing awareness towards the conservation of natural and cultural assets, both among locals and tourists". (UNWTO, 2002)

A growing body of literature has indicated that nature-based tourism possesses a strong potential to positively benefit the traveler's mental health and wellbeing. There are several studies that indicate a need for service providers and policymakers to collaborate and partner, in order to increase access to, and develop nature-based recreation and travel to reap more therapeutic benefits (Lackey et al., 2019).

Partnerships

A partnership refers to the coming together of many different stakeholders who will pool their resources, knowledge and expertise to produce tangible solutions to their shared problems (Graci, 2016; Adu-Ampong, 2017; Towner, 2018). Partnerships and collaboration can be a tool to help Indigenous communities and other traditionally disadvantaged groups achieve success in developing sustainable tourism in their communities (Carr et al., 2016). These benefits can include improved capacity for livelihood building and community engagement, feelings of accomplishment for the community, cultural pride, and recovery and entrepreneurial and creative confidence (Espeso-Molinero, Carlisle, & Pastor-Alfonso, 2016, p. 1333). Partnerships can also be valuable in Indigenous communities since forming partnerships may procure more resources to use for the development of tourism (Olsen, 2016). Governments and DMOs may be able to provide funding or resources to help Indigenous tourism businesses be established or develop. For example, partnering with a DMO to receive support in advertising may widen the reach of the tourism product or service which will generate more business. In addition, non-Indigenous partners will benefit from including Indigenous partners and gaining access to their perspectives and products which are increasingly in demand (ITAC, 2019).

The literature on Indigenous experience in partnerships advocate for more inclusive partnerships, appropriate and considerate representations of culture, "bottom-up" approaches to policy formulation and implementation, and increased benefits directly impacting Indigenous communities (Olsen, 2016; Espeso-Molinero et al., 2016). By making these partnership goals explicit, it is more likely that Indigenous communities will be able to benefit from an ecotourism partnership based on trust and respect rather than be exploited for their desirable tourism products and services.

Focus on sustainability

Indigenous ecotourism has a focus on sustainability from an environmental and resource perspective. Biodiversity and resource conservation are primary goals of Indigenous ecotourism in many cases in Canada. For example, many lodges such as the Spirit Bear Ecolodge in British Columbia and Cree Village Ecolodge in Ontario have been developed with sustainability goals in mind that focus on water, energy, biodiversity, waste, and supply chain. Low-impact forms of tourism consisting of animal and marine life viewing are conducted with minimal impact on the natural environment and promote education in regards to conservation and cultural practices. This can also lead to the development of conservation areas that can be managed by the Indigenous communities themselves.

Case studies

Given the close ties between Indigenous peoples and nature, ecotourism widely depends on the 'sustainability' aspect in regards to tourism development (Nepal, 2004). Sustainable development has become the buzzword in the tourism industry and particularly in tourism research. Indigenous ecotourism that puts sustainable concepts to practice provide the best micro-solutions (Wheeler, 1991). Best practices can help identify a potential tourism opportunity, the impact of tourism on communities and develop sustainable business models (Tides Canada, 2018). Two case studies will be discussed that exemplify best practices in Indigenous ecotourism: Tundra North Tours in Inuvik, Northwest Territories and Spirit Bear Ecolodge in Klemtu, British Columbia.

Tundra North Tours

Tundra North Tours (TNT) is an Indigenous-owned and -operated tour operator that provides authentic Indigenous tourism experiences in Inuvik, Northwest Territories. TNT was founded by Kylik Kisoun Taylor with a vision of preserving Indigenous culture, providing work opportunities for Indigenous peoples, and connecting resources to Indigenous lands. The owner of TNT believes that tourism is a gateway for non-Indigenous people to explore the Indigenous culture (K. Taylor, personal communication, 2018). Tundra North Tours conduct adventure tours with an educational focus. Their tourism is low impact and consists of immersing tourists in the way of life in the Arctic. Tourists learn to make Igloos, participate in traditional forms of food preparation, learn about Indigenous culture, and live on the land. Tundra North Tours is an example of an Indigenous ecotourism product owned by an Indigenous entrepreneur. The tourism product is focused on community wellbeing, capacity building, and sustainability.

Community based: Tundra North Tours has created an opportunity for Inuvialuit people to share their traditions through tourism. TNT created itineraries that maintain the safety and cultural integrity of the Indigenous community and provide a source of income for local residents seeking employment as a tour guide. Kylik Kisoun Taylor also believes that, "It is always easy to find a solution when it is an Indigenous owned company, because we are more community based and helps us more in a financial way. Sharing culture and promoting culture is the biggest benefit for Tundra north tours (K. Taylor, personal communication, 2018). Tundra North Tours inspires Inuvialuit youth and Indigenous youth across Canada to embrace their culture by showing that it is possible to be successful and make a living through sharing the Indigenous way of life. TNT instills a sense of pride and empowerment in knowing that people seek out and support Indigenous tourism.

Authenticity: Visitors experience authentic Indigenous cuisine, culture, and traditions as a result of the close-knit communities of the North. TNT and their guests are welcomed by the Inuvialuit community which allows for a local experience. TNT incorporates cultural experiences that are based off of the way of the land and the people residing on it. TNT has created blogs specifically describing the adventures many visitors can partake in. To ensure authenticity, participants are able to discover the history of the burial site of Albert Johnson known as the Mad Trapper of Rat River or taste different Indigenous delectable foods such as Muktuk often made from the skin and blubber of the beluga whale—a staple food of the Inuit diet as it contains healthy concentration of Vitamins C and D.

Nature based: TNT focuses on sustainability in their operations through having low impact tourism such as winter camping, compost toilets, and living off the grid. The tours include activities such as reindeer herding, winter camping, overnight camping in an Igloo, ice fishing, and other low-impact activities. There is a focus on having guests immerse themselves in the local culture and participating in traditional activities such as igloo building and traditional food preparation. Tourists live on the land at the camp and experience the way of life in the Arctic.

Partnership/Collaboration: TNT has partnered with many other Indigenous tourist companies to combine their tours and offer a larger variety of experiences. They have partnered with Indigenous Tourism Association of Canada, Spectacular Northwest Territories and World Indigenous Tourism Alliance (WITA) to combine tours and provide a wider product offering. Tundra North Tours has also partnered with universities to provide a transformative experience for students from Canada and internationally.

Focus on sustainability: Tundra North Tours ethos is focused on sustainability of both the culture and land. Tundra North Tours is incorporating food security into their current business plan with the development of a greenhouse and farm. The purpose of the farm is to grow fresh produce and herbs for the local community and to assist in the creation of a sustainable food system in the Arctic region. Tundra North Tours also practice sustainability initiatives at the camp with composting toilets, low impact accommodations and using locally sourced products. Tundra North Tours is an example of an Indigenous tourism business that is focused on capacity building, preservation of culture, and cross-cultural education.

Spirit Bear Lodge

Spirit Bear Lodge provides a unique wildlife experience with local Kitasoo guides (Destination Canada, 2019). Located in the remote coastal village of Klemtu, in the heart of the Great Bear Rainforest in British Columbia, this unique lodge takes visitors on an authentic cultural journey in search of the rare and beautiful Spirit Bear (ITAC, 2019). The Spirit Bear Lodge specialises in bear viewing and their experienced tour guides have an intimate knowledge of the behaviour and movement patterns of bears. In addition, Spirit Bear provides tours of culturally significant sites including the 'Big House' where visitors learn about the First Nations culture that has flourished in this region for thousands of years. In the 1990s, increasing media awareness of the Spirit Bear began to draw wildlife enthusiasts to this BC region. The ongoing interest in the bears prompted the local community to examine the viability of expanding and diversifying tourism and as a result, the Kitasoo Development Corporation (KDC) was created (ITAC, 2019).

Community-based: Spirit Bear Lodge is owned and operated by the community. The lodge is a way to increase economic development in the community as well as providing opportunities for capacity development and preservation of culture. Spirit Bear Lodge empowers youth through education and training. Through a local program, the lodge trains youth and provides local employment opportunities to high school graduates.

Authenticity: The Spirit Bear Lodge is owned and operated by the Kitasoo/Xai'xais First Nation. The exterior of the Spirit Bear Lodge pays homage to the traditional longhouses built by West Coast First Nations. The lodge is certified by the Aboriginal Tourism Association of BC as an Authentic Aboriginal tourism product. The certification is awarded to Aboriginal owned tourism programs that have made the commitment to excellence, quality, safety, hospitality, and cultural integrity (Coastal Funds, 2018).

Nature based: The Kitasoo developed environmental protocols, identified and protected areas, and implemented eco-based management philosophies into any extractive or non-extractive resource plans (Lemelin et al., 2015). With the help of conservation groups such as Valhalla Wilderness Society, Raincoast Conservation Fund, Greenpeace, and many others, the council and the people of Klemtu planned to protect the ecosystems and wildlife that make up one of the most biodiverse places on the planet (Lemelin et al., 2015). One of the programs currently in place is the Watchmen Program. Employing two full-time seasonal band members, the Watchmen Program is an initiative developed by BC's Coastal First Nations communities to control and participate in the stewardship and monitoring of their traditional lands and waters.

Partnership and collaboration: In 2010, the KDC invested over $1 million into the Spirit Bear Lodge to hire a general manager, expand the lodge by constructing an additional six rooms, purchase another vessel, and implement an international marketing plan (Coastal Funds, 2018). Spirit Bear Lodge also partners with conservation groups to ensure sustainable operations as well as several DMOs and tourism organisations to promote and market their business.

Focus on sustainability

Spirit Bear Lodge is now an integral part of the conservation economy in the Great Bear Rainforest (GBR). The Great Bear Rainforest is now recognised as a globally significant conservation model, and according to National Geographic, 'the wildest place in North America'. Spirit Bear Lodge is a showcase community tourism business in the Great Bear Rainforest, and a recognised best practice model for Indigenous community-based tourism in Canada as evidenced by their winning the Indigenous Adventure Award at the 2017 International Indigenous Tourism Conference hosted by the Indigenous Tourism Association of Canada. The award recognises best practice in Indigenous Adventure travel with a focus on responsibility and sustainability (Linking Tourism and Conservation, 2020).

These case studies exemplify that for Indigenous ecotourism to be successful it should incorporate community, authenticity, be nature based, focus on sustainability, and include partnerships and collaboration. These fall in line with the principles of ecotourism and Indigenous philosophies and values. Indigenous values that focus on honouring nature and culture, are in line with the concept of ecotourism.

Conclusion

Ecotourism is a natural complement to Indigenous tourism development. As ecotourism focuses on being nature based with a component of education, culture, and sustainability this is congruent with the values and beliefs of many Indigenous tourism operations in Canada. Indigenous ecotourism can lead to an increase in capacity building in a community, cultural connectedness, cross cultural relationships and education, conservation, an increase in biodiversity, and a strong economic recovery in many communities. In a world post-COVID-19, it is especially pertinent for Indigenous communities and entrepreneurs to develop tourism that is focused on the facets of ecotourism. The world needs global healing and Indigenous tourism that is authentic and nature based. Indigenous ecotourism is by nature small scale and sustainable. The benefits from Indigenous ecotourism are immense for both community, the planet, and the tourists. Indigenous ecotourism can be transformational, leading to a greater understanding of the need for environmental and cultural preservation and a greater understanding about the need for reconciliation. It also exemplifies the approach to tourism that is high quality and focused on yield rather then numbers. The case for Indigenous ecotourism is strong in terms of benefits and the principles of the Larrakia and Naut'sa mawt Declaration should be incorporated to continue to build a strong Indigenous tourism industry in Canada.

References

Adu-Ampong, E. (2017). Divided we stand: Institutional collaboration in tourism planning and development in the Central Region of Ghana. *Current Issues in Tourism*, *20*(3), 295–314.

Bostick, J. (2020, February 27). Ecotourism Summit Dives into FSU-PC. Retrieved from https://www.newsherald.com/news/20200227/ecotourism-summits-dives-into-fsu-pc.

Butler, R., & Hinch, T. (Eds.). (2007). *Tourism and Indigenous Peoples: Issues and Implications*. London: Routledge.

Carr, A., Ruhanen, L., & Whitford, M. (2016). Indigenous peoples and tourism: The challenges and opportunities for sustainable tourism. *Journal of Sustainable Tourism*, *24*(8–9), 1067–1079.

Ceballos-Lascurain, H. (1996). *Tourism and Protected Areas*. Gland, Switzerland: IUCNWorld Conservation Union.

Coastal Funds. (2018). The Success of Spirit Bear Lodge: How a Remote, Community-Led Business Became a Global Model for Ecotourism. Retrieved from https://coastfunds.ca/stories/the-success-of-spirit-bear-lodge/.

Coria, J., & Calfucura, E. (2012). Ecotourism and the development of indigenous communities: The good, the bad, and the ugly. *Ecological Economics*, *73*, 47–55.

Davis, M. (2008). Indigenous struggles in standard-setting: The United Nations Declaration on the rights of indigenous peoples. *Melbourne Journal of International Law*, *9*, 439.

Destination Canada. (2019). Canadian Signature Experiences Member List. Retrieved from https://www.destinationcanada.com/sites/default/files/archive/525-CanadianSignatureExperiencesMemberList/CSE_MemberList_EN_FULL-Nov2019.pdf.

Espeso-Molinero, P., Carlisle, S., & Pastor-Alfonso, M. J. (2016). Knowledge dialogue through indigenous tourism product design: A collaborative research process with the Lacandon of Chiapas, Mexico. *Journal of Sustainable Tourism*, *24*(8–9), 1331–1349.

Falak, S., Chiun, L. M., & Wee, A. Y. (2016). Sustainable rural tourism: An indigenous community perspective on positioning rural tourism. *Turizam: Međunarodni Znanstveno-Stručni Časopis*, *64*(3), 311–327.

Fennell, D. (1999) *Ecotourism: An Introduction*. London: Routledge.

Graci, S. (2010). The potential for aboriginal ecotourism in Ontario. in *Geography Research Forum*,*30*,33–148

Graci, S. (2016). Collaboration and partnership development for sustainable tourism. *Tourism Geographies*, *15*(1), 25–42.

Graci, S., Maher, P. T., Peterson, B., Hardy, A., & Vaugeois, N. (2019). Thoughts from the think tank: Lessons learned from the sustainable Indigenous tourism symposium. *Journal of Ecotourism*, 1–9.

Hall, C. M. (2007). Biosecurity and ecotourism. In J. Higham(Ed.), *Critical issues in ecotourism: Understanding a complex tourism phenomenon*(pp. 102–116).London: Routledge.

ITAC. (2016). National Guidelines: Developing Authentic Indigenous Experiences in Canada. Retrieved from https://indigenoustourism.ca/corporate/national-guidelines/.

ITAC. (2017) Indigenous Cultural Experiences Guide. Retrieved from https://indigenoustourism.ca/corporate/wp-content/uploads/2017/11/ITAC-Indigenous-Cultural-Experiences-Guide-web.pdf.

ITAC. (2019, May 17). Indigenous Cultural Experiences National Guidelines. Retrieved from https://indigenoustourism.ca/corporate/national-guidelines/.

Johnston, A. (2000). Indigenous peoples and ecotourism: bringing indigenous knowledge and rights into the sustainability equation. *Tourism Recreation Research*, *25*(2), 89–96.

Lackey, N. Q., Tysor, D. A., McNay, G. D., Joyner, L., Baker, K. H., & Hodge, C. (2019). Mental health benefits of nature-based recreation: a systematic review. *Annals of Leisure Research*, 1–15.

Linking Tourism and Conservation. (2020). Retrieved from https://www.ltandc.org.

Lemelin, R. H., Koster, R., & Youroukos, N. (2015). Tangible and intangible indicators of successful aboriginal tourism initiatives: A case study of two successful aboriginal tourism lodges in Northern Canada. *Tourism Management*, *47*, 318–328.

Liu, A. (2006). Tourism in rural areas: Kedah, Malaysia. *Tourism Management*, *27*(5), 878–889.

Menzies, C. R. (Ed.). (2006). *Traditional Ecological Knowledge and Natural Resource Management*. Lincoln, NE: University of Nebraska Press.

Nepal, S. K. (2004). Indigenous ecotourism in central British Columbia: The potential for building capacity in the Tl'azt'en Nations territories. *Journal of Ecotourism*, *3*(3), 173–194.

Notzke, C. (2004). Indigenous tourism development in southern Alberta Canada: Tentative engagement. *Journal of Sustainable Tourism, 12*(1), 78–91.

Olsen, L. S. (2016). Sami tourism in destination development: Conflict and collaboration. *PolarGeography, 39*(3), 179–195.

Paulauskaite, D., Powell, R., Coca-Stefaniak, J. A., & Morrison, A. M. (2017). Living like a local: Authentic tourism experiences and the sharing economy. *InternationalJournal of Tourism Research, 19*(6), 619–628.

Pacific Asia Travel Association (PATA) and World Indigenous Tourism Alliance (WINTA), (2014). Indigenous tourism and human rights in Asia and Pacific Region: Review, analysis & guidelines. Bangkok, Thailand: PATA.

Scheyvens, R. (1999). Ecotourism and the empowerment of local communities. *Tourism Management, 20*(2), 245–249.

Sirakaya, E., Sasidharan, V., & Sönmez, S. (1999). Redefining ecotourism: The need for a supply-side view. *Journal of Travel Research, 38*(2), 168–172.

Spirit Bear Lodge. (2019, September 24). Things To Do. Retrieved from https://indigenoustourism.ca/en/things-to-do/spirit-bear-lodge/.

Tides Canada. (2018). Indigenous Led-Ecotourism: A Source for Positive Community Impact: A Summative Report of the Indigenous Ecotourism Summit. Retrieved from https://coastfunds.ca/news/indigenous-ecotourism-offers-many-benefits-to-communities/.

Towner, N. (2018). Surfing tourism and local stakeholder collaboration. *Journal of Ecotourism, 17*(3), 268–286.

United Nations. (2007). UN Declaration on the Rights of Indigenous People. Retrieved from www.un.org/esa/socdev/unpfii/documents/DRIPS_en.pdf.

UNWTO. (2002). World Ecotourism Summit: Final Report. *World Ecotourism Summit.* https://www.e-unwto.org/doi/epdf/10.18111/9789284405503.

Wachtel, P. (1989). *The Poverty of Affluence: A Psychological Portrait of the American Way of Life.* Gabriola Island, BC: New Society Publishers.

Warren, D. M. (1996). Indigenous knowledge, biodiversity conservation and development. *Sustainable development in third world countries: Applied and theoretical perspectives,* 81–88.

Wheeler, B. (1991) Tourism's troubled times: Responsible tourism is not the answer. *Tourism Management, 12*(2), 91–96.

World Indigenous Tourism Alliance. (2012). Larrakia Declaration on Indigenous Tourism.Retrieved from www.winta.org/wp-content/uploads/2012/08/The-Larrakia-Declaration.pdf.

Zeppel, H. (2006). *Indigenous Ecotourism: Sustainable Development and Management* (Vol. 3). England:CABI.

11

THE CONNECTION BETWEEN NATURE AND SÁMI IDENTITY

The role of ecotourism

Cecilia De Bernardi

Introduction

As the concern for human-induced climate change is gaining attention with tourism at the forefront of the debate, the role of ecotourism gains new relevancy. For small companies in northern Sweden, tourism can be a good way to make a living or to integrate other work, such as for Sámi tourism entrepreneurs and one of their main occupations, which is reindeer herding (Leu, Eriksson, & Müller, 2018; Müller & Viken, 2017).

The Sámi are an Indigenous population of Europe, who have been traditionally living in an area called Sápmi. This region stretches from Norway to Russia (Om Sápmi - Samer.se, n.d.). Sámi people speak different languages (De samiska språken - Samer.se, n.d.) and the exact size of the population is not known; an estimation is about 100,000 Sámi people in the whole of Sápmi (Samerna i siffror - Samer.se, n.d.). The Sámi have operated in tourism differently time depending on the country of residence. For instance, from the late 1970s in Sweden and Finland, the recognition of Sámi culture and the establishment of Sámi institutions favoured funding and the support for creating tourism enterprises (Winsa, 2007).

Treatment of Sámi culture in the context of tourism has not been uniform, for example Finland is usually considered to be a negative example, where cultural exploitation has been particularly negative and prolonged long (e.g., Pettersson, 2006; Saarinen, 1999). Recently, the Finnish Sámi parliament published some ethical guidelines for the use of Sámi culture in tourism (Heith, 2019; Samediggi, n.d.). Other guidelines and certifications have been previously mapped by the Swedish Sámi Parliament (Sametinget, 2010) and the current situation has been charted by Olsen et al. (2019). The Sámi have been engaged in different activities for substance; some examples are fishing and reindeer herding (Näringar - Samer.se, n.d.). The latter is an exclusive right of the Sámi population in both Sweden and Norway, but not in Finland (Müller & Viken, 2017; Pettersson, 2006). As previously mentioned, there is a Sámi parliament and this institution is present in each of the Nordic countries. There are also other institutions that are specifically connected to Sámi matters (Organisationer - Samer.se, n.d.).

Tourism development can bring different issues related to marketing representations (e.g., de Bernardi, Kugapi, & Lüthje, 2017), linking indigenous populations to nature and primitivism (Fonneland, 2013; Olsen, 2006), and to a traditional identity (Pashkevich & Keskitalo, 2017). Another connection is nature stewardship (Koot, 2018), depending on the context. This link

 DOI: 10.4324/9781003001768-11

between Indigenous populations, such as the Sámi, and nature has been influenced by discourses on the North (Pashkevich & Keskitalo, 2017) and by a romantic view on how Indigenous peoples live (Condevaux, 2009; Stasch, 2014), and through strong historical dimensions (Valkonen & Valkonen, 2014).

The deep relation between nature and Sámi traditional activities is also part of the tourism experiences offered and it is also expressed in the acquisition of tourism certifications and labels. One example is the Nature's Best label from Sweden (Natures Best, n.d.). Other examples are the previously mentioned guidelines for tourism operations recently created by the Finnish Sámi parliament (Heith, 2019) and the Sámi tourism label that was created in Sweden called Sápmi Experience. Labels have been identified as a good way to protect Sámi culture and to provide a direction in Sámi tourism (Sametinget, 2010). Labels can also potentially support the tourism operations of the Sámi (de Bernardi et al., 2017). Sustainability has also been identified as an important aspect of Sámi-related labelling schemes (Sametinget, 2010; VisitSapmi – kriterierna, n.d.). Considering these factors, this chapter investigates the conceptualisation of nature and sustainability, together with other themes that emerged from the interviews with 16 entrepreneurs and other actors operating in both Sámi and other local tourism in Sweden and Norway. The aim is to understand the role of ecotourism and labels in Sámi tourism.

Ecotourism: A brief overview

Positive aspects of ecotourism include the involvement of the locals, environmental protection, and profits remaining in the area (Gale & Hill, 2009). However, there are negative effects to ecotourism (Das & Chatterjee, 2015), such as damage and disturbances (Gale & Hill, 2009), land appropriation (Bluwstein, 2017), and frustration with how ecotourism is defined, implemented and managed (Cobbinah, 2015). Ecotourists do not always visit because of nature protection, but to inflate their ego. Furthermore, "ecotourism encourages increased use of natural areas and greater penetration into sensitive environments, thereby putting the very future of indigenous tourism industries at risk" (Mihalic, 2000, as cited in Gale & Hill, 2009, p. 4). Despite these potential negative effects, ecotourism still has a strong potential, but this depends on how it is established and managed (Das & Chatterjee, 2015; Wang, Cater, & Low, 2016; Stronza, Hunt, & Fitzgerald, 2019). The concept has been debated for long and academic discourse may have contributed negatively to the formulation of an established definition (Fennell, 2001). However, ecotourism definitions seem to revolve around three main criteria (Weaver & Lawton, 2007, p. 170, as cited in Gale & Hill, 2009, p. 5):

1. Nature-based attractions
2. Focus on learning and education
3. Based on principles of economic, social, and environmental sustainability

Furthermore, a definition of ecotourism should also focus on the values of the visitors (Gale & Hill, 2009). The role of ecotourism experiences in promoting sustainability-related values have been described as very important (Higham & Carr, 2002; Walker & Moscardo, 2014), and could also be used to segment visitors or to adapt activities (Zografos & Allcroft, 2007). People also consider social values important in relation to community-based ecotourism experiences, such as feeling special and making a good impression on others (Kim & Park, 2017). Ecotourism can also be an incentive for conservation for locals in certain areas (Boley & Green, 2016; Stronza et al., 2019) and a tool for cultural protection (Masud, Aldakhil, Nassani, & Azam, 2017; Stronza et al., 2019). Jamrozy and Lawonk (2017) have shown that emotional values, novelty, and knowledge are important motivations related to ecotourism. Ecotourism criteria are a tool that can render this form of tourism more beneficial (Gössling, 2006).

Labels and ecotourism in Sámi tourism

Sámi tourism entrepreneurs work in different contexts with tourism, some are more culture-related while other enterprises focus more on nature (Leu et al., 2018). This is also related to the ACE framework (Adventure Culture Ecotourism), which implies that different companies operate in ecotourism to a higher or smaller degree, depending on the activities and how much these are related to either adventure or culture (Fennell, 2004, pp. 28–31). Considering these different dimensions of the ACE framework, it is important to discuss the role of labels to establish some criteria for ecotourism.

In Sweden there are different institutions and labels dealing with ecotourism. The most important is the "Swedish Ecotourism Association" (Ekoturismföreningen in Swedish), which created one of the oldest ecotourism labels in the world: Nature's Best (Naturturismföretagen – Nature's Best, n.d.; Sametinget, 2010). Nature's Best is also the basis for the Sámi tourism label Sápmi Experience (Olsen, 2016). The fact that Nature's Best is so old and connected to a Sámi tourism label is the reason why it was chosen as the focus for this chapter. The Swedish Ecotourism Association has about 80 members (Naturturismföretagen – Medlemmar, n.d.) and around 50 companies are certified with Nature's Best (Natures Best - Godkända företag, n.d.). The Ecotourism Association has recently changed structure and has become "Nature tourism companies" (Naturturismföretagaren in Swedish) (Redaktionen, 2018, July) to include a broader spectrum of companies and highlight its collective structure (Naturturismföretagen – Om oss, n.d.). This means that nature-based companies may be confused with ecotourism companies (Stronza et al., 2019), potentially undermining the efforts of ecotourism companies; Nature's Best has a key role in highlighting this distinction.

The Global Sustainable Tourism Council (GSTC) manages sustainable tourism criteria and certification bodies (GST Council, n.d.). In 2019, the GSTC made a revision of its criteria, which implied also a change in Nature's Best. For instance, a criterium for Sámi tourism was added (Naturturismföretagen – Nature's Best, n.d.). The label is based on the cooperation between a series of stakeholders: the tourism industry, landowners, and other institutions and is based on six basic principles which are: to contribute to the economy of rural areas, the adaptation to the environment for the whole experience, to actively protect nature and culture, the quality of guiding, quality assurance, and hosting (Naturturismföretagen – Nature's Best, n.d.). The Nature's Best has been criticised for missing third-party auditing (Haaland & Aas, 2010), which also highlights the main difference between a 'certification' and a 'label' (FAO, n.d.). In this case, the focus will be on labels.

Labels can have an important role for Sámi culture and for other indigenous cultures (de Bernardi et al., 2017) for instance as a way to renegotiate authenticity and indigeneity (Keskitalo, Schilar, Heldt Cassel, & Pashkevich, 2019). Labels are also used to distinguish indigenous handicrafts produced in a certain way (de Bernardi et al., 2017; Dlaske, 2014; Keskitalo et al., 2019). However, this kind of scheme can also be a tool for a company to evaluate its operations in light of environmental sustainability and respect for nature, as discussed later in the chapter. Labels can also be used to show to potential tourists and ecotourists the work that is being done (Minoli, Goode, & Smith, 2015) and to promote awareness (Pencarelli, Splendiani, & Fraboni, 2016), even though the appeal is usually for a smaller niche of engaged tourists (Karlsson & Dolnicar, 2016). Certifications and labels can also be used as marketing tools (Suzer, 2019; Testa, Iraldo, Vaccari, & Ferrari, 2015). Nonetheless, 'best practise' approaches are dependent on local context and on the support of policy (Fletcher, Pforr, & Brueckner, 2016).

Labelling schemes involving Indigenous culture should be flexible and relevant for the indigenous population, planning power should stay in their hands and certifications should be collaborative projects (Vivanco, 2007). Furthermore, education of visitors has been identified as an important factor in promoting sustainability through tourism in Indigenous contexts (Walker & Moscardo, 2016).

Methods

This study is based on the analysis of 16 interviews conducted with different actors operating in tourism in Sweden and Norway. The majority of the interviewees, 11, are women and half of the respondents were companies and the other half stakeholders such as DMOs. Companies and DMOs were interviewed with different interview guides. Due to several constrains, seven interviews were completed by phone, one on Skype without video, three on Skype with video and five in person, one of which was completed by a senior researcher working for the ARCTISEN project (Culturally Sensitive Tourism in the Arctic). This one interview was of course different from the others, but the consistency of the data was a source of additional trustworthiness. The findings are also congrous with other project's findings (Olsen et al., 2019). The ARCTISEN project is funded through the EU's Northern Periphery and the Arctic Programme. The aim is to support local and indigenous companies in the creation of sustainable tourism products (ARCTISEN, n.d.). This research was not funded by ARCTISEN, but there has been a close cooperation.

The respondents were both Sámi and non-Sámi and almost all of the interviews were in Swedish except for two that were in English and one that was a mix between English and Swedish. The respondents chose the language. The typical interview for a company was close to one-hour long and for stakeholders it was 40 minutes. The interviews are presented here as a numeric code and direct quotes have also been kept short (Davies, 2008, pp. 59–60; Hall, 2014), both measures are adopted to keep the participants anonymous. The interviews were semi-structured, and all were tape-recorded and, so far, partly transcribed.

In a previous study published by the author (de Bernardi, 2019a), different themes related to the tourism discourse and to how indigenous populations are framed in order to be interesting for the tourists (e.g., Pettersson, 2006) were identified (Table 11.1). The same themes were used as a coding matrix for the interviews with open coding for new themes to provide the

Table 11.1 The themes identified in marketing (from de Bernardi, 2019a)

Theme 1: Connection to nature/harmony with nature/peacefulness
• Connection to untouched nature
Theme 2: Connection to reindeer
Theme 3: Connection to the past
Theme 4: The use of friendly language
Theme 5: *Authenticity* as a noun or adjective
Theme 6:
• The Sámi costume
• Sámi huts or tents
• Traditional singing connected to indigenous populations or the traditional Sámi *yoik* chant
• *Handicrafts*
• *Food*
Theme 7: Joining the Sámi
Theme 8: Modernity and useful information for tourists

author with a guide for interpretation and to see if the same themes would be also present in the interviews. New themes emerged as the respondents focused on nature and its importance for tourism as well as its protection. This included the role of labels and therefore ecotourism was used as a frame of interpretation for the interview data.

The interview transcripts were analysed by thematic analysis informed by abduction, which implies a constant dialogue between a pre-existing theoretical framework and the data that is being analysed (Graneheim, Lindgren, & Lundman, 2017). This process was informed by an approach called Critical Realism, which is based on a single reality with multiple interpretations. The goal of an abductive approach is the ability to describe what is constitutive for a certain phenomenon (e.g., Danermark, Ekstrom, Jakobsen, & Karlsson, 2002).

Findings

During the coding process, the following codes were added (Table 11.2). The themes related to *handicrafts* and *food* were also added during the coding of the interviews and are therefore in italics in Table 11.1.

Table 11.3 shows in which interviews the themes were identified, just to provide a sense of the themes' frequency.

Themes 1–8 were previously identified (de Bernardi, 2019a) and there are a few differences with the interview data. Themes 4 and 7 were less prominent in the interviews, since both are more generally used as marketing tools (e.g., de Bernardi, 2019a). Theme 4 was mentioned while exemplifying how the tourists visit those areas "to meet us" (050108). Theme 8 relates to contemporary Sámi life, so it is very common when describing everyday operations.

Sustainability, nature, and the role of ecotourism

Ecotourism is related to many topics touched on by the respondents. Nature is described as a connection to the surroundings, and this is also closely related to reindeer herding (Theme 2) which is described as "part of the nature" (Int 030122) and also that the tourism operations of the Sámi "build on nature, so herding, the animals, how they move" (Int 050209, author's translation). Another example is that the use of the local environment is based on entrepreneurs being "out in nature [...] use it all the time. Animals and nature. The whole time. It is what is talked about, it is what is shown" (Int 050109, author's translation). Interviewees also describe that nature is very close to Sámi culture as "it has of course always lived very close to [...] nature" (Int 050110, author's translation). This connection to nature due to the way of life of the Sámi also connects to the theme of sustainability (Theme 11). For instance, one respondent thought that there was some positive tendency connected to being proud of Sámi culture and that entrepreneurs really want "to do Sámi tourism in a sustainable way and I think the sustainable is also very important" (Int 030122). Another respondent expressed that the way of living described for this particular company and family is "a sustainable way to live"

Table 11.2 The themes that emerged from the interview analysis

Theme 9: 'Fake' Sámi culture
Theme 10: Sustainability
Theme 11: The size of the company
Theme 12: Education

Table 11.3 Summary of the themes present in which interview

Theme	*Interview*
Theme 1: Connection to nature/harmony with nature/peacefulness • Connection to untouched nature	030102; 030122; 050107; 050108; 050209; 050109; 050110; 050120; 050121; 050123; 060103; 060111; 060119
Theme 2: Connection to reindeer	030102; 030122; 040124; 050107; 050108; 050209; 050109; 050110; 050120; 050121; 050123; 050130; 060111; 060119
Theme 3: Connection to the past	030122; 040124; 050107; 050108; 050209; 050109; 050110; 050120; 050121; 050123; 050130; 060103; 060111; 060119
Theme 4: The use of friendly language	050108
Theme 5: *Authenticity* as a noun or adjective	030122; 040124; 050108; 050110; 050121; 050123; 050130; 060103; 060111; 060119
Theme 6: • The Sámi costume • Sámi huts or tents • Traditional singing connected to indigenous populations or the traditional Sámi *yoik* chant • *Handicrafts* • *Food*	030122; 040124; 050107; 050108; 050209; 050109; 050110; 050120; 050121; 050123; 060111; 060119
Theme 7: Joining the Sámi	040124; 050120
Theme 8: Modernity and useful information for tourists	030102; 030122; 040124; 050107; 050108; 050209; 050109; 050110; 050120; 050121; 050123; 050130; 060103; 060111; 060119
Theme 9: 'Fake' Sámi culture	030102; 030122; 040124; 050108; 050209; 050109; 050110; 050121; 050123; 050130; 060111
Theme 10: Sustainability	030122; 040124; 050108; 050209; 050109; 050120; 050123; 050130; 060111
Theme 11: The size of the company	030122; 040124; 050108; 050120; 050121; 050123; 050130; 060103; 060111
Theme 12: Education	030102; 040124; 050107; 050108; 050109; 050209; 050110; 050120; 050121; 050123; 060119

Source: de Bernardi, C. (2019a). Authenticity as a compromise: A critical discourse analysis of Sámi tourism websites. *Journal of Heritage Tourism*, *14*(3), 249–262.

(Int 050123, author's translation). Sustainability is also something that reflects in how the operations are organised, and this respondent works "very much with sustainability and then it includes to meet at step one to be able to develop something" (Int 040124, author's translation).

Sustainability is also connected to labels. For instance, one of the respondents answered on a question related to the labels saying that it is about "commercialising the right and sustainable way" (Int 030122). Another respondent expresses how the company is certified as "an ecological company" and the label is also a "tool to think right and steer which phases you should work with and how you do" (Int 050209, author's translation). Many of the companies report to have acquired or to be working with Nature's Best (Int 050209, author's translation; Int 050110, author's translation). Labels are described as a "guarantee that that company [...] makes an effort [and that] one cares for what the certification represents" (Int 050110, author's translation). Another company has also gone through "a sustainability certification"

(Int 050123, author's translation). Labels are a way "to act in a very ethical way and to preserve environment" (Int 050130).

Other important themes related to sustainability from a cultural viewpoint are about how Sámi culture is presented. Several of the respondents mention the importance of education (Theme 12) and of transmitting knowledge about the Sámi (Int 050110; 050121; 050221; 050123, 060119). Labels are also a way to "get inspiration" and to "act sustainably" (Int 060111, author's translation) and to check one's operations (Int 050209; 050120; 050121; 050123; 060111). Labels are also a way to protect both tangible and intangible aspects of culture. Some respondents indicated that labels would be a good way to protect Sámi culture, for instance handicrafts (Int 050108).

Other themes

Many respondents used the word 'authentic' or the synonym 'genuine' (both in Swedish and in English) to describe how they see an enterprise representing Sámi culture ethically (Table 11.3), although there was no mention of it in the interview guide. Sámi culture is described as very old with a long history, yet modern. There are some recognisable markers such as the traditional Sámi costume, the reindeer and the traditional Sámi tent (Table 11.3). The new themes that emerged in the interview analysis are all intertwined. Many of the respondents referred to a 'fake' Sámi culture, which for them is embodied in the tourism industry of Rovaniemi, Finland; often considered a bad example of how Sámi culture is treated in tourism (Pettersson, 2006; Saarinen, 1999; Olsen et al., 2019). An example is one respondent saying that people "want genuine and authentic experiences. Not, not like Rovaniemi Sámi" (Int 030102, author's translation). The connection to presenting what many of the respondents consider to be 'real' ('authentic') Sámi culture is related to sustainability, to the information given and to the fact that tourists are educated in some way. The respondents also describe the operations as being usually small-scale, which in some cases are seen as a way in which the company can be authentic (Int 030102; 050123; 050130) and more sustainable (Int 040124; Sametinget, 2010).

Discussion

The interviews are a way to gain knowledge on the indigenous peoples' viewpoint on ecotourism and conservation (Higgins-Desbiolles, 2009). As previously mentioned, the connection to nature is a prominent theme when it comes to Sámi tourism marketing (e.g., Olsen, 2006; Pashkevich & Keskitalo, 2017; Prestholdt & Nordbø, 2015). Valkonen and Valkonen (2014) have described how nature in Sámi context is connected to two dimensions, one discursive and more related to worldview myths and one related to practices. Both of these dimensions are present in the interviews.

This connection is also related to the conceptualisation of wilderness, especially regarding the Arctic areas (e.g., Saarinen, 2005; Saarinen 2019) and the Indigenous populations living there (Shultis & Heffner, 2016). For instance, the discourses of conservation and traditional use. Furthermore, the Arctic is also described as a last frontier, far from civilisation where primitive populations live (Keskitalo, 2004) and this is also connected to a discourse of environmental protection (Haila, 1997; Shultis & Heffner, 2016). The 'Indigenous stewardship' discourse is also common and is grounded in the fact that Indigenous populations are caretakers of nature (Fennell, 2008; Pashkevich & Keskitalo, 2017). Wilderness is a conceptualisation of nature that can impede tourism conservation efforts by placing the focus on nature as separate from human activities, downplaying issues such as waste production or air travel (Brookes, 2001). Nonetheless, there can be synergy between wilderness and conservation (Saarinen, 2005). Wilderness was also mentioned by two interviewees as something people expect (Int 050110), when in reality these areas are 'bruksmark' in Swedish, which means agricultural land (Int 060111). Another respondent instead described the wilderness issue by saying: "yeah yeah wilderness, but what is wilderness?" (Int 050209).

The connection to nature has been related to a creation of discourses from within the northern areas in which actors such as Sámi tourism entrepreneurs are active participants (Keskitalo & Schilar, 2017). Even in the case of marketing representations, the Sámi exercise counter-discursive action by presenting their modernity for the potential tourists (de Bernardi, 2019a). The reason that certain elements of Sámi culture are emphasised in marketing is because such elements are part of Sámi culture (de Bernardi, 2019b) and not just a way to present Sámi culture that is in line with dominant discourses, as discussed by Heldt Cassel (2019). As previously mentioned, labels can protect both tangible and intangible elements of culture, and both examples are mentioned by respondents. Handicrafts and other cultural elements are examples. This implies that the connection to nature expressed by respondents is something that is closely connected to their identity and to Sámi identity (Valkonen & Valkonen, 2014). Nature is also the place where the Sámi have been living and operating for generations.

When it comes to presenting nature as wilderness, it has been shown that many people connect wilderness to something that is different from their everyday context (Conti & Heldt Cassel, 2019), and is therefore worth visiting. Nature is not wild, because nature is exploited one way or another (Keskitalo & Schilar, 2017; Keskitalo et al., 2019), which is why an approach based on ecotourism and certifications is important in this context. The protection of nature has been identified as an important priority in the context of Sámi tourism as well as a connection between tourism work and Sámi culture, which can be a good way to spread information (Sametinget, 2010). As previously mentioned, environmental values are important in the context of ecotourism (Higham & Carr, 2002; Walker & Moscardo, 2014). Values are then also connected to tourists' ethical approach in the context of ecotourism (Fennell, 2004, pp. 171–188). There is also an aspect of learning and education connected to ecotourism, which was an important theme identified in the interviews and was also mentioned by other companies and stakeholders interviewed during the ARCTISEN project (Olsen et al., 2019). Social sustainability is also important (Weaver & Lawton, 2007, p. 170, as cited in Gale & Hill, 2009, p. 5), particularly in the context of Indigenous tourism (Walker & Moscardo, 2016).

As previously mentioned, labels, especially related to ecotourism, do not mean that the tourists will choose a certified company rather than one that is not; it is important to concentrate on a specific segment (Hausmann et al., 2017) and to increase emotional ties and direct experience, especially with nature, in order to improve pro-environmental attitudes (Kazeminia, Hultman, & Mostaghel, 2016). Best practice approaches are also based on local context and on policy support (Fletcher et al., 2016) and should take into consideration several aspects related to Indigenous values (Vivanco, 2007). Sámi identity could be more competitive, especially in ecotourism (Palomino, 2012, as cited in Hägglund, Schilar, & Keskitalo, 2019, p. 61). Labels can also be a way to ensure that certain criteria are met to support conservation, as argued by Stronza et al. (2019), even though a third-party monitored certification would prove to be even a better tool to fulfil this purpose.

Conclusion

To conclude, the connection to nature in the context of Sámi tourism is stressed both in marketing material as well as in how companies and other tourism actors describe their operations. For people operating in interdependence with nature, both in tourism and in other activities, its conservation is an important part of everyday operations. In the case of Sámi culture specifically, a connection to nature is an important constitutive aspect and it has a historical characteristic. Labels can be used to protect Sámi culture from unethical use (de Bernardi et al., 2017), usually by tourism companies or other stakeholders.

In Sámi tourism operations it is important to ensure positive outcomes for the Sámi (Ween & Riseth, 2017) without undermining the cultural aspect with the natural one (Smed, 2017). In Indigenous tourism, principles of sustainable development have been considered particularly relevant. Indigenous control and empowerment are very important, while education helps promoting sustainability through tourism in Indigenous contexts (Walker & Moscardo, 2016).

The use of ecotourism labels or certifications can support these aspects and also spread information to tourists, both on Sámi culture and on nature conservation. Furthermore, labels can also be a good tool for companies to ensure more sustainable tourism operations and a balance in a company's activities based on the previously mentioned ACE framework (Fennell, 2004). It has been previously argued that the development of Indigenous-related ecotourism experiences would benefit both companies and visitors; the question is only how interested Sámi entrepreneurs are in such a development (Pettersson, 2006). The results of the interviews show that there is such an interest, but it needs backing by rigorous evaluation of ecotourism enterprises (Stronza et al., 2019).

References

ARCTISEN. (n.d.). Culturally Sensitive Tourism in the Arctic. Retrieved from http://sensitivetourism.interreg-npa.eu/ (accessed on 10.02.2020).

Bluwstein, J. (2017). Creating ecotourism territories: Environmentalities in Tanzania's community-based conservation. *Geoforum*, *83*, 101–113.

Boley, B. B., & Green, G. T. (2016). Ecotourism and natural resource conservation: The 'potential'for a sustainable symbiotic relationship. *Journal of Ecotourism*, *15*(1), 36–50.

Brookes, A. (2001). Doing the Franklin: Wilderness Tourism and the Construction of Nature. *Tourism Recreation Research*, *26*(1), 11–18. doi:https://doi.org/10.1080/02508281.2001.11081172.

Cobbinah, P. B. (2015). Contextualising the meaning of ecotourism. *Tourism Management Perspectives*, *16*, 179–189.

Condevaux, A. (2009). Māori Culture on Stage: Authenticity and Identity in Tourist Interactions. *Anthropological Forum*, *19*(2), 143–161. doi:https://doi.org/10.1080/00664670902980389.

Conti, E., & Heldt Cassel, S. (2019). Liminality in nature-based tourism experiences as mediated through social media. *Tourism Geographies*, 1–20.

Danermark, B., Ekstrom, M., Jakobsen, L., & Karlsson, J. C. (2002). *Explaining Society: Critical Realism in the Social Sciences*. London: Routledge.

Das, M., & Chatterjee, B. (2015). Ecotourism: A panacea or a predicament? *Tourism Management Perspectives*, *14*, 3–16.

Davies, C. A. (2008). *Reflexive Ethnography: A Guide to Researching Selves and Others*. London: Routledge.

de Bernardi, C. (2019a). Authenticity as a compromise: a critical discourse analysis of Sámi tourism websites. *Journal of Heritage Tourism*, *14*(3), 249–262.

de Bernardi, C. (2019b). A critical realist appraisal of authenticity in tourism: the case of the Sámi. *Journal of Critical Realism*, *18*(4), 437–452.

de Bernardi, C., Kugapi, O., & Lüthje, M. (2017). Sámi indigenous tourism empowerment in the Nordic countries through labelling systems: Strengthening ethnic enterprises and activities. In I. B. de Lima, & V. King (Eds.), *Tourism and ethnodevelopment inclusion, empowerment and self determination* (pp. 200–212). Abingdon, UK: Routledge.

De samiska språken – Samer.se. (n.d.). Språk, dialekt eller varietet? Retrieved from http://samer.se/1186 (accessed on 10.02.2020).

Dlaske, K. (2014). Semiotics of pride and profit: interrogating commodification in indigenous handicraft production. *Social Semiotics*, *24*(5), 582–598. doi:https://doi.org/10.1080/10350330.2014.943459.

FAO (Food and Agriculture Organization). (n.d.). The Concepts of Standards, Certification and Labelling. Retrieved from http://www.fao.org/3/y5136e/y5136e07.htm (accessed on 10.02.2020).

Fennell, D. A. (2001). A content analysis of ecotourism definitions. *Current Issues in Tourism*, *4*(5), 403–421.

Fennell, D. A. (2004). *Ecotourism: An Introduction*. London: Routledge.

Fennell, D. A. (2008). Ecotourism and the myth of indigenous stewardship. *Journal of Sustainable Tourism*, *16*(2), 129–149. doi:https://doi.org/10.2167/jost736.0.

Fletcher, C., Pforr, C., & Brueckner, M. (2016). Factors influencing Indigenous engagement in tourism development: An international perspective. *Journal of Sustainable Tourism*, *24*(8–9), 1100–1120.

Fonneland, T. A. (2013). Sami tourism and the signposting of spirituality. The case of Sami Tour: a spiritual entrepreneur in the contemporary experience economy. *Acta Borealia*, *30*(2), 190–208. Doi:https://doi.org/10.1080/08003831.2013.844422.

Gale, T., & Hill, J. (2009). Ecotourism and environmental sustainability: An introduction. In J. Hill, & T. Gale, *Ecotourism and environmental sustainability* (pp. 21–34). London: Routledge. doi:https://doi.org/10.4324/9781315578767.

Gössling, S. (2006). Ecotourism as experience-tourism. In *Ecotourism in Scandinavia: Lessons in theory and practice* (pp. 89–97). Wallingford, UK: CABI. doi:https://doi.org/10.1079/9781845931346.0089.

Graneheim, U. H., Lindgren, B. M., & Lundman, B. (2017). Methodological challenges in qualitative content analysis: A discussion paper. *Nurse Education Today*, *56*(May), 29–34. doi:https://doi.org/10.1016/j.nedt.2017.06.002.

GST Council (n.d.). About us. Retrieved from https://www.gstcouncil.org/about/about-us/ (accessed on 25.03.2020).

Haaland, H., & Aas, Ø. (2010). Eco-tourism certification - does it make a difference? A comparison of systems from Australia, Costa Rica and Sweden. *Scandinavian Journal of Hospitality and Tourism*, *10*(3), 375–385. doi:https://doi.org/10.1080/15022250.2010.486262.

Hägglund, M., Schilar, H., & Keskitalo, E. C. H. (2019). How is 'Sámi tourism' represented in the English-language scholarly literature? *Polar Geography*, *42*(1), 58–68. doi:https://doi.org/10.1080/1088937X.2018.1547327.

Haila, Y. (1997). "Wilderness" and the Multiple layers of Environmental Thought. *Environment and History*, *3*(2), 129–147. doi:https://doi.org/10.3197/096734097779555935.

Hall, S. M. (2014). Ethics of ethnography with families: A geographical perspective. *Environment and Planning A: Economy and Space*, *46*(9), 2175–2194.

Hausmann, A., Slotow, R., Fraser, I., & Di Minin, E. (2017). Ecotourism marketing alternative to charismatic megafauna can also support biodiversity conservation. *Animal Conservation*, 20(1), 91–100.

Heith, A. (2019). Nordic place branding from an indigenous perspective. In C. Cassinger, A. Lucarelli, & S. Gyimóthy (Eds.), *The Nordic wave in place branding*. Cheltenham, UK: Edward Elgar Publishing.

Heldt Cassel, S. (2019). Branding Sámi tourism: Practices of indigenous participation and place-making. In C. Cassinger, A. Lucarelli, & S. Gyimóthy (Eds.), *The Nordic Wave in Place Branding*. Cheltenham, UK: Edward Elgar Publishing.

Higgins-Desbiolles, F. (2009). Indigenous ecotourism's role in transforming ecological consciousness. *Journal of Ecotourism*, *8*(2), 144–160. doi:https://doi.org/10.1080/14724040802696031.

Higham, J., & Carr, A. (2002). Ecotourism visitor experiences in Aotearoa/New Zealand: Challenging the environmental values of visitors in pursuit of pro-environmental behaviour. *Journal of Sustainable Tourism*, *10*(4), 277–294.

Jamrozy, U., & Lawonk, K. (2017). The multiple dimensions of consumption values in ecotourism. *International Journal of Culture, Tourism and Hospitality Research*, *11*(1), 18–34.

Karlsson, L., & Dolnicar, S. (2016). Does eco certification sell tourism services? Evidence from a quasi-experimental observation study in Iceland. *Journal of Sustainable Tourism*, *24*(5), 694–714.

Kazeminia, A., Hultman, M., & Mostaghel, R. (2016). Why pay more for sustainable services? The case of ecotourism. *Journal of Business Research, 69*(11), 4992–4997.

Keskitalo, E. (2004). *Negotiating the Arctic*. New York: Routledge. doi:https://doi.org/10.4324/9780203508114.

Keskitalo, E. C. H., & Schilar, H. (2017). Co-constructing "northern" tourism representations among tourism companies, DMOs and tourists. An example from Jukkasjärvi, Sweden. *Scandinavian Journal of Hospitality and Tourism, 0*(0), 1–17. doi:https://doi.org/10.1080/15022250.2016.1230517.

Keskitalo, E. C. H., Schilar, H., Heldt Cassel, S., & Pashkevich, A. (2019). Deconstructing the indigenous in tourism. The production of indigeneity in tourism-oriented labelling and handicraft/souvenir development in Northern Europe. *Current Issues in Tourism*, 1–17.

Kim, K. H., & Park, D. B. (2017). Relationships among perceived value, satisfaction, and loyalty: Community-based ecotourism in Korea. *Journal of Travel & Tourism Marketing, 34*(2), 171–191.

Koot, S. (2018). The Bushman brand in southern African tourism: An Indigenous Modernity in a Neoliberal Political Economy. *Senri Ethnological Studies, 99*, 231–250.

Leu, T. C., Eriksson, M., & Müller, D. K. (2018). More than just a job: exploring the meanings of tourism work among Indigenous Sámi tourist entrepreneurs. *Journal of Sustainable Tourism, 26*(8), 1468–1482.

Masud, M. M., Aldakhil, A. M., Nassani, A. A., & Azam, M. N. (2017). Community-based ecotourism management for sustainable development of marine protected areas in Malaysia. *Ocean & Coastal Management, 136*, 104–112.

Minoli, D. M., Goode, M. M., & Smith, M. T. (2015). Are eco labels profitably employed in sustainable tourism? A case study on Audubon Certified Golf Resorts. *Tourism Management Perspectives, 16*, 207–216.

Müller, D. K., & Viken, A. (2017). Indigenous tourism in the Arctic. In A. Viken, & D. K. Müller (Eds.), *Tourism and Indigeneity in the Arctic* (pp. 3–15). Bristol, UK: Channel View Publications.

Näringar – Samer.se. (n.d.). Retrieved from http://samer.se/naringar (accessed on 10.02.2020).

Natures Best. (n.d.). About. Retrieved from https://naturesbestsweden.com/en/about-natures-best/ (accessed on 10.02.2020).

Natures Best – Godkända företag. (n.d.). Retrieved from https://naturesbestsweden.com/sv/certifierade-foretag/ (accessed on 10.02.2020).

Naturturismföretagen – Medlemmar. (n.d.). Retrieved from https://naturturismforetagen.se/bli-medlem/vem-ar-medlem/ (accessed on 10.02.2020).

Naturturismföretagen – Nature's Best. (n.d.). Retrieved from https://naturturismforetagen.se/natures-best/ (accessed on 10.02.2020).

Naturturismföretagen – Om oss. (n.d.). Retrieved from https://naturturismforetagen.se/om-oss/ (accessed on 10.02.2020).

Olsen, K. (2006). Making Differences in a Changing world: The Norwegian Sámi in the Tourist Industry. *Scandinavian Journal of Hospitality and Tourism, 6*(1), 37–53. doi:https://doi.org/10.1080/15022250600560570.

Olsen, K. O., Avildgaard, M. S., Brattland, C., Chimirri, D., De Bernardi, C., Edmonds, J., … & Institute, M. T. (2019). Looking at Arctic tourism through the lens of cultural sensitivity: ARCTISEN – a transnational baseline report. *Lapin yliopisto*. Retrieved from https://lauda.ulapland.fi/bitstream/handle/10024/64069/Arctisen%20-%20a%20transnational%20baseline%20report.pdf?sequence=1&isAllowed=y

Olsen, L. S. (2016). Sami tourism in destination development: conflict and collaboration. *Polar Geography, 39*(3), 179–195. doi:https://doi.org/10.1080/1088937X.2016.1201870.

Om Sápmi - Samer.se. (n.d.). Retrieved from http://samer.se/1002 (accessed on 10.02.2020).

Organisationer - Samer.se. (n.d.). Retrieved from http://samer.se/4369 (accessed on 10.02.2020).

Pashkevich, A., & Keskitalo, E. C. H. (2017). Representations and uses of indigenous areas in tourism experiences in the Russian Arctic. *Polar Geography, 40*(2), 85–101. doi:https://doi.org/10.1080/1088937X.2017.1303753.

Pencarelli, T., Splendiani, S., & Fraboni, C. (2016). Enhancement of the "Blue Flag" Eco-label in Italy: an empirical analysis. *Anatolia, 27*(1), 28–37.

Pettersson, R. (2006). Ecotourism and Indigenous People: Positive and Negative Impacts of Sami Tourism. In Stefan Gössling, & Johan Hultman (Eds.), *Ecotourism in Scandinavia: Lessons in Theory and Practice* (pp. 166–177). Wallingford, UK: CABI.

Prestholdt, R., & Nordbø, I. (2015). Norwegian landscapes: An assessment of the aesthetical visual dimensions of some rural destinations in Norway. *Scandinavian Journal of Hospitality and Tourism*, *15*(1–2), 202–222. Doi:https://doi.org/10.1080/15022250.2015.1014129.

Redaktionen. (2018, July). Ekoturismföreningen blir Naturturismföretagen. *Travel News*. Retrieved from https://www.travelnews.se/turism/ekoturismforeningen-blir-naturturismforetagen/.

Saarinen, J. (1999). Representation of indigeneity: Sami culture in the discourses of tourism. In J. N. Brown, & P. M. Sant (Eds.), *Indigeneity: Construction and Re/Presentation* (pp. 231–249). Commack, NY: Nova Science Publishers.

Saarinen, J. (2005). Tourism in the Northern Wildernesses: Wilderness Discourses and the Development of Nature-Based Tourism in Northern Finland. In C. M. Hall, & S. W. Boyd (Eds.), *Nature-based Tourism in Peripheral Areas: Development or Disaster?* (pp. 36–49). Bristol, UK:Channel View Publications.

Saarinen, J. (2019). What are wilderness areas for? Tourism and political ecologies of wilderness uses and management in the Anthropocene. *Journal of Sustainable Tourism*, *27*(4), 472–487. doi:https://doi.org/10.1080/09669582.2018.1456543.

Samediggi. (n.d.). Culturally Responsible Sámi Tourism. Retrieved from https://www.samediggi.fi/ongoing-projects/culturally-responsible-sami-tourism/?lang=en (accessed on 10.02.2020).

Samerna i siffror – Samer.se. (n.d.). Retrieved from http://www.samer.se/1536 (accessed on 10.02.2020).

Sametinget. (2010). *Samisk upplevelseturism - Definition, kartläggning och förutsättningar för utveckling av samisk turism*. Kiruna: Sametinget.

Shultis, J., & Heffner, S. (2016). Hegemonic and emerging concepts of conservation: A critical examination of barriers to incorporating Indigenous perspectives in protected area conservation policies and practice. *Journal of Sustainable Tourism*, *24*(8–9), 1227–1242. doi:https://doi.org/10.1080/09669582.2016.1158827.

Smed, K. M. (2017). Culture in nature: Exploring the role of 'culture'in the destination of Ilulissat, Greenland. In A. Viken, & D. K. Müller (Eds.), *Tourism and Indigeneity in the Arctic* (pp. 137–153). Bristol, UK: Channel View Publications.

Stasch, R. (2014). Primitivist tourism and romantic individualism: On the values in exotic stereotypy about cultural others. *Anthropological Theory*, *14*(2), 191–214. doi:https://doi.org/10.1177/1463499614534114.

Stronza, A. L., Hunt, C. A., & Fitzgerald, L. A. (2019). Ecotourism for Conservation?. *Annual Review of Environment and Resources*, *44*, 229–253.

Suzer, O. (2019). Analyzing the compliance and correlation of LEED and BREEAM by conducting a criteria-based comparative analysis and evaluating dual-certified projects. *Building and Environment*, *147*, 158–170.

Testa, F., Iraldo, F., Vaccari, A., & Ferrari, E. (2015). Why eco-labels can be effective marketing tools: Evidence from a study on Italian consumers. *Business Strategy and the Environment*, *24*(4), 252–265.

Valkonen, J., & Valkonen, S. (2014). Contesting the Nature Relations of Sámi Culture. *Acta Borealia*, *31*(1), 25–40. doi:https://doi.org/10.1080/08003831.2014.905010.

VisitSapmi - kriterierna. (n.d.) Retrieved from https://web.archive.org/web/20160823050133/https:/www.visitsapmi.org/kriterierna.html (accessed on 10.02.2020).

Vivanco, L. A. (2007). The prospects and dilemmas of indigenous tourism standards and certifications. *Quality Assurance and Certification in Ecotourism*, *13*, 218–240. doi:https://doi.org/10.1079/9781845932374.0218.

Walker, K., & Moscardo, G. (2014). Encouraging sustainability beyond the tourist experience: ecotourism, interpretation and values. *Journal of Sustainable Tourism*, *22*(8), 1175–1196.

Walker, K., & Moscardo, G. (2016). Moving beyond sense of place to care of place: the role of Indigenous values and interpretation in promoting transformative change in tourists' place images and personal values. *Journal of Sustainable Tourism*, *24*(8–9), 1243–1261. doi:https://doi.org/10.1080/09669582.2016.1177064.

Wang, C. C., Cater, C., & Low, T. (2016). Political challenges in community-based ecotourism. *Journal of Sustainable Tourism*, *24*(11), 1555–1568.

Ween, G. B., & Riseth, J. Å. (2017). Indigenous hospitality and tourism: Past trajectories and new beginnings. In A. Viken, & D. K. Müller (Eds.), *Tourism and Indigeneity in the Arctic* (pp. 205–221). Bristol, UK: Channel View Publications.

Winsa, B. (2007). Social capital of indigenous and autochthonous ethnicities. In L. P. Dana (Ed.), *International handbook of research on indigenous entrepreneurship* (pp. 257–286). Cheltenham, UK: Edward Elgar Publishing.

Zografos, C., & Allcroft, D. (2007). The environmental values of potential ecotourists: A segmentation study. *Journal of Sustainable Tourism, 15*(1), 44–66.

12
THE ROLE OF THE VISITOR IN STEWARDSHIP AND VOLUNTEERING IN TOURISM

James Malitoni Chilembwe

Introduction

The term *stewardship* is "a recognition of a duty of care to the present and future custodians of valuable resources of the earth" (May, 1991, p. 116). They are caring and protecting resources in the natural and cultural environment rest in all people inclusive of visitors, tourists, and local communities. The involvement has its purpose, to maintain the resources in the present status for the benefit of the present tourism consumers including the future visitors as well as the host communities in the form of sustainability. In order for sustainability to function properly, it requires responsible human actions to preserve the natural environment and its resource in their present form for the benefit of future generations. Sustainability should truly connect humans and the natural world (Fennell, 2019), and particularly of the marginalised communities (Acquino & Andereck, 2018). In order to exercise care of both the natural environment and local communities to sustainability, some visitors to various places of tourist interest take volunteering activities. Again, there is a growing demand for volunteerism and international growth in popularity for visitors, and tourists to travel for a volunteering activity (Guttentag, 2009; Wearing & McGehee, 2013). Despite the growing demand for volunteering, there is also questioning of the actual benefits of trekking to the host communities through volunteerism (Tomazos & Butler, 2012; Mostafanezhad, 2014). Sometimes foreign interests are put in the forefront at the expense of the desires of host communities (Palacios, 2010), and volunteers without the required skills can go a volunteer work, consequently hindering development projects and produce unwanted results (Guttentag, 2009). The true value and cost with regards to the triple bottom line are equally questionable (Goodwin, 2011; Wearing, 2001). Other criticisms facing volunteer tourism are profit-driving factors that largely benefit organisations. First, over-promising benefits that organisers and volunteers fail to accomplish; second, harming destinations as a result of lack of local knowledge and ethics; third, creating communities dissatisfaction due to lack of respecting host community cultural values (Benson & Henderson, 2011; Crossley, 2012; Simpson, 2004; Tomazos & Cooper, 2012; Lyons, Hanley, Wearing, & Neil, 2012). Also, a recent study by Acquino and Andereck (2018) acknowledges that volunteer tourism cannot always benefit the host community, beyond the volunteer's contribution mainly in marginalised communities.

DOI: 10.4324/9781003001768-12

Volunteer tourism brings together both local and international volunteers in tourism. Individuals partake on working holiday, volunteering their labour for causes that are meaningful to the volunteers and their related organisations (Tomazos & Butler, 2009a, 2009b), including being considered as a hero for volunteering in some causes (Tomazos & Butler, 2010). Volunteer tourism is promoted as a way of experience authenticity within the context of alternative tourism beneficial to a destination, leading to expectations of a responsible tourism ethos. Wearing (2001, p. 1) defines the purpose of volunteer tourism as "aiding or alleviating the material poverty of some groups in a society; the restoration of certain specific environments or research into aspects of society or environment", carried out alongside touristic activities. In volunteer tourism, it involves utilising discretionary time and income to travel out of the normal place, where most of the activities take place, to a completely new environment to assist others in need (McGehee & Santos, 2005). Others refer to the term *volunteer tourism* as 'voluntourism' or 'international volunteering' (Schwarz, 2018). Due to its dynamism use of the term *volunteer tourism*, and linking to its motivating factors and outcomes, it is becoming challenging to conceptualise the term by using a sole idea to interpret its meaning. Although, volunteering holidays contribute to the "reduction in barriers to travel, also an increase in the middle class in many developing countries, and the desire of that middle class to seek out more unusual travel experience" (Wearing & McGehee, 2013, p. 121). However, more increasingly are also other travel motivations that may clash with the requirements and consequences of volunteer activity (Palacios, 2010). Some are engaged in volunteer activity with the purpose of learning and acquiring a new experience (Sin, 2009). Given such motives, there are questioning their volunteering if they will adhere to issues of local, national and international ethics, which needs to be taken into consideration when doing their activities. Applying international ethics may be easy to obey due to internationalisation and industrialisation. However, most of the international volunteers in tourism find it challenging to adhere to local ethics. These are ethics applying to local cultures in many developing countries. Most of the volunteers in tourism can obtain basic training in some of the national ethics—general cultural practices that are applicable at a national level, for example; national language, greetings, some acceptable national dressing, and beliefs.

Nevertheless, it is difficult to train international volunteers in everything due to the given short time of their visit to a destination. Challenges may, as well as appear because of the various motives driving the volunteering activities. For example, the search for autonomous freedom, seeking out obligations and dependencies. For example, there is a need for mutual understanding and interest between guests and hosts. However, due to personal motives meetings places for activities can become places for both sides to gaze at each other (Krippendrf, 1987). There may also be personal beliefs, whereby volunteer tourism becomes a spiritual mission to spread philosophies to hosts whose cultural beliefs may differ. The behaviours of egoism may appear, whereby volunteer tourists make themselves feel better through an opportunity for superiority (Taillon & Jamal, 2010). Expectations by volunteers from the hosts may create some differences, particularly, when local cultural beliefs play a role, in which volunteer tourists ignore them or decide to understand from the perspective of another destination they had already volunteered. Another factor can be governance challenges to level the playing field between the host community and volunteers on the understanding of local ethical practices that sustainable tourism in the destination (Chilembwe & Mponda, 2016). Many volunteers in tourism programmes do not critically consider issues of poverty and inequalities when imposing development agenda (McGehee, 2012), for volunteering based on countries where volunteers originate and the kind of solutions, and they are to offer to benefit the host destinations. Also, volunteering in tourism should aim at empowering the local communities (Ong, King, Lockstone-Binney, & Junek, 2018), by providing them with the know-how knowledge, and skills to handle activities that sound technical and suitable for volunteer experts.

Stewardship and volunteerism managing approaches

The success of stewardship and volunteer tourism depends on the willingness of volunteers to engage in a travel experience that has an approach to assist a destination community and engage in work that more mainstream tourists may view as 'hardship' (Acquino & Andereck, 2018). Both stewardship and volunteer tourism should aim at making a legitimate contribution, create a positive impact on the host communities, especially in less developed countries. Volunteer tourism should also develop a mutually beneficial relationship between the visitor-tourists, visitor-tour operators, and the host community (Beirman, Upadhayaya, Pradhananga, & Deray, 2018; Butcher & Smith, 2010). Moreover, they must drive their community projects using a bottom-up approach and accommodate volunteer tourists to have their experiences to learn some of the local ethics through interactions. Therefore, there is a need to reconsider the nature of stewardship and volunteer tourism using development aid if working properly to the mutual understanding of local ethics and beliefs between the tourists and hosts. According to Wearing, Young, and Everingham (2017), volunteer tourism should move away from developmental aid to an environment where people can agree and exchange cultural values. A volunteer can be local or inter-organisation, non-governmental organisations (NGOs), charitable organisations, universities, conservation agencies, religious organisations, and a growing of non-profit organisations (Briggs, Peterson, & Gregory, 2010).

These organisations use a variety of models to manage volunteers in tourism and the wildlife environment. Some of the examples of management models are outsourcing volunteers, joint ventures with profits, and non-profit community organisations willing to work on various tourism projects. Outsourcing to a community is suitable for convenience and conservation. However, it may be low for the community model. Environment and conservation criteria are all important aspects of the choice of a management model. However, Spenceley, Snyman, and Eagles (2019) note, the model to contribute to biodiversity conservation is high with NGOs and insourcing, moderate with joint ventures and for-profits, and slow for community model. Using NGOs approaches is functional by using existing NGOs for delivery of services, such as operating a festival or special event. Second, the development of specialised NGOs dedicated to volunteer activities, and can also be called friends of the volunteers. It is important to utilise volunteers when managing the environment and conservation. Using NGOs could handle low visitor numbers, perhaps, due to the ability of NGOs to utilise volunteers, rather than paid staff, providing some buffer capacity in times of low tourism volume and low income. As Spenceley et al. (2019) indicate, the management of expectation is critical to the success of the long-term sustainability of the management model. However, environmental degradation could happen after a long time despite managing the environment and conservation by using NGOs to utilise small numbers of volunteers (Nyirenda & Chilembwe, 2015). Environmental degradation may happen fast or slow, that is, regardless of how small or large the number of stewards engaged at the place because they can also partake in other leisure activities in the same natural environment.

Stewardship and volunteer activities

Visitors can volunteer on several types of tourism services depending on their capabilities in areas of lodges, restaurants, campsites, horse trails, guided walking or hiking trails, fishing, filming, rafting, and boat transport or cruises, mountaineering, and rock climbing. They can also volunteer in the infrastructure that supports services such as roads, airstrips, electrical distribution, communication facilities, water supplies, waste management, and security.

The benefits to the host community and the promotion of personal learning primarily drive young adult volunteer tourists (Francis & Yasué, 2019). However, there could be frustrations in the case voluntourism provides limited benefits to host communities. On a positive note, despite such frustrations, the volunteer tourists acquire personal growth and learning as the decision of whether to withdrawal from volunteering takes place after a certain time. Several volunteer activities exist as categorised below into the following areas:

- volunteering requiring expertise, knowledge, and skills,
- volunteering not requiring expertise, knowledge, or skills,
- volunteering requiring local or community and national ethics, and
- volunteering not requiring local or community and national ethics.

Volunteer tourists or volunteer tourists seek travel experiences that are mutually beneficial for themselves and the host community (Francis & Yasué, 2019; Palacios, 2010; Woosnma & Lee, 2011). Participants provide free labour as well as direct and indirect economic stimulation to benefit the host community and contribute to sustainable development (Chilembwe & Mweiwa, 2019; Frilund, 2018; Wearing & McGehee, 2013). With such impacts on the host community, voluntourism may help educate volunteer tourists by fostering cross-cultural communication and understanding, raising consciousness about racism, and privilege about building knowledge, skills, and confidence (Bailey & Russell, 2010; Raymond & Hall, 2008). Such personal lessons and enhanced self-efficiency could lead to broader societal change if experiences spur greater civic engagement, activism, and pro-social or pro-environmental behaviours when the volunteer tourists return home (Brondo, 2015). They can have many reflections after a volunteer experience and understanding some dynamics of other societies. Some of the volunteers may have repeat visits organised by themselves, to check how the communities have changed over time, but also understand some areas of help needed to contribute something that may be meaningful in their society.

Volunteer tourists as a diverse group of people with different backgrounds, ages, and skills who are motivated to see a range of different types of experiences (Francis & Yasué, 2019; Wearing & McGehee, 2013). Volunteer tourists have philanthropical motivations, such as 'giving back' to the host communities and helping people with less privilege (Bailey & Russelll, 2012). Besides that, there are self-serving interests such as pleasure, adventure, recognition, skill training, personal growth, and making friends (Lo & Lee, 2011). The interaction between the visitors and the hosts, as well as their blend would likely encourage them to share the skills and learn more from those who have better experiences like environment and activities assigned to work.

Other studies of voluntourism indicate that volunteer tourists' partial motivation is to benefits the host community (Ballantyne, Packer, & Hughes, 2009; Curtin, 2010; D'Souza et al., 2019; Wearing & McGehee, 2013). However, there is also questioning whether voluntourism can benefit the host communities (McGloin & Georgeou, 2016). It appears many Western countries increasingly run voluntourism by for-profit businesses. On the contrary, in sub-Saharan Africa, volunteer tourism is to a larger extent based on culture, mainly on 'Ubuntu' concept (Kayuni & Tambulasi, 2012). Ubuntu means becoming a humane being, having a kind heart of helping others without expecting something back as a benefit.

The volunteering programs in tourism can focus on either short-term or medium-term activities. However, voluntourism is run by for-profit businesses, whose programs may prioritise the volunteers' self-interested motivations over the goal of designing the projects that have long-term benefits to the host communities (McGloin & Georgeou, 2016; Strzelecka, Nisbett, & Woosnam, 2017). McLennan (2014) argues that placements of unskilled people in the short term do not benefit local communities but rather harm the communities because of the culture or local context by the volunteer tourists. Besides, voluntourism may have limited educational benefits for volunteer tourists (Francis & Yasué, 2019), as they expect to learn more things in the process of volunteering activities. As such, they would add very little value that can benefit the local communities and the natural environment where the stewardship is taking place.

In order to contribute to a meaningful social change in society, volunteer tourism providers need to emphasise the broader intrinsic values of voluntourism such as helping less privileged people, protecting rare species, and contributing to positive social change in one's communities (Francis & Yasué, 2019; Bailey & Russelll, 2012; McGehee & Santos, 2005; Shaw & Miller, 2016). Volunteer tourists should prioritise ecotourism sustainability, by incorporating the socio-environmental aspects, particularly of personal responsibility for sustainability issues and actions, and developing an ethics of care (Jamal & Camargo, 2012; Mair & Laing, 2013; Miller, Rathhouse, Scarles, Holmes, & Tribe, 2010). Good stewardship takes into consideration of responsibility for the sustainability of the present and future valuable resources in the environment.

The International Ecotourism Society (The International Ecotourism Society [TIES], 2015) defines ecotourism as responsible travel to natural areas that conserve the environment, sustain the wellbeing of the local people, and involves interpretation and education. In ecotourism, voluntourism should accommodate both visitors and staff as well as volunteer tourists, based on the principles as follows:

- minimising physical, social, behaviour, and negative psychological impacts,
- building environmental and cultural awareness and respect,
- provides positive experiences for both visitors and hosts, including volunteers who could help provide sensitivity to host countries' political, environmental, and social climate, and
- recognise the rights and spiritual beliefs of the indigenous people in the community and work in partnership with them to create empowerment (TIES, 2015).

Local communities play a critical role in enhancing positive outcomes for the conservation of culture and its environment, working closely with the visitors and volunteer tourists in order to have their better experiences of the community. (Chilembwe, 2019). Also, local guides contribute to both the local communities' environment and visitor experiences (Chilembwe & Mweiwa, 2014). However, their contribution does not go beyond the narrow education focus goal of interpretation in ecotourism (Walker & Weiler, 2017). Therefore, the local guides should extend to assisting tourists and visitors, in general, to identify their role and responsibility in sustainability outcomes. There is a need for the guides to impact visitors' or tourists' attitudes and behaviour changed beyond their specific tourist experience (Ballantyne, Packer, & Falk, 2011; Packer & Ballantyne, 2013).

Again, there is a need to connect the interpersonal experience to visitor's personal life (Walker & Moscardo, 2014) as the local guides provide a critical role in fostering sustainable wildlife (Ballantyne et al., 2009; Curtin, 2010). As volunteering in tourism can be local or inter-organisations, NGOs, charitable organisations, and the growth of for-profit organisations. Wearing and McGehee's (2013) review of volunteer tourism research describes a strong relationship between volunteer tourism and ecotourism. The reports indicate that both share parameters such as similarity developed infrastructure, supplied marketing by the tourism industry, established environmental and cultural carrying capacities with strict monitoring, and environmentally sensitive behaviour and operations from both the tourists and operators. Volunteering in an ecotourism environment can help the formulation of symbolic relationships between tourists, Indigenous people, and natural areas (D'Souza et al., 2019). However, volunteering in ecotourism is not without obstacles when willing to achieve positive results in the communities. For example, volunteer tourists cannot adhere to the themes of commitment, communication, and cultural understandings at the forefront. There could also be language barriers between the volunteers and the host communities. As such, there is a need for special considerations to be given to the most vulnerable to minimise negative effects on their emotional well-being. Again, there should be cultural understanding and better social representation of volunteer tourism, as cross-cultural understanding does not happen automatically (Raymond & Hall, 2008). The social impact of volunteer tourism should also consider the ethics of medical volunteering, and not only those international volunteers who can work in orphanages but also medical volunteering in a wildlife setup. These are medical volunteers who have the know-how skills to detect various sicknesses and treat animals in the natural environment.

People associated with protected areas should have knowledge and skills that enable conservation, and cultural assets in protected areas and stewardship (McCool, Nkhata, Breen, & Freimund, 2013), as well as, in the protected area and asset stewardship (Jepson et al., 2017). Human assets, which comprise nature cultural attributes and dynamic involves the interaction between the protected areas and wider cultural practices and narratives to create a public identity for the protected areas (Jepson et al., 2017). The volunteers in tourism should take into account all aspects of sustainability, with a high priority on the social and cultural aspects of the local communities or cultural conditions of the communities that volunteer tourism projects design to influence the mind-set change between the visitors and the hosts.

Traditionally, volunteer tourism, when utilised correctly, is a powerful strategy for a sustainable alleviation form of tourism (Wearing, Beirman, & Grabowski, 2020). However, they also argue that the driving motives for people to have opportunities for volunteer travel holidays and engage in volunteer activities are disasters, for example, the December 2004 Indian Ocean Tsunami (Wearing et al., 2020). Examples in the conservation area are wildlife rescue and treatment, wildlife protection, and wildlife translocation (see also Table 12.1) for further areas of volunteering. Contrary to a sustainable approach, Subă (2017) and Ooi and Laing (2010) found out that the motives for people to volunteer in tourism abroad are travelling and leisure, experiencing a new culture and people, acquiring advantage at the labour market, gaining experience from abroad, and altruism. These motivations are to a larger extent for the self-benefiting as they provide little community benefits. The word 'self-benefiting' is somehow in confirmation to McGehee (2012), who indicated that most volunteer tourists demonstrate a sense of self-efficacy. They have a belief that one is capable of doing an activity and making a change in society through joining volunteering programmes. However, taking part in volunteering work simply strengthening the personal motives as they gain volunteering experience while working, learning from others, and also forming part of the personal and social development of volunteers (Chen & Chen, 2011). However, in some other circumstances, the kind of volunteers in tourism needed must be those who have highly specialised skills and knowledge in order to volunteer effectively. Even those that may want to gain experience through learning will find it hard to understand and follow up on what is

Table 12.1 Department and projects to attach volunteers

Wildlife department/section	*Volunteering project*
Wildlife rescue and welfare	• Wildlife reintroduction programmes • Clinical projects in wildlife health • Wildlife emergency response unit • Wildlife rescue projects
Wildlife enforcement	• Wildlife policy and law • Wildlife crime investigation • Wildlife justice • Wildlife detection dogs
Wildlife research	• Elephants research • Monkey research • Primate conservation research • Conservation medicine research • Illegal wildlife trade
Conservation education and advocacy	• Conservation education • Protected areas conservation education • Stop wildlife crime campaigns • Parliamentary conservation assembly lobby • Community projects

going on due to their given short time of attachments, as a combination of theories or classroom lessons and practices are critical to have a meaningful contribution as a steward.

Applying a case study: Conducting stewardship and volunteering activities

This case study of WLC Centre is useful to governments, government agencies, and non-governmental organisations (NGOs) interested in protecting natural environmental areas and conserve wildlife. It is also useful to the visitors or general public including tourism volunteers to act responsibly by avoiding engaging in wildlife illegally for trade and subject animals to a lifetime in captivity, injuries from snares, traps, spears, or bullets which lead to slow, painful deaths to animals. Finally, it may also be useful to those animal rights groups or defenders who are concerned with the decline of wildlife homes which has led to an increase in human-wildlife conflict. The centre was established in 2008 as a wildlife rescue, rehabilitation, and education facility to support the government's initiatives in fighting against wildlife crimes such as pets and bushmeat, and ivory trade (GoM, 2012).

The centre provides many activities that both visitors or volunteers may take part ranging from wildlife law enforcement, wildlife research, environmental education and training, and advocacy. Besides, international visitors and tourism volunteers, many schools and communities, both in urban and rural areas utilise the centre's programmes. They deliver awareness messages about the dangers of wildlife poaching. They also provide friendly solutions to income-generating activities like the production of green energy stoves. The centre stepped in after realising that the demand for products like charcoal and firewood is very high in Malawi, mainly in urban areas, which exerted more pressure on natural resources. It is part of the solution to reduce deforestation and loss of habitat for animals, hence subjecting danger to wildlife. While urban areas are expanding, there is also a demand for more land by the urban residents to use, leading to the human's activities conflicting with the remaining wildlife in the urban wildlife parks. They introduced several community projects targeting schools in urban areas, members of parliament, wildlife managers, city residents, media, communities surrounding protected areas, police officers, and airport staff in order to work together to stop wildlife crime. There is also community activism, with the help of volunteers; they organise awareness campaigns against poaching of wildlife animals like elephants and rhinos. They involved the community such as the youth, scouts, non-governmental organisations (NGOs), media, and members of parliament. They place billboards at strategic places as one way of sensitising both staff, passengers, and the public on the implication of trafficking ivory or other wildlife products. They train airport personnel on the new wildlife legislation policy and how to identify ivory which is in the process of being trafficked through airports after endangered species have been poached. These activities and approaches explained are easy for visitors and any tourism volunteer to take the park, provided there is an interest in the preservation of wildlife including in the urban areas. Volunteers who do not have skills and knowledge can easily learn from others as they interact with each other. However, issues bordering cultural differences cannot are hard to eliminate, though minimal, and as work expects to accommodate local, national, and international ethics.

In contrast, volunteering in other projects (see Table 12.1) requires expertise and technical know-how. For example, in wildlife rescue and welfare, there are projects such as clinical projects in wildlife health, wildlife rescue projects, and wildlife emergency response projects that need volunteers with some background knowledge in wildlife management. Likewise, all the projects under wildlife enforcement and wildlife research are technical requiring volunteers' skills and knowledge. Those volunteers with motives to learn and acquire experiences will likely find it hard to apply local, national, and international ethics when doing their work.

Furthermore, the centre created an innovative recycling bin which was designed exclusively for recyclable materials. As Lane (2018) notes, recycling is a widespread practice in developed countries; for example, Austria and the USA. However, the recycling practice in Malawi is still in an infancy stage. The centre introduced recycling practices and provides necessary training in collaboration with Environmental Foundation, and International Conservation and Clean up Management partners. They set up a collection point in the city. However, they attached recyclables collection from homes at a small fee per month. This development motivated some of the visitors and the communities who received training from the volunteers at the centre. Some local people can now afford to generate income through collecting recyclable products such as cans, papers, glasses, plastic bottles, and bags for sustainable construction projects. This initiative is likely to provide more opportunities for the local community at large who visit the centre to learn and practice practical waste management on a volunteer basis.

Further, the initiative is likely to contribute positively to behaviour change not only among the youth and students but also on several families and the general community, as many people learn from volunteers, to build their useful structures from recyclables. The practice expects to expand extensively and contribute to a cleaner and greener Malawi in the long-term process. Besides, the centre through tourism volunteers is also involved in the outreach and education initiative to assist women in the community in making fuel-briquettes. Fuel-briquette uses waste papers from offices, water, and sawdust. These are pounded into a gluey mixture and left to dry for 24 hours to make a brick-like fuel-briquette or green energy. They support women through training on how to make and use green energy stoves. Briquette is an excellent source of energy for local communities, it is cheaper as compared to charcoal, generates fewer emissions, and it is a healthier method of cooking. Besides that, it also engages the communities in forestation activities. They work jointly with schools and the local community to plant trees. They have planted over 20,000 trees and both local and international volunteers were engaged, see other activities Table 12.2 for volunteerism.

Table 12.2 Some of the activities for volunteerism in environment and conservation

Conservation	*Environmental education*	*Wildlife campaigns*
• Collaborate with schools in urban areas to convey messages about urban biodiversity, and how it supports ecosystem services like air and quality water • Community forest for household activities and income-generating activities • Tree planting activities • Providing stoves lessons	• Provides awareness to the communities on the dangers of illegal wildlife trade like ivory • Deliver programmes on specific issues that affect communities, for example, conflicts, wildlife welfare, and conservation • Delivery of practical recycling practices • Provides economic opportunities to the local community through the production of green energy (fuel-briquette)	• Community activism • Domestic and international travellers' sensitisation

These responsibilities are incorporated into the projects, depending on the goals and objectives of the organisation. For example, the centre decided to embark on programmes that are related to wildlife environment, education, and conservation to engage the youth, schools, and community at large as well as economically empowering the communities while disseminating conservation messages against wildlife crime and protecting both the wildlife and environment. They engage in wildlife-related activities related to the environment and maintain wildlife. It is supported by over 45 international organisations whose and most of their clients are interested in visiting protected areas and viewing wildlife. Similarly, the organisations that support wildlife centres' programmes connect most of their international volunteers to the centre. They worked jointly with international conservation organisations and local organisations; for example, members of parliament, local schools, legal firms, and community at large to deliver various initiatives effectively. Both local and international volunteers also play a significant role in the delivery of their projects to preserve the environment for wildlife sustainability.

The strength of the wildlife centre is that it operates from a private-sector perspective through a concession basis it received from the Malawi government. The centre wears a new face with better infrastructure, increased city wildlife, and activities, as well as visitors unlike when the government was operating the centre. It has established several network partnerships and donors who are supporting most of the projects. It also contributed tremendously to the improvement of wildlife policy change in Malawi by working with legal partners and Malawi Parliament to pass laws for wildlife protection. However, the centre is not without implications. First, the centre is mostly dependent on donors who provide financial and material support to their key projects including operations and management of all activities. It seems the centre does not generate adequate income on its own. As Moffett (2010) notes, wildlife conservation centres are generally capable of making business sense and generate more revenue and profits. The wildlife centre of this nature needs to generate substantial income and make profits for its sustainability.

Given that the donors and partners have decided to withdraw their financial support, it is likely that all projects that are currently running will face some challenges for continuity in the long-term process. Consequently, most of the international volunteers, who partake in activities due to some sort of partnership with a centre, will likely disappear. Similarly, illegal wildlife campaigns and fights for wildlife welfare are supposed to be carried out for several years to provide a proper transition from one generation to another in terms of best approaches or practices in wildlife conservation areas and waste recycling practices and management. The prolonged use of international volunteers, as well as local volunteers, makes them share a variety of ideas to move forward the awareness campaigns about the environment and conservation of wildlife.

To this end, some local ethics or beliefs liked to nature of the working environment in developing countries like Malawi. Stewards and volunteering in tourism should, therefore, understand the links in a wildlife set-up environment situation. Volunteers should have some knowledge about the variety of animal behaviours. Besides having skills and expertise may sometimes not be enough. It is critical also to make reflections on these questions when volunteering in a wildlife environment:

- can someone volunteer in wildlife set-up without knowledge in animal behaviours;
- are there any approaches to volunteers when defending from animals without causing any harm to wildlife;
- how to associate with animals like elephants and lions without being harmed in the cause of volunteering;
- what to do when faced with danger among wild animals; and finally
- can someone volunteer in a wildlife environment without having specific vaccinations about wild animals?

There is one of the widely known beliefs, which says 'do what Romans do when among them'. Similarly, a certain section of people who work in the natural wildlife environment has a belief to turn themselves into any wildlife animal when they are in danger among the wild animals. They act locally using local ethics, a defense mechanism approach by becoming part of the wildlife and return to normal human beings later when out of danger. As a local approach, it works for those who are well conversant with the practice working or surrounded by the natural environment. However, the approach may not be supported or applied as an international practice. As a result, international volunteers in tourism cannot know the practice and it is also challenging to learn and master the practice.

Conclusion and recommendations

The extent to which the role of a visitor in stewardship and volunteering in a host destination is practised depends so much on the values of both volunteer organisers and the local communities (hosts). The local ethics, cultural practices, and beliefs cannot be detached among the local communities. These local cultural attachments present some challenges to the role of visitors in stewardship and volunteering in tourism because of some gaps in knowledge and practices. While some practices are trainable and easy to adopt by the visitors or volunteers, some of them require more time to understand, given the limited time most of the volunteers spend on volunteering their work. As McGehee and Santos (2005) define, volunteering is a form of tourism activity, which aims to provide an alternative journey or visitation that can help tourism development and ecological restoration in a community. It is, therefore, a need for playing the role of a visitor's stewardship and volunteering ethically and responsibly, while exploring attractions and enjoying activities in a given cultural or natural context and contribute to the sustainability of the tourism environment. Volunteerism should provide a beneficial difference in a community and its stakeholders. Visitor stewardship with a strong relationship that builds on the understanding of local ethics has the potential power to change the mindset of the local communities through community activity engagement in projects and empowerment of the host community. It is, therefore, recommended that visitors willing to engage in tourism volunteering should be flexible to learn some of the national and local ethics and practices in tourism destinations where activities take place. While it is understandable that they cannot gain knowledge in all the local practices, but knowing some of the practices is crucial (practices across the country or region), helps to minimise some of the challenges that volunteers may face when interacting with the local people. It is also critical and advisable that first-time visitors or volunteers should go through some briefings and orientations before they can engage in meaningful volunteering activities. Finally, the case study applied is useful to governments who wish to attract responsible and commercially successful tourism investors into natural areas. It helps the community to benefit through tourism and be responsible for the conservation of natural resources and wildlife. The case demonstrates the type of impacts the private sector can apply in destination, on the natural environment, and society. The case also provides the basic enabling environmental requirements that require attracting the best volunteers to support the adoption of best practices under conditions of a natural environment. While motives for embarking on volunteering tourism vary, some of them do not require and skills and knowledge, as well as local or national beliefs. However, there should be an understanding that other volunteering activities require high skills and knowledge, in addition to adherence to local, national, and international ethics.

References

Acquino, J. F., & Andereck, K. (2018). Volunteer tourists' perception of the impacts on marginalised communities. *Journal of Sustainable Tourism*, *26*(11), 1967–1983.

Bailey, A. W., & Russell, K. C. (2010). Predictors of interpersonal growth in volunteer tourism: A latent curve approach. *Leisure Sciences*, *32*(4), 352–368.

Bailey, A. W., & Russelll, K. C. (2012). Volunteer tourism: Powerful programs or predisposed participants? *Journal of Hospitality and Tourism Management*, *19*(1), 123–132.

Ballantyne, R., Packer, J., & Falk, J. (2011). Visitors' learning for environmental sustainability: Testing short and long-term impacts of wildlife tourism experiences using structural equation modelling. *Tourism Management*, *32*, 1243–1252.

Ballantyne, R., Packer, J., & Hughes, K. (2009). Tourists' support for conservation messages and sustainable management practices in wildlife tourism experience. *Tourism Management*, *30*, 658–664.

Beirman, D., Upadhayaya, P. K., Pradhananga, P., & Deray, S. (2018). Napal tourism in the aftermath of the April/May 2015 earthquake and aftershocks: Repercussion, recovery and the rise of new tourism sectors. *Tourism Recreation Research*, *43*(4), 544–554.

Benson, A. M., & Henderson, S. (2011). A strategic analysis of volunteer tourism organisation. *The Service Industries Journal*, *31*(3), 405–424.

Briggs, E., Peterson, M., & Gregory, G. (2010). Toward a better understanding of volunteering for non-profit organisations: Exploring volunteers' pro-social attitudes. *Journal of Macromarketing*, *30*, 60–76.

Brondo, K. V. (2015). The spectacle of saving: Conservation voluntourism and new neoliberal economy on Ultila, Honduras. *Journal of Sustainable Tourism*, *23*(10), 1405–1425.

Butcher, J., & Smith, P. (2010). Making a difference: Volunteer tourism and development. *Tourism Recreation Research*, *35*(1), 27–36.

Chen, L.-J., & Chen, J. S. (2011). Motivations and expectations of international volunteer tourists: A case study of 'Chinese Village Traditions'. *Tourism Management*, *32*(2), 435–442.

Chilembwe, J. M. (2019). Nature tourism, wildlife resources, and community-based conservation in Malawi. In M. T. Stone, M. Lenao, & N. Moswete (Eds.), *Natural resources, tourism and community livelihoods in southern Africa: Challenges for sustainable development* (pp. 26–37). London, UK: Routledge.

Chilembwe, J. M., & Mponda, I. K. (2016). Tourism sustainable governance practices in Malawi as a tourist destination: The challenges and opportunities for tourism development. *Tourism Spectrum, 2*(1), 1–10.

Chilembwe, J. M., & Mweiwa, V. R. (2014). Tour guides: Are they tourism promoters and developers? Case study of Malawi. *International Journal of Research in Business Management, 2*(9), 29–46.

Chilembwe, J. M., & Mweiwa, V. R. (2019). Responsible travel and tourism adventure: Evidence from Malawi as a tourist destination: In A. Sharma (Ed.), *Sustainable tourism development: Futuristic approaches* (pp. 31–53). New York, US: Apple Academic Press.

Crossley, E. (2012). Poor but happy: Volunteer tourists' encounters with poverty. *Tourism Geographies, 14*(2), 235–253.

Curtin, S. (2010). Managing the wildlife tourism experience: The importance of tour leaders. *International Journal of Tourism Research, 12*, 219–236.

D'Souza, C., Taghian, M., Marjoribanks, T., Sullivan-Mort, G., Manirujjaman, M. D., & Singaraju, S. (2019). Sustainability for ecotourism, work identity and role of community capacity building. *Tourism Recreation Research, 44*(4), 533–549.

Fennell, D. A. (2019). Sustainability ethics in tourism: The imperative next imperative. *Tourism Recreation Research, 44*(1), 117–130.

Francis, D. A., & Yasué, M. (2019). A mixed-methods study on the values and motivations of voluntourists. *Tourism Recreation Research, 44*(2), 232–246. Doi:10.1080/02508281.2019.1594575.

Frilund, R. (2018). Teasing the boundaries of 'volunteer tourism'" local NGOs looking for global workforce. *Current Issues in Tourism, 21*(4), 355–368.

GoM. (2012) *Malawi Growth and Development Strategy*. Zomba: Government Press.

Goodwin, H. (2011). *Taking Responsibility for Tourism*. Oxford: Goodfellow Publishers.

Guttentag, D. (2009). The possible negative impacts of volunteer tourism. *International Journal of Tourism Research, 11*, 537–551.

Jamal, T., & Camargo, B. A. (2012). Sustainable tourism, justice and an ethic care: Towards the just destination. *Journal of Sustainable Tourism, 22*(1), 11–30.

Jepson, P. R., Caldecott, B., Schmitt, S., Carvalho, S. H. C., Correira, R. A., Gamara, N., ... & Ladle, R. J. (2017). Protected area asset stewardship. *Biological Conservation, 212*, 183–190.

Kayuni, M. H., & Tambulasi, R. I. (2012). Ubuntu and corporate social responsibility: The case of selected Malawian organisations. *African Journal of Economic and Management Studies, 3*(1), 64–76.

Krippendrf, J. (1987). *The Holiday Makes: Understanding the Impact of Leisure and Travel*. Oxford: Butterworth-Heineman.

Lane, H. (2018). *The Wildlife Centre Becomes a Recycling Hub*. Lilongwe: WLO.

Lo, A. S., & Lee, C. Y. S. (2011). Motivations and perceived value of volunteer tourists from Hong Kong. *Tourism Management, 32*(2), 326–334.

Lyons, K., Hanley, J., Wearing, S., & Neil, J. (2012). Gap year volunteer tourism: Myths of global citizenship? *Annals of Tourism Research, 39*(1), 361–378.

Mair, J., & Laing, J. H. (2013). Emerging pro-environmental behaviour: The role of sustainability focused events. *Journal of Sustainable Tourism, 21*(8), 1113–1128.

May, V. (1991). Tourism, environment and development: Values, sustainability and Stewardship. *Tourism Management*, 112–118.

McCool, S. F., Nkhata, B., Breen, C., & Freimund, W. (2013). A heuristic framework for reflecting on protected areas and their stewardship in the 21st century. *Journal of Outdoor Recreation and Tourism, 1–2*, 9–17.

McGehee, N. G. (2012). Oppression emancipation and volunteer tourism. *Annals of Tourism Research, 39*(1), 84–107.

McGehee, N. G., & Santos, C. A. (2005). Social change, discourse and volunteer tourism. *Annals of Tourism Research, 33*(3), 760–779.

McGloin, C., & Georgeou, N. (2016). Looks good on your CV: The sociology of voluntourism recruitment in higher education. *Journal of Sociology, 52*(2), 403–417.

McLennan, S. (2014). Medical voluntourism in Honduras: 'Helping' the poor? *Progress in Development Studies, 14*, 163–179.

Miller, G., Rathhouse, K., Scarles, C., Holmes, K., & Tribe, J. (2010). Public understanding of sustainable tourism. *Annals of Tourism Research, 37*(3), 627–645.

Moffett, R. (2010). What contribution can tourism make to biodiversity conservation? Presentation at the IUCN WCPA Tourism and Protected Areas Seminar on '*What Conservation can Tourism make to Biodiversity Conservation?*' 11 November, World Responsible Tourism Day, World Travel Market, London, UK.

Mostafanezhad, M. (2014). Volunteer tourism and the popular humanitarian gaze. *Geoforum, 54*, 111–118.

Nyirenda, B. M. C., & Chilembwe, J. M. (2015). Environmental impacts of tourism: Chose hill at Nyika National Park in Malawi. *International Journal of Research in Business Management, 3*(9), 17–24.

Ong, F., King, B., Lockstone-Binney, L., & Junek, O. (2018). Going global, acting local: Volunteer tourists as prospective community builder. *Journal of Recreation Research, 43*(2), 135–146.

Ooi, N., & Laing, J. H. (2010). Backpacker tourism: Sustainable and purposeful? Investigating the overlap between backpacker tourism and volunteer tourism motivations. *Journal of Sustainable Tourism, 18*(2), 191–206.

Packer, J., & Ballantyne, R. (2013). Encouraging reflective visitor experience in ecotourism. In R. Ballantyne, & J. Packer (Eds.), *International handbook on ecotourism* (pp. 169–177). Cheltenham: Edward Elgar.

Palacios, C. M. (2010). Volunteer tourism, development and education in postcolonial world: Conceiving global connections beyond aid. *Journal of Sustainable Tourism, 18*(7), 861–876.

Raymond, E. M., & Hall, C. M. (2008). The development of cross-cultural (mis) understanding through volunteer tourism. *Journal of Sustainable Tourism, 16*, 530–543.

Schwarz, K. C. (2018). Volunteer tourism and the intratourist gaze. *Tourism Recreation Research, 43*(2), 186–196.

Shaw, A. E., & Miller, K. K. (2016). Preaching to the converted? Designing wildlife gardening programs to encourage the unengaged. *Applied Environmental Education and Communication, 15*, 214–224.

Simpson, K. (2004). 'Doing development': The gap year volunteer-tourists and popular practice of development. *Journal of International Development, 16*(5), 681–692.

Sin, H. L. (2009). Volunteer tourism - "Involve me and I will learn"? *Annals of Tourism Research, 36*(3), 480–501.

Spenceley, A., Snyman, S., & Eagles, P. F. J. (2019). A decision framework on the choice of management model for park and protected area tourism services. *Journal of Outdoor Recreation and Tourism, 26*, 72–80. doi:https://doi.org/10.1016/j.jor.2019.03.004.

Strzelecka, M., Nisbett, G. S., & Woosnam, K. M. (2017). The hedonic nature of conservation volunteer travel. *Tourism Management, 63*, 417–425.

Subă, P. (2017). Motives for young people to volunteer abroad: A case study of AIESEC from the perspective of volunteer tourism. *African Journal of Hospitality, Tourism and Leisure, 6*(3). Retrieved from http://www.ajhtl.com/uploads/7/1/6/3/7163688/article_4_vol_6__3__2017_r.pdf.

Taillon, J., & Jamal, J. (2010). Understanding the voluntourist: A qualitative study. *Voluntourist Newsletter, 4*(1), 1–3.

The International Ecotourism Society (TIES). (2015). What Is Ecotourism? Retrieved from https://www.ecotourism.org/what-is-ectourism (accessed on 25.03.2020).

Tomazos, K., & Butler, R. (2009a). Volunteer tourism: The new ecotourism? *Anatolia: An International Journal of Tourism and Hospitality Research, 20*(1), 196–211.

Tomazos, K., & Butler, R. (2009b). Volunteer tourism: Working on holiday or playing at work. *Tourismos: An International Multidisciplinary Journal of Tourism, 4*(4), 331–349.

Tomazos, K., & Butler, R. (2012). Volunteer tourists in the field: A question of balance? *Tourism Management, 33*(1), 177–187.

Tomazos, K., & Cooper, W. (2012). Volunteer tourism: At the crossroads of commercialisation and service? *Current Issues in Tourism, 15*(5), 405–423.

Tomazos, S. R., & Butler, R. (2010). The volunteer tourist as 'hero'. *Current Issues in Tourism, 13*, 363–380.

Walker, K., & Moscardo, G. (2014). Encouraging sustainability beyond the tourism experience: Ecotourism, interpretation and values. *Journal of Sustainable Tourism, 22*(8), 1175–1196.

Walker, K., & Weiler, B. (2017). A new model for guide training and transforming outcomes: A case study in sustainable marine-wildlife ecotourism. *Journal of Ecotourism, 16*(3), 269–290. doi:10.1080/14724049.2016.1245736.

Wearing, S. (2001). *Volunteer Tourism: Seeking Experience That Make a Difference*. Wallingford: CAB International.

Wearing, S., Beirman, D., & Grabowski, S. (2020). Engaging volunteer tourism in post-disaster recovery in Nepal. *Annals of Tourism Research*, *80*, 102802. doi:https://doi.org/10.1016/j.annals.2019.102802.

Wearing, S., & McGehee, N. G. (2013). Volunteer tourism: A review. *Tourism Management*, *38*, 120–130.

Wearing, S., Young, T., & Everingham, P. (2017). Evaluating volunteer tourism: Has it made a difference? *Tourism Recreation Research*, *42*(4), 512–521. doi:http://dx.doi.org/10.1080/02508281/2017.1345470.

Woosnma, K., & Lee, Y. (2011). Applying social distance to voluntourism research. *Annals of Tourism Research*, *38*(1), 309–313.

13
ECOTOURISM IMPACT ON LIVELIHOODS AND WELLBEING

Ian E. Munanura and Edwin Sabuhoro

Introduction

Since the introduction of the ecotourism concept in the 1980s, it has experienced tremendous growth in application and scholarship (Fennell, 2008). Developing countries, in particular, have embraced ecotourism due to its ability to generate social, economic, and environmental benefits (Mbaiwa, 2015). The literature verifies that in Africa, over the years, ecotourism has arguably empowered most tourism countries (Backman & Munanura, 2017). The opportunities associated with ecotourism in developing countries have led to substantial investment in ecotourism-based livelihoods improvement programs in local communities, with the added expectation of benefits for wildlife conservation (Mbaiwa & Stronza, 2010). Some, however, have empirically challenged the actual realisation of the potential of ecotourism to substantially improve livelihoods and benefit wildlife (Archabald & Naughton-Treves, 2001). Specifically, Archabald and Naughton-Treves (2001) attribute this deficiency of actual realisation of the benefits of ecotourism to various constraints including mismanagement, inadequate tourism revenue, and inequity in benefit distribution. These constraints represent the operational limitations of ecotourism. There is yet another set of constraints distinct from operational constraints. Namely, constraints that are related to the complexities of livelihoods at the household level. Operational ecotourism constraints cannot be improved without enabling ecotourism to enhance the livelihoods of local communities and expect to achieve the full potential of ecotourism. Narrowing the gap between ecotourism operational constraints and livelihood constraints is a worthwhile endeavor.

This chapter addresses this gap by exploring the various aspects of the livelihoods concept which can be strengthened by ecotourism initiatives. This approach aims to demonstrate the diverse potential of ecotourism to sustain the wellbeing of people living in tourism destinations. Conceptually, the literature examines the impact of ecotourism using diverse theoretical frameworks: 1) sustainable livelihoods (Bennett, Lemelin, Koster, & Budke, 2012; Munanura, Backman, Hallo, & Powell, 2016); 2) capital assets (Duffy, Kline, Swanson, Best, & McKinnon, 2017); 3) community empowerment (Boley, McGehee, Perdue, & Long, 2014; Scheyvens, 1999); 4) social exchange theory (Faulkner & Tideswell, 1997); and other frameworks (e.g., political ecology, see Mathis & Rose, 2016). These frameworks have contributed to our understanding of ecotourism benefits from the traditional welfare economics perspective.

DOI: 10.4324/9781003001768-13

However, these frameworks lack much of the ecotourism impact opportunities embedded in the poorly understood livelihood processes. Current theoretical frameworks typically used in assessing the impact of ecotourism, such as the sustainable livelihoods frameworks, do not provide clarity on what constitutes the livelihood process. This missed clarity in turn creates a missed opportunity for realising the full potential of ecotourism. Moreover, most of the above frameworks, when applied to evaluating ecotourism constraints, focus on community-level analyses. However, De Sherbinin et al. (2008) argue that livelihoods are produced and reproduced at the household level. Thus, in our opinion, household-level analyses have the most potential to reveal underlying constraints for livelihoods. A conceptual model is proposed in this chapter to overcome the knowledge gap just stated. This model integrates multiple theoretical frameworks to broaden the concept of livelihoods and to provide diverse ways to view ecotourism opportunities for local communities.

Ecotourism opportunities for local communities

The definition of the ecotourism concept is still debatable (Sharpley, 2006). In this chapter, however, a working definition is one adapted from Fennell (2008), suggesting that ecotourism is a form of non-consumptive and low adverse impact tourism, which aims to advance tourists' experience and knowledge about nature while creating tangible and intangible opportunities for local communities. The substantial growth of ecotourism since the1990s (Sharpley, 2006), is arguably a result of increasing tourists' concern for the negative impact of their travel (Mieczkowski, 1995). The growth of ecotourism also results from its potential to generate economic opportunities for local communities without eroding their natural and cultural heritage (Mbaiwa, 2015). The potential of ecotourism to create opportunities for local communities is of interest in this chapter. This potential is illustrated in the literature. For example, according to Vivanco (2002), ecotourism empowers local communities to improve their living conditions. Scheyvens (1999), has argued that ecotourism has the potential to strengthen the local community's economic, social, political, and psychological conditions. Similarly, Simpson (2007) has indicated that ecotourism has enhanced local community livelihoods through the creation of opportunities for local communities to own, control, and influence tourism functions.

The empowerment framework suggested by Scheyvens (1999), however, enables observation of both tangible and intangible opportunities of ecotourism, and therefore it is the most inclusive approach to conceptualise ecotourism opportunities. According to the empowerment framework, local communities are likely to draw economic, social, psychological, and political opportunities from tourism (Scheyvens, 1999). The economic opportunities include employment of community residents in formal and informal sectors of ecotourism, which generates income for residents (Scheyvens, 1999). Economic opportunities also include revenue from ecotourism enterprises that are wholly or partially owned by local community residents. However, access to jobs and business opportunities are insufficient indicators of the tangible economic benefits of ecotourism. Therefore, additional indicators of economic opportunities from ecotourism include reliable income, recurring revenue, equity in income distribution, and compensation for the opportunity cost of protecting ecotourism resources. Together such indicators represent tangible benefits of ecotourism for local communities.

According to Scheyvens (1999), the intangible benefits of ecotourism occur in the form of psychological, social, and political opportunities. Psychological empowerment opportunities of ecotourism encompass the creation of cultural tourism activities and services linked to local traditions, enabling the active involvement of residents, which potentially generates pride in

Ecotourism Impact
Tangible impact:
Economic empowerment
Intangible Impact:
Social, psychological, political, and environmental empowerment

Figure 13.1 Potential ecotourism impact based on the empowerment framework (Scheyvens, 1999) and the ecotourism framework (Buckley, 1994)

culture and self-esteem (Scheyvens, 1999; Simpson, 2007). The extent to which cultural tourism activities are owned, controlled, and operated by local communities, determines the strength of their psychological empowerment (Simpson, 2007). According to Scheyvens (1999), social empowerment opportunities of ecotourism occur when ecotourism activities enable the creation and sustenance of active community groups, particularly the most disadvantaged groups such as the youth groups, women groups, artisan groups, cultural dancing groups, traditional healer's group, and tourist guides groups. It also occurs through the establishment of social infrastructure (e.g., schools, hospitals, roads) and other indicators, including the reduction of crime. Political empowerment opportunities occur when local community residents are represented at all levels of decision-making related to ecotourism. Indicators of political empowerment opportunities may include representation and active involvement of community groups such as grassroots community organisations and indigenous institutions in tourism planning and management. Lastly, Buckley (1994) indicates ecotourism's potential to improve environmental conservation education and awareness among local communities, which is likely to create positive attitudes toward, and support for, environment among local communities. (Figure 13.1)

Ecotourism and the UN Sustainable Development Goals

Ecotourism as a strategy to improve community wellbeing advanced with international recognition of the sustainable development concept (Honey, 2008). The sustainable development concept emerged in tourism literature following the 1987 Brundtland Commission report Our Common Future (World Commission on Environment and Development (WCED), 1987). The Brundtland Commission defined sustainable development as one that meets present needs without compromising future needs (World Commission on Environment and Development (WCED), 1987). According to Müller (1994), sustainable development encompasses growth that balances economic, social, and environmental goals. Given the potential of ecotourism to advance social, economic, and environmental benefits (Fennell, 2008), ecotourism has become an essential tool to advance sustainable development (Butcher, 2006). For example, the United Nations declared the year 2002 as the International Year of Ecotourism in recognition of its potential to facilitate sustainable development, especially in developing countries (Butcher, 2006). The importance of ecotourism in advancing sustainable development has been reaffirmed by the emergence of UN Sustainable Development Goals (SDGs), aimed to reduce poverty and sustain wellbeing globally (Griggs et al., 2013).

In 2015, the United Nations introduced SDGs, following the 2012 Rio+20 summit in Brazil (Griggs et al., 2013). The SDGs refocus the global community attention toward reducing poverty and sustaining social, economic, and environmental wellbeing by 2030 (Griggs et al., 2013). The SDGs emerged following the end of the Millennium Development Goals 2015 deadline (Griggs et al., 2013). The parallels of ecotourism goals in each of the 17 Sustainable

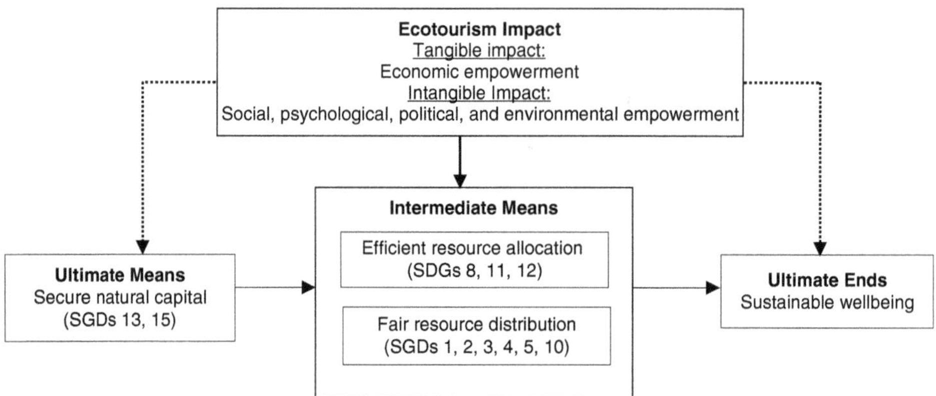

Figure 13.2 Ecotourism impact on SDGs and ultimate aim of sustaining wellbeing (Costanza et al., 2014, 2016)

Note: Dotted lines represent ecotourism impact, SDG 1 = End poverty in all its forms everywhere, SDG 2 = End hunger, achieve food security and improved nutrition, and promote sustainable agriculture, SDG 3 = Ensure healthy lives and promote well-being for all at all ages, SDG 4 = Ensure inclusive and equitable quality education and promote life-long learning opportunities for all, SDG 5 = Achieve gender equality and empower all women and girls, SDG 8 = Promote sustained, inclusive and sustainable economic growth, full and productive employment and decent work for all, SDG 10 = Reduce inequality within and among countries, SDG 11 = Make cities and human settlements inclusive, safe, resilient and sustainable, SDG 12 = Ensure sustainable consumption and production patterns, SDG 13 = Take urgent action to combat climate change and its impacts, SDG 15 = Protect, restore and promote sustainable use of terrestrial ecosystems, sustainably manage forests, combat desertification, and halt and reverse land degradation and halt biodiversity loss.

Development Goals (see Costanza et al., 2016) are striking. For example, some of the SDGs aimed to advance economic growth, improve health and education, and reduce inequity (Griggs et al., 2013), are also important goals of ecotourism (Fennell, 2008; Honey, 2008).

Considering the SDGs means-ends spectrum framework suggested by Costanza, McGlade, Lovins, and Kubiszewski (2014) and Costanza et al. (2016), the SDGs are means to achieve an overarching goal of sustainable wellbeing (see Figure 13.2). Ecotourism, therefore, not only facilitates the achievement of SDGs but also leads to SDGs' overarching goal of improving the sustainable wellbeing of communities in tourism-dependent countries. According to the means-ends spectrum framework (Costanza et al., 2014, 2016), achievement of sustainable wellbeing, the overarching goal of SDGs, is hierarchically shaped by: 1) sustainable scale shown in this chapter as securing natural capital or ecosystem services (e.g., SGDs 15 and 13—see Figure 13.2 for SDG statements); 2) efficient allocation of resources (e.g., SDGs 8 and 12); and 3) fair distribution of resources (e.g., SDGs 1 and 5). These examples of SDGs, based on the ecotourism literature (Scheyvens, 1999; Fennell, 2008), demonstrate that ecotourism and UN Sustainable Development Goals, share the overarching goal of improving sustainable wellbeing of communities. They also show that ecotourism is likely to substantively contribute toward the attainment of SGDs.

Ecotourism impact on means of Sustainable Development Goals

It has been indicated in Figure 13.2 that ecotourism has a positive impact on the ultimate means, intermediate means, and ultimate ends of SDGs, according to the hierarchical means-ends spectrum framework suggested by Costanza et al. (2014, 2016). The ecotourism literature has revealed the potential of ecotourism to strengthen natural capital and attainment of the ultimate means of SGDs (Buckley, 1994; Fennell, 2008; Stronza, 2007). Buckley (1994) provides a framework through which the impact of ecotourism on the ultimate means of SDGs can be viewed. For example, ecotourism has the potential to generate revenue, which provides direct and indirect financial contributions to the conservation of the environment (Buckley, 1994). The financial benefits of ecotourism have also been linked to improved political support for conservation (Gössling, 1999). Ecotourism also has the potential to improve environmental conservation awareness (Buckley, 1994). Environmental awareness is likely to create positive attitudes toward the environment and enables ecotourism service providers to minimise the adverse impact of their operations on the environment (Buckley, 1994; Fennell, 2008).

Figure 13.2 indicates that ecotourism has the potential to impact the intermediate means of SDGs, by contributing toward the efficient allocation of resources and fair distribution of resources, according to the means-ends spectrum framework (Costanza et al., 2014). However, efficient and fair resource allocations are secondary intermediary means, and it is argued in this chapter that access to livelihood resources regardless of whether access is fair and equitable represents primary intermediary means of SDGs. As indicated in Figure 13.3, primary and secondary intermediary means of SDGs interact to influence the SDGs overarching goal of sustainable wellbeing.

The literature reveals that the improvement of wellbeing is likely a function of access to livelihood assets that are vital for human life (Bebbington, 1999; Bennett et al., 2012). Bennett et al. (2012) have argued that access to livelihood assets is critical to a household's potential to produce and maintain livelihoods. According to the concept of capital assets, livelihood assets are diverse and encompass elements such as financial capital, social capital, human capital, natural capital, and built capital (Bebbington, 1999; Bennett et al., 2012). However, according to the sustainable livelihoods framework (Bebbington, 1999), access to such diverse livelihood assets is a means to improved wellbeing (see primary intermediate means in Figure 13.3). Further, according to the and the means-ends spectrum framework (Costanza et al., 2014), wellbeing is not sustainable, unless access to livelihood assets is fair and efficient (see intermediate means in Figure 13.3).

The potential for ecotourism to enable access to, and strengthen, livelihood assets, has been documented in the literature. Mbaiwa and Stronza (2010), for example, indicated that ecotourism, especially in developing countries, is perceived as a means to facilitate access to essential livelihood assets. Research shows that ecotourism facilitates access to financial capital (Scheyvens, 1999; Snyman, 2014). In most developing countries, ecotourism opportunities are created through employment, revenue, tax, philanthropic financial contributions, community development programs, as well as grants and loans which capitalise small-scale community-led tourism enterprises (Gössling, 1999). Social capital is strengthened through the creation of strong relational networks and cohesion in communities empowered by ecotourism activities (Scheyvens, 1999). Similarly, Scheyvens (1999) and Simpson (2007) indicate that tourism provides opportunities for people to collaborate in pursuit of shared interests, such as tourism revenue, and a strong sense of trust, belonging, and integrity. Ecotourism also strengthens natural capital through raising awareness of the opportunity cost for losing natural systems and strengthening environmental stewardship in host communities (Gössling, 1999). In fact, Gössling (1999) argues that economic gains from ecotourism opportunities typically help to create incentives for local people and governments to protect wildlife.

Further, ecotourism-supported programs are aimed to improve knowledge, skills, and competencies needed by the host communities to optimise benefits derived from ecotourism

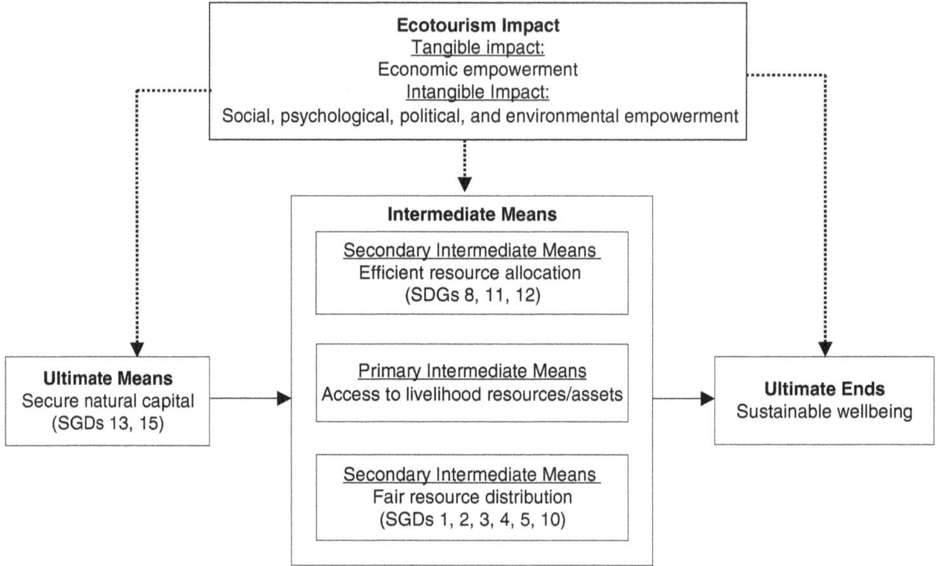

Figure 13.3 Ecotourism impact on livelihoods, SDGs, and sustainable wellbeing (Bennett, Lemelin, Koster, & Budke, 2012; Costanza et al., 2016)

Note: Dotted lines represent ecotourism impact, SDG 1 = End poverty in all its forms everywhere, SDG 2 = End hunger, achieve food security and improved nutrition, and promote sustainable agriculture, SDG 3 = Ensure healthy lives and promote well-being for all at all ages, SDG 4 = Ensure inclusive and equitable quality education and promote life-long learning opportunities for all, SDG 5 = Achieve gender equality and empower all women and girls, SDG 8 = Promote sustained, inclusive and sustainable economic growth, full and productive employment and decent work for all, SDG 10 = Reduce inequality within and among countries, SDG 11 = Make cities and human settlements inclusive, safe, resilient and sustainable, SDG 12 = Ensure sustainable consumption and production patterns, SDG 13 = Take urgent action to combat climate change and its impacts, SDG 15 = Protect, restore and promote sustainable use of terrestrial ecosystems, sustainably manage forests, combat desertification, and halt and reverse land degradation and halt biodiversity loss.

opportunities arguably strengthen human capital (Snyman, 2014; Stronza, 2007). As a result, ecotourism enhances access to built capital through direct and indirect financing of vital social infrastructures (Duffy et al., 2017). The social infrastructure programs, typically funded through tourism revenue or government investment are aimed to improve tourism services, such as roads, bridges, telecommunications, electricity, water supply, and sewage systems. Together, these arguments illustrate the potential of ecotourism to strengthen the ultimate and intermediate means of improving SDGs' overall aim of sustaining wellbeing (see Figure 13.3).

Ecotourism impact on ultimate ends of Sustainable Development Goals

The ultimate outcome of SDGs is the achievement of sustainable wellbeing (Costanza et al., 2016). The concept of wellbeing, however, is nuanced in the literature to render a clear understanding of links to ecotourism. Using the means-ends spectrum framework, Costanza et al. (2014, 2016), indicate that sustainable wellbeing is a function of ultimate means or access to natural capital, and intermediary means or efficient and fair access to productive resources such as financial capital (see Figure 13.4). The hierarchical nature of the means-ends spectrum approach used to conceptualise sustainable wellbeing is supported by the bottom-up situational

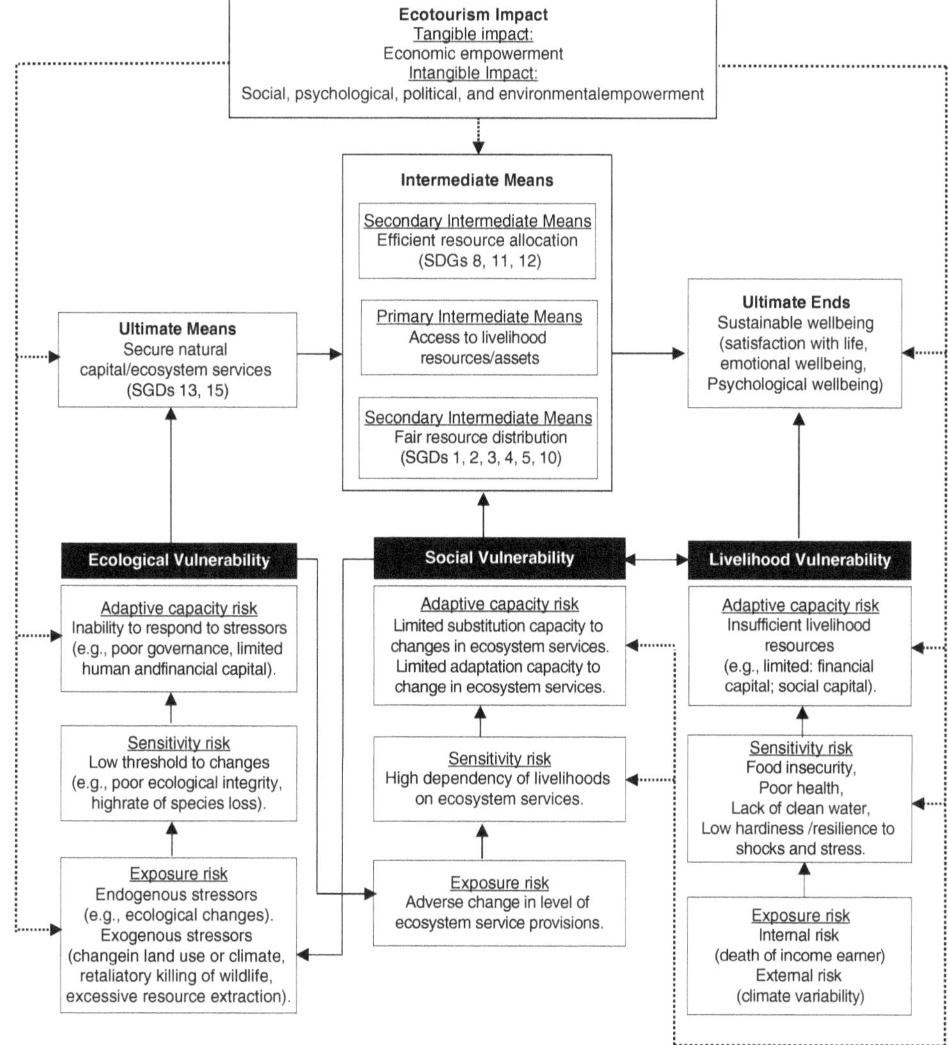

Figure 13.4 Ecotourism impact on vulnerabilities in livelihoods and SDGs processes (Adgers, 2006; Berrouet, Machado, & Villegas-Palacio, 2018; Hahn, Riederer, & Foster, 2009)

Note: Dotted lines represent ecotourism impact, SDG 1 = End poverty in all its forms everywhere, SDG 2 = End hunger, achieve food security and improved nutrition, and promote sustainable agriculture, SDG 3 = Ensure healthy lives and promote well-being for all at all ages, SDG 4 = Ensure inclusive and equitable quality education and promote life-long learning opportunities for all, SDG 5 = Achieve gender equality and empower all women and girls, SDG 8 = Promote sustained, inclusive and sustainable economic growth, full and productive employment and decent work for all, SDG 10 = Reduce inequality within and among countries, SDG 11 = Make cities and human settlements inclusive, safe, resilient and sustainable, SDG 12 = Ensure sustainable consumption and production patterns, SDG 13 = Take urgent action to combat climate change and its impacts, SDG 15 = Protect, restore and promote sustainable use of terrestrial ecosystems, sustainably manage forests, combat desertification, and halt and reverse land degradation and halt biodiversity loss.

influences approach suggested by Diener, Suh, Lucas, and Smith (1999). The bottom-up situational influences approach indicates that subjective wellbeing, or ones' perception of the quality of life, is influenced by bottom-up factors, which represent access to universal and basic needs of life (Diener et al., 1999). It is indicated that when the basic needs of life are attained, one's cognitive evaluation of the quality of life is positive (Diener et al., 1999). Most ecotourism studies evaluating the impact of ecotourism on wellbeing have paid more attention to the impact on objective wellbeing aspects, or the means to sustaining wellbeing (Snyman, 2014). This body of knowledge has demonstrated evidence of ecotourism's potential to strengthen environmental awareness and protection, economic vitality, and social capital in tourism destinations.

However, the literature indicates that wellbeing is a state of one's life, and it conceptually encompasses objective and subjective dimensions (Summers, Smith, Case, & Linthurst, 2012). It is also indicated that subjective wellbeing condition is a function of objective wellbeing (Kahneman & Deaton, 2010). Objective wellbeing encompasses access to basic material needs, and therefore represent means to sustainable wellbeing, as shown in Figure 13.4 (Summers et al., 2012). Subjective wellbeing, on the other hand, represents one's perception of the quality of life, that is, the ultimate ends, as shown in Figure 13.4 (Organisation for Economic Co-operation and Development, 2013). Therefore, understanding the impact of ecotourism on objective and subjective wellbeing aspects broadens understanding of links between ecotourism and wellbeing. In fact, the literature calls for integrating objective and subjective wellbeing in research models to fully grasp the impact of tourism on wellbeing (Uysal, Sirgy, Woo, & Kim, 2016). Research linking ecotourism to subjective wellbeing is emerging in the current literature (Uysal et al., 2016). This body of knowledge shows evidence that tourism positively impacts people's perceived quality of life.

The concept of subjective wellbeing encompasses three dimensions; evaluative, eudaimonic, and affective feelings (Organisation for Economic Co-operation and Development, 2013). The evaluative dimension represents an individual's perceived satisfaction with life (Organisation for Economic Co-operation and Development, 2013). According to Diener et al. (1999), life satisfaction is a cognitive evaluation of an individual's quality of life. It is a cognitive judgment process that is dependent on the comparison of an individual's life circumstances with what is perceived to be a desired standard of life (Diener et al., 1999). The eudaimonic dimension represents an individual's perceived psychological wellbeing condition (Diener et al., 2009; Organisation for Economic Co-operation and Development, 2013). Psychological wellbeing refers to an individual's perceived socio-psychological prosperity, encompassing psychological needs of competence, relatedness, self-acceptance, and other life fulfillment factors (Ryan & Deci, 2001). The affective dimension represents an individual's emotional wellbeing conditions that influence emotions and functioning (Diener et al., 2009). According to Kahneman and Deaton (2010), emotional wellbeing encompasses the emotional quality of experiences of joy, sadness, and other emotions that inform pleasantness or unpleasantness of one's life. Together, these aspects of subjective wellbeing are likely to demonstrate the extent to which ecotourism positively impacts diverse aspects of people's lives.

Socio-ecological vulnerability and the impact of ecotourism

The potential for ecotourism to strengthen SDGs' ultimate means (i.e., natural capital), intermediate means (i.e., access to livelihood resources), and ultimate ends (i.e., sustainable wellbeing) has been demonstrated earlier. However, the above suggested positive impact of ecotourism assumes that the system in which ecotourism interacts with livelihood processes is free from shock, stress, and turbulence. According to the literature, this assumption is not likely, and in fact, it is argued that livelihoods processes typically experience, and are constrained by, shocks and stresses. Devereux (2002) has argued that vulnerability to shocks and stresses is likely responsible for chronic poverty in the developing world. Therefore, notwithstanding the opportunities that ecotourism presents to SDGs' aim of sustaining well-being, one of the remaining limitations of SDGs is the socio-ecological vulnerability in livelihood processes.

Vulnerability refers to the extent to which a system is likely to be adversely affected by risk (Turner, 2010). As indicated in Figure 13.4, the vulnerability in livelihood processes can be viewed from two distinct but complementary perspectives. For example, in the initial phase of livelihood processes where natural capital or ecosystem services provision livelihoods, according to the means-ends spectrum framework (Costanza et al., 2014), vulnerabilities are shaped by risks within a coupled human-environment system (Turner, 2010). Within a coupled human-environment system, vulnerability, according to the vulnerability risk framework (Adger, 2006; Berrouet et al., 2018), encompasses social and ecological dimensions. Social vulnerability refers to the disruption of a social system (Adger, 2000) due to changes in ecosystem service provisions (Berrouet et al., 2018). The literature shows that the extent to which changes in the ecosystem services disrupt a social system depends on sensitivity risk (e.g., high level of people's dependency on the ecosystem service provisions), and adaptive capacity risk (e.g., limited capacity to adapt to changes in ecosystem services) (Adger, 2006; Berrouet et al., 2018).

Ecological vulnerability, on the other hand, refers to the extent to which an ecosystem fails to maintain its functions or experiences changes that diminish its capacity to provision ecosystem services (Berrouet et al., 2018). Ecological vulnerability, typically, is a function of exposure to endogenous risk factors (e.g., vegetation cover change) or exogenous risk factors (e.g., change in land use or excessive extractive use) (Adger, 2006; Berrouet et al., 2018). Exogenous risk factors are arguably the most immediate and impactful to environmental vulnerability (Baynham-Herd, Redpath, Bunnefeld, Molony, & Keane, 2018). For example, Adgers (2006) attributes ecological vulnerability to inadvertent or deliberate human activities that promote self-serving interests. Baynham-Herd et al. (2018) point out human actions such as, overuse of natural resources for subsistence and commercial interests, and retaliatory killing of wildlife, to be responsible for ecological vulnerability. Similarly, the level of the impact of exogenous and endogenous risk factors on the ecological system depends on the sensitivity risk (e.g., the low threshold at which the ecological system is unable to maintain ecosystem functioning when exposed to risk factors), and the adaptive capacity risk (e.g., poor wildlife governance and protection) (Adger, 2006; Berrouet et al., 2018).

The potential of ecotourism to mitigate ecological risk factors responsible for social vulnerability has already been demonstrated in the literature (Hornoiu, 2016; Gössling, 1999; Stronza, 2007; Stone & Nyaupane, 2017). For example, Hornoiu (2016) has indicated that ecotourism has strengthened resilience to climate change. Further, the literature has indicated the potential for ecotourism to create alternative livelihood opportunities for communities whose livelihoods are dependent on, and threaten, natural systems (Hornoiu, 2016; Mbaiwa & Stronza, 2010; Stronza, 2007). Stronza (2007) has also indicated that ecotourism has minimised illegal and unsustainable extractive forest use practices. Similarly, Sabuhoro, Wright, Munanura, Nyakabwa, and Nibigira (2017) have indicated that ecotourism opportunities have dissuaded poachers at Volcanoes National Park in Rwanda.

The literature also reveals potential of ecotourism to mitigate social risks linked to drivers of ecological vulnerability (Stronza, 2007; Fennell, 2008; Stone & Nyaupane, 2017). For example, in Botswana, ecotourism has contributed positively toward the preservation and protection of natural land, hence minimising ecological adaptive capacity risk (Stone & Nyaupane, 2017).

Similarly, the literature has shown the potential of ecotourism to strengthen human and institutional capacity for wildlife protection (Buckley, 1994; Fennell, 2008; Gössling, 1999). Ecotourism has also raised environmental awareness and support for conservation (Buckley, 1994). At the same time however, ecotourism is also likely to impact the environment negatively through: 1) increase in demand for agricultural land due to increased access to financial capital (Stone & Nyaupane, 2017), development of ecotourism in fragile ecosystems (Das & Chatterjee, 2015); 2) unfair distribution of socio-economic benefits where the most natural resource-dependent groups do not access ecotourism opportunities (Goodwin, 2002); and 3) loss of ancestral use rights (Das & Chatterjee, 2015). Such negative impacts of ecotourism are likely to increase sensitivity and adaptive capacity risks in the social system, which arguably leads to increased exposure of the ecological system to exogenous risk factors (e.g., excessive extractive use).

Livelihood vulnerability and the impact of ecotourism

According to Adgers (2006), the third aspect of vulnerability, distinct from the vulnerability of a coupled socio-ecological system, is poverty vulnerability. Poverty vulnerability is embedded in the sustainable livelihoods system, and framed in the sustainable livelihoods framework (Adgers, 2006; Hahn et al., 2009). In this chapter, poverty vulnerability is considered as livelihoods vulnerability for conceptual alignment with the livelihood's literature (Hahn et al., 2009). According to the sustainable livelihoods' framework, livelihood vulnerability occurs when an individual or household experiences shocks and stress and is unable to sustain livelihoods due to limitations in coping capabilities (Bebbington, 1999) and livelihood resources constraints (Hahn et al., 2009). Such risks to livelihoods can be observed at three levels: the level of exposure to risk, sensitivity to risk, and adaptive capacity to adverse livelihood changes (Adgers, 2006; Hahn et al., 2009).

Sensitivity and adaptive capacity risks to livelihood stressors are described in Hahn et al. (2009). For example, adaptive capacity risks are reflected in the socio-economic characteristics and constraints (Hahn et al., 2009). As shown in Figure 13.4, the adaptive capacity risk is likely linked to constraints within the SDGs' intermediary means. Such constraints, arguably, include limited access to financial and social capital. These adaptive capacity risks are also likely to influence ecological vulnerability risks (Gössling, 1999; Munanura, Backman, Hallo, Powell, & Sabuhoro, 2018). Sensitivity risks to livelihood shocks and stress encompass poor health, food insecurity, and limited access to clean water (Hahn et al., 2009). Poor health, coupled with limited access to food and water, are likely to increase livelihood vulnerability, and arguably limits the potential of attaining the SDGs' ultimate aim of sustaining wellbeing. Sensitivity risks are also likely to increase the ecological vulnerability risks, especially when people opt to rely on natural systems for food, water, and medicine (Hahn et al., 2009; Munanura et al., 2018).

The potential for ecotourism to mitigate both the sensitivity and adaptive capacity risks responsible for livelihood vulnerability has been explored in the ecotourism literature (Duffy et al., 2017; Fennell, 2008; Mbaiwa & Stronza, 2010; Scheyvens, 1999; Stronza, 2007; Snyman, 2014). For example, Scheyvens (1999) pointed out the potential of ecotourism to strengthen tangible means (e.g., financial capital), and intangible means (e.g., social capital) of producing livelihoods. Das and Chatterjee (2015) and Munanura et al. (2018) have also demonstrated that ecotourism has improved the livelihoods of communities near protected areas. Such links between livelihood constraints, ecological vulnerability risk, and the mitigation potential of ecotourism have also been consistently suggested in the literature (Fennell, 2008; Gössling, 1999; Das & Chatterjee, 2015; Mbaiwa & Stronza, 2010). Mbaiwa and Stronza (2010) have indicated that ecotourism has improved the livelihoods of communities in the Okavango Delta, and has reduced the environmental vulnerability risk attributed to livelihood constraints (Mbaiwa & Stronza, 2010).

Broadening the perspective of livelihood vulnerability and ecotourism impact potential

Exposure risk to livelihood shocks and stress, unlike sensitivity and adaptive capacity risks, has been narrowly examined in the livelihoods' literature. Yet, if exposure risk to livelihood shocks and stress is not clearly understood, the potential to strengthen the resiliency of livelihoods, and attain SDGs' ultimate goal of sustaining wellbeing will be unlikely. For example, Hahn et al. (2009) have conceptualised exposure risks to livelihood vulnerability as natural disasters and climate variability. However, from the family stress and resilience perspective, natural disaster and climate variability exposure risks represent external shocks and stress conditions responsible for adversity.

According to family stress theory (Boss, 2002), households are typically exposed to diverse adversity risk factors in the process of transforming and producing livelihoods. Such adversity risk factors are two-dimensional conceptually and include intra-household risk factors and external risk factors. The intra-household risk factors encompass stressful events such as the loss of a family income earner, divorce, chronic illness, death in the family, and other similar stressful events within a household (Boss, 2002). External adversity risk factors are typically beyond the control of households and include stressor events such as natural disasters (e.g., the 2005 hurricane Katrina), disease outbreak (e.g., the COVID-19 pandemic), human-made disasters (e.g., the 1994 genocide in Rwanda), economic instability (e.g., the 2008 economic recession), and climate variability (Hahn et al., 2009). Both internal and external adversity risk factors are likely to create exposure to livelihood shocks and stress, which could lead to livelihood vulnerability and increased human threat to the socio-ecological system (Das & Chatterjee, 2015).

The external and internal adversity risk factors introduced earlier, arguably give meaning to human response to adversity (McCubbin, 2001). According to family resilience theory, when exposure to adversity risks occurs, people respond by drawing from available protective factors to cope or overcome adversity (Benzies & Mychasiuk, 2009; Patterson, 2002). Such protective factors are typically drawn from adaptive capacity, that is, available livelihood resources such as financial capital (e.g., use of savings during financial strain) social capital (e.g., support from social network) (Benzies & Mychasiuk, 2009). In addition, the potential to cope with exposure to adversity risk factors, is also influenced by a household's sensitivity risks outlined in Hahn et al. (2009).

Further, potential to cope with exposure to adversity depends on a households' resilience capacity (McCubbin, 2001). For example, despite available resources, a household that is less hardy or resilient cognitively, and therefore less optimistic about life, is likely to maladapt and experience livelihood vulnerability when exposure to adversity occurs (Walsh, 2016). It is also argued that when resilience capacity exceeds adversity, livelihoods are likely to be resilient (Patterson, 2002). Therefore, in addition to exposure and sensitivity risk factors suggested by Hahn et al. (2009), this chapter suggests consideration of internal stress risk factors and broadening the scope of external stress factors beyond climate variability and natural disasters suggested by Hahn et al. (2009). According to Walsh (2016), a household's ability to overcome adversity is dependent on a household's belief systems (e.g., making sense of adversity), organisational patterns (e.g., connectedness), and communication processes (e.g., clarity of information on adversity). Other scholars have linked adversity coping capability to intra-household hardiness (Funk, 1992), and sense of coherence (Antonovsky, 1993). Research aiming to understand the positive impact of ecotourism on local community livelihoods is likely to benefit from considering the potential of ecotourism to strengthen resilience factors that may enable local communities to cognitively minimise the impact of exposure to adversity.

Summary

This chapter has integrated a diverse set of theoretical frameworks to propose a model that conceptually presents antecedents of livelihoods (referred to as the ultimate means of SDGs in Figure 13.4), livelihood resources (referred to as the intermediary means of SDGs in Figure 13.4), and sustainable wellbeing (referred to as ultimate ends of SDGs in Figure 13.4). The measures for the ultimate means can be adapted from the ecosystem services literature (e.g., Boyd & Banzhaf, 2007; Millennium Ecosystem Assessment, 2005). The measures for the livelihood resources may be adapted from capital assets literature (e.g., Akamani & Hall, 2015; Bennett et al., 2012). The measures for subjective wellbeing aspects, such as the perceptions of satisfaction with life, emotional wellbeing, and psychological wellbeing, can be adapted from the subjective wellbeing literature (e.g., Diener et al., 2009). In addition, the proposed model illuminates the process of producing livelihoods, which remains conceptually abstract in the literature. Informed by the vulnerability in the coupled socio-ecological systems (Adgers, 2006), and sustainable livelihood systems (Hahn et al., 2009), the model suggests consideration of vulnerability in livelihood processes and in the efforts to achieve the SDGs. The measures for livelihood vulnerability can be adapted from Hahn et al. (2009), measures for socio-ecological vulnerability can be adapted from Berrouet et al. (2018).

Drawing from the family stress and resilience theories, the model suggests consideration of intra-household and other external risk factors imbedded in the socio-political environment in which livelihoods are produced. The model indicates that inattention to resilience to adversity likely leads to livelihood vulnerability, which may have negative socio-ecological implications. These vulnerabilities in the livelihood processes, arguably, diminish the potential to attain SDGs. The measures for households' resilience capacity may be adapted from family resilience literature (e.g., Antonovsky, 1993; Bartone, 1995; Chew & Haase, 2016). Overall, the suggested model broadens the understanding of livelihood constraints for local communities in tourism destinations and argues for consideration of constraints and opportunities in the livelihood processes when designing ecotourism programs. Further, the model demonstrates the diverse potential of ecotourism to strengthen livelihoods beyond enabling access to livelihood assets such as financial capital. Until the complexity of livelihood processes is conceptually unpacked, the potential of ecotourism to strengthen local community livelihoods in tourism destinations will remain partially understood. Empirical evidence is needed to advance research on the links between ecotourism and vulnerability in the livelihood processes and SDGs. Hopefully, this chapter will initiate a scholarly discourse about diverse ways ecotourism can strengthen the livelihoods of local communities and facilitate the achievement of SDGs, particularly in developing countries.

References

Adger, W. N. (2000). Social and ecological resilience: Are they related? *Progress in Human Geography*, *24*(3), 347–364.

Adger, W. N. (2006). Vulnerability. *Global Environmental Change, 16*(3), 268–281.

Akamani, K., & Hall, T. E. (2015). Determinants of the process and outcomes of household participation in collaborative forest management in Ghana: A quantitative test of a community resilience model. *Journal of Environmental Management, 147*, 1–11.

Antonovsky, A. (1993). The structure and properties of the sense of coherence scale. *Social Science & Medicine, 36*(6), 725–733.

Archabald, K., & Naughton-Treves, L. (2001). Tourism revenue-sharing around national parks in Western Uganda: Early efforts to identify and reward local communities. *Environmental Conservation, 28*(2), 135–149.

Backman, K. F., & Munanura, I. E. (Eds.). (2017). *Ecotourism in Sub-Saharan Africa: Thirty Years of Practice.* London: Routledge.

Bartone, P. T. (1995). A short hardiness scale. Presentation at the Annual Convention of the American Psychological Society, New York.

Baynham-Herd, Z., Redpath, S., Bunnefeld, N., Molony, T., & Keane, A. (2018). Conservation conflicts: Behavioral threats, frames, and intervention recommendations. *Biological Conservation, 222*, 180–188.

Bebbington, A. (1999). Capitals and capabilities: A framework for analyzing peasant viability, rural livelihoods and poverty. *World Development, 27*(12), 2021–2044.

Bennett, N., Lemelin, R. H., Koster, R., & Budke, I. (2012). A capital assets framework for appraising and building capacity for tourism development in aboriginal protected area gateway communities. *Tourism Management, 33*(4), 752–766.

Benzies, K., & Mychasiuk, R. (2009). Fostering family resiliency: A review of the key protective factors. *Child and Family Social Work, 14*(1), 103–114.

Berrouet, L. M., Machado, J., & Villegas-Palacio, C. (2018). Vulnerability of socio—Ecological systems: A conceptual Framework. *Ecological Indicators, 84*, 632–647.

Boley, B. B., McGehee, N. G., Perdue, R. R., & Long, P. (2014). Empowerment and resident attitudes toward tourism: Strengthening the theoretical foundation through a Weberian lens. *Annals of Tourism Research, 49*, 33–50.

Boss, P. (2002). *Family Stress Management: A Contextual Approach.* Thousand Oaks, CA: Sage Publications.

Boyd, J., & Banzhaf, S. (2007). What are ecosystem services? The need for standardized environmental accounting units. *Ecological Economics, 63*(2–3), 616–626.

Buckley, R. (1994). A framework for ecotourism. *Annals of Tourism Research, 21*(3), 661–665.

Butcher, J. (2006). Natural capital and the advocacy of ecotourism as sustainable development. *Journal of Sustainable Tourism, 14*(6), 629–644.

Chew, J., & Haase, A. M. (2016). Psychometric properties of the family resilience assessment scale: A Singaporean perspective. *Epilepsy & Behavior, 61*, 112–119.

Costanza, R., Daly, L., Fioramonti, L., Giovannini, E., Kubiszewski, I., Mortensen, L. F.,... & Wilkinson, R. (2016). Modelling and measuring sustainable wellbeing in connection with the UN Sustainable Development Goals. *Ecological Economics, 130*, 350–355.

Costanza, R., McGlade, J., Lovins, H., & Kubiszewski, I. (2014). An overarching goal for the UN Sustainable Development Goals. *Solutions, 5*(4), 13–16.

Das, M., & Chatterjee, B. (2015). Ecotourism: A panacea or a predicament?. *Tourism Management Perspectives, 14*, 3–16.

De Sherbinin, A., VanWey, L. K., McSweeney, K., Aggarwal, R., Barbieri, A., Henry, S., ... & Walker, R. (2008). Rural household demographics, livelihoods and the environment. *Global Environmental Change, 18*(1), 38–53.

Devereux, S. (2002). Can social safety nets reduce chronic poverty? *Development Policy Review, 20*(5), 657–675.

Diener, E., Suh, E. M., Lucas, R. E., & Smith, H. L. (1999). Subjective well-being: Three decades of progress. *Psychological Bulletin, 125*, 276–302.

Diener, E., Wirtz, D., Biswas-Diener, R., Tov, W., Kim-Prieto, C., Choi, D., & Oishi, S. (2009). *New measures of well-being.* The Netherlands: Springer.

Duffy, L. N., Kline, C., Swanson, J. R., Best, M., & McKinnon, H. (2017). Community development through agro-ecotourism in Cuba: An application of the community capitals framework. *Journal of Ecotourism, 16*(3), 203–221.

Faulkner, B., & Tideswell, C. (1997). A framework for monitoring community impacts of tourism. *Journal of Sustainable Tourism, 5*(1), 3–28.

Fennell, D. (2008). *Ecotourism: An Introduction.* London: Routledge.

Funk, S. C. (1992). Hardiness: A review of theory and research. *Health Psychology, 11*(5), 335.

Griggs, David, Stafford-Smith, M. Gaffney, O., Rockström, J., Öhman, M. C., Shyamsundar, P., ... & Noble, I. (2013). Policy: Sustainable development goals for people and planet. *Nature, 495*(7441), 305.

Goodwin, H. (2002). Local community involvement in tourism around national parks: Opportunities and constraints. *Current Issues in Tourism, 5*(3–4), 338–360.

Gössling, S. (1999). Ecotourism: A means to safeguard biodiversity and ecosystem functions? *Ecological Economics, 29*(2), 303–320.

Hahn, M. B., Riederer, A. M., & Foster, S. O. (2009). The Livelihood Vulnerability Index: A pragmatic approach to assessing risks from climate variability and change—A case study in Mozambique. *Global Environmental Change, 19*(1), 74–88.

Honey, M. (2008) *Ecotourism and Sustainable Development: Who Owns Paradise?* Washington, DC: Island Press.

Hornoiu, R. I. (2016). Resilience capacity of local communities from protected areas under the impact of climate change and their strengthening through ecotourism. *The ASEAN Countries' Case. Quality-Access to Success, 17*(153).

Kahneman, D., & Deaton, A. (2010). High income improves evaluation of life but not emotional well-being. *Proceedings of the National Academy of Sciences, 107*(38), 16489–16493.

Mathis, A., & Rose, J. (2016). Balancing tourism, conservation, and development: A political ecology of ecotourism on the Galapagos Islands. *Journal of Ecotourism, 15*(1), 64–77.

Mbaiwa, J. E. (2015). Ecotourism in Botswana: 30 years later. *Journal of Ecotourism, 14*(2–3), 204–222.

Mbaiwa, J. E., & Stronza, A. L. (2010). The effects of tourism development on rural livelihoods in the Okavango Delta, Botswana. *Journal of Sustainable Tourism, 18*(5), 635–656.

McCubbin, M. (2001). Pathways to health, illness and well-being: From the perspective of power and control. *Journal of Community & Applied Social Psychology, 11*(2), 75–81.

Mieczkowski, Z. (1995) *Environmental Issues of Tourism and Recreation.* Lanham, MD: University Press of America.

Millennium Ecosystem Assessment (2005). *Ecosystems and Human Well-Being: Synthesis.* Washington, DC: Island Press.

Müller, H. (1994) The thorny path to sustainable tourism development. *Journal of Sustainable Tourism, 2*(3), 131–136.

Munanura, I. E., Backman, K. F., Hallo, J. C., & Powell, R. B. (2016). Perceptions of tourism revenue sharing impacts on Volcanoes National Park, Rwanda: A sustainable livelihoods framework. *Journal of Sustainable Tourism, 24*(12), 1709–1726.

Munanura, I. E., Backman, K. F., Hallo, J. C., Powell, R. B., & Sabuhoro, E. (2018). Understanding the relationship between livelihood constraints of poor forest-adjacent residents, and illegal forest use, at Volcanoes National Park, Rwanda. *Conservation and Society, 16*(3), 291–304.

Organization for Economic Co-operation and Development (2013). *Guidelines on Measuring Subjective Well-Being.* Paris: Organization for Economic Cooperation and Development.

Patterson, J. M. (2002). Integrating family resilience and family stress theory. *Journal of Marriage and Family, 64*(2), 349–360.

Ryan, R. M., & Deci, E. L. (2001). On happiness and human potentials: A review of research on hedonic and eudaimonic well-being. *Annual Review of Psychology, 52*(1), 141–166.

Sabuhoro, E., Wright, B., Munanura, I. E., Nyakabwa, I. N., & Nibigira, C. (2017). The potential of ecotourism opportunities to generate support for mountain gorilla conservation among local communities neighboring Volcanoes National Park in Rwanda. *Journal of Ecotourism*, 1–17.

Scheyvens, R. (1999). Ecotourism and the empowerment of local communities. *Tourism Management, 20*(2), 245–249.

Scoones, I. (1998). *Sustainable Rural Livelihoods: A Framework for Analysis* (Vol. 72). Brighton: Institute of Development Studies.

Sharpley, R. (2006). Ecotourism: A consumption perspective. *Article in Journal of Ecotourism, 5*(1–2), 7–22.

Simpson, M. C. (2007). Tourism, livelihoods, biodiversity, conservation and the climate change factor in developing countries. In B. Amelung, K. Blazejczyk, & A. Matzarakis (Eds.), *Climate change and tourism – assessment and coping strategies*. Maastricht – Warsaw-Freiburg: Institute of Geography and Spatial Organization-Polish Academy of Sciences, 190-208

Snyman, S. (2014). The impact of ecotourism employment on rural household incomes and social welfare in six southern African countries. *Tourism and Hospitality Research, 14*(1–2), 37–52.

Stone, M. T., & Nyaupane, G. P. (2017). Ecotourism influence on community needs and the functions of protected areas: A systems thinking approach. *Journal of Ecotourism, 16*(3), 222–246.
Stronza, A. (2007). The economic promise of ecotourism for conservation. *Journal of Ecotourism, 6*(3), 210–230.
Summers, J. K., Smith, L. M., Case, J. L., & Linthurst, R. A. (2012). A review of the elements of human well-being with an emphasis on the contribution of ecosystem services. *Ambio, 41*(4), 327–340.
Turner II, B. L. (2010). Vulnerability and resilience: Coalescing or paralleling approaches for sustainability science? *Global Environmental Change, 20*(4), 570–576.
Uysal, M., Sirgy, M. J., Woo, E., & Kim, H. L. (2016). Quality of life (QOL) and well-being research in tourism. *Tourism Management, 53*, 244–261.
Vivanco, L. A. (2002). Ancestral homes. *Alternatives Journal, 28*, 27–28.
Walsh, F. (2016). *Strengthening Family Resilience.* New York: The Guilford Press.
World Commission on Environment and Development (WCED). (1987). *Our Common Future.* Oxford: Oxford University Press.

14
FEMALE ENTREPRENEURSHIP AND ECOTOURISM

Ige Pirnar

Introduction

International tourism industry is changing rapidly towards more environmental approaches which increase the employment rate by positively impacting welfare of local communities. One of these approaches involves ecotourism applications which is very suitable for female entrepreneurs. Female entrepreneurship is a rising global trend in developed, developing, and even less developed nations. Within this regard, female ecotourism entrepreneurship matters deeply for communities and for nations since it is closely linked to income generation, increase in the quality of local life, higher employment rates, human capital accumulation and balanced wealth generation, while focusing on sustainability and environmental awareness. Due to the vitality of the topic, this chapter emphasises the understanding and the concept of female entrepreneurship, the characteristics, motivations, constraints, gender-related differences, and other related issues of the female ecotourism entrepreneurs.

Female entrepreneurship

Female established and managed businesses are one of the main growing and popular entrepreneurial trends (Brush & Cooper, 2012). The female entrepreneurship segment is growing fast and as a result, female entrepreneurship becomes an important economic factor since as The Global Entrepreneurship Monitor report states that nearly 274 million female-owned and/or established businesses exist in 74 world economies (Hechavarria, Bullough, Brush, & Edelman, 2019). Female entrepreneurship is a vital topic for all areas of managerial and research fields and tourism industry is one of them. Though the female entrepreneurs are rising in numbers, and the literature related to female entrepreneurship is extensive in parallel (Deng, Liang, Li, & Wang, 2020), female entrepreneurs are rather neglected in the tourism entrepreneurship research, even though the numbers in practice are significant (Page & Ateljevic, 2009).

Female entrepreneurship and woman entrepreneurship are both used interchangeably in the literature and have the same meaning. Other terms commonly used for female entrepreneurship are; female business founder and woman business owner (Achtenhagen & Welter, 2003). As for the definition; female entrepreneurs are business people who start up a new business and deal with the business process from the beginning to the end (Hughes & Jennings, 2012; Zapalska & Brozik, 2014). During the entrepreneurial process, they are involved in making independent

 DOI: 10.4324/9781003001768-14

business decisions related to the management (Serafimova & Petrevska, 2018). Thus, for Humbert & Brindley (2015) and Santos, Marques, and Ferreira, (2018), organising, leading, self-employment, having entrepreneurial traits like risk taking, and following innovational approaches are important issues which female entrepreneurship definition should include.

The importance of female entrepreneurship

Female entrepreneurship is important to individuals, to regions, and to countries due to its economic and social benefits. Though the number of the benefits and the levels of positive impacts change among developed and developing countries, female entrepreneurship positively contributions to all the societies. Thus, the main economic and social contributions of female entrepreneurship involve: increase in the employment rates, women empowerment, economic growth contribution to the prosperity and local wellbeing, increase in innovation and wealth creation in almost all economies (Minniti & Naudé, 2010; Serafimova & Petrevska, 2018). In addition, according to Berger and Kuckertz (2016, p. 5163), "increased levels of female entrepreneurship can contribute to a higher quality of entrepreneurship through conferring greater diversity" and female eco system starters' social contribution is impressive since they usually concentrate on social goals more than economic goals. Kearins and Schaefer (2017) support this factor by indicating that female entrepreneurs are found to be more environment focused and sustainability oriented in general. Female entrepreneurship usually results in better gender equality and decrease in the inequalities between genders in the business sector. Consequently, we may conclude that since gender equality is an issue which is in direct relationship with sustainability, female entrepreneurship positively impacts social sustainability and sustainable development (Dal Mas & Paoloni, 2019).

In addition to the stated ones, the other female entrepreneurship contributions include: improvement in women's social confidence, increase in female independence and security, reduction of local poverty and better distribution of earnings and wealth within the society (De Vita, Mari, & Poggesi, 2014). Further, Kevehazi suggests that (2016: 90) "the key element of social, economic and environmental sustainability is to extend the women's personal autonomy, their rights, to strengthen their ventures", which is a net outcome of female entrepreneurship.

Personal traits and characteristics of female entrepreneurs

Entrepreneurs, by definition, are people who start, manage, and run a new business where they figure out new markets and/or market needs and implement new ways to satisfy these needs. They are usually high risk takers, they are open to innovation, they are motivated by progress, and they tolerate changing environmental conditions (Pirnar, 2015). Some of the main entrepreneurial attributes, skills, and characteristics for all genders may be stated as (Brush, De Bruin, & Welter, 2009; Fitriati & Hermiati, 2011; Phelan & Sharpley, 2012):

- Commitment, determination, and perseverance
- Motivation for achievement,
- Orientation for opportunity,
- Internal locus of control
- Persistent managerial, planning, decision-making, and problem-solving skills
- Feedback search
- Tolerance for ambiguity, risk-taking, and failure
- Creativity and innovativeness
- Having a high mental and physical energy level

- Having a long-term and strategic vision
- Having a dynamic nature
- Self-confidence and independence
- Team building and communication abilities

Though the female entrepreneurs possess the similar stated entrepreneurial traits as male entrepreneurs, when compared they have special characteristics specific to their gender. The Global Entrepreneurship Monitor Report annually analyses these differences and presents the analysis results. As the report results indicate, the level and the areas of distinction between male and female entrepreneurs vary from country to country (De Vita et al., 2014). As a generalisation, female entrepreneurs are said to be more cautious, more open to the new ideas, more adaptive to changes, and more social needs oriented. In addition, it is understood that they possess more advanced training skills, they give importance to staff training, they encourage their staff more, and they have improved communication, observation, and problem-solving skills (Krueger, 2000; Ramadani, Hisrich, & Gërguri-Rashiti, 2015). The general comparative characteristics of the female entrepreneurs are summarised in Table 14.1.

It is possible to group the female entrepreneurial traits and characteristics under three sub-categories as characteristics related to entrepreneurship, characteristics related to socio-cultural values, and characteristics related to gender role as described further.

1. The characteristics related to entrepreneurship: These are the female entrepreneurs' common traits and characteristics which are usually similar with the male entrepreneurs. Examples of these characteristics include: self-confidence, risk taking nature, assertiveness, creativeness, innovativeness, competitiveness, and wish for economic independence.

Table 14.1 The general comparative characteristics of the female entrepreneurs

Experience in the job	Females often prefer to be more experienced than males when starting the business. After start-up and during operation, they tend to spend less time than male entrepreneurs on the job.
Aims	Females focus more on the job structure than profitability (when compered to males). Females focus more on revenue generation, whereas male entrepreneurs focus on ownership.
Potential age	In some studies, potential age is the same, for the studies with difference males are often ages 25–35, females are from the 35–45 age group.
Job factors amd entrepreneurial traits	Female entrepreneurs often prefer to start up smaller and more routine businesses (reason is indicated as environmental challenges). Female entreprenurs tend to be less willing to take risks and they do not favour uncertainity but they are good in autonomy and flexible to change.
Environmental factors	Female entrepreneurs tend to face more challenges when communicating with financial suppliers/intermediaries. Male entrepreneurs concentrate on production, export, advanced technology, and construction areas, whereas female entrepreneurs prefer to operate in trade, health, education, and food and beverage and tourism and hospitality.

Adapted from sources: Sexton & Bowman-Upton (1990); Woldie & Adersua (2004); Kepler & Shane (2007); Hughes & Jennings (2012); Ramadani et al. (2015); Hechavarria et al. (2017)

2. The characteristics related to socio-cultural values. These characteristics change from region to region and country to country. Though sometimes these characteristics differ as a specific gender, usually they are similar within the same cultural and social groups. In general, they consist of the level of responsibility sharing, attitude towards respectfulness, being protection and security oriented, being open to partnerships, and being good in teamwork and leadership.
3. Third group consists of the characteristics related to the specific gender role of females. Examples of this type of characteristic include: possessing good communication skills, being better in observation and empathy, having better human relations, and being sensitive to social needs.

Ecotourism and female entrepreneurship

Sustainability is a growing trend for general tourism industry and ecotourism is developed as a sustainable solution to the mass tourism's negative environmental and socio-cultural impacts (Wishitemi, Momanyi, Ombati, & Okello, 2015). Thus, ecotourism is an alternative tourism product that has a high growth potential with an increasing demand. In relation, ecotourism entrepreneurship is a new entrepreneurial area which is suitable to entrepreneurs dealing with agriculture and/or farmers located in rural areas far away from the seashores and beaches. Ecotourism entrepreneurship is promoted as a local solution and as an Indigenous tourism development to many rural region's socioeconomic problems (Fuller, Buultjens, & Cummings, 2005). Nowadays, be it developed or developing, in many parts of the world local people contribute to tourism activities as ecotourism entrepreneurs who are sensitive to sustainable, natural, and eco-friendly operations (Thompson, Gillen, & Friess, 2018). As ecotourism entrepreneurs, they market local products to tourists and by doing so they promote sustainability of local values and socio-cultural issues (Tekin & Kasalak, 2014).

"Whether it is defined as an investment opportunity, tourism experience, land-use practice or conservation tool, ecotourism is attractive for those interested in private conservation" (Serenari, Peterson, Wallace, & Stowhas, 2017, p. 1793). Hence, Hayombe, Agong, Mossberg, Malbert, and Odede (2012, p. 160) state that "eco-tourism is specific, delivered by small-scale enterprises involving responsible behaviour, contributes to the conservation of biodiversity, lowest possible consumption of non-renewable resources and is a learning experience". All together combined, some of the benefits of ecotourism applications may be stated as (Honey, 2008): improvement of quality of local life, protecting local values and cultures, socio-economic sustainability, environmental protection, biological conservation, community involvement and development, respect for local cultures, women empowerment, local participation, improvement in responsible tourism, increase in the number of tourism business ownership, infrastructure development, increase in employment and increase in tourist awareness and finally, "the community pride resulting from global recognition of local ecotourism" (Cobbinah, Amenuvor, Black, & Peprah, 2017, p. 39).

Benefits gained by female ecotourism entrepreneurs

When the characteristics of the female ecotourism entrepreneurs are examined, some literature emphasises that compared to their male counterparts some of the female entrepreneurs are motivated more by social causes and environmental protection issues and they prefer to be engaged in green and sustainable issues (Braun, 2010; Kearins & Schaefer, 2017; Dal Mas & Paoloni, 2019). This is a very important fact that makes ecotourism a very promising entrepreneurial investment area for female entrepreneurs. Ecotourism is also a suitable area for female empowerment through sustainability, local economic development, and rural development (Scheyvens, 2007; Dilly, 2003; Pleno, 2006; Honey, 2008; Wishitemi et al., 2015).

The individual benefits gained by female ecotourism entrepreneurs are many. The most important ones may be stated as (Scheyvens, 2007; Gentry, 2007; Cobbinah et al., 2017; Morgan & Winker, 2020) increased economic independence, additional income, self-realisation and increased social status, and women's participation in domestic level decision making like the householding spending and in community-level decision making. In addition, ecotourism can provide rural women new job opportunities, more control over resources, and it also helps their own self-development, self-improvement, and greater self-confidence (Belsky, 1999; Gentry, 2007). Female ecotourism applications often result in an increase in the sense of local cultural pride and high level of training (Scheyvens, 2007; Dilly, 2003; Pleno, 2006). Yet, in some rare cases ecotourism entrepreneurship gives results to declining birth rates. Female ecotourism empowerment is commonly found in female ecotourism entrepreneurship cases (see Table 14.2) and since female empowerment is one of the most important factors to obtain gender equality, it is possible to conclude that female ecotourism applications lead to gender equality resulting in narrowing the gender gap in the local community (Narwan & Mulia, 2019; Lohne, 2019). To optimise these benefits further, "gender equity planning in specific ecotourism projects may operationalise ecological goals of social movement and development". (Swain & Swain, 2004, p. 4). Thus, when the benefits of female ecotourism entrepreneurship is gathered, it may be concluded that there are psychological, political, social, and economic benefits that affect both the society and the individual as herself (Gil Arroyo, Barbieri, Sotomayor, & Knollenberg, 2019).

For better understanding, the benefits provided by pulled and pushed female ecotourism entrepreneurship may be grouped under three categories, as economic, social, and environmental positive impacts as Figure 14.1 indicates (Anup, 2017).

Motivations of female ecotourism entrepreneurs: 'push and pull' factors

There are many motivations for female ecotourism entrepreneurs to start and run a new business. According to their motivations, Tambunan (2009) groups female entrepreneurs into three groups as change entrepreneurs, forced entrepreneurs, and created entrepreneurs. Tambunan (2009) describes change female entrepreneurs as women who become entrepreneurs without serious planning. They are usually motivated by keeping occupied or started their business as a hobby. The second group of female entrepreneurs are called forced entrepreneurs who become entrepreneurs due to their situational factors. The death of a family member who is in charge of the business or succession approach in a family business are examples for this type where motivation of the female entrepreneur is mainly financial. Third type of female ecotourism entrepreneurs are called created entrepreneurs who are directed, developed, and encouraged to entrepreneurship. Their main motivations are being independent, having control over business and family time, self-achievement, owning her own business, self-realisation, and being her own boss (Kunjuraman & Hussin, 2017). Created female entrepreneurs also have social motivators like providing employment to others and being an inspiring role model to their children. They are also grouped under pulled entrepreneurs. Hughes (2003) grouped female entrepreneurs according to their motivations into two groups as affected by 'push or pull' factors. In this grouping distribution, pulled female entrepreneurs are attracted to start and establish their own business, whereas 'pushed' entrepreneurs are forced to their new and own businesses by environmental forces other than themselves. Some of these environmental factors may occur due to unemployment, underemployment, unfair, or unfavorable working conditions (Yetim, 2008), whereas monetary motivations like profit maximisation are also usually termed as a pull motivational factor for female entrepreneurs (Kirkwood, 2009).

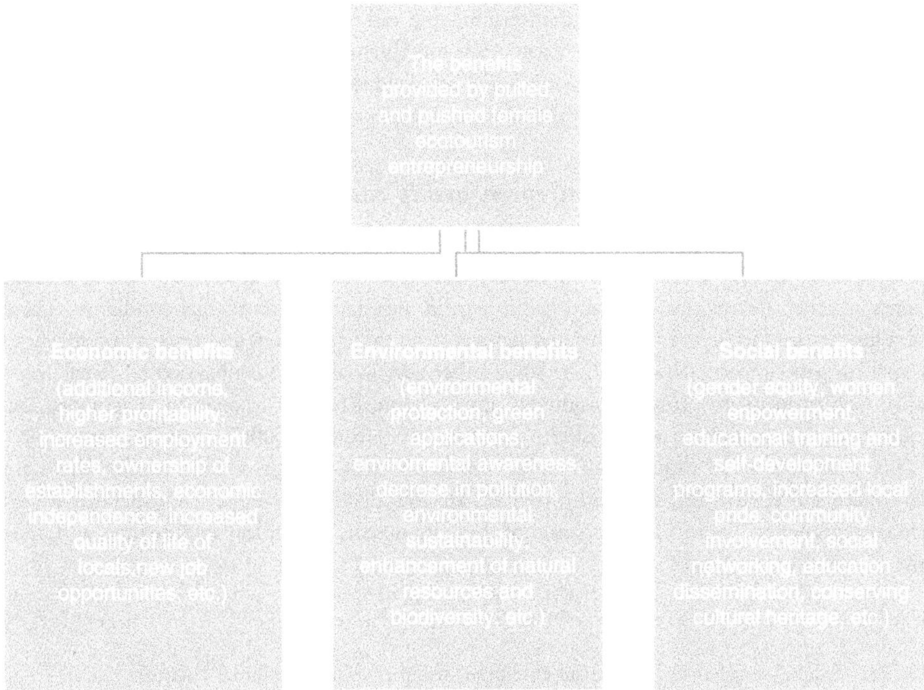

Figure 14.1 Benefits provided by pulled and pushed female ecotourism entrepreneurship

Altogether, the main motivations for both forced/pushed and created/pulled female ecotourism entrepreneurship may be stated as (Hughes, 2003; Dunn, 2007; Goyal & Parkash, 2011; Bakas, 2017; Deng et al., 2020):

- Self-identity, self-achievement, and self-realisation
- Support of family members, succession in family businesses
- Additional income
- Independency (economic and/or own time)
- Entrepreneurial success stories as role models
- Innovative thinking style and creative personality
- Filling up a pass time, as a hobby
- New opportunities and/or new challenges
- Financial entrepreneurial incentives for females
- Successful role models
- Having self-control over time/flexibility work hours
- Trying something new on one's own
- Self-satisfaction
- Setting a role model example to a friend, relatives, own children
- Social factors like providing job opportunities and employment for people

When the gender differences are examined, it is observed that both male and female entrepreneurs usually have the same important needs like new opportunities, creative and innovative ideas, personal entrepreneurial characteristics, and applicable and flexible business plans with suitable strategies. Thus, some motivators like additional income and economic and personal independence are the same for both genders; female ecotourism entrepreneurs are choosing to start their own business in order to balance their work and home time and responsibilities since working hours in tourism industry are usually quite long (McGowan, Redeker, Cooper, & Greenan, 2012).

Female ecotourism entrepreneurship cases and applications

Female ecotourism entrepreneurs are found to be quite flexible to changes. Many small businesses like small farms or agri-tourism accommodation enterprises have the ability to adapt to prompt changes and their flexible structure also helps them in uncertain environmental conditions. These establishments tend to be very creative and innovative in the crisis times and respond positively to external negative impacts (Bakas, 2017). Some other additional benefits obtained by female ecotourism entrepreneurs are greater community involvement, having a voice in local activities, and though rarely, sometimes new leadership roles in society (Stronza, 2008; Tran & Walter, 2014).

Due to all the stated motivational issues, positive factors, and provided benefits, there are many successful ecotourism applications which are founded, encouraged, and operated by female entrepreneurs. Table 14.2 lists some of these interesting examples, applications, and cases from all around the world.

Table 14.2 Examples of female ecotourism entrepreneurship cases from different countries

Female Ecotourism Entrepreneurship Cases and Their Location	*References (Authors)*
Case study of Guyanese Rain Forest	Dilly (2003)
Case Study in the Province of Bohol, Philippines	Pleno (2006)
LEELED community-based tourism project, Thailand	Dunn (2007)
Case research in the Peruvian Amazon	Stronza (2008)
Women's Entrepreneurship and Rural Tourism in Greece	Koutsou, Notta, Samathrakis, and Partalidou (2009)
Case of ecotourism project in Lombok, Indonesia	Schellhorn (2010)
Case of the Isecheno Women's Conservation Group (CBO) Isecheno, Kakamega, Western Provence, Kenya	Barry (2012)
Case of Barpak village at Gorkha District of Western Nepal	Acharya and Halpenny (2013)
Case of Beypazarı, Turkey	Kose (2014)
Case of Giao Xuan CBET project in Vietnam	Tran and Walter (2014)
Ecotourism development and female empowerment in Botswana	Lenao and Basupi (2016)
Case of women-centred ecotourism enterprises in Bunyoro, Uganda	Mwesigwa and Mubangizi (2016)
Case of Abai Village, Malaysia	Kunjuraman and Hussin (2016)
Case of Women Empowerment through Ecotourism Activities in Lower Kinabatangan Area of Sabah, East Malaysia	Kunjuraman and Hussin (2017)
Case of Bardia National Park, Nepal	Panta and Thapa (2018)
Case of 21 Village, province of Edirne, Turkey	Serinikli (2019)
Cultivating Women's Empowerment through Agritourism: Evidence from Andean Communities	Gil Arroyo et al. (2019)
Case study of a female-only ecotourism cooperative Orquideas project, Mexico	Morgan and Winker (2020)

As stated earlier, related literature shows that ecotourism entrepreneurship emphasises female empowerment (Swain & Swain, 2004; Lenao & Basupi, 2016). Local rural women perceive ecotourism projects as promoting socio-cultural empowerment of females and they become better educated with the help of provided educational training and self-development programs. Thus, cases and applications related to female ecotourism entrepreneurship indicate that there is a direct relationship between female empowerment and involvement in environmental protection and sustainability consciousness (Pleno, 2006).

The success factors established in successful female ecotourism entrepreneurial applications include: participation in community networking, increase in decision-making, local innovation, collective responsibility, successful environmental awareness, collective sharing of local resources, increased skills and knowledge on leadership and management.

Challenges of female ecotourism entrepreneurship

As there are many female ecotourism entrepreneurship benefits, there exists some challenges that women entrepreneurs should face. Some of the most common ones may be stated as: financial problems to start and run the new business, credit and loan obtaining problems, education and training problems, lack of technical skills and knowledge, and networking problems (Dilly, 2003; Mwangi, 2012). In addition, lack of managerial knowledge, need for special ecological and entrepreneurial training, and high competition between domestic and international establishments are also among the challenging technical issues that female ecotourism entrepreneurs have to overcome (Dunn, 2007; Soysal, 2013; Panta & Thapa, 2018). Moreover, the increase in female ecotourism entrepreneurship leads to increased competition for the ecotourism revenues, which may result in social cohesion and social disharmony in time (Scheyvens, 2007). This case is a proven social challenge since, already, some ecotourism entrepreneurs ask for the availability of official mechanisms for even and fair distribution of ecotourism outcomes (Cobbinah et al., 2017).

Furthermore, in some less developed and developing countries, the gender inequalities occurring from a manifestation of local customs and traditions result in female entrepreneurs' marginalisation (Gil Arroyo et al., 2019). Not only in less developed countries and developing countries, but also in developed countries, female entrepreneurs face already established social gender roles for domestic work and high family demands in childcare leading to the personal need to balance work and home time. Thus, challenging social gender pressures, discriminatory gender practices, and local prejudices for working women become quite common problems for many female ecotourism entrepreneurs from all over the world (Morgan & Winkler, 2020).

UN Sustainable Development Goals (SDGs) and the importance of female ecotourism entrepreneurship

Nowadays, the global world is facing many vital environmental, social, and economic problems and challenges. To overcome these problems and to find sustainable solutions to global challenges, the General Assembly of the United Nations adopted the 2030 Agenda for Sustainable Development in September 2015 (Gratzer & Keeton, 2017).

Sustainable Development Goals (SDGs) are defined as "a universal call to action to end poverty, protect the planet and ensure that all people enjoy peace and prosperity" United Nations (2014, 2015a). Further, the SDGs are building upon globally accepted Millennium Development Goals (MDGs) where sustainable development is defined as "development that meets the needs of the present without compromising the ability of future generations to meet their own needs" (Uitto, 2016). There are 17 Sustainable Development Goals (SDGs) taking place in the Agenda which need urgent action by all countries. They are stated as follows (United Nations, 2015b, 2018):

1. End poverty in all its forms everywhere.
2. End hunger, achieve food security and improved nutrition, and promote sustainable agriculture.
3. Ensure healthy lives and promote wellbeing for all at all ages.
4. Ensure inclusive and quality education for all and promote lifelong learning.
5. Achieve gender equality and empower all women and girls.
6. Ensure access to water and sanitation for all.
7. Ensure access to affordable, reliable, sustainable, and modern energy for all.
8. Promote inclusive and sustainable economic growth, employment, and decent work for all.
9. Build resilient infrastructure, promote sustainable industrialisation, and foster innovation.
10. Reduce inequality within and among countries.
11. Make cities inclusive, safe, resilient, and sustainable.
12. Ensure sustainable consumption and production patterns.
13. Take urgent action to combat climate change and its impacts.
14. Conserve and sustainably use the oceans, seas, and marine resources.
15. Sustainably manage forests, combat desertification, halt and reverse land degradation, and halt biodiversity loss.
16. Promote just, peaceful, and inclusive societies.
17. Revitalise the global partnership for sustainable development.

Global female ecotourism entrepreneurship applications play an important role in achieving some of the stated UN SDGs (Uitto, 2016; Nigar, 2018; Lohne, 2019). For example, Sustainable Development Goals (SDG5: gender equality) (United Nations Industrial Development Organisation (UNIDO), 2018) include women's economic empowerment and since women empowerment is one of the main benefits of female ecotourism entrepreneurship (Sarfaraz, Faghih, & Majd, 2014), it highly contributes to the achievement of SDG5 (gender equality) (Guney-Frahm, 2018; United Nations Population Fund, 2020). Indeed, female ecotourism entrepreneurship has a huge potential for women empowerment and with its related benefits as participation in social life, self-identity, self-achievement and self-realisation, independency, having a voice in local population and such, it helps to achieve SDG5 (Demartini, 2019).

In addition, by improvement in the daily life and local QOL, female ecotourism entrepreneurship contributes to SDG3 (promote wellbeing for all). Krasavac, Karamata, and Djordjevic (2019) state that female entrepreneurs promote sustainable practices in economy, social system, and ecology, which further indicates that female ecotourism entrepreneurs support the achievement of ecology related SDGs. To support this fact, Katila, de Jong, Galloway, Pokorny, and Pacheco (2017) emphasise that girls and women have an important role in sustainable different forest-based activities and sustainable ecotourism forestry. Gratzer and Keeton (2017) indicate that especially in mountain forests and mountain areas, ecotourism entrepreneurship is one of the most important "options for ensuring that local communities benefit from protected areas" which interacts with SDGs 1 and 2 (ending poverty and hunger) and SDG15 (land-based conservation). In addition, Minniti and Naudé (2010) and Alarcón and Cole (2019), mention that by providing jobs and promoting economic growth and development, female ecotourism entrepreneurship supports SDG8 (promote economic growth, employment).

Kanowski, Yao, and Wyatt (2019) suggest that education and training of ecotourism entrepreneurship may improve environmental awareness and entrepreneurial- and technology-related skills of female ecotourism entrepreneurs, while contributing to SDG4 (quality education) and SDG5 (gender equality). Lastly, with the benefits of cultural preservation and facilitating community partnerships (Movono & Hughes, 2020), female ecotourism entrepreneurship has a potential to positively impact SDG16 (peaceful and inclusive societies).

Conclusion

Ecotourism is a vital and promising area for women's empowerment, which directly and indirectly contributes to many of the UN's social development goals. Yet, women empowerment in ecotourism and female ecotourism entrepreneurship applications are rapidly increasing due to the provided local and individual sustainable, environmental, psychological, economic, and social benefits. Some of these benefits may be stated as:
increased economic independence, additional income, new job creation, increased social status, women's participation in domestic and/or community level in decision making, more female control over resources, self-improvement and greater self-confidence. In addition, female ecotourism applications often result in increase in the sense of local cultural pride, higher female training and gender equality. All these female ecotourism entrepreneurship benefits and motivations increase the popularity of local and international applications since benefits provided are important to individuals, local communities and countries. Thus, to optimise the associated benefits, it is important to understand not only the concept and motivations of female entrepreneurship, but the challenges and outcomes of the application cases. It is also important to realise the success factors established in successful female ecotourism entrepreneurial applications for better local QOL and rural development outcomes. Some of these success criteria may be stated as: participation in community networking, increase in decision-making, local innovation, collective responsibility, successful environmental awareness, collective sharing of local resources, increased skills, and knowledge on leadership and management.

Overall, this chapter focuses on female ecotourism entrepreneurs in terms of traits, characteristics, motivations, challenges, and gender-related distinctiveness. In addition, the interaction points to female ecotourism entrepreneurship cases and the UN's social development goals are determined.

References

Acharya, B. P., & Halpenny, A. E. (2013). Homestays as an alternative tourism product for sustainable community development: A case study of women- managed tourism product in rural Nepal. *Tourism Planning & Development*, *10*(4), 367–387. doi:https://doi.org/10.1080/21568316.2013.779313.

Achtenhagen, L., & Welter, F. (2003). *Female Entrepreneurship in Germany* (pp. 71–100). Greenwich, CT: Information Age Publishing.

Alarcón, D. M., & Cole, S. (2019). No sustainability for tourism without gender equality. *Journal of Sustainable Tourism*, *27*(7), 903–919. doi:https://doi.org/10.1080/09669582.2019.1588283.

Anup, K. C. (2017). Ecotourism in Nepal. *The Gaze: Journal of Tourism and Hospitality*, *8*, 1–19. doi:https://doi.org/10.3126/gaze.v8i0.17827.

Bakas, F. E. (2017). Community resilience through entrepreneurship: the role of gender. *Journal of Enterprising Communities: People and Places in the Global Economy*, *11*(1), 61–77. doi:https://doi.org/10.1108/JEC-01-2015-0008.

Barry, K. S. (2012). Women empowerment and community development through ecotourism. *Capstone Collection*, 2579. Retrieved from https://digitalcollections.sit.edu/capstones/2579.

Belsky, J. M. (1999). Misrepresenting communities: The politics of community-based rural ecotourism in gales point manatee, Belize 1. *Rural Sociology*, *64*(4), 641–666.

Berger, E. S., & Kuckertz, A. (2016). Female entrepreneurship in startup ecosystems worldwide. *Journal of Business Research*, *69*(11), 5163–5168. doi:https://doi.org/10.1016/j.jbusres.2016.04.098.

Braun, P. (2010). Going green: Women entrepreneurs and the environment. *International Journal of Gender and Entrepreneurship*, *2*(3), 245–259. Doi:https://doi.org/10.1108/17566261011079233.

Brush, C. G., & Cooper, S. Y. (2012). Female entrepreneurship and economic development: An international perspective. *Entrepreneurship & Regional Development*, *24*(1–2),1–6. doi:https://doi.org/ 10.1080/08985626.2012.637340.

Brush, C. G., De Bruin, A., & Welter, F. (2009). A gender-aware framework for women's entrepreneurship. *International Journal of Gender and Entrepreneurship*, *1*(1), 8–24. doi:https://doi.org/10.1108/17566260910942318.

Cobbinah, P. B., Amenuvor, D., Black, R., & Peprah, C. (2017). Ecotourism in the Kakum Conservation Area, Ghana: Local politics, practice and outcome. *Journal of Outdoor Recreation and Tourism*, *20*, 34–44. doi:https://doi.org/10.1016/j.jort.2017.09.003.

Dal Mas, F., & Paoloni, P. (2019). A relational capital perspective on social sustainability; the case of female entrepreneurship in Italy. *Measuring Business Excellence*, *24*(1), 114–130. doi:https://doi.org/10.1108/MBE-08-2019-0086.

De Vita, L., Mari, M., & Poggesi, S. (2014). Women entrepreneurs in and from developing countries: Evidences from the literature. *European Management Journal*, *32*(3), 451–460. doi:https://doi.org/ 10.1016/j.emj.2013.07.009.

Demartini, P. (2019). Why and how women in business can make innovations in light of the Sustainable Development Goals, *Administrative Sciences*, *9*(3), 64. doi:https://doi.org/10.3390/admsci9030064.

Deng, W., Liang, Q., Li, J., & Wang, W. (2020). Science mapping: A bibliometric analysis of female entrepreneurship studies. *Gender in Management*, *36*(1). doi:https://doi.org/ 10.1108/GM-12-2019-0240.

Dilly, B. J. (2003). Gender, culture, and ecotourism: Development policies and practices in the Guyanese rain forest. *Women's Studies Quarterly*, *31*(3/4), 58–75.

Dunn, S. F. (2007). *Toward Empowerment: Women and Community-Based Tourism in Thailand* (Doctoral dissertation, University of Oregon).

Fitriati, R., & Hermiati, T. (2011). Entrepreneurial skills and characteristics analysis on the graduates of the Department of Administrative Sciences, FISIP Universitas Indonesia. *Bisnis & Birokrasi Journal*, *17*(3).

Fuller, D., Buultjens, J., & Cummings, E. (2005). Ecotourism and indigenous micro-enterprise formation in northern Australia opportunities and constraints. *Tourism Management*, *26*(6), 891–904. doi:https://doi.org/10.1016/j.tourman.2004.04.006.

Gentry, K. M. (2007). Belizean women and tourism work opportunity or impediment? *Annals of Tourism Research*, (34), 477–496. doi:https://doi.org/10.1016/j.annals.2006.11.003.

Gil Arroyo, C., Barbieri, C., Sotomayor, S., & Knollenberg, W. (2019). Cultivating women's empowerment through agritourism: Evidence from Andean communities. *Sustainability*, *11*(11), 3058. doi:https://doi.org/10.3390/su11113058.

Goyal, M., & Parkash, J. (2011). Women entrepreneurship in India-problems and prospects. *International Journal of Multidisciplinary Research*, *1*(5), 195–207.

Gratzer, G., & Keeton, W. S. (2017). Mountain forests and sustainable development: The potential for achieving the United Nations' 2030 Agenda. *Mountain Research and Development*, *37*(3), 246–253. doi:https://doi.org/10.1659/MRD-JOURNAL-D-17-00093.1.

Guney-Frahm, I. (2018). A new era for women? Some reflections on blind spots of ICT-based development projects for women's entrepreneurship and empowerment. *Gender, Technology and Development*, *22*(2), 130–144. doi:https://doi.org/10.1080/09718524.2018.1506659.

Hayombe, P. O., Agong, S. G., Mossberg, L., Malbert, B., & Odede, F. (2012). Up scaling ecotourism in Kisumu City and its environs: Local community perspective. *International Journal of Business and Social Research*, *2*(7), 158–174.

Hechavarria, D., Bullough, A., Brush, C., & Edelman, L. (2019). High-growth women's entrepreneurship: Fueling social and economic development. *Journal of Small Business Management, 57*(1), 5–13. https://doi.org/10.1111/jsbm.12503.

Hechavarria, D. M., Terjesen, S. A., Ingram, A. E., Renko, M., Justo, R., & Elam, A. Taking care of business: The impact of culture and gender on entrepreneurs' blended value creation goals. *Small Business Economics, 48*, 225–257. doi:https://doi.org/10.1007/s11187-016-9747-4.

Honey, M. (2008). *Ecotourism and Sustainable Development: Who Owns Paradise?* (2nd ed.). Washington D.C., USA: Island Press.

Hughes, K. D. (2003). Pushed or pulled? Women's entry into self-employment and small business ownership. *Gender, Work & Organization, 10*(4), 433–454. doi:https://doi.org/10.1111/1468-0432.00205.

Hughes, K. D., & Jennings, J. E. (Eds.). (2012). *Global Women's Entrepreneurship Research: Diverse Settings, Questions, and Approaches*. UK: Edward Elgar Publishing.

Humbert, A. L., & Brindley, C. (2015). Challenging the concept of risk in relation to women's entrepreneurship. *Gender in Management: An International Journal, 30*(1), 2–25. doi:https://doi.org/10.1108/GM-10-2013-0120.

Kanowski, P., Yao, D., & Wyatt, S. (2019). SDG 4: Quality education and forests—'The golden thread'. In *Sustainable development goals: Their impacts on forests and people* (pp. 108–145). UK: Cambridge University Press.

Katila, P., de Jong, W., Galloway, G., Pokorny, B., & Pacheco, P. (2017). *Building on Synergies: Harnessing Community and Smallholder Forestry for Sustainable Development Goals*. Vienna, Austria: IUFRO.

Kearins, K., & Schaefer, K. (2017). Women, entrepreneurship and sustainability. *The Routledge Companion to Global Female Entrepreneurship*, 48–61.

Kepler, E., & Shane, S. (2007). *Are Male and Female Entrepreneurs Really That Different?* Washington, DC: Office of Advocacy, US Small Business Administration.

Kirkwood, J. (2009). Motivational factors in a push-pull theory of entrepreneurship. *Gender in Management: An International Journal, 24*(5), 346–364. doi:https://doi.org/10.1108/17542410910968805.

Koutsou, S., Notta, O., Samathrakis, V., & Partalidou, M. (2009). Women's entrepreneurship and rural tourism in Greece: Private enterprises and cooperatives. *South European Society and Politics, 14*(2), 191–209. doi:https://doi.org/10.1080/13608740903037968.

Kose, Z. (2014). Turizmde kadin istihdami ve kadin girisimciligi Beypazari orneği. Masters thesis, Antropology Department, Hacettepe University, Ankara.

Krasavac, B. C., Karamata, E., & Djordjevic, V. (2019). Innovative potential of environmentally motivated female entrepreneurship for sustainable development in the Republic of Serbia. *Economics of Agriculture, 66*(3), 721–735. doi:https://doi.org/10.5937/ekoPolj1903721C.

Krueger, D. (2000). Characteristics of the female entrepreneur. *Journal of Business and Entrepreneurship, 12*(1), 87.

Kunjuraman, V., & Hussin, R. (2016). Women participation in ecotourism development: Are they empowered? *World Applied Sciences Journal, 34*(12), 1652–1658. doi:https://doi.org/10.5829/idosi.wasj.2016.1652.1658.

Kunjuraman, V., & Hussin, R. (2017). Women empowerment through ecotourism activities in lower Kinabatangan area of Sabah, East Malaysia. *JGD, 13*(1), 135–148.

Lenao, M., & Basupi, B. (2016). Ecotourism development and female empowerment in Botswana: A review. *Tourism Management Perspectives, 18*, 51–58. doi:https://doi.org/10.1016/j.tmp.2015.12.021.

Lohne, L. (2019). *Recognition by Participation? Social Justice and Equality in Community-Based Ecotourism among the Hmong in Sa Pa, Viet Nam* (Master's Thesis, Centre for East and South-East Asian Studies, Lund University).

McGowan, P., Redeker, C. L., Cooper, S. Y., & Greenan, K. (2012). Female entrepreneurship and the management of business and domestic roles: Motivations, expectations and realities. *Entrepreneurship & Regional Development, 24*(1–2), 53–72. doi:https://doi.org/10.1080/08985626.2012.637351.

Minniti, M., & Naudé, W. (2010). What do we know about the patterns and determinants of female entrepreneurship across countries? *The European Journal of Development Research, 22*, 277–293. doi:https://doi.org/ 10.1057/ejdr.2010.17.

Morgan, M. S., & Winkler, R. L. (2020). The third shift? Gender and empowerment in a women's ecotourism cooperative. *Rural Sociology, 85*(1), 137–164. doi:https://doi.org/10.1111/ruso.12275.

Movono, A., & Hughes, E. (2020). Tourism partnerships: Localizing the SDG agenda in Fiji. *Journal of Sustainable Tourism*, 1–15. doi:https://doi.org/10.1080/09669582.2020.1811291.

Mwesigwa, D., & Mubangizi, B. C. (2016). Exploring the Best Practices of Women-Centred Ecotourism Enterprises in Bunyoro, Uganda. Retrieved from https://ir.lirauni.ac.ug/handle/123456789/53.

Mwangi, S. M. (2012). Psychosocial challenges facing female entrepreneurs in rural informal sector and their coping mechanisms: A case study of Gucha District, Kenya.*Research on Humanities and Social Sciences*, 2(2), 15–27.

Narwan, K., and Mulia, D. S. (2019). The impact of ecotourism on women empowerment, gender equality and community development in rural area. *Malaysian Journal of Social Sciences and Humanities*, *4*(6),139–151.

Nigar, N. (2018). Ecotourism for sustainable development in Gilgit-Baltistan: Prospects under CPEC. *Strategic Studies*, *38*(3), 72–85. doi:10.2307/48539385.

Page, S., & Ateljevic, J. (Eds.). (2009). *Tourism and Entrepreneurship: International Perspectives*. London: Routledge.

Panta, S. K., & Thapa, B. (2018). Entrepreneurship and women's empowerment in gateway communities of Bardia national park, Nepal. *Journal of Ecotourism*, *17*(1), 20–42. doi:https://doi.org/10.1080/14724049.2017.1299743.

Phelan, C., & Sharpley, R. (2012). Exploring entrepreneurial skills and competencies in farm tourism. *Local Economy*, *27*(2), 103–118.

Pirnar, I. (2015). The specific characteristics of entrepreneurship process in tourism industry. *Selcuk University the Journal of Institute of Social Sciences*, (34), 75–86.

Pleno, M. J. L. (2006). Ecotourism projects and women's empowerment: A case study in the province of Bohol, Philippines. *Forum of International Development Studies*, (32), 137–155.

Ramadani, V., Hisrich, R. D., & Gërguri-Rashiti, S. (2015). Female entrepreneurs in transition economies: insights from Albania, Macedonia and Kosovo. *World Review of Entrepreneurship, Management and Sustainable Development*, *11*(4), 391–413. doi:https://doi.org/10.1504/WREMSD.2015.072066.

Santos, G., Marques, C. S., & Ferreira, J. J. (2018). A look back over the past 40 years of female entrepreneurship: Mapping knowledge networks. *Scientometrics*, *115*(2), 953–987.

Sarfaraz, L., Faghih, N., & Majd, A. A. (2014). The relationship between women entrepreneurship and gender equality. *Journal of Global Entrepreneurship Research*, *4*(1), 6.

Schellhorn, M. (2010). Development for whom? Social justice and the business of ecotourism. *Journal of Sustainable Tourism*, *18*(1), 115–135.

Scheyvens, R. (2007). Ecotourism and gender issues. *Critical Issues in Ecotourism*, 185–213.

Serafimova, M., & Petrevska, B. (2018). Female entrepreneurship in tourism: A strategic management perspective in Macedonia. *Journal of Applied Economics and Business*, *6*(1), 21–32.

Serenari, C., Peterson, M. N., Wallace, T., & Stowhas, P. (2017). Private protected areas, ecotourism development and impacts on local people's well-being: A review from case studies in Southern Chile. *Journal of Sustainable Tourism*, *25*(12), 1792–1810. doi:https://doi.org/ 10.1080/09669582.2016.1178755.

Serinikli, N. (2019). The attitudes of the micro female entrepreneurs in the rural area towards becoming a cooperative: Agro-tourism woman cooperative. *Journal of Entrepreneurship and Development*, *14*(1), 45–57.

Sexton, D. L., & Bowman-Upton, N. (1990). Female and male entrepreneurs: Psychological characteristics and their role in gender-related discrimination. *Journal of Business Venturing*, *5*(1), 29–36. doi:https://doi.org/10.1016/0883-9026(90)90024-N.

Soysal, A. (2013). Women's entrepreneurship in rural areas: An assessment for Turkey. *Eskişehir Eskişehir Osmangazi University Journal of Economics and Administrative Sciences*, *8*(1), 163–190.

Stronza, A. (2008). Hosts and hosts: The anthropology of community-based ecotourism in the Peruvian Amazon. *Napa Bulletin*, *23*(1), 170–190. doi:https://doi.org/10.1525/napa.2005.23.1.170.

Swain, M. B., & Swain, M. T. B. (2004). An ecofeminist approach to ecotourism development. *Tourism Recreation Research*, *29*(3), 1–6. doi:https://doi.org/10.1080/02508281.2004.11081451.

Tajeddini, K., Ratten, V., & Denisa, M. (2017). Female tourism entrepreneurs in Bali, Indonesia. *Journal of Hospitality and Tourism Management*, *31*, 52–58. doi:https://doi.org/10.1016/j.jhtm.2016.10.004.

Tambunan, T. (2009). Women entrepreneurship in Asian developing countries: Their development and main constraints. *Journal of Development and Agricultural Economics*, *1*(2), 027–040.

Tekin, M., & Kasalak, M. A. (2014). In regional development role of ecotourism entrepreneurship. *Selcuk University Journal of Institute of Social Sciences*, *32*, 129–136.

Thompson, B. S., Gillen, J., & Friess, D. A. (2018). Challenging the principles of ecotourism: Insights from entrepreneurs on environmental and economic sustainability in Langkawi, Malaysia. *Journal of Sustainable Tourism*, *26*(2), 257–276. doi:https://doi.org/10.1080/09669582.2017.1343338.

Tran, L., & Walter, P. (2014). Ecotourism, gender and development in northern Vietnam. *Annals of Tourism Research*, *44*, 116–130. doi:https://doi.org/10.1016/j.annals.2013.09.005.

Uitto, J. I. (2016). The environment-poverty nexus in evaluation: Implications for the Sustainable Development Goals. *Global Policy*, 7(3), 441–447. doi:https://doi.org/10.1111/1758-5899.12347.

United Nations. (2014). Sustainable Development Goals. Retrieved from http://www.un.org/sustainabledevelopment/gender-equality/ (accessed on 07.09.2020).

United Nations. (2015a). Transforming Our World: The 2030 Agenda for Sustainable Development. Retrieved from https://sustainabledevelopment.un.org (accessed on 04.09.2020).

United Nations. (2015b). The Millennium Development Goals Report 2015. Retrieved from https://www.un.org/millenniumgoals/2015_MDG_Report/pdf/MDG%202015%20rev%20(July%201).pdf (accessed on 07.09.2020).

United Nations. (2018). The 17 Goals. Retrieved from https://sdgs.un.org/goals (accessed on 07.09.2020).

United Nations Industrial Development Organization (UNIDO). (2018). UNIDO Event at SPIEF 2018 to Focus on Increasing Women's Contribution to Economic Growth. Retrieved from https://www.unido.org/news/unido-event-spief-2018-focus-increasing-womens-contribution-economic-growth (accessed on 07.09.2020).

United Nations Population Fund. (2020). Gender Equality. Retrieved from https://www.unfpa.org/gender-equality (accessed on 05.09.2020).

Wishitemi, B. E., Momanyi, S. O., Ombati, B. G., & Okello, M. M. (2015). The link between poverty, environment and ecotourism development in areas adjacent to Maasai Mara and Amboseli protected areas, Kenya. *Tourism Management Perspectives*, *16*, 306–317. doi:https://doi.org/10.1016/j.tmp.2015.07.003.

Yetim, N. (2008). Social capital in female entrepreneurship. *International Sociology*, *23*(6), 864–885. doi:https://doi.org/10.1177/0268580908095913.

Zapalska, A. M., & Brozik, D. (2014). Female entrepreneurial businesses in tourism and hospitality industry in Poland. *Problems and Perspectives in Management*, *12*(2), 7–13.

THEME 3

Change, conflict, and consumption

15

ECOTOURISM AND ACCESSIBILITY FOR PERSONS WITH DISABILITIES

Brian Garrod

Introduction

Access to ecotourism by persons with disabilities has rarely been the subject of academic research; nor has a great deal of guidance been developed in this regard for organisations responsible for the planning, management and delivery of ecotourism. The assumption seems to be that ecotourism is similar enough to other forms of tourism for research recommendations and practical guidance to be transferred directly from the general to the specific. As other chapters in this volume clearly demonstrate, however, the defining characteristics of ecotourism require that decisions about its management, planning, and delivery cannot be based simply upon an understanding of general tourism, no matter how sophisticated this might be. Indeed, Fennell (2020) defines ecotourism as:

> travel with a primary interest in the natural history of a destination. It is a non-invasive and participatory form of nature-based tourism that is built around learning, sustainability (conservation and local participation/benefits), and ethical planning, development and management. (Fennell, 2020, p. 20)

Ecotourism is thus different in its focus, its operation, and its intent. Sustainability is clearly a central goal. While this definition does not explicitly state that ecotourism should consider the needs and wants of people with disabilities, the condition that ecotourism should be ethically planned, developed, and managed might reasonably be taken to imply such. Indeed, as this chapter will argue, there are strong ethical reasons for ensuring that access to ecotourism is provided for people with disabilities.

Defining disability is also something that has greatly exercised the minds of researchers. Language is crucial when discussing disability issues, particularly in respect of the terminology employed (e.g., Alén, Domínguez, & Losada, 2012). Gillovic, McIntosh, Darcy, and Cockburn-Wootten (2018, p. 615) argue that we all have the power to "(re) produce oppression through language that maligns and misrepresents, or to (re) conceptualise and (re) construct the world we live in with liberating language that facilitates positive social change". The range of terms that has been used to describe people with disabilities has in the past included terms such as 'handicapped', 'wheelchair-bound', 'retards', 'cripples', and 'crazies':

DOI: 10.4324/9781003001768-15

terms that are today considered to be not only outdated but also condescending, stigmatising and discriminatory.

Much has changed, however, in the last quarter of a century. This has been ascribed to the gradual transition from the medical model of understanding disability to the social model. The now largely outmoded medical model assumes that disability is something opposite of what is 'normal' that should somehow be 'cured' or 'fixed' (Gillovic et al., 2018). The social model, in contrast, proposes that disability is a social construct. As such, disability only exists because society puts barriers in the way of people doing things. Were these barriers to be removed, a person considered disabled would no longer effectively be disabled because they could fully access every aspect of life. The social model therefore proposes that disability is, always and everywhere, a socially reproduced phenomenon. Research following the social model of disability therefore tends to focus on identifying the barriers to disabled access and then determining suitable strategies for overcoming those barriers (Randle & Dolnicar, 2019).

The terminology used in this chapter will follow the social model by referring to the reduction of function a person may experience as an 'impairment', while the term 'person with disability' (PwD) will be used to describe someone who has one or more impairments. This reflects a preference not to label people with disabilities by using the term as an adjective (as in 'disabled person') but rather to acknowledge that they are people first and foremost. Similarly, rather than describing people without disabilities as 'normal' or even 'able-bodied', this chapter will use the term 'non-disabled people'.

Following the World Tourism Organization (UNWTO) (2016), the terms *mobility, sensory,* and *intellectual disability* will be used in the chapter to describe the various impairments of PwDs. Mobility disability refers to a restriction in a person's ability to freely move about. Such individuals may or may not use some form of aid to assist their mobility (for example a wheelchair, walking frame, or sticks). Mobility disability may include those who lack full use of their arms and/or hands. Sensory disability relates to an impairment relating to the use of one's senses, these being chiefly visual and hearing, although it is quite possible that impairment of other senses may affect a person's tourism experience. Intellectual disability relates to a person's intellectual functioning, which may be delayed relative to people of their same age. This includes the development of conceptual skills (e.g., those relating to money, time, or use of language), social skills and everyday-life skills. Mental illnesses may also be included in this group. Such distinctions are, however, largely artificial and for the convenience of researchers and policy makers, with many PwDs having combinations of impairments from one or more groups. The UNWTO also argues that people who would not normally be considered PwDs may have wants and needs that overlap with those who would, for example, as people age, they tend to have reduced capacities in all three categories of impairment. Also included would be people who have young children in prams or pushchairs, pregnant women, those with temporary disabilities (such as a broken limb), people who are obese, those of very tall or short stature, and indeed children (see Figure 15.1).

None of these forms of impairment should, or indeed necessarily do, prevent PwDs from taking tourism trips. By the same token, PwDs with many different impairments are likely to wish to engage in ecotourism (Chikuta, du Plessis, & Saayman, 2018). It is widely acknowledged that tourism organisations should not simply aim to provide special access to PwDs but to ensure that their operations allow full, universal access to all (Darcy, Cameron, & Pegg, 2010; World Tourism Organization, 2013). The term 'accessible tourism' is therefore often used to denote research and practice in this area (Altinay, Saner, Bahçelerli, & Altinay, 2016; Buhalis & Darcy, 2010; Darcy, 2010). As this chapter will shortly argue, there are good reasons why this

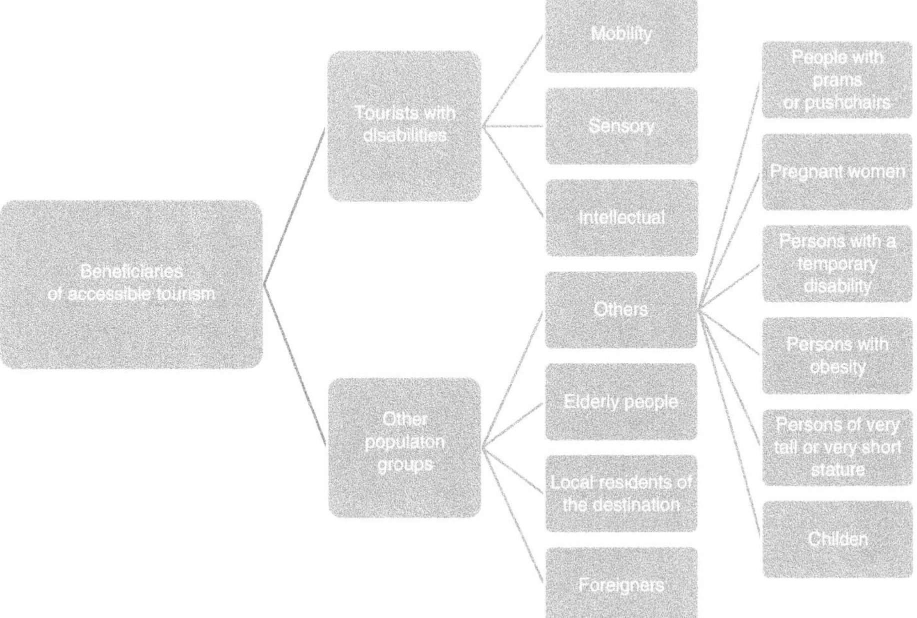

Figure 15.1 Classification of beneficiaries of accessible tourism

Source: Adapted from World Tourism Organization (2016)

principle should also be applied to ecotourism: not only ethical reasons (which are important in themselves) but sound commercial reasons as well.

There has been a tendency for the primary focus of disability concerns to be on physical disabilities. This stereotypical thinking tends to be based on simple visual cues (such as a person's use of a wheelchair) when deciding who might require additional assistance to access a location or use a service. This is reminiscent of the medical model of disability. With the greater acceptance of the social model of disability, however, this (often unconscious) bias is becoming less prevalent in society. Even so, the literature on accessible tourism has been slow to respond to this social shift, with most published research focusing on mobility disabilities, sometimes including some forms of sensory disability such a visual impairment. Few studies of disability in the tourism context have considered intellectual disabilities. Exceptions include Sedgley, Pritchard, Morgan, and Hanna (2017), Hamed (2013) and Dattolo, Luccio, and Pirone (2016), who focus on tourism by people with autism spectrum disorder, Page, Innes, and Cutler (2015) and Connell, Page, Sheriff, and Hibbert (2017), who examine the development of dementia-friendly tourism, and Rix, Lowe, and The Heritage Forum (2010), who study the inclusion of people with learning difficulties at heritage sites. None of these focus on ecotourism, and none seems to recognise the possibility that PwDs may have combinations of impairments. The extent of impairment is also largely overlooked in such studies.

The market for accessible tourism

Providing access to tourism by PwDs has traditionally been viewed as an ethical issue rather than a business opportunity. One reason for this is the introduction of legislation in many countries to ensure that tourism organisations make minimum provision for PwDs. Relevant legislation includes the Equality Act of 2010 in the United Kingdom, the Americans with Disabilities Act of 1990 in the USA, and the Canadians with Disabilities Act of 2015 in Canada, all of which support and enforce the rights of PwD to undertake tourism. At the international level, meanwhile, there is the UN Convention on the Rights of People with Disabilities (2006), Article 30 of which concerns the rights of PwDs to participate in cultural life, recreation, leisure and sport. The European Union (EU) adopted this convention in 2010, which was the first time the EU formally accepted an international human rights treaty. It has since developed a framework for its implementation. There is also a wide range of guidelines and advice available, for example from global institutions such as World Tourism Organization (n.d., 2013, 2016), national destination marketing organisations such as VisitBritain (2020), and charities such as Tourism for All (2020). The need for such laws, conventions, guidelines, and advice tends to reflect the ethical perspective that PwDs have the same right to participate tourism as anyone else, and that any lack of provision on the part of the tourism industry must be corrected.

This ethical view is also based on the presumption that providing access to PwDs will not be profitable for tourism businesses or represent an acceptable use of taxpayers' money in the case of public-sector organisations. As Darcy et al. (2010) note, in the accommodation sector there has been a persistent belief that investment in 'rooms for the disabled' is unwarranted due to low expected market returns. This is because the accessible tourism market is thought to be inherently small and highly specialist in character, making the cost of the necessary adaptations prohibitive. Tourism providers also tend believe that PwDs travel infrequently and have lower disposable incomes (Darcy et al., 2010).

These presumptions are, however, largely incorrect. While there are currently no reliable statistics available on the number of PwDs taking general tourism trips (let alone ecotourism trips specifically), all the indications are that the potential market is not only substantial but also rapidly growing. For example, the World Tourism Organization (n.d.) notes that there are presently around one billion PwDs in the world, representing 15% of the global population. Furthermore, disability is strongly associated with age (Chikuta et al., 2018), and the world's population is aging rapidly. This means that there will be approximately 2 billion people aged over 60 by 2050, representing more than 20% of the global population (World Tourism Organization, n.d.). In most developed countries, the proportion of PwDs in the overall population tends to range from around 15% to 20% (Alén et al., 2012).

Studies do tend to suggest that the propensity to take a tourism trip is lower for PwDs than in the non-disabled population. A study conducted in Australia estimated that PwDs made up 11% of the total number of tourists in Australia; meanwhile a similar study in the UK found that PwDs made up 12% of domestic tourist numbers (World Tourism Organization, 2013). Studies show that while PwDs tend to travel less often than non-disabled people, a considerable number would take more tourism trips if there were fewer barriers to doing so (Alén et al., 2012). Research also demonstrates that PwDs are prepared to stay longer, spend more per day, and purchase more services than other tourists. Many choose to travel outside the high season to avoid crowds (Chikuta et al., 2018). They also generally prefer to travel with family members or carers, who also pay their way (Alén et al., 2012). This suggests that accessible tourism is not the inherently unprofitable market segment that many in the tourism industry seem to believe.

The accessible ecotourism market is also likely to be significant scale. A report by EBSCO (2009) suggested that even a decade ago, ecotourism made up between 5 to 7% of overall tourism. The report also suggested that ecotourism was one of the fastest sectors in the industry. While ecotourism statistics for ecotourism are notoriously difficult to validate, particularly because of the inherent disagreements about its proper definition, the market for accessible tourism might have grown considerably over the past decade. A figure of 7% of overall tourism is likely, therefore, to be a conservative estimate. This would imply that in the United Kingdom, for example, the accessible ecotourism market might be in the region of 8.5 million trips.

Tourism providers nevertheless tend to treat accessible tourism as a legal obligation rather than a commercial opportunity (Chikuta et al., 2018). It is difficult to explain why this disparity remains. Possibly it is because industry leaders are not persuaded by the available market research, and there is certainly a need for more and better-quality market research to be undertaken, particularly with respect to ecotourism. Another explanation might simply be that tourism providers are unaware of the wants and needs of tourists with disabilities (Chikuta et al., 2018).

Motivations and expected benefits

Little research has been undertaken into the motivations and expectations of PwDs taking tourism trips. The few studies that do exist tend, however, to emphasise certain overarching themes. The first is that PwDs generally want to be treated as equals to non-disabled tourists (Chikuta et al., 2018). Unfortunately, this is still not the experience of many PwDs, despite the tourism industry in many countries having done much to provide additional facilities for them. Sometimes, PwDs need to adopt extreme coping strategies as a result. Yau, McKercher, and Packer (2004), for example, report that people who use wheelchairs often find that the toilets in airplanes are too restricted in terms of space, and that asking for assistance is embarrassing. Some therefore take to deliberately starving and dehydrating themselves before the flight so they do not need to use the toilet. As such, it is not enough for PwDs to have their needs met: they would like their wants to be met also. This means that they should be able to make full use of any facilities any activities that are offered by the tourism provider.

The second theme is that PwDs, just like non-disabled tourists, are often looking for challenge. This includes activities such as exploring nature and going to inaccessible places such as wilderness areas, so it is probable that many ecotourism offerings would be of interest (Ray & Ryder, 2003). Sometimes PwDs seek out tourism activities that are even more challenging relative to their abilities compared to non-disabled tourists (Chikuta, du Plessis, & Saayman, 2017). The trip may be seen an opportunity to challenge themselves, to increase their self-confidence (Chikuta et al., 2018; Yau et al., 2004), and to build positive self-image and self-esteem (Lord & Patterson, 2008). Burns and Graefe (2007) argue that PwDs often undertake outdoor recreation to prove a point to themselves and/or to others. Related to this, Chikuta et al. (2018) argue that some PwDs are also motived to undertake challenging tourism activities while they still can, particularly if they do not know how much harder it might be to undertake them in the future if their disabilities worsen with age.

A third theme is learning and personal development. A study of the motivations for PwDs to engage in physically active leisure by Lord and Patterson (2008), for example, found that PwDs believe that being involved in such activities helps them build stronger relationships with family and peers, as well as to increase their personal competencies in social, decision-making and communication skills. Such activities may also provide learning experiences. While Lord and Patterson's study focuses on active leisure participation, rather than tourism, it is likely that these benefits are sought by PwDs who undertake tourism trips.

Seeking healing and wellbeing benefits is a fourth theme. It is now widely accepted that spending time in natural environments can have positive health and wellbeing benefits to people in general (Buckley et al., 2019). There is no reason to expect these benefits not to apply equally to PwDs, if not more so. A study by Chikuta et al. (2017), for example, found that seeking healing in nature was a strong motivation for PwDs to travel to national parks.

Knowledge of the motivations and expectations of PwDs to undertake tourism trips is still, however, quite limited. One thing that can be taken as given is that PwDs with different types, combinations, and degrees of severity of impairment are likely to have different motivations and expectations. Very little research has been undertaken, however, to investigate these likely differences. One exception is the study by Chikuta et al. (2017), which found that people with hearing disabilities were statistically more likely to seek adventure- and enrichment-related benefits than those with physical or visual disabilities. In terms of escape-related motivations, meanwhile, no statistically significant relationship was found. Another study in the context of physically active recreation found that people with high support needs felt it was more important to challenge their abilities and work towards a goal than those with low support needs (Lord & Patterson, 2008). Perhaps contrary to expectations, however, they found no statistical differences between men and women in any of the motivations they studied.

It is also important to note that, by definition, ecotourists are likely to have different motivations and expected seek different benefits to general tourists. These include being physically active, escape from the home routine, challenging oneself, enjoying undisturbed nature, learning new things, and spending time with family and friends (Chikuta et al., 2017). They are also likely to exhibit different behaviours. Research tends to suggest that ecotourists often travel alone or with a partner/spouse, stay somewhat longer in the destination, spend as much as other tourists, and prefer more basic accommodation. The same is likely to be true of ecotourists with disabilities.

Barriers to access by persons with disabilities

Much of the recent research on accessible tourism has focused on the barriers that PwDs encounter (Yau et al., 2004). This is doubtless the product of the growing use of the social model of disability. Table 15.1 presents a summary of barriers that may be faced by PwDs in the course of their tourism trip. As per the social model, these are divided into three distinct categories: attitudinal, informational, and physical.

The first group, attitudinal barriers, relate to the assumptions people make about PwDs. This might include, for example, making stereotypical assumptions about what PwDs can and cannot do, overlooking or ignoring them (including speaking to them indirectly through a companion or carer), making disparaging remarks, and offering them inappropriate assistance. Research suggests that PwDs simply wish to be treated like everyone else: as valued and respected customers (Darcy, 2010; Randle & Dolnicar, 2019). While they are often overlooked, attitudinal barriers can be more influential than physical or informational barriers. Indeed, a study by Zhang and Cole (2016) suggested that staff attitudes can be the most important variable in terms of tourist accommodation service performance; often being the factor that solely determined the satisfaction of guests with disabilities.

The second group of barriers are informational: they limit or prevent PwDs from gaining information about the provision available to tourists and, in particular, those with disabilities (Mills, Han, & Clay, 2008). PwDs often use the Internet to research their trips in order to ensure that their needs and wants can be met (Ray & Ryder, 2003) and to predict any problems that they might encounter so that they can develop coping strategies. This includes both website searches and the use of the social media (Altinay et al., 2016). PwDs often need to put considerable time and effort into such research before they feel confident taking a trip (Buhalis & Michopoulou, 2011). The range, detail, comprehensiveness, and correctness of such information are all therefore important factors. More importantly, however, PwDs will need to

Table 15.1 Social model of disability applied to tourism

	Barriers	*Strategies*
Attitudinal	• Treating disabled visitors differently to others • Good service is not seen as a right for all • Lack of awareness of needs/wants of visitors with disabilities • Lack of use of appropriate terminology when talking about disability issues • Assumptions and stereotyping about people with disabilities • Overhelpfulness or overprotectiveness • Visitors hiding disability and/or misrepresenting their abilities	• Training to improve knowledge and competencies of staff members • Recognise essential differences in wants and needs of different disability groups • Adopt a compliance-plus approach to disability issues • Include accessibility in all organisational policies and strategies • Develop a range of accessibility policies • Employ people with a range of disabilities • Make use of people with disabilities in advertising and promotion
Informational	• Websites that lack usability by people with different level of sensory ability • Lack of detailed, up-to-date, comprehensive, qualitative information about facilities • Websites where it is difficult to find information specific to particular needs and wants • Accessible room and/or parking space cannot be reserved in advance • Lack of signage that accessible to people with disabilities • Guides not available who have knowledge and competency to serve needs and wants of disabled patrons	• Provide simple and easy-to-use information on websites • Include photos of facilities and spaces on websites • Design information provision that is effective regardless of ambient conditions and user abilities • Use of captions and voiceovers on videos • Use social media to assist in the spread of information about facilities • Develop a clear accessibility policy and published it on the Internet • Encourage visitors with disabilities to ask questions, and provide clear answers in an accessible way • Consult people with disabilities in the design of information • Use profiling and personalisation features to tailor information to individual website users
Physical	• Non-accessible public transport, e.g., lack of suitable seating, restraints, arrangements for boarding and disembarkation, etc. • Lack of accessible, well-marked parking at the point of entry • Presence of steps and staircases • Narrow doorways, passages, gates, etc. • Lack of level access to waterside for water-based activities	• Prioritise the less costly but most influential improvements • Apply universal design principles to facilities to accommodate all preferences and abilities • Ensure that the most-used assets are the most accessible • Ensure that hazardous elements are eliminated, isolated, shielded, or minimised.

(*Continued*)

Table 15.1 (Continued)

	Barriers	*Strategies*
	• Trails with hazards, e.g., bridges with steps up to them • Toilets, showers and changing rooms that have not been suitably adapted • Lack of suitable aids, e.g., amphibious chairs and crutches to assist with swimming • Helpers not available to provide assistance when needed • Loss or damage to equipment (e.g., wheelchairs) in transit • Facilities for people with disabilities inferior in quality or availability	

Sources: Buhalis and Michopoulou (2011); Chikuta et al. (2018); Darcy et al. (2010); Poria, Reichel, and Brandt (2011); Randle and Dolnicar (2019); Ray and Ryder (2003); World Tourism Organization (UNWTO) (n.d., 2013, 2016); Yau et al. (2004); Zhang and Cole (2016).

be able to access that information and those with sensory impairments may find that particularly difficult, particularly if the websites do not provide alternative formats such as large font sizes to assist those with visual impairments or captioned videos for those with audio impairments (Randle & Dolnicar, 2019).

The third group of barriers relates to the features of the built or natural environment that prevent equal access. Architectural and design features are the most common sources of such barriers, including diverse factors such as room to manoeuvre wheelchairs in hotel rooms; sufficient width of doors and gateways; the heights of room furniture; the availability of suitably adapted toilets, showers, and changing facilities; the availability of lifts and ramps to avoid steps or stairs; provision for carers and service dogs; and accessible parking spaces (e.g., Zhang & Cole, 2016). PwDs will often find it difficult to navigate environments with which they are unfamiliar (Randle & Dolnicar, 2019). Such barriers are relatively well researched, having long been the subject of disability research under the medical model. They are not, however, necessarily the most influential in terms of restricting access by PwDs to tourism; nor is the relative importance of the different barriers generally well understood (Darcy, 2010; Poria et al., 2011).

The content of Table 15.1 is intended to be indicative rather than comprehensive. It nevertheless demonstrates the benefits of the social model in identifying a wide range of barriers tourists with disabilities face. This range clearly extends beyond the physical barriers that are so often focused upon. It should also be noted that tourists with different types, combinations, and severity of impairment are likely to have different emphases when it comes to the relevance and influence of each of the different barriers. They will therefore have different priorities when it comes to their wants and needs being addressed, and these will need to be fully taken into account by the organisations making those interventions.

Strategies for increasing to access for people with disabilities

The social model of disability is considered useful as a means not only of identifying the barriers to participation in a particular activity by PwDs but also of developing strategic interventions to reducing or, ideally, eliminate them (Randle & Dolnicar, 2019). The strategic interventions shown in Table 15.1 are therefore classified into the same three groups: attitudinal, informational and physical. These examples of strategic intervention are intended to be illustrative rather than comprehensive. The first group, which relates to attitudinal barriers, focuses on the strategies that tourism organisations can implement to challenge the ways of thinking and behaviours of staff towards customers with disabilities. Such training should be done across the organisation and at all levels of seniority. Some of the biases PwDs experience will be unconscious on the part of the staff member, so it is important that they uncover their biases and learn how to correct them. A key message is that PwDs deserve and expect the same level of service as any other customer.

Another area where strategic interventions can address the attitudinal barriers to disability is in the development of organisational policies. Organisations should ensure that disability issues are included in all of their policies and practices. Such policies should aim to go beyond simply meeting current disabilities legislation to adopt a compliance-plus approach. Indeed, as Poria et al. (2011) note, if customers with disabilities feel that the tourism organisation has gone beyond what is necessary, they will feel more welcome and that they have been extended due hospitality.

The second group of strategies relate to addressing informational barriers. Such strategies have tended to be applied to the 'old' media, such as paper-based and broadcast media, but increasingly the focus is on their use in 'new' media such as websites and the social media (Altinay et al., 2016). Many of the strategies that could be implemented apply, however, to both old and new media. For example, alternative formats will normally be required for access by PwDs with sensory disabilities: large print, Braille, or audio alternative formats for those with visual disabilities; written versions and signed/subtitled visual material for those with hearing disabilities. The overarching principle is that information should be universally accessible. It is important to note that PwDs are, if anything, more likely to use the Internet to research and book their trips (Buhalis & Michopoulou, 2011). Access by users with various disabilities is already a component of best practice in website design and following the guidelines that area already available in this respect would certainly help address the informational barriers to tourism.

Buhalis and Michopoulou (2011) also argue that tourism providers could make better use of the customisation and personalisation features of websites in order to market their services to PwDs more effectively. They argue that many tourism providers typically segment the accessible tourism market according to type of disability (be it mobility, sensory, or intellectual). Modern website functionality can, however, enable individual PwDs to express their own needs and wants, so that tourism providers can understand and serve them more precisely on an individual basis.

Thirdly, there are strategies to address the physical barriers to tourism. Perhaps understandably, these have tended to be the focus of a great deal of the effort of tourism organisations, often to the neglect of even recognising the attitudinal and information barriers that may exist alongside them. Indeed, physical barriers tend to be the most visible and apparently most easily solved of the three types, often relying on the application of technical 'fixes' such as the modification of buildings or the installation of equipment. They have also been the area upon which existing legislation has tended to focus, as they are easily verified by inspection. Most of guidance available to organisations thus focuses on physical barriers to access. The guidelines developed by the World Tourism Organization (n.d., 2013, 2016), for example, are based heavily on the concept of universal design (see Figure 15.2). This is defined as "The design of products, environments, programmes, and services to be usable by all people, to

Principle 1 Equitable use
The design is useful and marketable to people with diverse abilities

Principle 2 Flexibility in use
The design accommodated a wide range of individual preferences and abilities

Principle 3 Simple and intuitive use
Use of the design is easy to understand, regardless of the user's experience, knowledge, language skills or current concentration level

Principle 4 Perceptible information
The design communicates necessary information effectively to the user, regardless of ambient conditions or the user's sensory abilities

Principle 5 Tolerance of error
The design minimizes hazards and the adverse consequences of accidental or unintended actions

Principle 6 Low physical effort
The design can be used efficiently and comfortably and with a minimum of fatigue

Principle 7 Size and space for approach and use
Appropriate size and space is provided for approach, reach, manipulation and use, regardless of the user's body size, posture or mobility

Figure 15.2 Seven principles of universal design
Source: World Tourism Organization (2013)

the greatest extent possible, without the need for adaptation or specialized design" (World Tourism Organization, 2016, p. V49). Such principles relate well to the physical barriers, and to some extent to the informational barriers, but they far less relevant to addressing the attitudinal barriers to accessible tourism. They also tend to relate to the built rather than the natural environment. Indeed, the World Tourism Organisation's (2013) guidelines to implementing the principles include full sections for the urban environment, transport, and accommodation, for example, but relegate 'natural areas' to a 'miscellaneous' category along with excursions and sport.

Fundamental to the problem is that the principles of universal design presuppose that it is possible (and indeed desirable) to design the environment in which tourism is taking place. With the built environment, design is arguably a more practical possibility, as tourism facilities are often purpose-built and can thus be designed from scratch. The same cannot, however, be said in the case of ecotourism. While it is realistic to envisage the application of universal design rules to some elements of the ecotourism experience, for example the design of information boards, waymarking signs, safety announcements or interpretation in visitor centres, these are largely peripheral elements of the experience. When thinking about the core elements of ecotourism provision, however, it is much more difficult to envisage how the principles of universal design might best be applied, not only because there is often less scope to do so but because doing so may not be considered desirable for sustainability reasons. As Chikuta et al. (2018) suggest, proponents of sustainability often argue that any change to existing ecological conditions in order to accommodate the needs and wants of PwDs may serve to compromise the quality of the natural environment. For example, PwDs may require use of motorised vehicles to visit an ecotourism site, which may necessitate the building of roads and could generate noise in an otherwise tranquil wilderness environment. Ironically, the qualities of the natural environment that are being threatened are the very ones that all ecotourists want to experience, whether they have disabilities or not. Access for all may inadvertently bring ruin to all.

The above discussion exposes a clash of ethics. On the one hand, it seems only right that PwDs should have equal access to nature, and that it would be an infringement of their human rights to deny them this, particularly in national parks that have been established for the common good (Chikuta et al., 2018). On the other hand, there is also a strong ethical argument that the protection of these natural resources for the benefit of future generations should supersede considerations about who should or should not be allowed to use them. This reinforces the suggestion already made in this chapter than much more research specifically on how best to address physical accessibility concerns in the ecotourism context is urgently needed.

Conclusion

Research into accessible tourism remains limited, and that which does exist tends to focus on the physical barriers to access and to lack sophistication in terms of recognising the essential differences in motivations, expectations, barriers, and intervention strategies that are appropriate to people with different types, combinations, and degrees of impairment. Research specifically on ecotourism, meanwhile, remains elusive.

While it is not possible to state the number of global ecotourists with any great confidence, what we do know is that ecotourism is a specialist market segment. As such, it is likely that the research that has been undertaken on general tourism, and the various guidelines that have been produced to assist providers in meeting the requirements of tourists with disabilities, is not likely to be well-suited to the ecotourism context. There is thus a risk that the strategic interventions that are implemented to address them may be ineffective. Even worse, there is potential for such measures to conflict with the broader Sustainable Development Goals that are considered fundamental to ecotourism.

This is not to suggest that ecotourism providers should do nothing. There are strong ethical reasons for ensuring that ecotourism meets the needs and wants of PwDs, which are complemented by sound commercial reasons related to the size and growth rate of the market. The key will be to adopt a precautionary approach and start by implementing measures that are known to have few negative implications for sustainability. Best practices can be identified as model for the development of future provisions. Addressing the attitudinal and informational barriers to access may be a good place to start as they tend to pose fewer risks to the natural environment than the physical barriers. In the meantime, more research needs to be undertaken on how to achieve better access for PwDs in the specific context of ecotourism. This will produce more robust guidance to ecotourism providers in respect of how best to address the physical barriers to access in the longer term.

References

Alén, E., Domínguez, T., & Losada, N. (2012). New opportunities for the tourism market: Senior tourism and accessible tourism. In M. Kisimoglu (Ed.), *Visions for global tourism industry: Creating and sustaining competitive strategies* (pp. 139–166). Rijeka: Intech.

Altinay, Z., Saner, T., Bahçelerli, N. M., & Altinay, F. (2016). The role of social media tools: Accessible tourism for disabled citizens. *Journal of Educational Technology & Society, 19*(1), 89–99.
Buckley, R., Brough, P., Hague, L., Chauvenet, A., Fleming, C., Roche, E., ... & Harris, N. (2019). Economic value of protected areas via visitor mental health. *Nature Communications, 10*(1), 1–10.
Buhalis, D., & Darcy, S. (Eds.). (2010). *Accessible Tourism: Concepts and Issues.* Bristol: Channel View Publications.
Buhalis, D., & Michopoulou, E. (2011). Information-enabled tourism destination marketing: Addressing the accessibility market. *Current Issues in Tourism, 14*(2), 145–168.
Burns, R. C., & Graefe, A. R. (2007). Constraints to outdoor recreation: Exploring the effects of disabilities on perceptions and participation. *Journal of Leisure Research, 39*(1), 156–181.
Chikuta, O., du Plessis, L., & Saayman, M. (2017). Nature-based travel motivations for people with disabilities. *African Journal of Hospitality, Tourism and Leisure, 6*(1), 1–16.
Chikuta, O., du Plessis, E., & Saayman, M. (2018). Accessibility expectations of tourists with disabilities in national parks. *Tourism Planning & Development, 16*(1), 75–92.
Connell, J., Page, S. J., Sheriff, I., & Hibbert, J. (2017). Business engagement in a civil society: Transitioning towards a dementia-friendly visitor economy. *Tourism Management, 61*, 110–128.
Darcy, S. (2010). Inherent complexity: Disability, accessible tourism and accommodation information preferences. *Tourism Management, 31*(6), 816–826.
Darcy, S., Cameron, B., & Pegg, S. (2010). Accessible tourism and sustainability: A discussion and case study. *Journal of Sustainable Tourism, 18*(4), 515–537.
Dattolo, A., Luccio, F. L., & Pirone, E. (2016). Web accessibility recommendations for the design of tourism websites for people with autism spectrum disorders. *International Journal on Advances in Life Sciences, 8*(3–4), 297–308.
EBSCO. (2009). Sustainability Watch: Ecotourism. Retrieved from https://ebscosustainability.files.wordpress.com/2010/07/ecotourism.pdf.
Fennell, D. A. (2020). *Ecotourism* (5th ed.). London: Routledge.
Gillovic, B., McIntosh, A., Darcy, S., & Cockburn-Wootten, C. (2018). Enabling the language of accessible tourism. *Journal of Sustainable Tourism, 26*(4), 615–630.
Hamed, H. M. (2013). Tourism and autism: An initiative study for how travel companies can plan tourism trips for autistic people. *American Journal of Tourism Management, 2*(1), 1–14.
Lord, E., & Patterson, I. (2008). The benefits of physically active leisure for people with disabilities: An Australian perspective. *Annals of Leisure Research, 11*(1–2), 123–144.
Mills, J. E., Han, J. H., & Clay, J. M. (2008). Accessibility of hospitality and tourism websites: A challenge for visually impaired persons. *Cornell Hospitality Quarterly, 49*(1), 28–41.
Page, S. J., Innes, A., & Cutler, C. (2015). Developing dementia-friendly tourism destinations: An exploratory analysis. *Journal of Travel Research, 54*(4), 467–481.
Poria, Y., Reichel, A., & Brandt, Y. (2010). The flight experiences of people with disabilities: An exploratory study. *Journal of Travel Research, 49*(2), 216–227.
Poria, Y., Reichel, A., & Brandt, Y. (2011). Dimensions of hotel experience of people with disabilities: An exploratory study. *International Journal of Contemporary Hospitality Management, 23*(5), 571–591.
Randle, M., & Dolnicar, S. (2019). Enabling people with impairments to use Airbnb. *Annals of Tourism Research, 76*, 278–289.
Ray, N. M., & Ryder, M. E. (2003). "Ebilities" tourism: An exploratory discussion of the travel needs and motivations of the mobility-disabled. *Tourism Management, 24*(1), 57–72.
Rix, J., Lowe, T., & The Heritage Forum (2010). Including people with learning difficulties in cultural and heritage sites. *International Journal of Heritage Studies, 16*(3), 207–224.
Sedgley, D., Pritchard, A., Morgan, N., & Hanna, P. (2017). Tourism and autism: Journeys of mixed emotions. *Annals of Tourism Research, 66*, 14–25.
Tourism for All. (2020). Tourism for All: Making Accessible Tourism Better. Retrieved from https://www.tourismforall.org.uk/.
United Nations. (2016). UN Convention on the Rights of People with Disabilities. Retrieved from https://www.un.org/development/desa/disabilities/convention-on-the-rights-of-persons-with-disabilities.html.
VisitBritain. (2020). Make Your Business Accessible. Retrieved from https://www.visitbritain.org/business-advice/make-your-business-accessible.

World Tourism Organization (UNWTO). (n.d.). Messages of the World Committee on Tourism Ethics on Accessible Tourism. Accessible Tourism for All: Promoting Universal Accessibility in Tourism. Retrieved from http://cf.cdn.unwto.org/sites/all/files/docpdf/wctemessagesonaccessibletourism.pdf.

World Tourism Organization (UNWTO). (2013). *Recommendations on Accessible Tourism for All*. Madrid: UNWTO.

World Tourism Organization. (2016). *Manual on Accessible Tourism for All: Principles, Tools and Best Practices – Module I: Accessible Tourism – Definition and Context*. Madrid: UNWTO. Retrieved from http://cf.cdn.unwto.org/sites/all/files/docpdf/moduleieng13022017.pdf.

Yau, M. K. S., McKercher, B., & Packer, T. L. (2004). Traveling with a disability: More than an access issue. *Annals of Tourism Research*, *31*(4), 946–960.

Zhang, Y., & Cole, S. T. (2016). Dimensions of lodging guest satisfaction among guests with mobility challenges: A mixed-method analysis of web-based texts. *Tourism Management*, *53*, 13–27.

16
ECOTOURISM AND CLIMATE CHANGE

Jonathon Day and Steve Noakes

Climate change will have increasing impacts on both our natural world and society as the century progresses. While there is increasing understanding of how climate change will directly impact destinations in which ecotourism takes place, the indirect impacts of climate change are less clear. In this chapter we will consider the impacts of climate change on the tourism system and explore the ways ecotourism may respond to a changing world. Ecotourism represents a broad range of activities, from guiding in the arctic to ecolodges in tropical rainforests, and responses will vary across locations and business types. Nevertheless, in this chapter we will examine insights into the anticipated impacts of climate change and how ecotourism organizations and destination communities are dealing with these changes.

Climate change and the tourism system

Climate change represents one of the most significant challenges facing humanity, and its implications for tourism are only part of a much larger story. The discussion of the impacts of climate change on tourism requires application of systems thinking. Tourism is a complex adaptive system (Farrell & Twining-Ward, 2005; A. Morrison, Lehto, & Day, 2018). It is embedded within two other complex, interrelated systems: human society and the natural ecosystem. Each of these systems are impacted by and have an impact on climate itself, which is also a complex system. Any discussion of the impacts of climate change on ecotourism must consider broader issues of climate change on society, and climate change on tourism in general. Each of these phenomena are complex systems, sometimes embedded, but always deeply interrelated with each other. The challenge of climate change and its impact on tourism, and specifically ecotourism, can best be described as a set of wicked problems (Rittel & Webber, 1973). While generalisations can be made about the likely impacts of climate change, each destination community or organisation will respond in its own way. According to E. Morrison, Hutcheson, Nilsen, Fadden, and Franklin (2019), "There is no simple solution to a wicked problem. Proposed solutions are neither wholly right nor completely wrong, and each problem is unique" (p. 6). Every ecotourism destination and every ecotourism product will face unique challenges as climate change impacts accelerate. As each destination tackles the challenge, it is important that we learn from our collective experience.

 DOI: 10.4324/9781003001768-16

Climate change

Climate change will have a significant impact on natural and human systems. The Intergovernmental Panel on Climate Change (IPCC) notes that "changes in climate caused impacts on natural and human systems on all continents and across the oceans" (IPCC, 2014a, p. 6). It is very likely that heat waves will occur more often and last longer, and that extreme precipitation events will become more intense and frequent in many regions. The ocean will continue to warm and acidify, and the global mean sea level will rise" (IPCC, 2014a, p. 10). Nicholls (2014) notes that climate change will have broad implications for destinations, as well as the operations of tourism businesses. At a destination community level, the direct impacts of climate change will lead to rising sea levels, extreme weather including more severe storms, ocean acidification, and rising temperatures. Nicholls (2014) also notes that tourism businesses will face operational challenges because of climate change. These will include reduced water availability, extreme weather events, difficulties in security insurance, and new costs to reduce carbon.

Climate change is becoming an increasingly important topic in tourism literature. Studies of climate change to date have focused on some of the most vulnerable destinations, including mountain tourism—particularly ski-related tourism—and small island tourism. Nevertheless, climate change will impact all parts of the world, and significant impacts are expected in destinations with relatively little change.

Ecotourism and climate change

Ecotourism is an important form of sustainable tourism that has demonstrated an ability to protect and conserve natural ecosystems, provide economic benefits for local communities, and enrich travelers with greater appreciation of nature and culture. While the concept of ecotourism is broadly embraced, definitions vary. Our definition, based on Fennell (2001), recognises ecotourism as tourism that tends to take place in natural areas, is concerned with conservation, is respectful of local culture, and endeavors to ensure benefits of tourism accrue to local people. Ecotourism embraces principles of sustainable tourism with a focus on natural spaces and local culture. It has been a means of conserving natural spaces and biodiversity. It has also been an effective tool in sustainable tourism development around the world, particularly in developing countries. Ecotourism products are frequently important in developing countries, providing opportunities for economic and social development.

Climate change will have a disproportional impact on ecotourism. Ecotourism is broadly distributed across the globe, so while climate change impacts can be expected, the impacts will be varied. There is growing documentation of the impacts of climate change on ecotourism, with studies identifying issues from Jordan (Jamaliah & Powell, 2018), to Australia, to Nepal, to New Zealand (Kutzner, 2019).

Nevertheless, as our ecosystems change because of the changing climate, ecotourism's reliance on nature-based activities will require adaptations. Ecotourism businesses are often small scale and less able to adapt to changing circumstances or disaster, such as floods or fires. Many ecotourism products are in developing countries where societal stresses caused by climate change are expected to be most acute. The significance of climate change is acknowledged by ecotourism organisations. For instance, the Asian Ecotourism Network has recognized the importance of tackling climate change and became a signatory of the "Tourism Declares a

Climate Emergency" (AEN, 2020). In addition, Ecotourism Australia has incorporated climate change as a core element of their certification process.

The nexus of human and natural systems that is fundamental to ecotourism means that ecotourism is closely tied to the changes climate change is predicted to create. The direct impacts of climate change are coming into focus in many parts of the world. While there is some variation in models adopted to project climate change, there is growing clarity on the likely direct impacts of it. In 2019, Galudo Beach Lodge, Mozambique—recognised as an important example of effective ecotourism—was decimated by a major storm. While storms are not uncommon, the intensity of this storm was exacerbated by warming sea temperatures caused by climate change. The indirect impacts of climate change are far less clear and could have significant impacts on many tourism destinations. Indirect changes are those that are the result of a changing climate. These can range from changing biodiversity, to changes in health and wellbeing in destination communities, to decreased consumer demand in some locations (Nicholls, 2014).

In complex systems, the impacts of change are often nonlinear, so it is difficult to predict how climate change will impact ecotourism destinations. The use of impact chains as a tool for systematically observing change is becoming more common (Arabadzhyan et al., 2020). In their assessment of the effects of climate change on coastal tourism, Arabadzhyan et al. (2020) identify a broad range of impacts, including the loss of attractiveness of touristic marine environments, loss of comfort due to reduced beach availability, loss of attractiveness due to deterioration in land environment, increased danger of forest fires, increased thermal stress, increased health issues due to emergent diseases, decrease in available water for the tourist industry, and loss of attractiveness due to cultural heritage loss.

Climate change, natural systems, and ecotourism

Ecotourism, by definition, intersects with the natural environment. As a result, ecotourism products will be affected by a range of indirect impacts of climate change. These changes will be many, and they will vary depending on the ecosystem and the level of change. For example, in many places forest composition will change, and with it, the types of animals inhabiting the forest will change. These changes may change the visitor appeal of these locations. Climate change's impact on vegetation and water will significantly impact sub-Saharan Africa with up to 40% of species in national parks facing extinction due to inability to migrate (Nicholls, 2014); there is no doubt this will impact Africa's nature-based ecotourism sector. Aquatic ecosystems will be impacted as well. Fishing tourism companies on the Great Lakes of North America are already changing the fish they seek as a result of the changing availability of species brought on by climate change (Chin, Day, Sydnor, Prokopy, & Cherkauer, 2019). Additionally, birdwatching tours in New Zealand, faced with reduced numbers of yellow-eyed penguins due to climate change, are adding marine animals to their programs (Kutzner, 2019). Rising water temperatures are causing increased acidification of ocean water, which is placing reef ecosystems in jeopardy (GRMPA, 2019). Ecotourism operators on the Great Barrier Reef are responding by adopting climate action strategies, including mitigation and adaptation.

Proximity to nature is both an advantage of ecotourism and a risk. As temperatures rise and droughts become more prevalent, risks for ecotourism products will increase. In 2019, Australia experienced one of the worst bushfire seasons on record. These fires were exacerbated by a changing climate. The fires impact ecotourism products across the country, from Kangaroo Island, South Australia, to Queensland.

Climate change, human systems, and ecotourism

Climate change will also impact human systems in a variety of ways, including social disruption. The United States Department of Defense acknowledges that

> climate change poses another significant challenge for the United States and the World at Large … The pressures caused by climate change will influence resource competition while placing additional burdens on economies, societies, and governance institutions around the world. These effects are threat multipliers that will aggravate stressors abroad such as poverty, environmental degradation, political instability and social tensions. (USDoD, 2014, p. 8)

These impacts are expected to be most acute in developing countries, many of which rely on ecotourism for economic development.

The changing climate will have impacts on many aspects of human life. There will be different demands for energy in some areas as the need for cooling increases, new pressures on water management systems as droughts reduce available water or increased rain leads to flooding, and new demands for infrastructure to meet changing needs (IPCC, 2014a). Another important indirect impact of climate change will be those affecting human health. The impact of increased heat and humidity on agricultural workers has been well documented, and tourism workers will likely suffer similar issues (Filippelli et al., 2020). Climate change is anticipated to increase the range of disease, with tropical diseases like zika and dengue expanding to new locations.

In warmer climates, climate change is expected to impact the activities that are enjoyable, comfortable, or even safe. Most tourism takes place in a so-called comfort zone, where temperatures range from 60 to 80° F. As temperatures rise in traditional destinations, changes in travel patterns must be expected. The impacts of ecotourism products in the tropics may be substantial.

Climate change will also have impacts on heritage and cultural locations. For instance, Brooks, Clarke, Ngaruiya, and Wangui (2020) note that archeological sites across Africa face destruction as a result of direct and indirect impacts of climate change.

Changing consumer behaviors

One important indirect impact of climate change may be fluctuations in consumer demand. While travel has been considered a right, and social capital from sharing stories and images from travels to remote locations has been a driver of travel, there is growing evidence of a countertrend. Flight shaming has been introduced into the lexicon (Gössling, Humpe, & Bausch, 2020; Thornhill, 2019). The impact of a major change in what is socially acceptable or even socially desirable can be seen in the current pandemic, where travel shame is a factor suppressing demand. On the other hand, "last-chance tourism," where tourists seek to visit places before they are succumbed to biodiversity loss or other climate impacts, is creating opportunities for some destinations—at least in the short term (Denley et al., 2020; Eijgelaar, Thaper, & Peeters, 2010; Lemelin, Dawson, & Stewart, 2012).

Mitigation, adaptation, vulnerability, and resilience

Responses to climate change can be divided into two broad categories: mitigation and adaptation. The IPCC describes mitigation and adaptation as "complementary strategies for reducing and managing the risks of climate change" (IPCC, 2014a, p. 17).

Mitigation

Mitigation is described as "actions that reduce the amount of greenhouse gas emissions" (Reinhardt & Toffel, 2017, p. 104). Tourism is a significant contributor to climate change and has an important role in reducing GHG emissions. Assessments from 2005 suggest tourism contributes between 3.7 and 5.4% of global CO_2 to GHG emissions (UNWTO & UNEP, 2008, p. 177). Transportation, particularly air travel, and buildings are the largest contributors of GHG. Ecotourism products have taken a leading role in mitigation. By definition, ecotourism organisations have a commitment to the environment and to work to lower environmental impacts of their operations and monitor the behavior of their guests (Hornoiu, 2015). Many ecotourism products take care to ensure that their operations minimise greenhouse gas production. They utilise architectural and design features to reduce energy, and they are among the first to adopt regenerative principles in architecture. These processes not only minimize GHG, but also take carbon from the atmosphere.

Ecotourism takes place within the broader tourism system, and effective climate change programs require support from a variety of organizations. Industry associations, like Ecotourism Australia, encourage the adoption of mitigation strategies. Ecotourism Australia's "Climate Heroes" program promotes mitigation strategies and recognises organisations that are reducing their GHG emissions. Government programs designed to increase energy efficiency and reduce reliance on nonrenewable energy sources can make significant contributions to reducing GHG. There is also increasing evidence that ecotourists are prepared to support the mitigation strategies.

A systems approach to ecotourism reveals the irony that although many ecotourism operations are low impact and strive to minimise GHG, they tend to be distant from source markets and require travelers to fly long distances, thus generating GHG in their travel to the destinations.

Adaptation, vulnerability, and resilience

Adaptation is described as actions that "make an organization more resilient in the face of ongoing and forecasted changes in the earth's systems" (Reinhardt & Toffel, 2017, p. 105). As the impacts of climate change become more acute, ecotourism organisations will need to adapt. Responding to the need for adaptation to climate change requires a recognition of the issues, an understanding of the impacts, a willingness to act, and the ability to respond. Even when operators recognise the impacts of climate change, action may be minimal. For some ecotourism operators, climate change can be seen as "slow moving and low risk" (Chin, 2015). Studies in Indiana (Chin et al., 2019) and Australia (Turton et al., 2010) show that even though operators may have concerns about climate change, they rarely have clear plans for adapting to it. Hernandez and Ryan (2011) find that most operators are in the early planning stages of climate adaptation. While some impacts of climate change take place gradually, many ecotourism products—and the destinations in which they are located—are vulnerable to sudden shocks caused by climate change. Floods, droughts and fires, and storms are all likely to increase with climate change, and tourism businesses' preparedness for these types of disasters are limited (Sydnor-Bousso, Stafford, Tews, & Adler, 2011). As extreme weather impacts, from floods to storms to droughts, weigh more heavily on ecotourism products, resilience will be a necessary capacity of ecotourism businesses.

Resilience and vulnerability, are becoming critical issues in understanding climate change impacts. The Intergovernmental Panel on Climate Change (IPCC) describes resilience as

> the capacity of social, economic, and environmental systems to cope with a hazardous event or trend or disturbance, responding or reorganizing in ways that maintain their essential function, identity and structure, while also maintaining the capacity for adaptation, learning and transformation. (IPCC, 2014b, p. 127)

Sydnor-Bousso et al. (2011) describe resilience as the ability to withstand shocks and rebuild. It is the ability of the hospitality industry to return after disasters to pre-disaster levels of functioning or better (Sydnor-Bousso, 2009). As threats from extreme weather increase, ecotourism businesses must build their resilience by planning for disaster recovery. This presents a significant challenge as many tourism organizations fail to plan for disaster adequately (Sydnor-Bousso et al., 2011), and many ecotourism businesses exist on extremely tight budgets and few rainy-day resources. If operator level responses are still somewhat limited, there is emerging evidence that destination communities reliant on ecotourism are beginning to recognise the importance of building adaptative capacity and resilience. In a study in the Dana Biosphere Reserve in Jordan, Jamaliah and Powell (2018) note that a resilience in tourist destinations was dependent on a range of interconnected and interdependent social, economic, governance, and environmental factors. Their study highlights that while destinations may be strong in some components of resilience, they may need to build capacity in others. The awareness of climate change issues in destinations varies. Santos-Lacueva, Ariza, Romagosa, and Saladié (2019) note that factors such as the development of the destination, the policy framework, the presence of extreme meteorological events, and the dependency on natural resources all play a potential role. While there is some value in considering the destination community's resilience and the businesses' resilience separately, it is important to recognise the important interactions between ecotourism products and their destination communities. Effective destination governance can improve the resilience of ecotourism products. Similarly, ecotourism may contribute to the resilience of destination communities (Hornoiu, 2016).

Appreciating the vulnerability of ecotourism organisations and destinations is critical for effective risk-management. Vulnerability is "the propensity or predisposition to be adversely affected. Vulnerability encompasses a variety of concepts and elements including sensitivity or susceptibility to harm and lack of capacity to cope with or adapt" (IPCC, 2014b, p. 128). Understanding the vulnerability of destination communities is an emerging area of climate change research (Calgaro, Lloyd, & Dominey-Howes, 2013; Moreno & Becken, 2009). These studies have examined communities from the edge of protected areas in the south of New Zealand (Espiner & Becken, 2013), as well as agro-ecotourism in the Philippines (Hidalgo, 2015), Fiji (Moreno & Becken, 2009), Riviera Maya and the Spanish Mediterranean (Santos-Lacueva et al., 2019), and the Nepalese Himalayas (Nyaupane & Chhetri, 2009).

One of the great challenges of adaptation for ecotourism is that, in the short term, consumers have a greater ability to adapt quickly than ecotourism businesses, particularly those with physical structures (Day et al., 2018; Simpson, Gossling, Scott, Hall, & Gladin, 2008). Travelers can easily choose different destinations as climate changes. There is growing evidence of consumer awareness of climate change. In two studies conducted in 2019, consumers identified the possibility that their travel plans may be impacted by climate change. The "Portrait of the American Traveler" shows that 48% of travelers agree that climate change will have a significant influence on what destinations they choose to visit in the next five years (MMGYGlobal, 2019). In a different study, Destination Analysts' "State of the American Traveler" (2019) shows that 50% of American leisure travelers expect that climate change will impact their travels at some point in the next five years. Further, Destination Analysts reports that 28.5% of travelers expect climate change will affect the timing of trips, 20.9% will change the destinations they choose, and 15.2% will adjust how they travel.

The impact of some changes, particularly in the short term, may be minimal from the visitors' perspective. Tourists may be unaware of some changes caused by climate change or may make minor behavioral changes. For example, travelers to the Great Barrier Reef were unaware of changes in hard coral, an impact of climate change, when they visited (Ramis & Prideaux, 2013).

Lessons from Binna Burra Lodge: Queensland, Australia

On Sunday, 8 September 2019, bushfires destroyed Binna Burra Lodge (BBL), one of Australia's iconic ecotourism properties. An examination of the disaster, the recovery, and the broader context of the event provide important insights for ecotourism destinations and ecotourism operators preparing for the impacts of climate change. Australia's ecotourism sector differs from ecotourism in many parts of the world. Australia is a wealthy, developed country and has a long history of ecotourism and nature-based tourism. Ecotourism and nature-based tourism are an important part of the Queensland tourism portfolio and are actively promoted by state and national Destination Marketing Organisations (DMOs). Industry associations, including Ecotourism Australia (EA) and Queensland Tourism Industry Association, provide a range of support materials and resources to ecotourism throughout the state.

BBL is an ecotourism lodge in the Lamington National Park in Southern Queensland. Established in 1933, the lodge has been a leader in ecotourism in Australia. The Binna Burra complex included a lodge with some pioneering local timber accommodations, a number of apartment-style "Sky Lodges," and a Rainforest Campsite and Safari Tents. The Lamington National Park, one of the national parks in the Gondwana Rainforests World Heritage Site, is a temperate rain forest. It is described as "outstanding examples of major stages of the Earth's evolutionary history, ongoing geological and biological processes, and exceptional biodiversity" (UNESCO, 2020). The Binna Burra Cultural Landscape is listed on the State Government's Heritage Register a place that has "cultural heritage significance to the people of Queensland" (Queensland Government, 2016).

Binna Burra has a long tradition of commitment to ecotourism principles. It first achieved "Advanced Ecotourism" certification with EA in 1997 and has remained committed to ecotourism performance management and improvement. EA, established in 1991, inspires environmentally sustainable and culturally responsible tourism. EA is a membership-based organisation representing over 500 ecotourism operators, 1700 products, and a growing number of destinations. EA designs and delivers a range of certification programs promoting sustainable tourism. The "Advanced Ecotourism" certification is awarded to Australia's leading ecotourism products that have demonstrated a commitment to operate with minimal impact on the environment and provide opportunities to learn about the environment with operators that are achieving best practices, using resources wisely, and helping local communities. Ecotourism Australia's certification process incorporates a commitment to regular risk assessment and crisis preparations, as well as a range of mitigation activities, including a commitment to energy efficiency and reduction.

Disaster recovery is a complex activity. "The decisions of an entity impact the decisions and recovery of others. These decisions include rebuilding, replacing, repairing, resuming, and restoring everything from income sources to buildings to social networks to businesses to infrastructure" (Marshall & Schrank, 2014, p. 597) and this is the case for Binna Burra Lodge. Disaster recovery is best considered a process that takes place over time, and recovery times can be different for different organisations. Binna Burra's efforts to rebuild following the fires have been recognised as exemplary, and it is worthwhile noting some of the key elements of the recovery.

Pre-disaster period

Lamington National Park is part of the Queensland National Parks system. The park is well managed, and rangers from the National Park Service work closely with BBL management. In the weeks prior to the fire, controlled burns were undertaken as part of the ongoing forest management of the park.

BBL was prepared for the disaster, with established emergency plans and fire safety training drills that were conducted on a regular basis. In the days before the fire struck the lodge, BBL made several important decisions. Evacuations began two days before the fire reached the lodge, and all guests and staff were successfully evacuated before the fire blocked the escape routes. Binna Burra is a wilderness location, and the only land access is by a single road to the lodge. This critical infrastructure was damaged in the fires and, if evacuation had been delayed in the hope that the fire would not reach BBL, lives would have been placed in great danger. Cotterell and Gardiner (2019) note that the decision to "not let business get in the way of the early evacuation" (p. 7) avoided the possibility of emergency helicopter evacuations and the possibility of not being able to successfully get everyone to safety. Binna Burra staff, together with Emergency Services staff, were deployed to ensure that no curious visitors could reenter the area.

Perhaps more importantly, Binna Burra has established strong ties to the community and government—local, state, and national. The organisation is active in tourism industry activities and has established itself as a good corporate citizen and a positive contributor to the social, environmental, and economic wellbeing of the region. This investment over the years in relationship building with a wide variety of stakeholders represents significant social capital.

The fire

Early in the morning of September 8, 2019, the wildfire destroyed Binna Burra's heritage-listed buildings, including the original lodge accommodation and dining room. At the time of the devastation, there were concerns that the iconic ecolodge would not recover from the damage of the bushfires. *The Guardian* ran an obituary for the lodge (MacColl, 2019). The bushfires that impacted Binna Burra were at the early part of a six-month season of bushfires that covered the whole of Australia, with over 17 million hectares burned across New South Wales, Victoria, Queensland, ACT, Western Australia, and South Australia (Richards, Brew, & Smith, 2020). Bushfires are an indirect result of climate change. While climate doesn't cause fires, it does create the conditions in which bushfires are more likely. In 2019, Australia experienced its driest year (CSIRO, 2020) on record, and the severity of the bushfire season is partially attributed to the impacts of climate change (Thompson, 2020).

Throughout the process, staff followed a predetermined emergency plan (Cotterell & Gardiner, 2019). Even as the fire approached, a remote headquarters was established and a crisis communication plan was implemented with the BBL chairman, Steve Noakes, the most senior member of the management team, providing a single source of consistent messaging to media and stakeholders. Other important crisis management lessons included the importance of being able to operate remotely during the crisis and the importance of having a "go-kit" of information necessary to keep the business operating (Cotterell & Gardiner, 2019).

Immediate post-disaster recovery: #Bringbackbinnaburra

Binna Burra quickly committed to returning to business. Communication activities with media and key stakeholders focused on recovery. "Bring Back Binna Burra" became a rally theme with communications and fundraising efforts. A GoFundMe page generated over A$140,000. By quickly changing their website to focus on recovery efforts, Binna Burra was able to capitalise on publicity generated from the media coverage that reached as far as London and New York. Stakeholder communication remained a priority in the months following the fires.

Government commitment to rebuilding infrastructure was almost immediate. In the days following the fire, state, and federal government expressed willingness to collaborate to #BringBackBinnaBurra. The state government immediately established a Binna Burra Recovery Taskforce to support the restoration of infrastructure including water and telecommunications (Cotterell & Gardiner, 2019). The day after the fire, the taskforce also committed to supporting displaced staff if necessary (Dick, 2019). By Wednesday, the first taskforce meeting was convened and was visited by the Queensland Minister for State Development, Manufacturing, Infrastructure and Planning, and by the Regional Council Mayor. By the end of the week (September 13), the Prime Minister of Australia and the Premier of Queensland had visited the site.

Ongoing post-disaster recovery—the first year after the fire

In the weeks and months following the fire, additional business challenges were addressed by the BBL team. One significant challenge was organizing refunds for over 4000 forward bookings for accommodation and on-site venues. This issue was not only administratively challenging, but it also placed strains on cashflow and financial resources. Another ongoing management issue was the continued support of the staff. With 90% of the 65 staff having been made redundant, BBL arranged a "staff transition event with job and training offers to staff. Counselling and financial advice was also provided" (Cotterell & Gardiner, 2019).

BBL established an internal reconstruction council to manage the recovery process. Working groups provided information and supported five key sets of activities. These activities included coordination and government liaison, corporate issues, business operations, (stakeholder) relationships and communications, and people (including human resources and volunteer coordination). While working through their recovery, BBL was able to apply the framework developed for the Queensland State Disaster Plan. They also received guidance from the National Institute for Disaster Resilience (Schultz & Barnett, 2020).

Binna Burra was committed to re-open as soon as possible, but the lack of road access resulted in an eight-month delay until the single road access could be made safe for heavy equipment to commence the demolition work. In early 2020 BBL open a small off-site local café which traded for just four weeks before being closed down due to the global COVID-19 pandemic. It reopened mid-year for a short time, but was closed down when the primary Binna Burra site reopened for business in September 2020, one year after the devastating bushfires. On 1 September 2020, BBL welcomed its first guests back into the repaired Sky Lodges accommodation and opened a new 'Bushwalker's Bar' at the historic Grooms' Cottage. The focus on re-opening is critical as research shows that the speed to re-open is an important success factor in successfully surviving a disaster (Marshall & Schrank, 2014).

While BBL worked to recover, the repair of infrastructure supporting the operation began. The reopening of the road to Binna Burra on August 31, 2020, was an important milestone in the recovery. The repairs, funded by Australian and Queensland government, represented a $35 million investment (Dick, 2020). Walking trails in the Lamington National Park also needed repairs at a cost of $1 million. In addition to these critical repairs, the state government is currently finalising the construction of a cliff-climbing course, Via Ferrata (iron way), at Binna Burra. The course "will allow visitors to safely scale cliff faces usually only accessible to experienced rock climbers through a system of iron steps fixed in the rock face supported by a safety cable" (Schultz & Barnett, 2020, p. 44).

It is important to appreciate the relationship between BBL and the local community, with BBL playing an important role in the livelihood of the region. An example of the connection between BBL and the community for building the capacity for the long-term success is how the Binna Burra Kitchen continued support for local food suppliers. An innovation during the pandemic has been the development of a "Scenic Rim Farm Box." This initiative was born out of the COVID-19 lockdown as a collaboration between local farmers, the local government authority, Scenic Rim Regional Council, and distribution partners such as Binna Burra. The Farm Box connects Scenic Rim producers with consumers who were unable to visit the region during the health crisis, with a 'farm to you' home delivery business with the region's best produce, products and locally produced wine and beer. The marketing message was that 'it makes it easy to eat local *every* week' (Cunningham, 2020; Scenic Rim Farm Box, 2020).

As BBL rebuilds from the ashes of the bushfire, one of the critical elements of its recovery has been the support from their clients and friends. "Solidarity tourism," or the commitment to support those businesses impacted by disaster, has emerged from the experiences of the bushfires. According to Noakes:

> Solidarity tourism prioritizes positive attitudes to nature as well as the traditions and the interests of the local communities. It gives locals and visitors the opportunity to connect and support a place like Binna Burra following the difficulties and challenges that have resulted from the bushfire destruction in our community and at the lodge within the Lamington National Park. (Noakes, 2020, p. 1)

The next steps of recovery: Reset, reimagine, recreate

BBL knows that the bushfire will remain an important part of their history, and they recognise the challenge of maintaining progress over the long term. Informed by recovery science, they are committed to avoiding the disillusionment sometimes experienced by disaster survivors. BBL has focused efforts on a "Reset, Reimagine, Recreate" approach that incorporates a long-term vision, a comprehensive master plan, and significant new initiatives to begin immediately. These initiatives are based firmly in BBL's commitment to sustainable development. As they proceed, BBL recognizes the likely impact of climate change on their future. As Steve Noakes, the Chairman of Binna Burra Lodge, says, "Our responsibility now is to have a vision that is crafted on the knowledge and understanding of the climate as it will impact the tropical and subtropical rainforest" (Schultz & Barnett, 2020, p. 51).

BBL is well positioned to leverage the support of state and national industry associations that are already addressing climate change. BBL will be guided and supported in their climate strategies by the work of two organisations in particular—the Queensland Tourism Industry Council (QTIC) and Ecotourism Australia. Tourism operators in Queensland, and across Australia, have been encouraged by government and NGOs to prepare for the impacts of climate change over the past 15 years. From pioneering work conducted by CSIRO and Tourism Queensland, to the development of an industry-wide climate change plan—The Queensland Tourism Climate Change Sector Adaptation Plan (Becken, Montesalvo, & Whittlesea, 2018), developed by the Queensland Tourism Industry Council—Queensland tourism providers have been exposed to principles of climate change preparation. EA also plays an important role in promoting effective climate change adaptation and mitigation strategies. BBL has committed to undertaking the "Climate Action" certification with EA.

Lessons from Binna Burra Lodge

In some ways, Binna Burra Lodge represents a unique case study. Certainly, few ecotourism organizations could expect to receive the level of support BBL received as it approached its rebuilding process. This has confirmed its position as a significant nature-based tourism enterprise within a World Heritage listed national park accessible to all. Nevertheless, there are important lessons for all ecotourism products from the experience of BBL. Those key lessons are as follows:

- Build social capital: The support and goodwill of the community, industry, and government is an important asset following a disaster. Building those relationships over time is an important investment in resilience.
- Be prepared: Climate change is increasing the likelihood of a range of natural disasters. Being ready for the challenges of maintaining business continuity will assist in recovery. Recognising risks and preparing for them is critical for long-term success.
- Get back to business as soon as possible: Research shows getting back to business as quickly as possible is an important factor for recovery.
- Leverage available resources: There are a wide array of resources available to ecotourism organizations. These resources include tools and training on preparation for disaster. They also include resources to support recovery following the event. By leveraging these resources, ecotourism organisations can apply best practices even in the most difficult times.

There are other lessons from the BBL case as well. Despite following best-practice ecotourism, BBL was not as focused on the issues of climate change as they are today. In preparing for the future, they have committed to pay greater attention to the issues of climate change and its impact on their operations. The case also highlights the systems nature of recovery. BBL is one element in a system of stakeholders that includes the local community, parks management, the government, local suppliers, staff, community and friends, and past visitors. Each plays a role in the recovery process. Finally, the BBL incident highlights the dynamic nature of risk and vulnerability. Bushfires are rare in temperate rain forests, and the likelihood of the risk of fire at BBL has been relatively low. Nevertheless, as climate change affects conditions, ecotourism products must be mindful of the changing risk profile of their operations.

Conclusion

Climate change will have significant impacts on socioenvironmental systems in the coming years. This is true for tourism in general, and ecotourism in particular. Ecotourism's proximity to a natural world that is responding to a warming world means that our sector of the industry will be on the front lines of change. The direct changes are becoming increasingly certain, while the indirect changes are less predictable.

Tourism must commit to mitigation as part of a universal commitment to reducing GHG. Even while reducing GHG, we must adapt and build resilience to change. As Reinhardt and Toffel (2017) note:

> thirty years ago, mitigation and adaptation could have been viewed as substitutes: If we had invested in more aggressive mitigation then, we may not have to invest so much in adaptation now. But that window has shut. [Mitigation] efforts simply cannot obviate the need for extensive adaptation. (pp. 105/106)

Unfortunately, the commitment to GHG mitigation and the impacts of climate change are asymmetrical. Reducing GHG production does not insulate ecotourism products from the impacts of climate change. While ecotourism products must work to mitigate their impacts and reduce their GHG emissions, they must also adapt to anticipated changes in the environment and build their capacity for resilience.

It has been suggested that concern for climate change may overwhelm other legitimate sustainability concerns (Weaver, 2011). This approach overlooks the intersectionality of climate change with many of the other challenges facing the world. Natural space preservation, biodiversity conservation, and Indigenous rights—to name just three critical issues—intersect with climate change. Those committed to ecotourism principles must recognize how each of these issues are interrelated and must work to solve these issues at the same time.

References

AEN. (2020). AEN Declares a Climate Emergency. Retrieved from https://www.asianecotourism.org/tourismdeclares.

Arabadzhyan, A., Figini, P., García, C., González, M. M., Lam-González, Y. E., & León, C. J. (2020). Climate change, coastal tourism, and impact chains – a literature review. *Current Issues in Tourism*, 1–36. doi:10.1080/13683500.2020.1825351.

Becken, S., Montesalvo, N., & Whittlesea, E. (2018). *Building a Resilient Tourism Industry: Queensland Tourism Climate Change Response Plan*. Australia: Queensland Government.

Brooks, N., Clarke, J., Ngaruiya, G. W., & Wangui, E. E. (2020). African heritage in a changing climate. *Azania: Archaeological Research in Africa, 55*(3), 297–328. doi:10.1080/0067270X.2020.1792177.

Calgaro, E., Lloyd, K., & Dominey-Howes, D. (2013). From vulnerability to transformation: A framework for assessing the vulnerability and resilience of tourism destinations. *Journal of Sustainable Tourism*, 1–20. doi:10.1080/09669582.2013.826229.

Chin, N. (2015). *Dissertation Exploring the Potential Impacts of Climate Change on North America's Laurentian Great Lakes tourism sector* (PhD Dissertation, Purdue University).

Chin, N., Day, J., Sydnor, S., Prokopy, L. S., & Cherkauer, K. A. (2019). Exploring tourism businesses' adaptive response to climate change in two Great Lakes destination communities. *Journal of Destination Marketing & Management, 12*, 125–129. doi:10.1016/j.jdmm.2018.12.009.

Cotterell, D., & Gardiner, S. (2019). *Bushfire at Binna Burra Lodge: A Case Study*. Gold Coast, Australia: Griffith Institute for Tourism, Griffith University.

CSIRO. (2020). The 2019–2020 Bushfires: A CSIRO Explainer. Retrieved from https://www.csiro.au/en/Research/Environment/Extreme-Events/Bushfire/preparing-for-climate-change/2019-20-bushfires-explainer.

Cunningham, S. (2020). New Partnership to Market Fresh Produce Is Out of the Box! Retrieved from https://www.binnaburralodge.com.au/news/new-partnership-to-market-fresh-produce-is-out-of-the-box/.

Day, J., Widhalm, M., Chin, N., Dorworth, L., Shah, K., Sydnor, S., & Dukes, J. (2018). *Tourism and Recreation in a Warmer Indiana: A Report from the Indiana Climate Change Impacts Assessment.* West Lafayette, Indiana: Purdue University.

Denley, T. J., Woosnam, K. M., Ribeiro, M. A., Boley, B. B., Hehir, C., & Abrams, J. (2020). Individuals' intentions to engage in last chance tourism: Applying the value-belief-norm model. *Journal of Sustainable Tourism, 28*(11), 1860–1881. doi:10.1080/09669582.2020.1762623.

DestinationAnalysts. (2019). State of the American Traveler: Destination Edition. Retrieved from https://mk0destinationajcrrq.kinstacdn.com/wp-content/uploads/2019/03/SATS-Winter-2019small.pdf.

Dick, C. (2019). Taskforce to be Established to help Binna Burra Lodge Recover [Press release]. Retrieved from https://statements.qld.gov.au/statements/88323.

Dick, C. (2020). Binna Burra Road Reopening Marks Key Recovery Milestone [Press release]. Retrieved from https://statements.qld.gov.au/statements/90603.

Eijgelaar, E., Thaper, C., & Peeters, P. (2010). Antarctic cruise tourism: The paradoxes of ambassadorship, "last chance tourism" and greenhouse gas emissions. *Journal of Sustainable Tourism, 18*(3), 337–354. doi:10.1080/09669581003653534.

Espiner, S., & Becken, S. (2013). Tourist towns on the edge: Conceptualising vulnerability and resilience in a protected area tourism system. *Journal of Sustainable Tourism*, 1–20. doi:10.1080/09669582.2013.855222.

Farrell, B., & Twining-Ward, L. (2005). Seven steps towards sustainability: Tourism in the context of new knowledge. *Journal of Sustainable Tourism, 13*(2), 109–122.

Fennell, D. A. (2001). A content analysis of ecotourism definitions. *Current Issues in Tourism, 4*(5), 403–421. doi:10.1080/13683500108667896.

Filippelli, G., Freeman, J., Gibson, J., Jay, S., Moreno-Madrinan, M., Ogashawara, I., ... & Wells, E. (2020). Climate Change Impacts on Human Health at an Actionable Scale: A State-Level Assessment of Indiana, USA. *Climatic Change.* Retrieved from https://link.springer.com/article/10.1007/s10584-020-02710-9.

Gössling, S., Humpe, A., & Bausch, T. (2020). Does 'flight shame' affect social norms? Changing perspectives on the desirability of air travel in Germany. *Journal of Cleaner Production, 266*, 122015. doi:10.1016/j.jclepro.2020.122015.

GRMPA. (2019). Threats to the Reef: Climate Change. Retrieved from http://www.gbrmpa.gov.au/our-work/threats-to-the-reef/climate-change.

Hernandez, A., & Ryan, G. (2011). Coping with climate change in the tourism industry: A review and agenda for future research. *Tourism and Hospitality Management*, 17(1), 79–90.

Hidalgo, H. A. (2015). Vulnerability assessment of agri-ecotourism communities as influenced by climate change. *International Journal on Advanced Science, Engineering and Information Technology, 5*(6), 379. doi:10.18517/ijaseit.5.6.553.

Hornoiu, R.-I. (2015). Assessing climate change percepttion of ecotourism stakeholders from protected areas. *Calitatea, 16*(147), 68–70.

Hornoiu, R.-I. (2016). Resilience capacity of local communities from protected areas under the impact of climate change and their strengthening through ecotourism. The ASEAN Countries' Case. *Calitatea, 17*(153), 70–73.

IPCC. (2014a). *Climate Change 2014: Synthesis Report. Contribution of Working Groups 1,2,and 3 to the Fifth Report of the Intergovernmental Panel on Climate Change.* Geneva, Switzerland: IPCC.

IPCC. (2014b). IPCC, 2014: Annex II: Glossary. In K. Mach, S. Planton, & C. von Stechow (Eds.), *Climate Change 2014: Synthesis Report. Contribution of Working Groups I, II and III to the Fifth Assessment Report of the Intergovernmental Panel on Climate Change* (pp. 117–130). Geneva, Switzerland: IPCC.

Jamaliah, M. M., & Powell, R. B. (2018). Ecotourism resilience to climate change in Dana Biosphere Reserve, Jordan. *Journal of Sustainable Tourism, 26*(4), 519–536. doi:10.1080/09669582.2017.1360893.

Kutzner, D. (2019). Environmental change, resilience, and adaptation in nature-based tourism: Conceptualizing the social-ecological resilience of birdwatching tour operations. *Journal of Sustainable Tourism, 27*(8), 1142–1166. doi:10.1080/09669582.2019.1601730.

Lemelin, H., Dawson, J., & Stewart, E. J. (2012). *Last Chance Tourism.* London: Routledge.

MacColl. (2019, 9 September). An Obituary for Binna Burra Lodge, Which First Moved Me When I Was a Teenager. *The Guardian.* Retrieved from https://www.theguardian.com/commentisfree/2019/sep/09/an-obituary-for-binna-burra-lodge-which-first-moved-me-when-i-was-a-teenager.

Marshall, M. I., & Schrank, H. L. (2014). Small business disaster recovery: a research framework. *Natural hazards (Dordrecht), 72*(2), 597–616. doi:10.1007/s11069-013-1025-z.

MMGYGlobal. (2019). MMGY GLOBAL'S Portrait of American Travelers Survey Reveals Significant Shifts in What is Influencing Travelers Most. Retrieved from https://www.mmgyglobal.com/news/mmgy-globals-portrait-of-american-travelers-survey-reveals-significant-shifts-in-what-is-influencing-travelers-most/.

Moreno, A., & Becken, S. (2009). A climate change vulnerability assessment methodology for coastal tourism. *Journal of Sustainable Tourism, 17*(4), 473–488. doi:10.1080/09669580802651681.

Morrison, A., Lehto, X., & Day, J. (2018). *The Tourism System* (8th ed.). Dubuque, Iowa: Kendall Hunt.

Morrison, E., Hutcheson, S., Nilsen, E., Fadden, J., & Franklin, N. (2019). *Strategic Doing: Ten Skills for Agile Leadership.* Hoboken, New Jersey: John Wiley and Sons.

Nicholls, M. (2014). Climate Change: Implications for Tourism - Key Findings from the Intergovernmental Panel on Climate Change 5th Assessment. Retrieved from https://www.cisl.cam.ac.uk/business-action/low-carbon-transformation/ipcc-climate-science-business-briefings/tourism.

Noakes, S. (2020). Solidarity Tourism. https://www.binnaburralodge.com.au/news/solidarity-tourism-for-binna-burra/

Nyaupane, G. P., & Chhetri, N. (2009). Vulnerability to climate change of nature-based tourism in the Nepalese Himalayas. *Tourism Geographies, 11*(1), 95–119. doi:10.1080/14616680802643359.

Queensland Government. (2016). Binna Burra Cultural Landscape. Retrieved from https://apps.des.qld.gov.au/heritage-register/detail/?id=601899.

Ramis, M., & Prideaux, B. (2013). The importance of visitor perceptions in estimating how climate change will affect future tourist flows to the Great Barrier Reef. In M. V. Reddy, & K. Wilkes (Eds.), *Tourism, climate change and sustainability.* London: Routledge.

Reinhardt, F. L., & Toffel, M. W. (2017). Managing climate change: Lessons from the U.S. Navy. *Harvard Business Review, 95*(4), 102–111. Retrieved from http://search.ebscohost.com/login.aspx?direct=true&db=bth&AN=123739249&site=bsi-live.

Richards, L., Brew, N., & Smith, L. (2020). 2019–20 Australian Bushfires—Frequently Asked Questions: A Quick Guide. Retrieved from https://parlinfo.aph.gov.au/parlInfo/download/library/prspub/7234762/upload_binary/7234762.pdf.

Rittel, H., & Webber, M. (1973). Dilemmas in a general theory of planning. *Integrating Knowledge and Practice to Advance Human Dignity, 4*(2), 155–169. doi:10.1007/BF01405730.

Scenic Rim Farm Box. (2020). About. Retrieved from https://scenicrimfarmbox.com.au/pages/story.

Schultz, T., & Barnett, B. (2020). *Binna Burra Strategic Visions, Masterplan, and RAP Background Summary Report.* Binna Burra, Queensland:United Nations Environment Program.

Simpson, M., Gossling, S., Scott, D., Hall, C., & Gladin, E. (2008). *Climate Change Adaptation and Mitigation in the Tourism Sector: Frameworks, Tools and Practices.* Paris, France.

Sydnor-Bousso, S. (2009). *Assessing the Impact of Industry Resilience as a Function of Community Resilience: The Case of Natural Disasters* (Doctoral Dissertation, The Ohio State).

Sydnor-Bousso, S., Stafford, K., Tews, M., & Adler, H. (2011). Towards a resilience model for the hospitality and tourism industry. *Journal of Human Resources in Hospitality and Tourism, 10*(2), 195–217.

Thompson, A. (2020). Yes, Climate Change Did Influence Australia's Unprecedented Bushfires. *Scientific America.* Retrieved from https://www.scientificamerican.com/article/yes-climate-change-did-influence-australias-unprecedented-bushfires/.

Thornhill, J. (2019). Flight Shaming Is Taking Off – Can Travel Be More Ethical? *The Guardian.* Retrieved from https://www.theguardian.com/money/2019/jun/09/flight-airline-travel-rail-family-environment.

Turton, S., Dickson, T., Hadwen, W., Jorgensen, B., Pham, T., Simmons, D., … & Wilson, R. (2010). Developing an approach for tourism climate change assessment: Evidence from four contrasting Australian case studies. *Journal of Sustainable Tourism, 18*(3), 429–447. doi:10.1080/09669581003639814.

UNESCO. (2020). World Heritage List: Gondwana Rainforests of Australia. Retrieved from https://whc.unesco.org/en/list/368/.

UNWTO, & UNEP. (2008). Climate Change and Tourism – Responding to Global Challenges. Retrieved from https://www.e-unwto.org/doi/abs/10.18111/9789284412341.

USDoD. (2014). Quadrennial Defense Review 2014. Retrieved from https://archive.defense.gov/pubs/2014_Quadrennial_Defense_Review.pdf.

Weaver, D. (2011). Can sustainable tourism survive climate change? *Journal of Sustainable Tourism*, *19*(1), 5–15. doi:10.1080/09669582.2010.536242.

17

ANIMALS CAUGHT IN THE CROSSFIRE

Humanitarian efforts and responsible tourism opportunities

Nicholas Wise

Introduction

War and conflict bring about significant change to the local environment, and this has an impact on animals and biodiversity. Hanson et al. (2009) conducted a comprehensive study of warzones around the globe and found that 80% of conflicts occur in areas rich in biodiversity. Because war, conflict, and civil unrest is a battle over resources, there is often no mercy for what lies in the path by those trying to assume territory. Human life is a tragedy or war, with death tolls commonly conveyed to onlookers in the media and through research reports; but what can be distant from the headlines, narratives, and reporting of war is the destruction of nature and the death of wildlife (Dudley, Ginsberg, Plumptre, Hart, & Campos, 2002). This chapter is concerned with discussing and presenting examples of animals and nature that were caught in the crossfire and the practical responses that ensued following a time of conflict whereby conservation and tourism play a role in future preservation. Much published academic research focuses on how conflict, tragedy, and security issues affect tourism destinations (e.g., Butler & Suntikul, 2013; Henderson, 2000; Winter, 2008; Wise, 2019). Some research has considered the role of ecotourism in war-torn destinations (e.g., Azimi, 2012; McNeilage, 1996) and wildlife decline (e.g., Daskin & Pringle, 2018), but very few sources have considered the impact of war on animals (indirectly) in the tourism literature (e.g., Shackley, 1995). Countries designate nature reserves so they can showcase unique wildlife, but conflict disrupts both the management of these sites and the existing ecological value that attracts visitors (Ospina, 2006). This can have a disruption on tourism for years to come because negative associations such a war (which produced a negative destination image) can deter visitors.

Animals play an important role in their ecosystems and natural settings. When we consider tourism and ecotourism, animals can be a niche attraction—and are thus important to the promotion of particular destinations (Bulbeck, 2005). People have psychological, emotional, or even imaginative innate connections with wild animals, or biophilia (see Curtin, 2005; Fennell, 2012; Wilson, 1984). When animals are threatened, people seek ways to respond and protect them, may these animals be unique to a particular place, an endangered species, or a household pet. While the disturbance of war can distract people from ongoing conservation, but when

DOI: 10.4324/9781003001768-17

stories of animals under threat arise, people embark on humanitarian efforts to restore a sense of protection and normality to animals, may they be in the wild, in zoos/nature centres, or domesticated. However, we must not also rule out that tourism can also be a threat to animals in the wild, even though the examples in this chapter look at tourism opportunities as a response to create awareness opportunities for locals. Tourism requires development and infrastructures to access remote areas, and this can, and does, impact local habitats (see Perkins & Grace, 2009). When people seek animals in the wild, this can alter or disrupt their biological clocks as many wild animals are nocturnal. Gorillas, for instance, are primates closely related to humans, and because people can pass diseases directly to them, being in close proximity can threaten the health of gorillas. Sometimes people do not act respectively towards animals, and this concern leads to a change in behaviour (see Chakraborty, 2019; Mlozi & Pesämaa, 2019). Bulbeck (2005) informs and encourages us to critically consider the impact of habitat loss and human encroachment, but considers this alongside the rapid increases of nature and ecotourism development that brings people into direct contact with animals.

This chapter is concerned with addressing how animals are impacted (differently) by war, conflict, and active disputes over the territory in which they reside and then looking at what local or international responses have been to address the impact on animals. Insight from research from a range of disciplines across the natural and social sciences is considered, as well as media reporting when providing examples of animals caught in the crossfire later in this chapter. Jones (2015) argues we overlook animal rights and social justice issues, but animals succumb to violence during times of conflict, just as people do. Examples in this chapter include the loss of giraffe and elephant herds during Mozambique's Civil War; post-conflict efforts to protect endangered gorillas in the Democratic Republic of the Congo (DRC) and neighbouring countries in the Virunga Mountains area; people rescuing zoo animals in Aleppo, Syria; and how war deters poaching, allowing wildlife to thrive in Kashmir, but the opposite was seen in Afghanistan. Before discussing these examples, this chapter will consider literature on the impact of war and conflict on tourism and ensuing humanitarian conservation that is an important component of contemporary ecotourism.

War and conflict, ecotourism and conservation

War and conflict is the result of contested power relations, which involve violent and aggressive disputes over geography (Agnew, 2009). While war is often a dispute over territory, conflicts will ultimately disrupt any ongoing conservation efforts that might be ongoing in that given territory (Toft, 2014). Wildlife caught in the crossfire are not equipped to deal with drastic change to their living environments. Thus, either fall victim as casualties of war, they must find a way to disperse or relocate to unfamiliar environments where they face other pressures of a new landscape that they may not belong to (or know how to adapt to), or become threatened by other predators because they are entering another animal's territory. In this sense, both humans and animals are territorial.

Governments play a central role in protecting natural environments that contain animals and wildlife. Many countries can learn lessons from countries such as Costa Rica, which is observed as a benchmark for ecotourism development (Miller, 2012) and there is a history of humanitarian and non-governmental involvement in promote ecotourism to preserve wildlife for future generations and to benefit future tourism opportunities (see Butcher, 2007). However, in places around the world where damaging disputes over territory are occurring, or where competition to acquire land and resources leads to war and violence, the attempt to conserve is a challenge. Thus in disputed territories, conservation can be contested and political, especially where longstanding quarrels and complex power relations in protected areas exist (Ceballos-Lascurain, 1996; Mathis & Rose, 2016).

Times of active war, or directly following the aftermath of tragedy, affected regions usually have a demand for humanitarian need/aid (Cochrane & Cooke, 2016). In some places, a new humanitarian economy evolves to protect people and the environment in an attempt to de-escalate stresses (UNEP, 2017). Humanitarian efforts are also about raising awareness—often with the help of the media or non-governmental organisations. In addition, raising funds to protect people, animals, and the environment in an initiative alongside implementing safety and security measures help protect people and sustain natural biodiversity (Madianou, 2013). However, despite said opposition, Lopez, Bhungalia, and Newhouse (2015) also highlight that humanitarianism is a response to violence, as people seek ways to protect, preserve, or return normalcy to those affected—to aid people and/or animals and help preserve their local natural environment. A key challenge, however, is the disruption of conservation efforts during times of war given access and circumstances; but a re-emergence of humanitarian relief efforts immediately following times of war is an attempt to restore a natural setting. In some cases, there are attempts to promote tourism as a means to generate income in areas as a way of helping to preserve and protect unique wildlife (Ballantyne, Packer, & Hughes, 2009). To build a tourism industry in war-torn areas, destination managers need to ensure the destination is safe, conservationists need to be and feel safe, and that tourism and conservation efforts can take place. Animals such as gorillas (an endangered species) attract visitors who pay high fees to visit and experience these unique animals in the wild. DRC and Afghanistan that have unique wildlife, and ongoing conflict and violence remains an issue, significantly limiting conservation and tourism efforts (see Ospina, 2006). Visitor fees contribute to relief efforts to protect and conserve animal populations so that threats against them are minimal and so communities see regular income generation, which is happening in Uganda and Rwanda where there exist gorilla populations (Maekawa, Lanjouw, Rutagarama, & Sharp, 2013). A regular flow of income from an emerging ecotourism economy can allow local communities to become more aware of how conservation efforts can result in local economic sustainability to improve livelihoods. Tourism and the visitor economy can also help educate local people about the significance and value that unique wildlife have in a particular locale (Gadd, 2005). However, within such a framework, there is a need for business solutions that align with and support conservation efforts (see Andonova & Carbonnier, 2014).

Humanitarianism aligned with wildlife conservation is not a new concept, and some of the modern and contemporary thought and approaches align with perspectives going back to the early 1900s. Mighetto (1988) notes three highlighted points of wildlife conservation from the perspective of former United States president Theodore Roosevelt: 'the true sportsman', 'the nature lover', and 'the humanitarian'. For decades the literature aligned with conservation and outdoor recreation, but more recently as attempts to conserve focus on saving wildlife from the threat of war, the focus on humanitarianism is gaining presence (see Mlozi & Pesämaa, 2019). It is also possible to argue that humanitarianism efforts align with the global rise of tourism, especially the demand among visitors to see and experience animals in their natural settings and habitats opposed to seeing exotic animal species in zoos or other enclosed nature parks (Safina, 2018).

The topic of animal ethics is explored in the tourism literature ethical consumption standpoints, whereby animals are used for entertainment and lack any sort of naturalness (Fennell, 2012). Cochrane and Cooke (2016), focusing on animal rights, ask how "intervention in a state by external agents to prevent, halt or minimise violations of basic animal rights ('humane intervention') can be justified" (p. 106). It is more widely accepted that we can justify intervention efforts by militaries, human rights groups, or charities to help people, but as Cochrane and Cooke (2016) argue when it comes to acting and intervening for animals the justification and cause is not only much more difficult but even impossible. This is especially true in remote areas. Intervention is also a challenge because of additional stress (forced interactions with animals) and horrors (based on the trust animals have with humans) put on animals during war and conflict. Cochrane and Cooke (2016) note "all states have the responsibility to massively transform their relationship with non-human animals, and to build international institutions to oversee the proper protection of their most basic rights" (p. 106). In some places, building institutions is a challenge depending on financial resources. While institutions are dedicated societies to protect animals, what can follow are ecotourism opportunities to complement animal protection policies and build a new local economy—such as what is happening in gorilla sanctuaries in Uganda and Rwanda given the protective measures to ensure the animals service and thrive.

A critical note concerning wildlife tourism is that animals are simply a resource, or exploited commodities (McNeilage, 1996). There do exist ethical issues here, especially surrounding the commodification of animals in the wild, but others argue tourism endeavours are an attempt to create awareness to help protect and conserve wildlife in areas threatened by conflict and territorial dispute. The argument is allowing tourists to view animals in their natural habitats is more ethical than capturing and allocating wildlife to zoos and kept in captivity (Safina, 2018). While scholars have focused on how zoos affect animal behaviour (Hosey, 1997), zoos remain popular visitor attractions for those who may not have the resources to go to remote corners of the Earth. Moreover, there is a demand among nature based or environmental enthusiasts to see and experience wildlife in their natural settings. Again, the challenge with this is many of these natural settings are not only in very remote parts of the world, but some are in conflict-ridden areas. Frost (2011) argues that while zoos are controversial, they are an attempt to protect animals, and depending on this location educate about the need to conserve species. Animal populations in their natural can only be sustained so long they are kept safe and away from harm (this includes the impact of war on natural environment and poaching). For endangered species living in active areas of conflict, they risk extinction because humanitarian assistance is not always possible, often impossible. This was a major concern for the wild gorilla population in the DRC. Like any other resource, once it is depleted, or extinct, the resource no longer has economic value and thus different disputes over territory may emerge. These debates help us think about the critical challenges of tourism, especially in places suffering from war and conflict.

To enhance understandings going forward, there is a need for interdisciplinary cooperation to address challenges. For instance, Carbonnier (2016) argues that economists need to play a more central role to help scholars and local residents understand, reflect and address on humanitarian challenges. Jones (2015, p. 467) adds "the literature on social justice, and social justice movements themselves, routinely ignore nonhuman animals as legitimate subjects of social justice." Such a focus can allow future researchers to help local people and animals impacted by war to inform future conservation efforts. This way the protection of animals, wildlife and local community is based on mutual understandings and awareness of tragedy opposed to just a response to tragedy. Focusing on local communities living in fragile environments with unique wildlife need to understand how to maintain tourism, so that sustainable solutions can be realised environmentally, socially, and economically—for both the purpose of humanitarianism and ecotourism. The next section will address the impact of war and conflict on wildlife, and will look at examples and humanitarian efforts, tourism opportunities, and responses to aid and assist animals caught in the crossfire.

Animals caught in the crossfire: Examples and responses

To highlight the issues and impacts of conflict on nature and animals, this section will discuss case examples to address the impact on animals during times of war and conflict. These examples here build on the points highlighted above and include discussions of humanitarian efforts to spare animals deemed 'innocent victims of war' (see Marijnen & Duffy, 2018). The first two examples are in Africa, and emphasise the conclusion by Marijnen and Duffy (2018) that "more than 70% of Africa's national parks have been affected by war in recent decades, and wildlife has suffered as a result".

For several decades, mountain gorillas in DRC faced threats due to ongoing war, with accounts of gorillas shot and killed by rebel fighters. Hatcher (2012) reported that a quarter of the world's then 800 gorillas living in the wild at the time resided in Virunga National Park (in eastern DRC). Virunga National Park saw rebels seize control of the area in 2007, resulting in conservationists fleeing the park. Without conservationists and park rangers, gorillas "have been caught in the cross fire of pitched gun battles and have sometimes been attacked directly by rebels seeking to steal and sell infant gorillas" (Walsh, 2013). There were reports of 10 gorillas killed during the early years of the conflict and some even fleeing the area (Sanders, 2008). While rebel groups may not have targeted gorillas directly, rebel occupation of this territory halted conservation efforts and tourism here ceased (Sanders, 2008).

Reports of 40,000 people displaced were noted by Hatcher (2012) and 140 conservationists were killed (Rosen, 2014) in DRC, but the number of gorillas displaced and other wildlife were not calculated. While parks are 'safe' sanctuaries, gorillas still face the threat of poaching. Another important focus here is how locals perceive gorillas, given the threat of poaching. Poachers see gorillas as immediate profit, so a response in Uganda and Rwanda was to employ former poachers to run conservation/wildlife tours because they are familiar with the forest and can navigate through the thick vegetation (Feltner, 2009). There are also efforts to educate local residents in the surrounding communities about the value of gorillas to the tourism economy. Because they are an endangered species, tourists desire to see them in the wild, and protecting gorillas is a chance for local communities to realise sustained economic impacts from tourism (Haines, 2018). For instance, an initiative of Bwindi Impenetrable National Park in Uganda is to promote awareness of the impact of human conflict and poaching on wildlife, and to teach locals the importance of wildlife so that they can gain from a sustained tourism economy. The African Wildlife Foundation (2020) have adopted a partnership initiative in Uganda and Rwanda working with governments, private operators and local communities, so that tourism can thrive and gorillas can be sustained in their natural environment, and maintain efforts to keep gorillas safe in DRC. Conservation methods include "on the ground through ranger-based monitoring, transboundary collaboration, community and tourism development, anti-poaching activities, and habitat conservation" (African Wildlife Foundation, 2020).

For many countries in Africa, wildlife is important to the economy (see Price, 2017). In DRC, the occupation of territory, active conflict, and poaching has taken a toll on the gorilla population, in Mozambique the civil war that ravaged the country from 1977 to 1992 resulted in the significant loss of giraffe and elephant herds. Potenza (2018) reported that one park lost 90% of its wildlife during the civil war in the country. The direct killing of animals was a means of survival for many during the 15-year civil war due to widespread food shortages. Some would hunt wildlife for food and for others poaching was an attempt to make money (selling ivory). Shivni (2018), reports that in Mozambique it remains "unclear whether gunfire or bombs hurt any animals during this time" and that "indirect causes include disruption from humans fleeing violence and seeking shelter in animal habitats, food shortages or fewer opportunities to reproduce". Knock-on effects from unintentional compromises is what Daskin and Pringle (2018) highlight as a main cause for decline, especially when animal reproduction cycles and mating seasons are directly affected, as this impacts migration patterns as well. What is key is the post-war response efforts, and the commitment to conservation. Following the civil war in Mozambique, the conservation of Gorongosa National Park became a primary focus. Here, "former soldiers on opposite sides of the conflict were put to work together as rangers to protect the country's wildlife" (Potenza, 2018). Such a response was an attempt to build and work towards mutual understandings, whereby preserving Mozambique's bountiful natural resources and wildlife could unite people once in opposition. Thus, conservation efforts and building an ecotourism economy became a focus to distracting people from war tensions and societal opposition and put emphasis on rebuilding its society, natural environment, and economy to attract visitors (see Dondeyne, Kaarhus, & Allison, 2012).

In the examples noted to this point from Africa, a key threat to animals in war-torn areas has been poaching and people hunting for a means of survival. War also leads to the displacement or abandonment of animals in zoos and household pets (Guynup, 2017; Pleitgen, 2014). For example, Gouillou (2017) highlights that the ongoing conflicts in Syria and Palestine focus heavily on human death tragedy as well as the destruction of tangible and intangible cultural heritage, with little insight on how these conflicts are taking a toll on animal populations, as the author deems animals: forgotten war victims. Without ongoing conservationists working in active warzones, staff are unable to care for animals in zoos—the case in Aleppo and Gaza. Direct hits on a Gaza zoo during conflict between Israel and Hamas led to limited capacity for care staff and zookeepers—resulting in no water and food shortages (Pleitgen, 2014). Traumatised and weak animals had limited/no ability to flee, but even if zoo animals could flee they would lack survival skills in an unfamiliar (and likely unwelcome) environment. It was uncertain if the zoo attacks were direct or were collateral damage, but given the human death toll and displacement in this area, the focus on animals kept in captivity was not a priority when it came to receiving resources (Pleitgen, 2014). In Aleppo, Amir Khalil (who has experience saving animals from warzones) organised an effort to save 13 surviving animals from Magic World (Guynup, 2017). This effort resulted in the relocation of these surviving animals to a zoo in neighbouring Jordan. The challenge here was humanitarian intervention, and still this required diplomatic talks to allow people to not only collect the animals, but then to transport them overland through an active warzone. For household pets or zoo animals, the impact of war is different, but the result is also tragic. For wild animals, disruptions to their natural environment may force them to alter more traditional paths or migration routes, if only temporarily. As addressed in the examples noted in this chapter, many animals impacted by war and active conflict have little chance of recovery without humanitarian intervention and strict conservation policies/efforts, as this applies to animals in the wild, in zoos and household pets. However, for household pets and zoo animals, these animals are dependent on humans for survival, and without care, their survival chances are minimal. Animals that do survive face trauma due to abandonment and starvation (Lynne, 2019).

Reports from Kashmir show a different situation concerning the impact of war on wildlife. The conflict in Kashmir between India and Pakistan had a very different impact on certain animals, as the war brought about a time of safety for leopards, snow leopards, bears, and the hangul (closely related to the reindeer, native only to Kashmir). Because locals had to surrender their firearms, this stopped people from wandering the forests—thus deterring poaching (see Joshi, 2006). Despite conflict, having a positive impact on the snow leopard population in Kashmir, in nearby Afghanistan the impact on this same species was very different. Lynne

(2019) reports "an estimated 10 million landmines have been scattered throughout Afghanistan threatening the elusive snow leopard. Afghan soldiers also hunt goats and sheep for food in the remote mountainous areas of the country, reducing the snow leopard's access to prey". The negative impact on snow leopards in Afghanistan is twofold, with these animals killed by landmines or starving because their natural food supplies diminished from hunting.

In some warzones there has been speculation that animals were transporting supplies, as noted "during the Vietnam war, the United States routinely bombed Asian elephants as they were believed to be used to transport supplies for the opposition" (Lynne, 2019). For such places with a history of mines and casualties, an approach similar to Cambodia whereby efforts are about educating visitors of the impact of mines on people (see Winter, 2008), and in places such as Afghanistan in the future education and conservation efforts can highlight the wider impact on animal populations and the upsetting of a natural balance.

Conclusions and future research

As outlined in the many examples above, ecotourism and nature-based tourism can play a central role when it comes to post-war conservation efforts. Such tourism opportunities aim to protect wildlife, and to sustain local/regional economies so local communities in these destinations can generate income. The emphasis needs to be the long-term value of nature and wildlife opposed to short term survival, which is the mentality of poachers that we see in the example of how DRC, Uganda, and Rwanda are embracing the gorillas in the wild for sustained tourism futures in the Virunga Mountains. However, we must not surpass the issue of commodification noted earlier in this chapter. In Mozambique, the civil war was seen as the most tragic example of wildlife animals caught in the turmoil of conflict, whereas the very opposite was seen in Kashmir where wildlife were able to find protection because of conflict. In Aleppo, humanitarians risked their lives to save stranded animals during a time of active war.

According to Potenza (2018), "every country, and every conflict, is different", and this represents a challenge to assessing animals caught in the crossfire. Conducting research in war- and conflict-ridden areas is difficult due to safety concerns and limited access to these areas; in reality, researchers cannot begin to assess such the impact of animals in warzones until years after. Recent work assesses the impact of war on destinations by employing content analysis (e.g., Wise, 2019) or conducting thorough archival research (e.g., Daskin & Pringle, 2018). Field-based research proves a challenge for researchers given the difficulty to access sites and destinations due to safety restrictions, not to mention challenges with gaining ethical approval and funds to travel to destinations experiencing active conflict, or recovering from it. Additionally, when war or conflict consume a country or destination, cease-fires are not always immediate, as a resurgence is possible. The challenge then, is acquiring funding to carry out conservation efforts and investment in subsequent ecotourism opportunities (including infrastructure to access areas, tourism resources, training staff, and paying employees). War or conflict sees funds diverted away from social or natural agendas, and post-conflict, economic situations may be dire so trying to divert funds to conservation might not always be (initially) feasible. In many cases, conservation efforts will require awareness and investment from the outside with the assistance of humanitarian groups who have the ability to raise funds externally to support efforts to save wildlife. This has been a trend going back decades, for instance, the African Wildlife Foundation (2000) noted:

> Throughout the civil conflicts, IGCP has helped pay stipends to park staff members, replace equipment lost or stolen during the hostilities, restore damaged facilities, and

> train guards in gorilla-tracking techniques. Rangers from the three countries have been taught how to collect invaluable data on the mountain gorillas. They use global positioning systems and other methods to chart the ranging habits of gorilla groups. The collected information helps park wardens manage protection efforts.

Each of these activities require adequate funding and human resourcing to build facilities and monitor wildlife, which goes beyond the immediate needs and priorities that war-torn countries face, and which is often needed to repair and restore basic infrastructures essential to rebuilding a national economy.

Protecting animals following a time of conflict and building a new ecotourism economy is about creating an awareness to attract tourists and educating locals about conservation. Conflict, as mentioned, is a dispute over territorial control, but these territories contain populations of people recovering and also animals and wildlife seeking normality in their natural living environment. Liu, Li, and Pechacek (2013) stress the needs for a set of guidelines for managing wildlife ecotourism, but what is missing is the complexity associated with disputes over territory where animals are unfortunately caught in the crossfire. War devastates people, places, and wildlife, and "animals that do survive can rebuild their populations, so conservation efforts in war-torn areas are incredibly important" (Potenza, 2018). Importantly, this takes co-ordination efforts of people to restore normality to animal sanctuaries and natural living environments. Also important is educating local communities about the interplay between conservation and responsible tourism, and conservation efforts in warzones have been successful (such as in Mozambique). For animals in zoos and household pets that are abandoned, this remains a key concern, due to their limited ability to adapt.

The impact of animals caught in the crossfire is an underdeveloped area of research in the field of ecotourism and tourism development in general. Going forward, future work should endeavour to better understand both the protection animals and efforts to educate local communities—as these points together offer useful insight on the overlaps of sustainable community planning and natural resource protection. Thus, integrating such 'dark', 'tragic', or 'violent' tourism histories alongside ecotourism initiatives is important to inform visitors of the challenges of nature conservation, and strategies for overcoming conflict-related issues, given the physical, social, and psychological impacts this has on people, animals, conservation efforts, social development, and destination image.

References

African Wildlife Foundation. (2000). Conservation in War Zones. Retrieved from https://www.awf.org/news/conservation-war-zones (accessed on 20.04.2020).

African Wildlife Foundation. (2020). Mountain Gorilla. Retrieved from https://www.awf.org/wildlife-conservation/mountain-gorilla (accessed on 20.04.2020).

Agnew, J. (2009). Review: Killing for cause? Geographies of war and peace. *Annals of the Association of American Geographers, 99*(5), 1054–1059.

Andonova, L., & Carbonnier, G. (2014). Business-humanitarian partnerships: processes of normative legitimation. *Globalizations, 11*(3), 349–367.

Azimi, M. (2012). *Ecotourism in Afghanistan* (in Farsi). Kabul: Seraj and Sobhe Omid.

Ballantyne, R., Packer, R., & Hughes, K. (2009). Tourists' support for conservation messages and sustainable management practices in wildlife tourism experiences. *Tourism Management, 30*, 658–664.

Bulbeck, C. (2005). *Facing the Wild: Ecotourism, Conservation and Animal Encounters*. London: Routledge.

Butcher, J. (2007). *Ecotourism, NGOs and Development*. New York: Routledge.

Butler, R., & Suntikul, W. (2013). *Tourism and War*. New York: Routledge.

Carbonnier, G. (2016). *Humanitarian Economics: War, Disaster and the Global Aid Market.* London: Oxford University Press.
Ceballos-Lascurain, H. (1996). *Tourism, Ecotourism, and Protected Areas: The State of Nature-based Tourism around the World and Guidelines for its Development.* Gland: IUCN.
Chakraborty, A. (2019). Does nature matter? Arguing for a biophysical turn in the ecotourism narrative. *Journal of Ecotourism, 18*(3), 243–260.
Cochrane, A., & Cooke, S. (2016). 'Humane intervention': The international protection of animal rights. *Journal of Global Ethics, 12*(1), 106–121.
Curtin, S. (2005). Nature, wild animals and tourism: an experiential view. *Journal of Ecotourism, 4*(1), 1–15.
Daskin, J. H., & Pringle, R. M. (2018). Warfare and wildlife declines in Africa's protected areas. *Nature, 553*, 328–332.
Dondeyne, S., Kaarhus, R., & Allison, G. (2012). Nature conservation, rural development and ecotourism in central Mozambique: Which space do local communities get? In I. Convery, G. Corsane, & P. Davis (Eds.), *Making sense of place: Multidisciplinary perspectives* (pp. 291–302). Suffolk: Boydell Press.
Dudley, J. P., Ginsberg, J. R., Plumptre, A. J., Hart, J. A., & Campos, L. C. (2002). Effects of war and civil strife on wildlife and wildlife habitats. *Conservation Biology, 16*(2), 319–329.
Feltner, M. (2009). Turning Poachers into Conservationists in Rwanda. *National Geographic.* Retrieved from https://www.nationalgeographic.com/travel/intelligent-travel/2009/05/13/turning_poachers_into_conserva/ (accessed on 20.04.2020).
Fennell, D. (2012). *Tourism and Animal Ethics.* London: Routledge.
Ferreira, S., Strydom, J., Kriel, M., & Gildenhys, S. (2015). Tourism and development after civil war in Malange province, Angola. *South African Geographical Journal, 97*(2), 158–182.
Frost, W. (2011). *Zoos and Tourism: Conservation, Education, Entertainment?* Bristol: Channel View Publications.
Gadd, M. E. (2005). Conservation outside of parks: Attitudes of local people in Laikipia, Kenya. *Environmental Conservation, 32*(1), 50–63.
Gouillou, N. (2017). Forgotten Victims of the War. *Medium.* Retrieved from https://medium.com/@aetaqid/forgotten-victims-of-the-war-595d28ce4474 (accessed on 20.04.2020).
Guynup, S. (2017). See How Syrian Zoo Animals Escaped a War-Ravaged City. *National Geographic.* Retrieved from https://www.nationalgeographic.com/news/2017/10/wildlife-watch-rescuing-animals-aleppo-syria-zoo/ (accessed on 20.04.2020).
Haines, G. (2018). How Africa Is Winning the Battle to Save Its Mountain Gorillas. *Telegraph.* Retrieved from https://www.telegraph.co.uk/travel/destinations/africa/articles/how-east-africa-is-winning-the-battle-to-save-mountain-gorillas/ (accessed on 20.04.2020).
Hanson, T., Brooks, T. M., da Fonseca, G. A. B., Hoffmann, M., Lamoreux, J. F., Machlis, G., ... & Pilgrim, J. D. (2009). Warfare in biodiversity hotspots. *Conservation Biology, 23*(3), 578–587.
Hatcher, J. (2012). Gorillas Caught Between the Lines of New Congo War. *Independent.* Retrieved from https://www.independent.co.uk/environment/nature/gorillas-caught-between-the-lines-of-new-congo-war-7791468.html (accessed on 20.04.2020).
Henderson, J. C. (2000). War as a tourist attraction: The case of Vietnam. *International Journal of Tourism Research, 2*(4), 269–280.
Hosey, G. R. (1997). Behavioural research in zoos: Academic perspectives. *Applied Animal Behaviour Science, 51*(3/4), 199–207.
Jones, R. C. (2015). Animal rights is a social justice issue. *Contemporary Justice Review, 18*(4), 467–482.
Joshi, B. (2006). Insurgency Benefits Kashmir Wildlife. *BBC News.* Retrieved from http://news.bbc.co.uk/1/hi/world/south_asia/6169969.stm (accessed on 20.04.2020).
Liu, C., Li, J., & Pechacek, P. (2013). Current trends of ecotourism in China's nature reserves: A review of the Chinese literature. *Tourism Management Perspectives*, 7, 16–24.
Lopez, P. J., Bhungalia, L., & Newhouse, L. S. (2015). Geographies of humanitarian violence. *Environment and Planning A: Economy and Space, 47*(11), 2232–2239.
Lynne, B. (2019). War Can Have a Devastating Impact on Wildlife Populations Across the Globe. *Earth.com.* Retrieved from https://www.earth.com/news/war-devastating-impact-wildlife/ (accessed on 20.04.2020).
Madianou, M. (2013). Humanitarian campaigns in social media. *Journalism Studies, 14*(2), 249–266.
Maekawa, M., Lanjouw, A., Rutagarama, E., & Sharp, D. (2013). Mountain gorilla tourism generating wealth and peace in post-conflict Rwanda. *Natural Resources Forum, 37*(2), 127–137.

Marijnen, E., & Duffy, R. (2018). Animals Are Victims of Human Conflict, So Can Conservation Help Build Peace in Warzones? *The Conversation*. Retrieved from https://theconversation.com/animals-are-victims-of-human-conflict-so-can-conservation-help-build-peace-in-warzones-90045 (accessed on 20.04.2020).

Mathis, A., & Rose, J. (2016). Balancing tourism, conservation, and development: a political ecology of ecotourism on the Galapagos Islands. *Journal of Ecotourism, 15*(1), 64–77.

McNeilage, A. (1996). Ecotourism and mountain gorillas in the Virunga Volcanoes. In V. J. Taylor, & N. Dunstone (Eds.), *The exploitation of mammal populations* (pp. 334–344). Dordrecht: Springer.

Mighetto, L. (1988). Wildlife protection and the new humanitarianism. *Environmental Review, 12*(1), 37–49.

Miller, A. (2012). *Ecotourism Development in Costa Rica*. Lanham: Lexington.

Mlozi, S., & Pesämaa, O. (2019). Antecedents of national park knowledge in Tanzania. *Journal of Ecotourism, 18*(3), 199–220.

Ospina, G. A. (2006). War and ecotourism in the national parks of Colombia: Some reflections on the public risk and adventure. *International Journal of Tourism Research, 8*(2), 241–246.

Perkins, H., & Grace, D. A. (2009). Ecotourism: Supply of nature or tourist demand? *Journal of Ecotourism, 8*(3), 223–236.

Pleitgen, F. (2014). Gaza's Zoo Animals Caught in Crossfire of Israel-Hamas Conflict. *CNN Middle East*. Retrieved from https://edition.cnn.com/2014/08/18/world/meast/gaza-zoo-destroyed/index.html (accessed on 20.04.2020).

Potenza, A. (2018). When Humans Kill Each Other in War, Wildlife Dies Too. *The Verge*. Retrieved from https://www.theverge.com/2018/1/10/16871120/war-wildlife-population-decline-africa-conservation (accessed on 20.04.2020).

Price, R. A. (2017). *The Contribution of Wildlife to the Economies of Sub Saharan Africa*. K4D Helpdesk Report. Brighton, UK: Institute of Development Studies.

Rosen, J. (2014) The Battle for Africa's Oldest National Park. *National Geographic*. Retrieved from https://www.nationalgeographic.com/news/special-features/2014/06/140606-gorillas-congo-virunga-national-park/ (accessed on 20.04.2020).

Safina, C. (2018). Where are zoos going—or are they gone? *Journal of Applied Animal Welfare Science, 21*(1), 4–11.

Sanders, E. (2008). A Gorilla War in Congo. *Los Angeles Times*. Retrieved from https://www.latimes.com/archives/la-xpm-2008-dec-06-fg-gorillas6-story.html (accessed 20.04.2020).

Shackley, M. (1995). The future of gorilla tourism in Rwanda. *Journal of Sustainable Tourism, 3*(2), 61–72.

Shivni, R. (2018). In War-Torn Areas of Africa, Wildlife Is a Major Casualty. *PBS NewsHour*. Retrieved from https://www.pbs.org/newshour/science/in-war-torn-areas-of-africa-wildlife-is-a-major-casualty (accessed on 20.04.2020).

Toft, M. D. (2014). Territory and war. *Journal of Peace Research, 51*(2), 185–198.

UNEP. (2017). Placing the Environment at the Heart of Humanitarian Response. *United Nations Environment Programme*. Retrieved from https://www.unenvironment.org/news-and-stories/story/placing-environment-heart-humanitarian-response (accessed on 20.04.2020).

Walsh, B. (2013). Saving Congo's Gorillas: A Refuge for Species Under Threat. *TIME*. Retrieved from https://time.com/3803631/saving-congos-gorillas-a-refuge-for-species-under-threat/.

Wilson, E. O. (1984). *Biophilia*. Cambridge, MA: Harvard University Press.

Winter, T. (2008). Post-conflict heritage and tourism in Cambodia: The burden of Angkor. *International Journal of Heritage Studies, 14*(6), 524–539.

Wise, N. (2019). Narrating the scars of Sarajevo: Reminiscent memories of war and tragedy in the landscape. In R. Isaac, E. Çakmak, & R. Butler (Eds.), *Tourism and hospitality in conflict ridden destinations* (pp. 201–215). London: Routledge.

World Peace Foundation. (2015). Mozambique: Civil War. *Mass Atrocity Endings*. Retrieved from https://sites.tufts.edu/atrocityendings/2015/08/07/mozambique-civil-war/ (accessed on 20.04.2020).

18
ANTI-ECOTOURISM
The convergence of localism and way of life

Joe Pavelka

Introduction

Ecotourism has consistently been associated broadly with the way of life of local people. Case studies report ecotourism's application has supported a particular way of life, while others claim it has compromised or ruined a particular way of life that existed prior to its implementation (Das & Chatterjee, 2015). A post-hoc analysis of ecotourism links the case studies and has been applied and judged. The purpose of this chapter is to examine ecotourism within a different context. In this analysis, an application of ecotourism is attempted in the Big Horn region of Alberta Canada but rejected due to a combination of forces that are best described as localism and way of life. I argue here that it is as important to understand ecotourism's critiques and case study results, as it is to gain insight into particular settings in which it is not likely to be embraced.

Anti-ecotourism refers to a context in which ecotourism is attempted to be applied but is ultimately rejected by a community. The preconditions of anti-ecotourism are shrouded in a layered tension, most likely found in the global north, and especially western North America where there are economic alternatives to ecotourism. Anti-ecotourism suggests that ecotourism is rejected in concept, reputation, and association because of what it proposes to bring in terms of change and threats to a way of life. The application of ecotourism is framed by a climate of tension, change, and disruption to a way of life that views its relationship to the land in traditional instrumentalist manner.

It is important for us to better understand the pre-implementation conditions of ecotourism. Ecotourism has long been associated with conflict, but we know little about the conflict specific to ecotourism at its early stages of development (Connell, Hall, & Shultis, 2017). Hunt and Stronza (2014) argue that there is a need for two additional stages to be added prior to Butler's Life Cycle Model's *exploration* phase. These stages would consist of "the absence of tourism" and "the arrival of early tourists." These phases reflect the conditions of a community in which a tourism industry has yet to occur; its residents are naïve or unaware, do not have concerns and/or are not hesitant to accept tourism development (Hunt & Stronza, 2014). It is useful to learn how ecotourism's critiques may manifest in a pre-implementation context as communities look to it as a potential economic development strategy. Rural and forestlands in western North America have experienced a general decline in the productivity of staple resource economies.

DOI: 10.4324/9781003001768-18

Many have and continue to look to nature-based tourism and amenity migration for revitalisation which may produce intra-community conflict (Gosnell & Abrams, 2011).

Finally, at the core of anti-ecotourism is a fundamental tension related to the urban-rural cultural divide and the clash of way of life that Gosnell and Abrams (2011) describe. However, it is likely to increase in significance as a constraint to the implementation of urban-inspired nature-based tourism in rural settings. These factors suggest that it is important for us to understand the meaning and conditions of anti-ecotourism which include the concepts of localism and way of life.

Localism in tourism and recreation studies is often discussed in relation to surf tourism. It is defined as a situation whereby local surfers need to defend their territory from tourists or others wishing to take advantage of the resource (Usher & Gomez, 2016). Localism is not restricted to surf settings and may be applied to any recreation resource context. Localism links with anti-ecotourism because locals may believe that ecotourism will bring tourists who will clutter or displace them from their cherished recreation areas. This may be especially true for recreation that is not known to be compatible with mixed use, such as motorized off-road recreation that, due to its noise pollution and impact on the land is perceived to dominate to the exclusion of other activities, creating an intense winner-loser scenario (Miller, Vaske, Squires, & Olson, 2017).

Anti-ecotourism is more nuanced than simple localism. Localism is only an option for those who can afford to reject the economic promise of ecotourism. Those who cannot will often make the trade-off of improved economic conditions for the presence of tourism. One of ecotourism's tenets is that it can and should lift the economic wellbeing of communities (Fennell, 2008). This is a common aim in developing countries where ecotourism is often viewed as an economic savior. In the West, resource economies have dominated for decades and have created an entrenched but precarious way of life that provides decent wage employment therefore resistance to ecotourism is more likely. As Brouder (2012) claims, tourism is often viewed as a last resort for Western rural communities in economic trouble.

Way of life is a complex phenomenon that does not easily lend itself to precise definition. In fact, it is a concept seldom pondered except when it is perceived to be under attack, consequently it becomes an overriding and important issue. However, way of life is defined as the behaviour and habits that are typical of a particular person or group or is chosen by them (Collins Dictionary, 2020). It may include an important activity or resource that affects all parts of someone's life (Merriam-Webster, 2020). For example, the sea or mountains contribute to form a way of life.

Early on, Wirth (1938) defined urbanism as a way of life that is the sum of many (urban) factors converging to shape a daily pattern of life that is different from rural life. A way of life is linked to individual or group identity. Sorge (2008) examines localist and cosmopolitan identities in highland Sardinia, Italy. He claims these identities are shaped by the epistemology of modernity and divisible by practices, values, and beliefs that lean toward the modern or traditional. He argues that both modern and traditional identities form separate ways of life that may coexist in the same community, but not without tensions. An understanding of way of life must also include the economy that allows for a particular style of patterns of daily life. The application of ecotourism within a community may involve a similar tension that arises from a 'new way of doing things' that supposes the old ways to be out of step with modern values.

Ecotourism itself had a similar effect on tourism when in its early days it rose as a critique of mass tourism. Ecotourism advocated moving toward a modern intent for the industry that focused on the wellbeing of local people over the interests of multi-national corporations. Ecotourism essentially involves "responsible travel to areas that conserves the environment and sustains the wellbeing of local people..." (TIES, 2015). Ecotourism has a greater chance of success when the offering of a conservation-oriented economic development strategy based on inbound travel is viewed as enhancing the wellbeing of local people. We assume most communities that embody nature are amenable to ecotourism's aims. But when a particular way of life does not prioritise an economy of inbound visitors and foreign notions of environmental conservation, ecotourism's chances of implementation lessen.

This chapter presents literature that frames the tension surrounding ecotourism from the perspective of what residents may expect if it were to be implemented in their community. Later, I present a case study of an attempt to implement the Bighorn Country Provincial Park in Alberta, Canada. It was an attempt to implement a park system in a rural forested region of Alberta to enhance conservation and build a nature tourism economy. The case study embodies the concept of anti-ecotourism; its key features can be found in many other parts of the world. I close the chapter with a set of criteria to define the concept of anti-ecotourism.

Ecotourism and tension

The application of ecotourism can result in a myriad of tensions that span social, economic, and environmental issues. Ecotourism proposes a reorganization that will be viewed differently by individuals based on their values and beliefs within a meeting ground of hosts and guests characterised by tension (Schweinsberg, Wearing, & Darcy, 2012). Hosts often feel the threat of displacement, which can result in territorial resistance and manifest outwardly as localism. These concepts, along with the economic considerations that underpin a resource-based way of life, are discussed in the following section. In keeping with the concept of anti-ecotourism, the reader should consider the following concepts from the perspective of a community that is presented with the prospect of adopting ecotourism.

Place attachment

Place attachment is a term that widely captures various elements of the emotive aspects of the human-environment relationship. For example, Stokowski (2002) describes place attachment when referring to the feelings of attachment to particular settings, that are based on attentiveness and emotion. From a recreation perspective, Kyle, Bricker, Graffe, and Wickham (2004) define place attachment as, "The extent to which an individual values and identifies with a particular natural setting." Brehn, Eisenhauer, and Krannich (2004) states that place attachment can be especially evident within communities with high amenities, but it consists of two key dimensions involving "social ties particularly in high amenity settings, it may also involve attachment focused on physical attributes" (Brehn et al., 2004, p. 407).

Kyle et al. (2004) claim that for the individual recreationist, place attachment is largely a function of place dependence and place identity. Place dependence refers to the importance of a particular place or resource in providing the amenities necessary for the pursuit of a particular activity. Place identity is defined as "Those dimensions of the self that define the individuals' personal identity in relation to the physical environment by means of a complex pattern of conscious and unconscious ideas, beliefs, preferences feelings, values, goals, behavioural tendencies and skills relevant to this environment (Proshanshsky, 1978, cited in Kyle et al., 2004, p. 124). Additionally, place dependence can lead to place identity over time (Proshanshsky, 1978).

Stokowski (2002) agrees that individuals give meaning to, and gain identity from, the subjective interpretation of place. However, she argues for a circular type relationship whereby individuals are attracted to a place that holds potential to support one's identity. Once there, a mutually understood identity of a place is generated through social discourse and action. It is the social creation of place that reinforces personal identity. Stolowski's emphasis on the social aspect of place creation, recreation, and attachment is widely shared (Kyle et al., 2004; Brehn et al., 2004; Williams & McIntyre, 2002). Olive (2015) describes a similar phenomenon within the surf culture where local surfers develop a special relationship with the place that extends to stewardship, and believe they know what is best for the place. She refers to it as 'Ecological Sensibilities,' which is an ecocentric sensibility that transcends the narrow sense of self and human superiority.

Tension can arise when guests embody attitudes that conflict with their hosts. Ecotourism has been described as a conspicuous form of life politics that reflect a certain sensibility among Western tourists to make a difference to the societies visited. It is presented as a form of ethical consumption as an extension of the identity of the tourist, and presumed to be ethically better for the host region (Butcher, 2008). Tension may arise if the host community is not keen to have ethically minded nature-based tourists endeavoring to better their lives. It invokes a sense of superiority similar to the 'modernisation theory' critique of ecotourism.

Modernisation theory posits that there is a dichotomy between the modern and traditional, and that Indigenous people are pressured to abandon their traditional ways for the modern. It also contends that the route to change for developing nations is the route that developed nations have undertaken in the past, and that development will proceed in a linear manner following similar stages of development (Regmi & Walter, 2017). In essence, the application of ecotourism limits the development options for the community and diminishes its traditions. Modernisation theory is criticized for its ethnocentric focus whereby the practices that may seem odd and strange within economically poor non-Western nations are viewed as primitive and traditional, and therefore need to be corrected. Regmi and Walter (2017) present these criticisms in the context of developing nations, but its application to Western rural-urban contexts are equally valid.

Displacement

Displacement is understood to be about being 'displaced' from a place, economy, or position, but its disruptiveness on one's way of life cannot be overlooked. It may present an unmooring of the self from daily patterns and relationships and from opportunities of a future that the individual believes they are entitled to. Way of life is not simply a pattern of activity or presence of a significant resource; it is an expectation of benefit from one's environment. In that regard, the threat of displacement can be ever more significant.

Ecotourism-related displacement can occur when community residents are forced or coerced off their land generally to make way for a park or protected area (Cernea & Schmidt-Soltau, 2006). It can also occur when residents are displaced in absolute or partial terms due to rezoned protection status of key resources attributed to the ecotourism economy (Carrier & MacLeod, 2005) or when shut out of key business opportunities and jobs in the new economy (Das & Chatterjee, 2015). Marginalisation within the planning process by outsides or in-fighting is another form of displacement (Lane & Corbett, 2005). For example, outside groups may enter into community planning in a disingenuous manner that is predominately focused on achieving an environmental goal over community economic wellbeing (Butcher, 2005).

Another form of displacement involves a restriction of activities deemed too consumptive for inclusion in ecotourism. Melitis and Campbell (2007) argue that ecotourism is defined as non-consumptive limiting the direct use or removal of wildlife through activities such as hunting. They find this limitation problematic for a variety of reasons that involve cultural preservation or moral superiority and it can lead to increased tensions between locals and tourists when areas are regulated for exclusively for non-consumptive use.

Territoriality and localism

Place attachment suggests that an individual has a strong bond to a place based on emotive or cognitive dimensions Kyle et al. (2004). Territoriality is a form of social control of that place when an individual perceives it to be at risk (Usher & Gomez, 2016). Territoriality is based on three dimensions including ownership, boundaries, and regulation. It is an informal, citizen-driven phenomenon that can be linked to the identity of an entire community, not just a small group of individuals (Beaumont & Brown, 2016).

Ownership is thought to be psychological. For example, a group of dog walkers assume ownership over a particular part of a park, or motorbike users over specific trails. Boundaries can be understood informally as traditional uses of the land, whereby it is well known that certain areas are dedicated to a particular activity such as equestrian trails or logging roads that have served recreational uses. Regulation is the informal, or non-sanctioned, defense of the space from outsiders. In the surf community, it may include rules to show respect to local surfers, such as waiting on the shoulder of the break, it is also where localism may be observed. Usher and Gomez (2016) state that localism is the territorial behaviour of resident surfers over surf breaks, and it may manifest differently in each surf destination. Localism is the manifestation of territoriality and it is experienced from the mild to the extreme. Researchers have noted that localism in the surf community can be as benign as yelling profanities at outsiders through to extremely violent behaviours. However, localism is also demonstrated as environmental protection and stewardship (Usher & Gomez, 2016).

Economic tensions

In the West, rural communities have felt a threat to their way of life due to global economic restructuring. That threat has often led to the pursuit of forms of nature-based tourism to shore up or reinvent their rural economy (Gosnell & Abrams, 2011). Nature based tourism in Western rural contexts is often pursued as a last resort and sometimes portrayed as a panacea for economic issues (Brouder, 2012). In other cases, it is pursued as a strategy for economic diversification in resource-based communities (Connell et al., 2017) while others welcome tourism and amenity migration as a way to transition low value production land into higher value positional goods (Gosnell & Abrams, 2011). In each case there are tensions associated with the new economy.

Connell et al. (2017) examine the prospect of introducing ecotourism to the Robson Valley of British Columbia. The Robson Valley is a peripheral area that relies on a forestry staple-based economy. A peripheral area is geographically remote endowed with natural resources that lacks political and economic control over major decisions regarding its wellbeing and economic linkages. A staples-based economy refers "an economic relation within a region that historically depended upon the export of bulk, largely unprocessed commodities" (p. 176). It is based on a precarious dual dependency on the natural resource to produce raw goods and distant markets to purchase the commodities produced.

Carson and Carson (2011) examine contexts in which tourism may not be supported in the western United States. They characterize staples-based peripheral communities as having a government that tends to invest in the staples export to a point of overconcentration coupled with a reluctance to promote other industries. Large external investors tend to manage and control many of important functions such as transport and storage and discourage the formation of smaller internal community-based economic linkages. It results in a diminished capacity to manage economic change, resiliency, and diversification. They create a culture of reliance on external corporations for the commercialization of the final products, and lesson entrepreneurial capacity within the region. They tend to import high-skilled labour, leaving the local workforce to remain low-skilled and consequently develop attitudes that favour low-skilled employment in staples production over general skills and education. They become locked into traditional occupations with little capacity for mobility and foster few incentives to obtain higher education. On the other hand, they position the position the tourism-ready community to be one that possesses an educated entrepreneurial community with productive internal partnerships.

Urban-based amenity migrants come with their own life politic regarding the environment and recreating in the outdoors resulting in a clash of rural and urban ways of life. While local residents can maintain a semblance of gatekeeping, amenity migrants tend bring capital which can skew the community toward change (Gosnell & Abrams, 2011). Tourism remains an important tool in the development and reconstruction of rural communities, but Brouder (2012) aptly states that tourism may appear to occupy a neutral space, but it does not.

Bighorn Country Proposal, Alberta Canada

In 2018, Alberta's provincial government proposed the development of Bighorn Country as one way of diversifying Alberta's economy through nature-based tourism. The Bighorn park proposal would be this government's second major park development within a two-year period. The first was the Castle Wildland Provincial Park in the southern end of the province. Castle Park is set in an area dominated by ranching and crown land or unprotected land held by the government for a variety of economic purposes. Castle Park was presented as ecotourism-driven economic diversification and eventually approved but not without controversy. Some local residents protested additional protections and regulations on land they had come to rely on for motorised recreation (Weber, 2017). The Big Horn park process would be even more contentious and ultimately be rejected.

The stakeholders in the Big Horn park proposal included a provincial government keen to create additional parks prior to the end of its mandate. The New Democrat Party (NDP) government was viewed as occupying the left or center-left end of the political spectrum in a province long dominated by conservative governments. In the peripheral area, there is a community of rural towns that are reliant on a staple economy based on resource extraction, a small group of nature-based tourism operators and First Nations groups. Additionally, Yellowstone to Yukon (Y2Y) and the Canadian Parks and Wilderness Society (CPAWS) were two environmental, non-governmental organisations (ENGOs) working alongside the government. The Big Horn area rests between two iconic national parks, Jasper National Park to the north and Banff National Park to the south and could complete an unprecedented contiguous space of protected habitat making it especially attractive to the ENGOs. The designation of Big Horn as protected space would also boost the provincial government's total protected area from 15% to its target of 17%. The area had been a priority area for protection ENGOs (Riley, 2019). There were several previous attempts at an overarching land use recreation strategy for the area dating back to the 1970s. The 2018, process was initiated by the provincial government and actively supported by the ENGOs, but they soon encountered opposition from vocal elements of the local community (Legualt, 2020).

The Bighorn Country park proposal encompassed 5000 square km of crown land along the Eastern Slopes of the Rocky Mountains. It is a beautiful landscape where the foothills meet the mountains. It contains rivers, lakes, forests, and habitat for numerous species, including rare animals such grizzly bears, wolverines, and Athabasca rainbow and bull trout. Bighorn Park could easily transition into a world-class nature or ecotourism destination. It is located about three hours from international airports in the cities of Calgary and Edmonton (Figure 18.1).[2]

Like many similar landscapes in the west, the Bighorn area is a contested space. It includes the headwaters of the North Saskatchewan River that provides drinking water for more than a million people in the Edmonton region. According to the provincial government, the management intent of Bighorn Country was to protect the headwaters and watershed integrity, conserve and maintain biodiversity, recognise Indigenous peoples' rights and traditional uses, provide high-quality outdoor recreation opportunities, and support economic diversification and increase tourism opportunities (Alberta Government, 2018).

The management aim of Bighorn Country closely aligns with most definitions of ecotourism that emphasize environmental conservation, and social and economic benefit to local people. Where it differs is that ecotourism is to provide local people with control of the scale and benefit of the tourism development. Given that it was a top-down initiative led by the provincial government, it was never clear as to who would define and control the scale and benefit of any proposed ecotourism.

Alberta crown land has a long history of supporting the staple economies of timber and oil and gas extraction, along with the social role of supporting local residents' recreation. In essence, it supports a local way of life that includes well-paying resource jobs and forest-based recreation. The historic absence of strict regulation meant that forest recreation leaned toward activities the government and ENGOs deemed unnecessarily destructive to the environment. The most contentious activities included motorised recreation involving ATV, motorbike and snowmobile use, and random camping (Riley, 2019). The latter refers to camping that generally involves RV campers and motorized off-road vehicles parked along a random river or lake area that is not a park or campground. It is randomly situated and as such it does not contain garbage, sewage, or water services. Random campers may camp for a weekend or longer. Tension would ensue if the park proposal was seen to restrict the economic and recreational patterns of local residents. In other words, if the park would challenge the way of life of Bighorn residents.

Bighorn Park was to be set in different zones that included provincial parks, wildland parks, recreation areas and two types of land use zones. Provincial parks and wildland parks focus on conservation with varying degrees of allowable human recreation use. Provincial Recreation Areas tend to be smaller and essentially cater to overnight camping and local tourism opportunities. A Public Land Use zone allows for the management of natural resources along with motorized recreation, as does the Public Land Recreation Area. The distinction of what uses were to be permitted and not permitted should have framed given the potential for conflict. If many old uses were to be eliminated, than it would be reasonable to conclude that local residents would rebel against the Bighorn Country proposal.

Nature-based tourism could be developed in all of the designated areas. Pre-existing uses such as off-road ATV and snowmobile use would be permitted in all zones on designated trails. Commercial forestry, and oil and gas exploration would continue to be permitted in both classifications of Public Lands. New petroleum exploration would be allowed in public lands and in all park zones, but with no surface access. Grazing would be allowed in all zones. Hunting and equestrian use in all zones would be permitted, with limitations

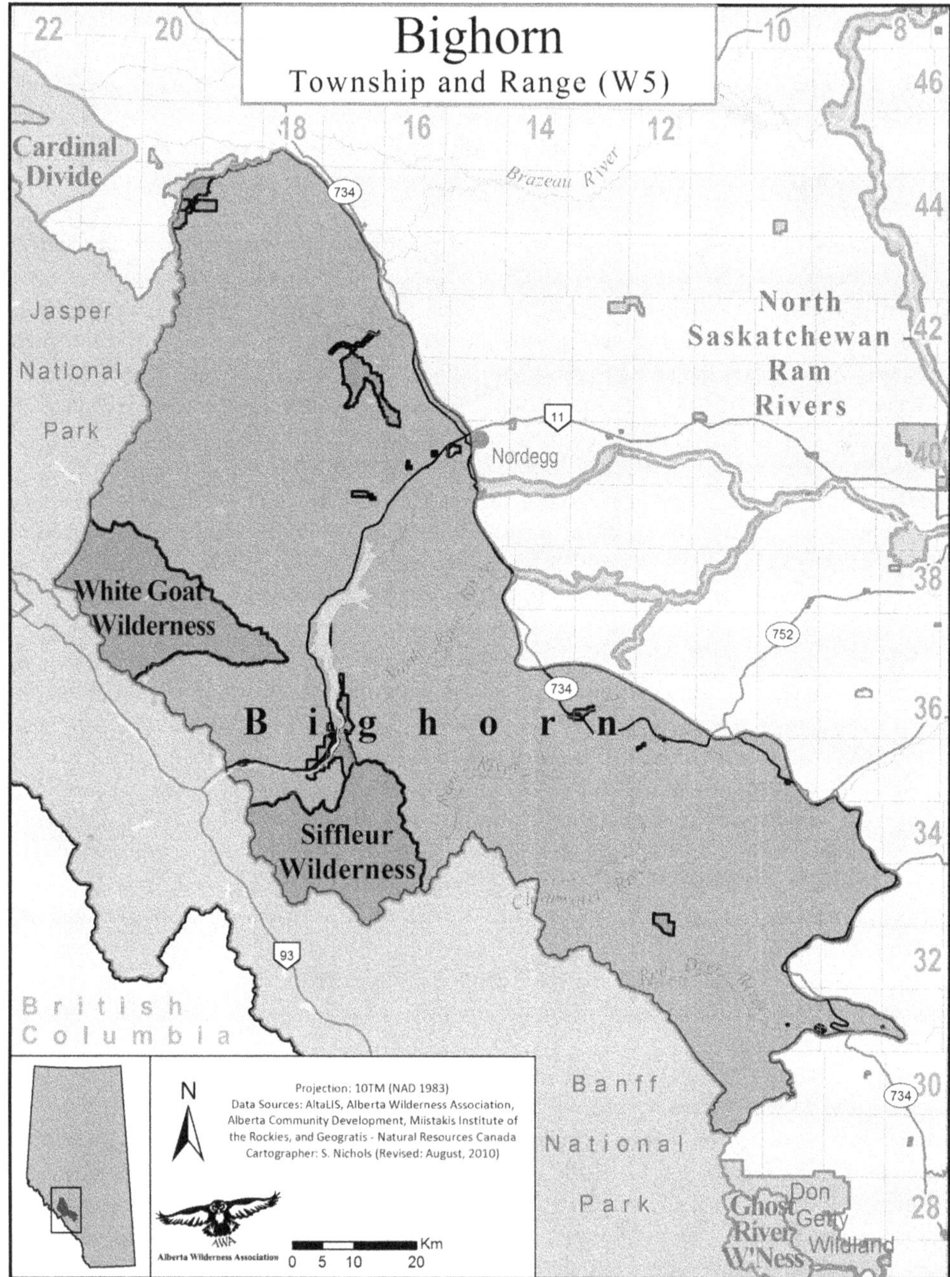

Figure 18.1 Bighorn area of concern; Courtesy of Alberta Wilderness Association

in provincial parks and recreation areas, as they are nearer to human use (Alberta Government, 2018).

Economic activity such as commercial forestry, coal and metallic mineral extraction, and oil and gas would be allowed in both land use zones. Existing oil and gas, free hold mineral, and sand and gravel extraction would be permitted to continue in all areas. Grazing and commercial trail riding would continue in all areas with permits. Recreational OHV activity and winter snowmobile activity would be permitted on designated trails in all zones. Auto access camping (or random camping) would be permitted all zones, but restricted to designated areas and prohibited in the wildland park zones. Hunting could continue in all areas with restrictions. Non-motorised recreation such as hiking, fishing, and climbing could occur in all areas, along with mountain biking and backcountry with permit. Overall economic activity would continue basically unchanged. Motorised recreation, including auto access camping would continue but restricted to designated trails and areas. Land uses would not be eliminated, but would be regulated. The park proposal included $40 million to develop visitor and tourism-based infrastructure (Alberta Government, 2018).

The geographic region in the proposed Bighorn Park is anchored by the communities of Rocky Mountain House, Nordegg north to Drayton Valley, and three First Nations groups in the region. Rocky Mountain House is the largest of the communities with a population of about 8000. The region is not homogenous, but in general it aligns with the Robson forestry community presented in Connell et al. (2017) and relies heavily on forestry and oil and gas extraction. Jobs are relatively well paying. The population is educated, but less so than the provincial average. Its main employers are the local government, forestry, transportation, and oil and gas companies. In 2014, the oil and gas industry throughout Alberta experienced a bust in its boom and bust cycling. It has not returned to pre-2014 levels. Global projections suggest it will not rise to a boom and it is now viewed as undergoing a structural shift, hence the imperative surrounding economic diversification.[1]

The park process was marred with controversy and tension. It is not feasible to provide a detailed account, but the summary is as follows. The park plan was introduced to the community as a proposal in 2018, while the controversial Castle Park plan process in the south of the province was still under way. From the start, many local residents rejected the park proposal. The *Globe and Mail*, a reputable national newspaper, quoted a longtime trapper as saying, "Nobody wants another Banff and Jasper ... referring to the national parks to the south and north, respectively. Tourism is great, but Albertans need a place to go." This was a consistent sentiment throughout the process. Residents felt the area belonged to them and tourism would disrupt that relationship. The government and ENGOs emphasized the conservation value of the proposed park including its value to the Y2Y habitat connectivity. The tone of the consultation changed by the end of 2018, when it was reported that a local opposition provincial Member of the Legislative Assembly (MLA) asserted that "the plan was a part of a foreign-funded plot to wall off the back country to Albertans who the region is their home." Many viewed this type of rhetoric as flaming tensions through unsubstantiated conspiracy theories, but others embraced the ideas (Lewis, 2019). The process deteriorated when accusations of intimidation toward park supporters and government officials were made public. Shannon Phillips, then Minister of Environment for the presiding government, cancelled four in-person public consultation sessions due to those concerns. How much intimidation actually occurred was debated in the media, but a member of the community who owns a small ecolodge and supported the plan, indicated she was the target of intimidation on several occasions, including to the point of having cars of men park in front of her rural house (Zapach, 2020). Shortly thereafter, the provincial government lost the general election and a new United Conservative Government was elected. Following the election, Jason Nixon, former opposition MLA for area who strongly opposed the park, became the new Minister for Parks, and announced that the Bighorn Park plan was dead and would not go forward. He cited numerous concerns with the plan, but inadequate consultation was viewed as the primary reason the park plan was halted (Deroworiz, 2019).

Discussion

In this section, I examine the Bighorn case study in light of the literature and then return to the concept of anti-ecotourism. The Bighorn park initiative existed as a document of the plan and what it represented to the community. The plan focused on environmental protection but was careful not to introduce significant changes to local recreation and resource access. It essentially wrote in regulations that people were already doing. To many locals, the Bighorn park proposal is a symbol of curtailed future liberties more than present. The aim of the pro-park group including the former government and ENGOs was in keeping with a globalist approach to environmental protection. This approach plainly acknowledges that environmental protection is the priority and nature-based tourism is the way to support it (Wolf, Croft, & Green, 2019).

The aim of the anti-park group made up of dominant community voices, was to maintain their existing way of life, with little interest in environmental protection as presented to them by the pro-park group (Legualt, 2020). Ethnocentric sensibilities can be observed in the approach of the pro-park group similar to modernisation theories' criticism that ecotourism may presume the ways of the community to be old and in need of modernising (Regmi & Walter, 2017). Media reported that local residents believed they were already good stewards of the land. Whether local people were or were not good steward's matters less than that they believed it to be true, and more so was the perception that they were being told by outside ENGOs how to properly care for their land. At a rally during the process, the director of the Alberta Off Highway Vehicle Association stated that people were not opposed to the preservation of the landscape, but to the process and how "you are being treated" ... "you guys are being vilified because you simply disagree and that's completely unacceptable in today's society" (Grant, 2019). Olive (2015) noted a similar situation whereby through their deep place attachment, local surfers believed themselves to be the best stewards of the coastal area. Pro- and anti-park initiatives represent divergent life politics and what Sorge (2008) noted in the Highlands of Sardinia as the battle between localist and cosmopolitan identities. Shannon Phillips, then environment minister for the government referred to the conflict as nothing less than a culture clash (Phillips, 2020).

Local concerns that parks, regulation, and a burgeoning ecotourism economy would curtail future freedoms and access reflect practical concerns of displacement. Residents worried that access to their livelihood resources and recreation areas would be compromised due to regulation and ecotourism would bring crowds both well understood forms of displacement (Carrier & MacLeod, 2005). They argued the consultation process was flawed and their views were not adequately heard (Lane & Corbett, 2005) and though hunting was not eliminated, it was mentioned as a threatened right in light of the association of a park and its protections (Melitis & Campbell, 2007). The 2018 Bighorn process was a top-down initiative originating from the government. This is generally discouraged in the literature in favour of a grassroots, bottom-up approach (Butcher, 2005). Or alternatively, in the literature of a Lazy Approach (Pavelka, 2020), whereby community projects are only initiated once considerable trust has developed between partners, and the community itself defines the intent and scope of the project. However, bottom-up community-based tourism projects may not be equitable with dominant local actors silencing less powerful local actors (Lane & Corbett, 2005). This was the case with Big Horn, as dominant local anti-park actors shut out locals who supported the plan.

Territoriality and localism are generally directed to tourists accessing the recreation resource (Beaumont & Brown, 2016). In the case of Bighorn, tourists had not yet arrived en masse, so it was directed to the proponents of the plan. It manifested in denigrating the plan for its content and process and the potential for future restrictions, as well as in the intimidation of members of the pro-park group (Legualt, 2020). Implied herein is that residents understood that if the area opened to ecotourism's sensibilities of quiet non-consumptive recreation it would not mix with the local preference for motorized vehicles and random camping recreation.

Economically, the Big Horn is a resource-based economy similar to Carson and Carson's (2011) description of the staple-based economy. It relies heavily on forestry and oil and gas extraction and thus far, generates enough local wealth to fend off a desperate or last resort shift to tourism as is common in other contexts (Brouder, 2012). This is a key factor that separates it from many other ecotourism contexts. Developing countries have little choice but to embrace any prospective development strategy that may lift it from poverty. However, during the planning process, the pro-park group commissioned an economic feasibility study to determine if a nature-based tourism economy could replace the existing resource economy as a mode of wealth generation. The studies' results showed that it could equal current economic output, but it would require a shift toward entrepreneurial employment, which is not likely in an entrenched staple-based economy (Legault, 2020; Carson & Carson, 2011).

Anti-ecotourism occurs when ecotourism is presented to a community but is rejected prior to implementation based on protecting a way of life viewed as incompatible with the tenets of ecotourism. The Bighorn park proposal demonstrates this phenomenon in three important ways. The first is that the pro-park group heavily emphasised environmental protection and the anti-park group interpreted that as restrictions on their way of life. Second, the Bighorn staple-based economy still allows residents adequate financial means to maintain their way of life thereby eliminating the last resort context for accepting ecotourism. Lastly, Bighorn represents a contemporary concern of growing divisions of urban and rural ways of life compounded by a lack of trust in institutions. The plan was introduced by a progressive government and regardless of merit, it was not supported or trusted. These are significant barriers for ecotourism to overcome.

Conclusion

The purpose of this chapter was to present and demonstrate the concept of anti-ecotourism, not to discourage ecotourism, but to support its thoughtful implementation. Rejection of tourism by a community is not a new idea. Carson and Carson (2011) have demonstrated that there are places in the American rural midwest which are not amenable to it. Schellhorn (2010) demonstrated that ecotourism did not fit with the Senaru people in Lombok, Indonesia, because it was fundamentally not compatible with local *adat* customs, despite being encouraged by various NGOs. The concept of anti-ecotourism should be used practically by ecotourism's proponents in government and civil society as a conceptual checklist prior to engaging.

Many years ago, I was invited to Fort Chipewyan, Alberta, to present workshops on ecotourism. Fort Chipewyan is a First Nations area in the heart of Alberta's Oil Sands region on Lake Athabasca. It is small and isolated and suffers from the environmental effects of being downriver of the Oil Sands, but it has also benefited economically from it. I was confused as to why they would want to develop ecotourism given their financial position. At a large public meeting, I asked that question and an elderly man responded by saying, "We want to tell our story." That makes sense; ecotourism is a good storyteller. Later, after working with other small resource-based, non-indigenous communities in Alberta also contemplating a shift toward tourism, I concluded that such communities would welcome tourism if it could achieve three aims; to support local quality of life foster a minor tourism economy akin to economic diversification, it must tell a community's story.

Notes

1 The mayor of Rocky Mountain House was contacted for an interview, but she did not respond.
2 The map is used with permission of Alberta Wilderness Association permission given March 12, 2020.

References

Alberta Government. (2018). Bighorn Country Proposal. Retrieved from https://open.alberta.ca/dataset/5535789d-4cdd-439d-b274-8988f082dab9/resource/e416bacd-e431-4821-9b8b-2e467c840188/download/bighorncountryproposal-nov2018.pdf.

Bascomb, B., & Taylor, M. (2008). Ecotourism and sustainability in a Q'eqchi' Maya Community, Guatemala. *FOCUS on Geography*, Winter.

Beaumont, E., & Brown, D. (2016). 'It's not something I am proud of but its … just I feel': Local surfer perspectives of localism. *Leisure Studies, 35*(3), 278–295.

Brehn, J. M., Eisenhauer, B. W., & Krannich, R. S. (2004). Dimensions of community attachment and their relationship to well-being in the amenity rich rural west. *Rural Sociology, 69*(3), 405–429.

Brouder, P. (2012). Tourism development against the odds: The tenacity of tourism in rural areas. *Tourism Planning & Development, 9*(4), 333–337.

Butcher, J. (2005). The moral authority of ecotourism: A critique. *Current Issues in Tourism, 8*(23), 114–124.

Butcher, J. (2006). The United Nations International Year of Ecotourism; A critical analysis of development implications. *Progress in Development Studies, 6*(2), 146–156.

Butcher, J. (2008). Ecotourism as life politics. *Journal of Sustainable Tourism, 16*(3).

Carrier, J., & MacLeod, D. (2005). Bursting the bubble: The socio-cultural context of ecotourism. *Royal Anthropological Institute, 11*, 315–334.

Carson, D. A., & Carson, D. B. (2011). Why tourism may not be everybody's business: The challenge of tradition in resource peripheries. *The Rangeland Journal, 33*, 373–383.

Cernea, M. M., & Schmidt-Soltau (2006). Poverty risks and national parks: Policy issue in conservation and resettlement. *World Development, 34*(10), 1818–1830.

Collins Dictionary. (2020). Way of Life. Retrieved from https://www.collinsdictionary.com/dictionary/english/way-of-life (accessed on 26.03.2020).

Connell, D. J., Hall, J., & Shultis, J. (2017). Ecotourism and forestry: A study of tension in a peripheral region of British Columbia Canada. *Journal of Ecotourism, 16*(2), 169–189.

Das, M., & Chatterjee, B. (2015). Ecotourism: A panacea or predicament? Tourism *Management Perspectives, 14*, 3–16.

Deroworiz, C. (2019). Calling it flawed, UCP scraps Bighorn parks proposal touted by NDP Canadian Press in Calgary Herald, https://calgaryherald.com/news/local-news/calling-it-flawed-ucp-scraps-bighorn-parks-proposal-touted-by-ndp

Fennell, D. (2008). *Ecotourism* (3rd ed., pp. 8–19). New York, NY: Routledge.

Gosnell, H., & Abrams, J. (2011). Amenity migration: Diverse conceptualizations of drivers, socio-economic dimensions, and emerging challenges. *GeoJournal, 76*, 303–322. doi:https://doi.org/10.1007/s10708-009-9295-4.

Grant, R. (January 31, 2019). Hundreds Rally Against Bighorn Park Proposal in Red Deer. *Red Deer Advocate*. Retrieved from https://www.ponokanews.com/news/hundreds-rally-against-bighorn-park-proposal-in-red-deer/.

Hunt, C., & Stronza, A. (2014). Stage-based tourism models and resident attitudes towards tourism in an emerging destination in the developing world. *Journal of Sustainable Tourism, 22*(2), 79–298.

Kyle, G., Bricker, K., Graffe, A., & Wickham (2004). An examination of recreationists' relationships with activities and settings. *Leisure Sciences, 26*, 123–142.

Lane, M. B., & Corbett, T. (2005). The tyranny of localism: Indigenous participation in community-based environmental management. *Journal of Environmental Policy and Planning*, 7(2), 141–159.

Legault, S. (2020. January 21). Telephone Personal Interview.
Lewis, J. (2019). On the Rockies' Edge, Frictions Form Over Alberta's Plan for New Provincial Park. *The Globe and Mail Environment Reporter*. Retrieved from https://www.theglobeandmail.com/canada/alberta/article-on-the-rockies-edge-frictions-form-over-albertas-plan-for-new/.
Melitis, Z., & Campbell, L. (2007). Call it consumption! Re-conceptualizing ecotourism as consumption and consumptive. *Geography Compass*, *1*(4), 850–870.
Merriam-Webster. (2020). Way of Life. Retrieved from https://www.merriam-webster.com/dictionary/way%20of%20life (accessed on 26.03.2020).
Miller, A. D., Vaske, J. J., Squires, J. R., & Olson, L. E. (2017). Does zoning winter recreationists reduce recreation conflict? *Environmental Management*, *59*, 50–67.
Olive, R. (2015). Surfing, localism, place-based pedagogies and ecological sensibilities in Australia. In *Routledge International Handbook of Outdoor Studies*.
Pavelka, J. (2020). The Deep Field School: A model to support sustained international service learning, teaching and scholarship. In E. Sengupta, B. Patrick, & M. Mandla (Eds.), *Humanizing higher education*. New York , NY: Emerald Publishing Editors.
Phillips, S. (2020). Personal Telephone Interview.
Pookhao, N. (2014). Community-Based Ecotourism: The Transformation of Local Community. *SHS Web of Conference* 12, 01033 EDP Sciences.
Proshansky, H. M. (1978). The city and self-identity.*Environment and Behavior,* 10(2), 147–169. doi: 10.1177/0013916578102002
Regmi, K., & Walter, P. (2017). Modernization theory, ecotourism policy and sustainable development for poor countries on the global South: Perspectives from Nepal. *International Journal of Sustainable Development & World Ecology*, *24*(1), 1–14.
Riley, S. (2019). It Can't Be a Free-for-All Anymore. *The Narwahl*. Retrieved from https://thenarwhal.ca/it-cant-be-a-free-for-all-anymore-the-battle-for-bighorn-country/.
Schellhorn, M. (2010). Development for whom? Social justice and the business of ecotourism. *Journal of Sustainable Tourism*, *18*(1), 115–135.
Schweinsberg, S. C., Wearing, S. L., & Darcy, S. (2012). Understanding communities' views of nature in rural industry renewal: The transition from forestry to nature-based tourism in Eden Australia. *Journal of Sustainable Tourism*, *20*(2), 195–213.
Sorge, A. (2008). Divergent visions: Localist and cosmopolitan identities in highland Sardinia. *Journal of Royal Anthropological Institute (N.S.)*, *14*, 808–824.
Stokowski, P. (2002). Languages of place and discourses of power: Constructing new senses of place. *Journal of Leisure Research*, *34*, 4.
TIES. (2015). What Is Ecotourism. Retrieved from https://ecotourism.org/what-is-ecotourism/ (accessed on 22.02.2020).
Usher, L. E., & Gomez, E. (2016). Surf localism in Costa Rica: Exploring territoriality among Costa Rican and foreign resident surfers. *Journal of Sport Tourism*, *20*(3/4), 195–216.
Walker, G. J., & Virden, R. J. (2005). Constraints to outdoor recreation. In E. L. Jackson (Ed.), *Constraints to leisure* (pp. 201–220). State College, PA: Venture Publishing Inc.
Weber, B. (2017). Alberta to expand Castle area parks, phase out off-highway use, Calgary Herald https://calgaryherald.com/news/local-news/alberta-to-expand-castle-area-parks-phase-out-off-highway-vehicles
Williams, D. R., & McIntyre, N. (2002). Where Heart and Home Reside: Changing Constructions of Place and Identity. Presentation at the 5th recreation and tourism trends symposium, East Lansing, MI, Department of Parks and Recreation and Tourism Resources, Michigan State University.
Wirth, L. (1938). Urbanism as a way of life. *American Journal of Sociology*, *44*(1), 1–24.
Wolf, I., Croft, D., & Green, R. (2019). Nature conservation and nature based tourism: Paradox? *Environments*, *6*, 104.
Zapach, M. (2020). Personal Telephone Interview.

19
SOCIALISATION

How it augments ecotourists' experiential satisfaction during ecotrips and after (in social media aided virtual settings)

Sudipta Kiran Sarkar

Introduction

Ecotourists, identified as the ones who travel to nature-based areas, both undisturbed and modified, and engage in experiences involving moderate to extreme levels of difficulty, constitute a major segment of green consumerism (Wearing, Cynn, Ponting & Mathew, 2002). Some of the principal motivations of ecotourists observed over the years have been nature-seeking, learning and education on environmental aspects, knowledge sharing, and last but not the least, socialisation (Hughes & Morrison-Saunders, 2005; Beaumont, 2011; Lee, Lee, & Lee, 2014). These largely distinctive motivations have separated ecotourists from mainstream tourists, and tourists related to consumption of other aspects of tourism. A wide range of previous literature focused on ecotourists' behaviour have suggested nature-seeking and environmental learning and education (and knowledge sharing) as the most dominant motivations of ecotourists. Though socialisation among ecotourists has been recognised as a major experiential and motivational aspect of ecotourists, it did not seem to generate much zeal for a wider academic discussion in the body of knowledge in ecotourism so far.

Socialisation, in the context of ecotourists, refers to engaging in shared experiences with likeminded ecotourists via social interactions, sharing past experiences, views and knowledge during ecotrips. Quality socialisation opportunities have been found to enhance the value and satisfaction ecotourists derive from trips to nature-based settings (Chan & Baum, 2007; Lu & Stepchenkova, 2012). Given the essentiality of socialisation for ecotourists, the possibilities for them to engage in it beyond real ecotourism contexts demands academic probing. Though largely absent in previous literature, a few studies have explored the possibilities of socialisation among ecotourists beyond real ecotourism settings particularly in the context of social media. These studies identified the causal influences of satisfying socialisation experiences for ecotourists upon knowledge sharing, a key behavioural trait, as well as the epistemological and ecological significance of socialisation for ecotourists, in social media contexts.

Online socialisation via social media can lead to the possibilities of collective awareness and actions or 'online citizenship' on ecological and responsible travel issues for ecotourists, supplementing beyond the sharing and exchange of nature-based travel knowledge. Such engagement via social media by ecotourists enhances the value in online communities, as well as

DOI: 10.4324/9781003001768-19

leads to collective knowledge building and to a new dimension of ecotourist behaviour as 'green consumers'. However, socialisation activities by social media platforms in the form of sharing of pictorial and audiovisual content have in many recent instances proliferated, and hence, led towards overwhelming influxes of tourists to sparsely populated and ecologically sensitive locations. Social networking platforms in social media—namely Instagram and Snapchat—where brief forms of online social interactions (short-length texts, short videos, and photos) are only possible, have merely contributed towards widespread popularisation of locations, without providing the necessary space for engaging in well-informed dialogue about responsible visitations.

Given this premise, this chapter will endeavour to engage in a conceptual exploration of socialisation as a major facet of ecotourists' behaviour, in both real and virtual (social media) settings, and its implications with relation to both theory and praxis.

Ecotourists' motivation, as we understand it

Ecotourists being 'green consumers' are primarily driven by desires to engage with nature-based experiential activities, actuated by their needs to learn about the ecological aspects and ways to sustainably bond with the wilderness of the sites they visit. The learning needs are largely realised by their eagerness to acquire knowledge about the geographical, the biodiversity in particular, and the anthropological elements they encounter as a part of their visitation experiences in nature-based settings (Beaumont, 2011; Wearing & Neil, 2009). Hence, the pedagogical elements involved in the consumption process become a vital aspect of ecotourists' gaze (Chan & Baum, 2007). Various studies have also indicated as learning through interpretation facilities at nature-based settings, as an imperative source of knowledge as well as a stimulus for ecotourists' to visit nature-based destinations (Beaumont, 2011).

Acquiring knowledge via interpretation not only encompass having insightful understanding of different elements of ecology in the natural surroundings, but also issues of conservation of biodiversity, green practices at ecotourism sites, and the compatibility of the ecotourism operations in terms of sustainability (Hughes & Morrison-Saunders, 2005; Okech, 2009). Interpretation comes in a range of forms—texts as signages, information boards being the most traditional ones; however, the role of talks by nature-tour guide on ecology as a mode of interpretation is a more interpersonal process involving engaging social exchanges (Hughes & Morrison-Saunders, 2005; Hill, Woodland, & Gough, 2007; Powell & Ham, 2008). Ecotourists who seek insightful knowledge on ecology may not be experts themselves in areas of biodiversity and environment, and hence, the extent and reliability of the knowledge they receive at ecotourism sites are vital for them.

Therefore, the imperativeness of knowledge acquired is central to the experiences of interpretation ecotourists go through. It is something that affects the cognitive process of ecotourists who endeavour to reflect on nature-based experiences at a post-trip stage, or the 'recollection' stage in which the experience of the visitation is recalled (Clawson & Knetsch, 1966). Interpretation that effectively disseminates insightful knowledge in various engaging forms enhances the satisfaction levels of ecotourists as they are able to retain such knowledge and provides them a fulfilling experience (Dodds & Joppe, 2001; Okech, 2009; Hill et al., 2007).

While seeking learning and pedagogical benefits of ecotourism brought about by interpretation is an important motivating factor for ecotourists, their desire to seek nature precedes it. Nature seeking desires of ecotourists are driven by their ecocentricism; need to reach out to the inherent spiritual value emanating from close human-nature exchanges (Cini, Leone, &

Passafaro, 2010). Such spiritual value can be enhanced by the recreational and therapeutic value derived from nature, particularly forests—sometimes referred to as 'forest bathing', alluring ecotourists towards nature-based settings (Cunningham, 2016). Besides these aesthetic experiential factors, ecotourists seek nature for their commitment towards environmental protection, and sustainable behaviour by means of deeper understanding of flora and fauna species, and making their consumption of nature compatible with such commitment and insightful understanding of the wilderness (Beaumont, 2011; Higham & Luck, 2002; Harlow & Pomfret, 2007; Lee & Moscardo, 2005; Zografos & Allcroft, 2007; Wearing & Neil, 2009). The degree of such commitment and understanding of ecological aspects in turn, demand a need for knowledge that creates further curiosity for ecotourists to learn about nature.

However, it is interesting to note, as pointed out by Lee et al. (2014) that knowledge attainment is brought about by ecotourists meeting and socialising with others who share similar interests about ecology, and certain aspects of ecotourism consumption. Ecotourists have been observed to associate with like-minded groups that are instrumental in building their personal identities as nature seekers via the means of the ecological knowledge and awareness they receive from such groups (Lee et al., 2014; Kim, Kim, Park, & Guo, 2008). The value derived from the process of association (and socialisation) taking place in these groups can lead towards ecotourists deciding to visit an ecotourism site or the nature of responsible ecotourism consumption they should engage in (Lee et al., 2014; Kim et al., 2008).

Socialisation: A less heeded experiential trait of ecotourists

Engaging in social interactions with peer ecotourists and loved ones during ecotrips has been an inextricable part of the experiential moments of ecotourists (Chan & Baum, 2007; Holden & Sparrowhawk, 2002, Kim et al., 2008). Ecotourists have been observed to have considerable levels of socialising intent during their ecotourism experiences in various studies. Trekkers to Annapurna, Nepal, were found to exercise significant levels of socialisation among their groups that brought them enhanced experiential satisfaction and better bonding among themselves (Holden & Sparrowhawk, 2002). Socialisation plays a catalytic role in bringing about a stronger peer influencing and social bonding among ecotourists in the process of their experiential interactions with their natural surroundings (Harlow & Pomfret, 2007). Such social bonding via socialisation can further become a driving factor towards ecotourist groups developing a shared commitment towards the surrounding natural living environment, resulting in a 'sense of community' (Harlow & Pomfret, 2007, p. 199). The social bondages among ecotourists also enriches their hedonic desires derived during their social exchanges of affective elements. While engaging with immersive wildlife experiences and nature-based adventure experiences, the thrill and excitement derived from it is often reflected in their social exchanges with each other (Chan & Baum, 2007). Ecotourist communities found in the forms like voluntary environmental organizations or birding communities accomplish their collective environmental and experiential goals by harnessing from the deep social bondages and shared hedonic contentment enabled by socialisation (Eubanks Jr., Stoll, & Ditton, 2004). Ecotourists belonging to communities engaging with specialized nature-based activities like birding have found socialising with fellow organisational members considerably instrumental in their intention to participate in such activities (MacFarlane, 1996). Socialisation in such communities entails acquiring existing and new knowledge about the specific nature-based recreational activity, and the establishment of distinct community identity via the sharing of "similar attitudes, beliefs, and ideologies", "similar behaviour", as well as sharing of "a sense of group identification" (MacFarlane, 1996, p. 37). It also establishes a set of vital community norms to adhere to for members of these communities; for which interactions via socialisation is further necessary to keep the structures of such community norms intact (MacFarlane, 1996). Therefore, for communities of ecotourists, engaging with a range of specialised nature-based activities, socialisation has been instrumental in bringing cooperative actions, community norms, and shared values as well as sense of kinship in turn, leading towards enhancing the levels of gratifications derived from nature-based experiences (Lee, Graefa, & Li, 2007; Moore, Scott & Moore, 2008; Kuentzel & Heberlein, 2006; Eubanks Jr. et al., 2004).

Socialisation particularly plays a much stronger role in the experiences of softer ecotourists. Socialisation opportunities in the form of social interactions with peer ecotourists, as well as with ecotourism staff and local communities at destinations, tends to bring increased levels of gratification to them from their ecotourism experiences (Cini et al., 2010; Kim et al., 2008). In the context of ecolodge experiences of ecotourists, socialisation with peers with identical trip interests, as well as with friends, and family members have been found to play a significant role in their selection of ecolodges as well as a major source of satisfaction enhancing the quality of their ecolodge experiences (Chan & Baum, 2007; Lu & Stepchenkova, 2012). Chan and Baum (2007) found socialising opportunities brought about by socio-psychological factors to ecotourists were highly instrumental in their visitation to the ecolodges in Sukau, East Malaysia. Similarly, ecotourists in their online social interactions about ecolodge experiences in Costa Rica were evidently indicative of socialisation as the principal gratifying factor that augmented the quality of their ecolodge experiences in that country (Lu & Stepchenkova, 2012).

Satisfying experiences of socialisation with peer ecotourists during ecotrips, can also enable ecotourists towards the fulfilment of their personal development desires (Harlow & Pomfret, 2007; Ryan, Hughes, & Chirgwin, 2000). The different facets of socialisation that an ecotourist goes through during an ecotourism experience; sharing and exchange of knowledge, ideas, emotions as well as the shared moments with peer ecotourists, friends, and family members, can be educational for them (Ryan et al., 2000). Personal development desires enabled by socialisation can lead towards gaining spiritual benefits from the natural environment, as well as shared sensitivity towards the environment, and collective attitudes towards conservation and sustainable behaviour (Harlow & Pomfret, 2007). It can, therefore, be elevating especially for certain sections of ecotourists who are driven by motives of having a scholarly desire for deeper understanding of nature during an ecotourism experience. Moreover, their intentions to engage more in similar ecotourism experiences are also brought by such personal development benefits (Galley & Clifton, 2004; Harlow & Pomfret, 2007).

Socialisation for ecotourists: The online context

Socialisation can be referred to as the collective exercise by like-minded individuals to associate with each other through social interactions for a shared purpose (Kesebir, Uttal, & Gardiner, 2010). In other words, it is "a dynamic and constructive process embedded in the practices of social interactions" (Kesebir et al., 2010, p. 97). According to Nonaka and Takeuchi (1995, p. 62), socialisation can be defined "as a process of sharing process experiences and thereby creating tacit knowledge such as shared mental models and technical skills". Socialisation tends to be more effective depending on the aptness of the settings that aid in the intensity of the exchange process (Kesebir et al., 2010).

Socialisation among ecotourists also entails the important dimension of knowledge sharing. This emanates from their learning and pedagogical needs in terms of the ecological aspects of the natural settings they visit. Socialisation among them is therefore, not merely about kinship and togetherness or sharing of hedonic moments during an ecotour, but also sharing of expert knowledge of different ecological aspects during their encounters with nature. This dimension therefore aids in the development of environmental values of ecotourists, as it encompasses the levels of their deeper understanding and admiration for the ecology as well as the sustainability issues relating to the nature-based settings they visit. Knowledge shared by ecotourists through their socialising activities during an ecotour further enhances the quality and their contentment of their shared experiences of nature-based areas (Lu & Stepchenkova, 2012; Sarkar, Au, & Law, 2017). Therefore, given the imperativeness of the socialisation process for ecotourists, its continuity during the post-trip stage is essential to understand ecotourists' shared behaviour.

Social media platforms which can provide dyadic or many-to-many online social interactions can become a medium of socialisation experiences for ecotourists before and after they engage with ecotourism activties (Sarkar et al., 2014, 2017). Social media platforms that include social networking sites (SNS), carry the potential in initiating dyadic socialisation between ecotourists, knowledge sharing and trip research (Charters, 2009). Based on User-generated content (UGC), SNS sites have the ability to provide satisfying socialisation and knowledge-sharing experiences to ecotourists (Lu & Stepchenkova, 2012). Though relatively sporadic, few empirical studies have inquired into the role of social media leading towards socialisation (and knowledge sharing) among ecotourists. Studies by Sarkar et al. (2014, 2017), examined the social exchange factors embedded in social media enabled social interactions among ecotourists in bringing rewarding socialisation experiences for them, while the study by Cheng, Wong, Wearing, and McDonald (2016) determined how social media enabled social interactions among ecotourists was instrumental in their trip motivations as well as in bringing about awareness towards sustainability and pro-environmental behaviour. The role of social exchange factors of cooperation, reputation, altruism, and trust reflected via the concept of Value in Online communities (VOC) were specifically examined Sarkar et al. (2017), as to how they had a significant causal effect on social media–enabled socialisation that in turn, led towards knowledge sharing intentions. Moreover, the studies by Sarkar et al. (2014, 2017) both confirmed that for ecotourists, both hard and soft, it is imperative for ecotourists to continue socialising and sharing knowledge online via social media and derive similar satisfaction as in the case of face-to-face socialisation during actual ecotrips.

While these studies ascertained the importance of online socialisation and knowledge sharing by ecotourists, Cheng et al. (2016) ascertained that building of collective knowledge about environmental awareness, and pro-sustainable behaviour and ecotourism consumption can be brought about by via social media–enabled socialisation among ecotourists. Such collective knowledge can enrich ecotourists' online experiences and motivate them towards visiting ecotourism sites (Cheng et al., 2016). The study was also indicative thatsocial media–enabled communication, dyadic and many-to-many, among ecotourists can lead towards effective ecotourism promotion in China (Cheng et al., 2016). In addition, Fennell (2017) explored the videos shared on video-sharing platform, Vimeo, by individuals engaging in the nature-based recreational activity of fishing. One of the key observation of this study was the tendency of fly anglers to share personal experiences online via Vimeo, and interact with a wider community of likeminded individuals having similar interests in fishing. Moreover, it was through this sharing and interactions, these individuals were able to establish mutually acceptable and unified view of how fly angling should be practised (Fennell, 2017).

A wider significance of socialisation via social media and ecotourists; ecological citizenship

The socialisation process that ecotourists engage with via social media involves a co-creative activity, which can lead towards collective consciousness and responsible behaviour towards environment. Such collective behaviour is not only mutually beneficial for them, but also results in collective online actions pertaining to the welfare of the ecological elements, and sustainable practices. One such collective online action is found in the form of ecological citizenship. Ecological citizenship (or sustainable citizenship) entails an intense dialogical process relating to pro-environmental issues carried out by ecologically conscious individuals via online social networking platforms, resulting in actual collective actions and practices necessary for the welfare of the environment (Rokka & Moisander, 2009). These ecologically conscious individuals (who can also be nature-based travelers) belong to online communities in social media that are responsible in engaging in collaborative and dialectical online conversation about environmental sustainability (Rokka & Moisander, 2009). Such conversations also encompass issues like responsible tourism consumption and environmentally compatible ecotourism practices, resulting in enhanced commitment towards contributing to sustainable wellbeing of local communities and ecosystems (Rokka & Moisander, 2009). Hence, such conversations via social media becomes central to the establishment of ecological citizenship. Social media–enabled ecological citizenship can have a major influence on ecotourists as green consumers and the ethical and responsible behaviour that comes with it (Budeanu, 2007). Furthermore, ecological citizenship also demands altruism among ecotourists to jointly articulate the importance of ecological responsiveness in social media platforms, in the forms of audiovisual, textual and pictorial content, (Rokka & Moisander, 2009).

Therefore, the environmental citizenship evident among eco-conscious travelers via online socialisation, is clearly indicative of its consistency with the ecotourists' motivations of learning, nature seeking, and engaging in socialisation during actual ecotrips. The ecological awareness and advocacy-related initiatives of ecotourists evident in these motivations are clearly reflected in their online ecological citizenship. The pro-environmental behaviour and desire for responsible ecotourism consumption that underlie as factors in their socialisation process during actual ecotourism experiences is markedly reflected in their equally ecologically conscious online socialisation process. Therefore, this could be indicative of the fact that the gratifying outcome of online socialisation of ecotourists could be compatible with the outcomes of satisfying socialisation experienced face-to-face with peer groups, friends, and family in an actual ecotour.

Conclusion

Socialisation, both as a construct and a critical experiential component, in ecotourists' consumption process now appears more protruded as demonstrated by this study. Given this distinctiveness it carries, the academic discourse on socialisation as a construct in ecotourism is conspicuously absent in the recent literature. In particular, and as evident in the discussion in this study, the pursuit of online socialisation by ecotourists' enabled by social media, and the collective and environmentally constructive actions it can produce, has received very little attention. Therefore, it puts forward a scholarly duty on future pundits researching ecotourists to ascertain firstly, the underlying reasons for the presence of only few studies on online socialisation given its imperativeness in a face-to-face context. Secondly, future studies need to probe into the wider evidence of socialisation in social media in newer platforms like Instagram, and the resulting collective actions; pro-environmental online behaviour and advocacy, as well as the shared recreational benefits they derive from it. Finally, future scholars may also examine whether online socialisation activities and the resulting collective actions translate into actions in the real world, with relation to responsible consumption of ecotourism and environmental advocacy. Furthermore, it would also be interesting to observe the consistencies and inconsistencies between online socialisation and face-to-face socialisation of ecotourists and the satisfaction derived from them.

Finally, scholars and practitioners, DMOs, and local communities, have to be mindful of online socialisation via popular social networking platforms like Facebook, Instagram, and YouTube leading towards undesirable outcomes. The boundless, expansive coverage of social media and the resulting unprecedented popularity of its usage by tourists are making destinations, particularly ecotourism sites, increasingly experience overcrowding, visual (or aesthetic) pollution, and irresponsible digital behaviour (i.e., selfie-taking at nature-based sites considered sacred by local communities, and sharing of such selfie photos on social media platforms resulting in disrespectful behaviour). Such situations would pose major challenges for stakeholders involved in responsible ecotourism, and therefore, online socialisation initiated by both, hard and soft ecotourists, needs to be driven by their core motivations of learning, knowledge sharing, and nature-seeking.

References

Beaumont, N. (2011). The third criterion of ecotourism: Are ecotourists more concerned about sustainability than other tourists? *Journal of Ecotourism, 10*(2), 135–148.

Budeanu, A. (2007) Sustainable tourist behavior - a discussion of opportunities for change. *International Journal of Consumer Studies, 31*(5), 315–326.

Chan, J. K. L., & Baum, T. (2007). Ecotourists' perception of ecotourism experience in Lower Kinabatangan, Sabah, Malaysia. *Journal of Sustainable Tourism, 15*(5), 574–590.

Charters, T. (2009) Climate change response- back to the future? *World Ecotourism Conference 2009*, July 15th–17th, 2009, Vientiane, Laos PDR.

Cheng, M., Wong, I. A., Wearing, S., & McDonald, M. (2016). Ecotourism social media initiatives in China. *Journal of Sustainable Tourism, 25*(3), 416–432.

Cini, F., Leone, L., & Passafaro, P. (2010) Promoting ecotourism among young people: A segmentation strategy. *Environment and Behaviour, 44*(1), 87–106.

Clawson, M., & Knetsch, J. L. (1966). *Economics of Outdoor Recreation*. Baltimore: John Hopkins Press.

Cunningham, E. J. (2016). (Re) creating forest natures: Assemblage and political ecologies of ecotourism in Japan's central highlands. In M. Mostafanezhad, R. Norum, E. Shelton, & A. Thompson-Carr (Eds.), *Political ecology of tourism: Community, power and the environment* (pp. 169–187). London: Routledge.

Dodds, R., & Joppe, M. (2001). Promoting urban green tourism; the development of the other map of Toronto. *Journal of Vacation Marketing*, 7(3), 167–261.

Eubanks Jr., T. L., Stoll, J. R., & Ditton, R. B. (2004). Understanding the diversity of eight-birder subpopulations: Socio-demographic characteristics, motivations, expenditures and net benefits. *Journal of Ecotourism, 3*(3), 151–172.

Fennell, D. (2017). Fly-fishing on Vimeo: Cas(t)ing the channel. *Humans Dimensions of Wildlife; An International Journal, 22*(1), 18–29.

Galley, G., & Clifton, J. (2004). The motivational and demographic characteristics about research ecotourists: Operation Wallacea volunteers in Southeast Sulawesi, Indonesia. *Journal of Ecotourism, 3*(1), 69–82.

Harlow, S., & Pomfret, G. (2007). Evolving environmental tourism experiences in Zambia. *Journal of Ecotourism, 6*(3), 184–209.
Higham, J., & Luck, M. (2002) Urban ecotourism: A contradiction in terms? *Journal of Ecotourism, 1*(1), 36–51.
Hill, J., Woodland, W., & Gough, G. (2007) Can visitor satisfaction and knowledge about tropical rainforests be enhanced through biodiversity interpretation and does this promote a positive attitude towards ecosystem conservation? *Journal of Ecotourism, 1*(6), 75–85.
Holden, A., & Sparrowhawk, J. (2002). Understanding the motivations of ecotourists: The case of trekkers in Annapurna, Nepal. *International Journal of Tourism Research, 4*, 435–446.
Hughes, M., & Morrison-Saunders, A. (2005). Influence of on-site interpretation intensity on visitors to natural areas. *Journal of Ecotourism, 4*(3), 161–177.
Kesebir, S., Uttal, D. H., & Gardiner, W. (2010). Socialization: Insights from social cognition. *Social and Personal Psychology Compass, 4*(2), 93–106.
Kim, S. S., Kim, M., Park, J., & Guo, Y. (2008). Cave tourism: Tourists characteristics, motivations to visit, and the segmentation of their behavior. *Asia Pacific Journal of Tourism Research, 13*(3), 299–318.
Kuentzel, W., & Heberlein, T. A. (2006). From novice to expert? A panel study of specialization progression and change. *Journal of Leisure Research, 38*, 496–512.
Lee, S., Graefe, A., & Li, C (2007). The effects of specialization and gender on motivations and preferences for site attributes in paddling. *Leisure Sciences, 29*(4), 355–375.
Lee, S., Lee, S., & Lee, G. (2014). Ecotourists' motivation and revisit intention: A case study of restored ecological parks in South Korea. *Asia Pacific Journal of Tourism Research, 19*(11), 1327–1344.
Lee, W. H., & Moscardo, G. (2005). Understanding the impact of ecotourism resort experiences on tourists' environmental attitudes and behavioral intentions. *Journal of Sustainable Tourism, 13*(6), 546–565.
Lu, W., & Stepchenkova, S. (2012). Ecotourism experiences reported online, classification of satisfaction attributes, *Tourism Management, 33*(3), 702–712.
MacFarlane, B. L. (1996). Socialization influences of specialization among birdwatchers. *Human Dimensions of Wildlife, 1*(1), 35–50.
Moore, R. L., Scott, D., & Moore, A. (2008). Gender-based differences in birdwatchers' participation and commitment. *Human Dimensions of Wildlife, 13*, 89–101.
Nonaka, I., & Takeuchi, H. (1995). *The Knowledge-Creating Company: How Japanese Companies Create the Dynamics of Innovation*. UK: Oxford University Press.
Okech, R. N. (2009). Developing urban ecotourism in Kenyan cities: A sustainable approach, *Journal of Ecology and Natural Environment, 1*(1), 1–6.
Powell, R. B., & Ham, S. H. (2008). Can ecotourism interpretation really lead to pro-conservation knowledge, attitudes and behaviour? Evidence from the Galapagos island. *Journal of Sustainable Tourism, 16*(4), 467–489.
Rokka, J., & Moisander, J. (2009). Environmental dialogue in online communities: Negotiating ecological citizenship among travellers. *International Journal of Consumer Studies, 33*(2), 199–205.
Ryan, C., Hughes, K., & Chirgwin, S. (2000). The gaze, spectacle and ecotourism. *Annals of Tourism Research, 27*(1), 148–163.
Sarkar, S. K., Au, N., & Law, R. (2014). Analyzing ecotourists' satisfaction in socialization and knowledge sharing intentions via social media. In Z. Xiang, & I. Tussyadiah (Eds.), *Information and communication technologies in tourism 2014* (pp. 313–326). Switzerland: Springer International Publishing.
Sarkar, S. K., Au, N., & Law, R. (2017). Analyzing the effect of value in online communities on satisfaction in online socialization and knowledge-sharing intentions of eco – tourists. In P. O. de Pablos (Ed.), *Managerial strategies and solutions for business success in Asia*. Pennsylvania:IGI Global.
Wearing, S., Cynn, S., Ponting, J., & McDonald, M. (2002) Converting environmental concern into ecotourism purchases: A qualitative evaluation of international backpackers in Australia. *Journal of Ecotourism, 1*(2/3), 133–148.
Wearing, S., & Neil, J. (2009). *Ecotourism; Impacts, Potentials and Possibilities* (2nd ed.). Oxford, UK: Elsevier.
Zografos, C., & Allcroft, D. (2007). The environmental values of potential Ecotourists: A segmentation study. *Journal of Sustainable Tourism, 15*(1), 44–66.

20

VIETNAMESE ECOTOURISTS

Ecotourists from an unconventional market

Huong H. Do, David Weaver, and Laura Lawton

Introduction

Ecotourism is one niche segment of the larger tourism sector. As such it conforms to all core characteristics of tourism that Goeldner and Richie (2003, p. 5) define "as the processes, activities, and outcomes arising from the relationships and the interactions among tourists, tourism suppliers, host communities, and surroundings environments that are involved in the attracting and hosting of visitors". Therefore, ecotourists travel some threshold distance to destinations in order to find multiple psychological and physical experiences, employ tourism services, and seek satisfaction. However, this particular tourist segment involves the complex circumstances of relatively fragile environments and needs to be considered differently from the popular commercial tourism destinations such as theme parks and casinos.

In 1987, Laarman and Durst coined the concept of the hard-soft dimension of ecotourism by differentiating between dedicated or casual interest respectively and the physical rigor of the experience (Laarman & Durst, 1987). By examining national park visitors' behaviours, Weaver and Lawton employed a typology technique to prove and expand this theory. They found three groups of harder, softer, and structured ecotourists, in which a newly found structured ecotourists combine characteristics of both harder and softer ecotourists in the same trips (Weaver & Lawton, 2002).

Understanding about behaviours of ecotourists from conventional markets, particularly harder and softer ecotourists is well developed. In terms of ecotourism sites, for harder ecotourists, the destinations are wilderness or otherwise relatively undisturbed settings (Acott, La Trobe, & Howard, 1998; Valentine, 1993) as well as the efforts exerted to access those remote areas. In contrast, softer ecotourists prefer more infrastructure and services to make their trip more comfortable and less risky. Softer ecotourists (or 'occasional ecotourists') sometimes combine their ecotourism trip with other leisure purposes (Laarman & Durst, 1987; Wallace, 1993; Weiler & Richins, 1995). In addition, preference for big city attractions was used as a segmentation variable as it can differentiate between ecotourists and non-ecotourists (Eagles, 1992).

In addition, research on learning and education purposes displays consistent findings about harder ecotourists (Weaver, 2002b). Learning new things about nature and wildlife occurs on-site or before the trip through prepared documents and reading (Lindberg, 1991).

DOI: 10.4324/9781003001768-20

However, with softer ecotourists, learning activities are more passive as expressed by greater reliance on interpretation services (Meric & Hunt, 1998), the presence of interpretation facilities (Blamey & Hatch, 1998), and seeing wildlife (Lindberg, 1991; Ryan, Hughes, & Chirgwin, 2000).

The third ecotourism criterion comprised efforts to minimise negative impacts (or 'footprints') in the destination and improve conservation outcomes. Harder ecotourists require only basic accommodation and services (Laarman & Durst, 1987) and are also more likely to want to leave the destination in a better condition than when they arrive (Acott et al., 1998; Diamantis, 1999). Harder ecotourists make efforts to influence other people to not have a negative impact on the site. They are also more concerned with being ethical visitors (Wight, 1993). The softer negative impacts of ecotourists are well described in the literature (Duff, 1993; McClung, Seddon, Massaro, & Setiawan, 2004; Müllner, Eduard, & Wikelski, 2004; Orams, 2002). Approaching wildlife is one aspect in this dimension.

This discussion of the hard-to-soft ecotourism spectrum is important because of its variable implications for impacts, settings, and behaviour. If for example one adheres to a hard perspective, then ecotourism constitutes perhaps 0.1% or less of global tourism activity and as such is essentially negligible despite the wider spatial diffusion, and higher per capita income and expenditure of participants. Conversely, the generous parameters of soft ecotourism, with its masses of short-term, marginally engaged protected area visitors, potentially implicate 15% or even 20% of global tourism, representing a major segment of the industry and a powerful economic force. The previous discussion reflects a Eurocentric construct of ecotourism that associates with demand in Western countries—popularly known as conventional ecotourist market, and only marginally engages emergent unconventional ecotourist market share in that appear to emulate this Western model. Given the size of the Asian population and its expected dominance of future tourism growth, any discourse on ecotourism identities and trends is grossly incomplete without a more systematic consideration of emerging Eastern 'ecotourist' markets.

Ecotourists from unconventional markets

Ecotourists are mostly categorised by region, sub-region, and country. The nineties of the last century saw the popularisation of ecotourism among conventional markets such as Australia, North America (United States of America, Canada), and Europe (United Kingdom, Germany, Scandinavia) (Hall & Weiler, 1992; Wearing & Neil, 1999).

Notwithstanding the continuing dominance of Western sources, there is increasing evidence of ecotourist-type markets appearing in non-conventional source regions such as East Asia and Latin America, as indicated previously. Cochrane (2006) reported an emerging market from East Asia that includes Japan, Korea, Taiwan, China, and Hong Kong while Weaver (2002a) defined the emerging market of Asia as including East Asia and South East Asian countries such as Thailand, Malaysia, and Indonesia. Research in a Thai national park shows a 91:9 ratio of Thai residents to foreign visitors (Hvenegaard & Dearden, 1998) and supported patterns observed in Indonesian protected areas (Cochrane, 2006). Eagles and Higgins (1998) insisted that Japan is a rapidly developing market and may become a dominant player in the near future. It should be noted that each year, there are over 300 million visitors to Japan's national parks system (Suntherland & Britton, 1980) compared with the approximately 60 million annual visitors to the United States' much vaunted nature-based national parks system (National Park Service, 2000–2009). Since the number of inbound visitors to Japan was less than five million in 2000, it is presumed that the vast majority of these national park visitors are domestic and probably soft ecotourists or an East Asian variant of same. This is in part because the Japanese national park system accommodates landscapes that are much more modified by human activity than their U.S. counterparts, as per Categories IV to VI of the IUCN typology (Weaver, 2002a) and the Korean research of Lee, Lawton, and Weaver (2013).

Statistics found in other Asian countries also support the emerging trends of domestic ecotourism in Asia, such as the 20 million visitors to Korean national parks in 1999 (Kim, Lee, & Klenosky, 2003) and the 11.5 million visitors to Thailand's national parks in 1994 (Hvenegaard & Dearden, 1998). Huge flows of domestic park visitation in Asia are argued to indicate 'mass ecotourism' if identified against the core criteria (Weaver, 2002a). It is also possible for this large-scale ecotourism to contribute even more positively to the sustainable development in protected areas in Asia if appropriate visitor education and stricter environmental codes of conduct are implemented (Lück, 2002). Presumably, ecotourists, or at least a type of tourist that incorporates ecotourism-related characteristics, appear in other parts of the world but are not yet adequately captured in the English literature.

In the East, the concept of ecotourism was introduced in China in the mid- to late-1990s (Wen & Ximing, 2008). From their surveys of visitors to national parks, bio-reserve areas, and ecologically scenic spots, two prevalent types of ecotourists have been described. They are 'elite' and 'mass' in which the former includes a small number of international hard ecotourists and local highly educated elites, and the latter contains a much larger number of general and mostly domestic tourists interested in nature (Wen & Ximing, 2008). The respective similarities with hard and soft ecotourists are clear, notwithstanding the inclusion of a Western visitor component. Many of the Chinese elite have absorbed Western precepts of environmental education and roughly emulate the Western consciousness of ecotourism. Among them, students from Yunnan University, China exemplify these criteria of 'quasi-hard' ecotourists (Wen & Ximing, 2008) citing (Xue, 2005). The majority of domestic ecotourists (over 90%), however, are positioned as mass or soft ecotourists, according to evidence from Mount Taibai National Forest Park, Shanxi, China (Zhang & Zhao, 2005; cited in Wen & Ximing, 2008). There is no empirical evidence in this literature of structured ecotourists or their equivalent.

Evidence from other Asian countries is almost non-existent in the English language literature, and therefore not accessible to the researcher. So it is a question of whether the Chinese model of soft-hard ecotourists is applicable to other Asian countries because of the common influence of Confucianism or not applicable because of variations among Asian cultures and economic backgrounds. Responding to this shortage, Lee et al. (2013) showed that ecotourism in Korea can be superficially considered as soft ecotourism after taking into consideration the three core criteria. Through a review of 206 articles about ecotourism and field trips to four protected areas in Korea, the research also gave evidence as to how culture and human beings together constitute 'nature'. Accordingly, it is highly acceptable for 'nature-based' protected areas to accommodate monuments, gardens, and other highly visible manifestations of material human culture. Since there is no study about Vietnamese ecotourists undertaken to date, this chapter will examine the portrait of Vietnamese domestic ecotourists in Cat Tien National Park according to the core ecotourim criteria.

Cat Tien National Park

Cat Tien, a RAMSAR recognised national park, was selected as a site for this study. CTNP has an undeniable advantage as an accessible ecotourism destination compared with other national parks in Viet Nam due to its desirable location. The park is 150 kilometres by highway from Ho Chi Minh City (three hours driving) that is the economic and foreign trade capital of Viet Nam and home to about 9 million residents. The city airport, Tan Son Nhat airport, is the largest international flight hub in Viet Nam and makes the park easily accessible to international visitors (see Figure 20.1). However, the park is remote enough from Ho Chi Minh City that it does not function merely as a weekend sightseeing site for mass tourists from that city that has a serious shortage of leisure parks and green spaces. Located in the East of the Mekong Delta, CTNP is also on the Highway to Da Lat, a famous tourist town in the Highlands (four hours from the park), and is connected to Mui Ne's beach resorts (three hours). Thanks primarily to CTNP's location in the ecotone between two major geophysical regions, the fauna and flora of wet tropical forests and wetlands are both represented, giving rise to a high level of biodiversity.

Figure 20.1 Map of location of CTNP (map courtesy of Bui Huu Manh, 2021)

Flora—There are 1610 species, 724 genera, 162 families, and 75 vegetation orders in CTNP. Thirty-eight species from 13 families are listed in the Viet Nam Red Book that lists rare and endangered species of fauna and flora native to Viet Nam that need to be protected, recovered, and developed. These include red wood (*Afzelia xylocarpa*), rose wood (*Dalbergia bariensis*), narra padauk (*Pterocarpus macrocarpus*), and sindoer sepertir (*Sindora siamensis*). In addition, 22 species from 12 families are endemic local plants such as *Telectadium dongnaiensis, Telectadium edule*, and species from the Asclepiadaceae, Orchiadaceae, Moraceae, and Anacardiaceae families (Vietnam Forest Creatures, 2012).

Fauna – CTNP is one of the most important sites in Viet Nam for the conservation of large mammals. Many species are not only found in the Viet Nam Red Book, but also in the IUCN Red List of threatened species. Among the confirmed large mammal species are the Asian elephant (*Elephas maximus*), Indochinese tiger (*Panthera tigris corbetti*), Sun bear (*Ursus malayanus*), Eurasian wild pig (*Sus scrofa*), Sambar deer (*Cervus unicolor*), and Gaur (*Bos gaurus*). The latter three species reportedly occur at high densities relative to other areas in Viet Nam (Cat Tien National Park, 2010). The park was well known for its Javan rhino (*Rhinoceros sondaicus annamiticus*, also known as the lesser one-horned rhinoceros), which sadly became extinct in 2012.

Market segmentation study

The quantitative study surveyed visitors to Cat Tien National Park (CTNP) in order to extract ecotourists from nature-based tourists by motivations, preferred activities, behaviours, environmental, and socio-economic attitude. Twenty-two behaviour, 11 attitudinal, and 20 motivation scale items from the ecotourism literature resulted from the thorough review of literature about Western and Asian ecotourists upon which this selection was made examined the three core criteria of ecotourism, the distinction between harder and softer ecotourists.

Using random stratified sampling method, before implementing check-out procedure, the CNTP receptionists asked every second visitor whether they were willing to participate in the survey and then collected completed questionnaires from the visitors. A total of 1532 paper based questionnaires were distributed to all visitors who satisfied the participant criteria when they checked out of the park during the six-month dry season in 2010–2011. Of these, 1267 visitors returned the questionnaires, resulting in a response rate of 82.7%. After the treatment of missing data, 1082 questionnaires remained (46.2% Vietnamese, 53.8% Western). Ward's hierarchical clustering method and one-way ANOVA were utilised to cluster visitors into distinctive groups and t-test helped to compare the differences between Vietnamese and Western visitors in each group (Hair, Black, Babin, & Anderson, 2010).

The cluster analysis identified six groups of visitors: Classic Western ecotourists (n = 171), Service shunners (n = 89), Service seekers (n = 92), Typical CTNP visitors (n = 283), Sociable wildlife engagers (n = 208), and Unenthusiastic visitors (n = 239). The proportion of Vietnamese and Western visitors in each cluster was calculated in which three clusters tend to have a majority of Western visitors, i.e., 'Classic Western ecotourists' (91%), 'Service seekers' (90%), and 'Service shunners' (90%). On the other hand, Vietnamese visitors accounted for a large proportion in two clusters 'Unenthusiastic visitors' (64%) and 'Sociable wildlife engagers' (84%). Among them, 'Typical visitors' had a relatively balanced proportion of visitors with 49% Vietnamese and 50% Westerners.

All six visitor groups conform to the core expectations of the ideal ecotourism type as articulated by Blamey (1997, 2001), i.e., preference of being close to wild nature, commitment to a certain level of learning about nature, awareness of environmental protection, and striving to be responsible travellers in tandem at least with the neutral option of status quo sustainability.

Crucially, the commitment level to each criterion varies across and within the six groups, thereby corroborating the theoretical and empirical literature. In order to understand how rigorous the domestic ecotourism is, the three groups that comprise majority of Vietnamese visitors in the sample (93.4%) are examined according to the soft-hard dimension.

Unenthusiastic visitors = 'Softer ecotourists'

The 'unenthusiastic visitors' group that resembles 'soft ecotourists' in all aspects is suggestive of a soft ecotourism experience. It is noteworthy that even when visiting a national park, they only marginally view national parks as wilderness settings (M = 3.66). This is because their experiences are limited to the headquarters area and do not involve deeper forest environments. Although visiting a national park, they have had little contact with 'relatively undisturbed natural areas' (Ceballos-Lascuráin, 1996). In terms of sustainability, their visits adhere to the steady-state position by being responsible visitors (M = 3.80) and being ambivalent about eating wild animal meat (M = 2.97) at the same time. Moreover, these visitors have a correspondingly moderate level of pro-environmental awareness.

Typical visitors and sociable wildlife engagers = 'Structured ecotourists'

The group known as 'typical visitors' is a balanced mix of the Vietnamese and the Western visitors who tend to rate relatively lower on the scale in the category of anthropocentrism. For example, they travelled in medium-sized groups of 10 people on average. To some extent, this group might be regarded as 'structured' as per Weaver and Lawton (2002) that is, they are as likely as harder ecotourists to be physically active in order to see unique wildlife (M = 3.97). However, failure to see wildlife is not a problem for them (M = 3.29). Striving for sustainability is observed in this group through their lower desire for the park to have comfortable facilities and services (M = 2.76), and their intention to leave the destination in a better condition than when they arrive (M = 3.96). However, they resemble softer ecotourists in their motivation to learn about nature through the interpretation materials (M = 4.01) and tour guides (M = 4.10). It is noteworthy that their enhancing of sustainability corresponds with a moderate level of environmental awareness, rather than with the expected high awareness. Perhaps the sustainable behaviours that they practise on-site are contextual and as a consequence of the on-site environmental education program.

Almost all the Vietnamese visitors from the group 'sociable wildlife engagers' show an extremely high motivation to learn about nature (M = 4.35) that is similar to the harder ecotourists, though with a concomitantly high preference for learning through interpretation (M = 4.20). In terms of concerns about sustainability, these visitors (M = 3.68) expect facilities and services at approximately the same level as the 'service seekers' (M = 3.85) and the 'classic Western ecotourists' (M = 3.62). As with the 'typical CTNP visitors', they have a strong awareness for the need for environmental protection and tend to try to enhance sustainability through their visits (M = 4.44). To some extent, aspects of their visits comply with "non-damaging, non-degrading, ecologically sustainable" motivations (Valentine, 1993, p. 108).

Perhaps the main characteristic of this group is their especially high anthropocentric tendency. They prefer to be in a large group, a preference that is in alignment with previous studies showing that this choice may originate from the ecotourist's alleged need for socialising (Ballantyne, Packer, & Sutherland, 2011), entertaining (Singh, Slotkin, & Vamosi, 2007), and coping with the uncertainty of being in a wilderness setting. Even though they are in large groups, they manage to keep a certain distance from other group members inside the trekking trails in order to enjoy and learn about the environment. Their high interest in touching beautiful plants/flowers (M = 4.08) and cute wild animals (M = 3.85) is different from findings from Kerstetter, Hou, and Lin (2004) about 'touching the fauna or flora in order to have fun' by Taiwanese ecotourists (M = 2.96).

The use of the theory of harder-structured-softer typology of ecotourists (Weaver & Lawton, 2002) assists in identifying Vietnamese softer and structutured ecotourists. Implications suggest suitable ecotourism management and products for domestic ecotourist markets and effective interpretation strategies for enhancing learning/education.

Qualitative study

The qualitative study comprises two stages of total 24 in-depth face-to-face interviews. Stage one (I) comprises 14 intervews in a field trip to the CTNP in early June 2010. Fourteen Vietnamese informants were selected by the researcher for their representation of harder as well as softer ecotourism proclivities, agreed to participate in in-depth interviews. The respondents were simultaneously and invited along the trekking trails or in the night wildlife watching tour. They were questioned about their motivations, behaviours, attitude, and experiences during their trips in CTNP and their overall view towards nature and conservation. Three visitors were interviewed just after their trip in the park and the other 11 were approached in Ho Chi Minh City not later than one month after the trip.

From the quantitative study, Principal Component Analysis also revealed the factor 'anthropocentrism' (α = .714) which entails preferences for touching wild animals and beautiful flowers and plants; to be in a larger group of four persons or more; avoiding wilderness areas because of safety concerns; and big city attractions over national parks. Among six visitors groups, three groups comprising majority of Vietnamese visitors in the sample (93.4%), one group resembles 'soft ecotourists' and two groups as 'structured ecotourists'. The other three majority Western visitor groups pertain to harder, structure, and softer ecotourists. It is noteworthy that the three majority Vietnamese groups have higher 'anthropocentrism' than other three groups. The qualitative interview component of the research was designed to not only explain the new concept of 'anthropocentrism' among Vietnamese ecotourists but also explore their distinctive characteristics in relation to the three core ecotourism criteria.

In stage two (II), qualitative data consists of nine face-to-face in-depth semi-structured interviews and one telephone interview, was collected during a one-month period in early 2012. The informants were selected to represent each of the six groups revealed by the cluster analysis. Data collection occurred in Ho Chi Minh City after one week or maximum one month after their trips to CTNP. Most encounters took place in quiet coffee shops, except for one participant who invited the researcher to her home. Each interview took approximately 60 to 90 minutes. Ten interviews with Vietnamese informants were audio recorded and then transcribed. In addition, the participants were encouraged to provide explanatory materials, such as photos, objects, and other personal things relating to their behaviour in Cat Tien National Park. Structured according to the three ecotourism core criteria, the following sections report the relationship between nature and human in which concept of 'anthropocentrism' is explained, then introduced the nature of learning and sustainability from Vietnamese ecotourist perspective.

Nature and human

The Vietnamese people share a similar worldview with other East Asian people in embracing the *unity between humans and nature* (Sofield & Li, 2007; Wen & Ximing, 2008), and in regarding humans as an integral part of nature, as per the ancient influence of Confucianism and Taoism (Weller, 2006). As evidenced in the data, wildlife, landscapes, and humans harmonise as a complete entity. Inseparability is one of the most popular themes raised by almost all Vietnamese participants. This finding corresponds to Lee et al. (2013), who describe the reciprocal relationship between humans and the physical environment in Korean culture.

The existence of both unity and tension between nature and culture corresponds with the negative yin and positive yang of Taoism in which two opposite things exist reciprocally and paradoxically (Brunn & Kalland, 1995; Saso, 1972). Therefore, it was not surprising that besides unity between nature and humans, nature was also perceived as being *distant from people* so that luxurious nature appeared only in cyber space. "The nature I know is only from Discovery or Geographic channels" (C6V65f).[1] In addition, C2V68f similarly described her perception of the distance between humans and nature that she has experienced:

> I mean nowadays, people gather in cities too much. It means that they compete to live here, even myself, competitive to find a job. They completely don't have time to go there (forest) and visit like we, students, do now. The majority of employees work from Monday to Saturday. I know a girl who just wants to sleep on Sunday. Asking her to go shopping, or walking, she never goes because working is too tiring, 8 hours a day. There are people who work in offices. There are people who work extra time. They even work on Sunday. The opportunities for them to visit and connect [with nature] are rare. I think nowadays, they [human and nature] are separated.

Touching nature

Touching, therefore, enabled people to have contact with and brought participants closer to nature: "I felt like I was close to them, I saw them, I touched them" (PilotG, female[2]). The desire to touch also resulted from a desire to feel that nature still persists:

> I have a feeling that nature is close. It's close. It appears very completely. It's not a state that it is being destroyed. It's still untouched, wild. It's not faded. I think so. Once touching, I feel that it still exists...I respect it. I'm afraid that I don't have a chance to touch. Actually last time, I couldn't see any (C1V156m).

In addition, the cultural proclivity for touching is also evident in the Dong Ho folk woodcut paintings that were very popular in almost every household during the seventeenth to twentieth centuries (Dao, 2013). The focus of these paintings was on domestic animals such as pigs, chickens, cats, buffalo, and fish (that is tamed nature), as well as rural life, agriculture, and other cultural activities (that is human landscapes). Close relationships between humans and tamed nature were embodied and clearly represented in these paintings through images of humans holding and touching animals (Figure 20.2).

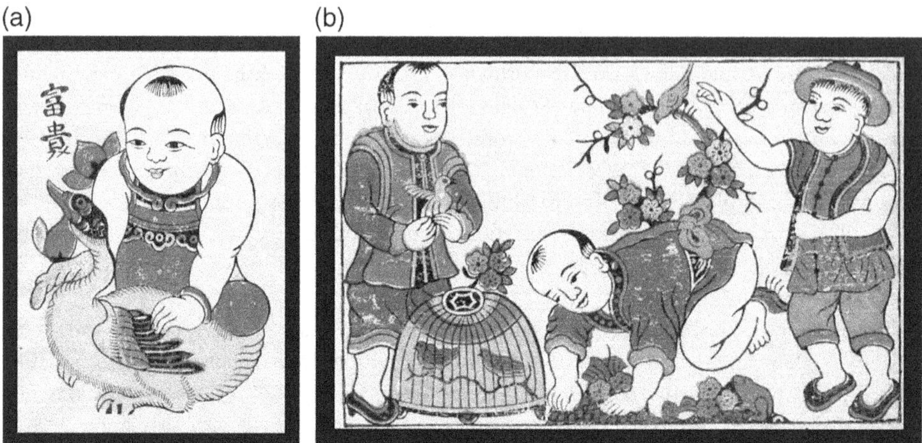

Figure 20.2 Dong Ho folk woodcut paintings (seventeenth–twentieth century), Lao Dong Publishing House, 2012

The alien concept of walking alone in the wilderness

Data distillation exposes three cultural constraints that demotivate visitors from being alone in the forest. Mystery is defined by Oxford Advanced Learner's Dictionary as something that is difficult to understand or to explain, "In general, forest is a place of mystery. It has inside something that we don't know" (C1V232m). However, people still believe and have a fear of myths, for example sacred forests in Cam Din Chin that curse and punish intruders (Nguyen & Nguyen, 2012). This animistic tendency is also found in Japan and India where curses associated with specific places or specific types of wildlife prevent people from approaching (Gadgil & Vartak, 1976; Sasaki, Sasaki, & Fox, 2010). As a result, Pilot N, male was "scared of being alone".

Ghosts are another issue, discouraging people from being alone in the darkness of the night. "But at night, I am afraid of ghost … not really … walking alone in the forest at night, I have a feeling of someone following me … and I hate that feeling", C6V65f. This fear is not confined to females. One male respondent who travels extensively to Cat Tien has also experienced the same feeling. Belief in ghosts is a reflection of local beliefs mixed with Taoist and Buddhist cultural characteristics (Peng, 2007). In the case of C4V198m, such spiritual aspects originated from Buddhism and prevent a harder ecotourist from walking in the night time when the 'yin' (dead people) are thought to appear; "I am a true follower of Buddhism, and there are yin people and yang people". Although the existence of ghosts is not scientifically proven, its impact as a temporal barrier hold true in Cat Tien and has also been reported in the case of tsunami-hit destinations in Thailand (Rittichainuwat, 2011).

The preference to be a part of larger groups

The fear of animistic spirits and ghosts aside, sociable interaction is an important consideration for Vietnamese people in protected area trips. Without accompanying people, the experience in the park would cause much distress for many Vietnamese visitors, with one respondent (Pilot N, male) equating solitude with an act of suicide: "Going to such a wild place needs friends. If I go alone, I might jump into Crocodile Lake". Less dramatically, C1V232m expressed the fundamental need for having others to communicate and enjoy leisure with: "Along the road, we have people to talk with, and make fun with".

When travelling with partners, visitors feel safer because they help and support each other as a way to cope with various constraints, such as being lost, a lack of knowledge, a lack of travelling experience, and a lack of survival skills. In addition, socialising is also a way to mobilise the collective expertise of the group: "If we meet dangerous animals, more heads can find out solutions" (C1V232m). This finding supports the high collectivism and low individualism of Vietnamese culture as measured by Hofstede's 2001 cultural survey, wherein Viet Nam ranks equally with the sister East Asian cultures of China, Thailand, and South Korea (score IDV 20/100). The individualism score is much lower than in Western cultures such as Australia (90), UK (89), France (71), and Germany (67). Besides the direct impact on group size, Vietnamese collectivism is also argued to be associated with the alien concept of being alone in the wilderness, in sharp contrast to Western culture where many protected area visitors cherish solitude in wild or semi-wild settings.

Big cities as safer than national parks

Some constraints that may affect domestic visitors travelling to protected areas are presented in this section. In addition to the above dimensions of mystery and the supernatural, forests have also traditionally been regarded as unfavourable and risky destinations to visit. Dangerous animals are the most prevalent risk mentioned by both harder and softer ecotourists. People are usually warned about dangerous animals such as snakes and tigers when travelling into forests, and can easily mention anecdotes about unpleasant encounters from friends, family, or other acquaintances. PilotL, female, shared her story; "I don't know which animals are inside the forest. They might harm me, don't they? People told me that there must have many dangerous animals, because it's the forest!". Visitors are also afraid of poisonous plants that are rumoured to be easily found in the forests (Thien Nhien, 2013). Protected areas, moreover, not only house hostile wildlife and plants but pose the risk of encountering illegal intruders. Illegal loggers, illegal traders, and wildlife poachers are all perceived to be security risks for the visitor, especially if they are by themselves. "While walking in the forest, people said that illegal loggers might attack ... so I've already prepared some self-protection tools" (PilotG, female).

In order to better understand this issue, the social context of Vietnam's increasingly urban population, which dominates the domestic portion of the sample, should be examined. Urbanisation in Viet Nam is a recent phenomenon affiliated with the *doi moi* policy of economic liberalisation, though city dwellers still accounted for only 30% of the population in 2010, or 26.3 million individuals. This situation contrasts with the figure of 85% in Australia, and 80.7% in the United States (Berg, 2012; Stevenson, 2003; World Bank, 2011), but indicates a major increase from the 11.8 million in 1986. At least one-half of current Vietnamese urban citizens, therefore, constitute people with deep rural roots who had to quickly adapt to new urban lives. Having experienced all features of early-, late-, and post-modernity within a single generation (Zukin, 1998), this new 'bourgeois' expects improved amenities, luxury brands, shopping arcades, easily accessible entertainment, golf courses, and of course travel to 'dream destinations'. National parks perhaps are the alternative destinations that satisfy a desire for novelty, but cosmopolitan and iconic tourist destinations are still preferred for general domestic tourists.

Learning

A desire to learn is indicated by the specific types of plants and animals that visitors wish to encounter. Question 6 in the survey questionnaires (Stage 1) asked respondents to list the names of wild animals and/or plants that they wanted to see in Cat Tien. Among the 374 Vietnamese visitors who responded to this inquiry, the coding shows that megafauna are most frequently listed. These include mammals such as rhino (82), deer (72), gaur (72), elephant (54), boar (16), and fox (9). Cited carnivores include bear (72), tiger (62), panther (35), weasel (8), cat (2), and civet (1). Primates comprise monkey (44) and gibbon (28). Listed amphibians are crocodile (62) and lizard (3) while reptiles are represented by snakes (14). Bird species include birds in general (73) and more specifically peacock (19) and pheasant (10). Other small species, such as leech (6), insect (3), and butterfly (3), are poorly represented. In terms of plants, giant trees (60) and wild flowers (39), in particular orchids (22), are of interest to visitors. It should be noted that 67 visitors, accounting for 18% of respondents to this question, were able to list specific types of species under families or scientific names. The highest number of fauna and flora listed by any single visitor was eight. Such high levels of familiarity, pending further testing, are likely associated with the preponderance of university students and others with university qualifications who may further reflect on the erosion of traditional taboos against 'dangerous' wild animals. Figure 20.3

Figure 20.3 Visitor activies in CTNP (a) Top left: Company colleagues boating in Crodile Lake (Photo courtesy: Duong Truong Son, 2011). (b) Top right: University students visiting Bear Rescue Centre (author's photo). (c) Bottom left: Group tour listening to tour guide interpretation (author's photo). (d) Bottom right: Group tour planting trees in a volunteer reforestation program where their names will be given to that tree (Photo courtesy: Cao Tuan Dung, 2014)

This salient Vietnamese preference for seeing megafauna is very much different from other studies alleging an East Asian (China, Japan, South Korea) focus on charismatic *micro*-flora and *micro*-fauna (Lee et al., 2013). The author's field observations confirm that the East Asian "blossom and waterfall" prototype (Weaver & Lawton, 2002) is not applicable in Viet Nam ecotourism. As can be seen from the aforementioned *Dong Ho* folk paintings, Vietnamese imagery is different from East Asian paintings that focus on panoramic views of landscapes (context-oriented) as well as small details. *Dong Ho* paintings simply present "human scale" images of subjects that share a certain similarity with the Western attention to major objects, and completely omit the background (that is, they are object-oriented) (Masuda, Gonzalez, Kwan, & Nisbett, 2008; Petersen, 1995).

The different style of learning that characterises Vietnamese culture is also pertinent. All three majority Vietnamese clusters adhere to passive learning, unlike Western cultures, and therefore express support for interpretation, tour guides, and other forms of attraction mediation. However, there is also awareness of how individual touching and other physical contact can facilitate the learning process. Through touching, people can gain some knowledge about the subjects, as emphasised by one informant: "I think that by touching trees, leaves, animals, they inspire us to love and understand nature" (C1V232m). What transpires in this exchange is that wildlife receives respect, empathy, and tenderness from humans, who in turn are inspired to love and respect nature.

Sustainability

Given the aforementioned context about forests as an unfamiliar and hostile environment for the majority of Vietnamese people, it is not surprising that the historical and contemporary human artefacts that are commonly encountered in the protected areas of China, Japan, and South Korea (Lee et al., 2013) are almost unknown in Viet Nam. Tombs are sometimes encountered, but these belong to indigenous people (Thanh, 2012). An important consideration here is how the agricultural culture still defines the relationship between nature and humans in the mindset of the many people with ongoing rural connections (Thomas, 2002). As a result, the Vietnamese proclivity was to protect forests only if they perceived them to be a direct benefit for themselves (Anh, 2013). One participant alluded to this proclivity in advocating more tourism for protected areas: "I saw another value ... if we put our efforts to protect it, and employ it for tourism, we will have much benefit from it" (PilotG, female).

Recent developments, however, indicate some change in these attitudes. As this research neared its completion in late 2013, it appeared for the first time that a civil environmental movement protecting CTNP from the threat of two hydro-electric power plants had gained momentum. The Save Cat Tien group was established and sent a letter to the Chairman of the State Council. The information was disseminated online and obtained over 4700 signatures on an online petition, according to savingcattiennationalpark.blogspot.com.au. It is the first time that the Vietnamese people lobbied to protect a forest not for their direct benefit but for the earth and for future generations.

This study finds that the biggest contribution of ecotourism experiences to sustainability is changing visitor awareness and traditional perceptions about wildlife protection and the role of conservation in improving the quality of life for an increasingly urban population. There is considerable evidence of an adherence to a 'steady-state' sustainability awareness that respects the environment by 'leaving no trace'. However, there is also evidence of a vanguard that evinces awareness of 'enhancive sustainability' in which human actions deliberately try to improve the condition of the environment. These "agents of change" remind other people to behave correctly and respect nature by clearing existing rubbish, inspiring other people, and being environmental models in their own behaviour. These environmental models, notably, often publicise their trip experiences through social media, calling for others to join the trips "to help people to obtain a right view of nature" (C4V198m). The degree to which this 'right view' assimilates Western environmental sensibilities is as yet unclear, since the transition is still early and the advocates still too few.

Last but not least is the awareness of eating wild meat. In a metaphor for a changing Viet Nam, one participant confessed that he is struggling with himself about whether or not to eat wild meat. Like many other respondents, he selected the Neutral option for his answer to this question in the survey. In terms of the reasons why most participants insist that they don't like to eat wild meat, some admit that they would like to eat the meat of wild animals at least once. Curiosity is the reason most often given: "If the meat is special, I am open to trying it – trying it once just to know" (PilotN, male). This curiosity may be related to perceptions that wild meat has good flavour and texture. The harder ecotourist (C4V198m) is aware of both sides of this issue and finds himself struggling with it:

> Desire for new things is a basis for being human. We all eat pork which is contaminated by chemicals so we want to try wild animal meat to see how it is by comparison. It's just the "like", but eating them is illegal, so we shouldn't do that. I hesitate, struggling with myself.

Confounding this issue is the observation that eating wild animals is not contradictory to a world view that sees animals and humans as part of the same unity of nature, though the logic of 90 million Vietnamese all participating in this activity is not conducive to harmony and balance because it dictates against the long term survival of wild animals. Socialising is another factor associated with eating wild meat. People just find it hard to resist eating it when they were invited to an event where wild meat was being served. They may feel bad about it, but they need to be diplomatic and keep a harmonious relationship with the hosts and other guests. This mismatch between attitude and behaviours is also observed in the hunting and consumption of wild meat on the Caribbean island of Trinidad (Waylen, McGowan, & Milner-Gulland, 2009). Often marketed as a local specialty, in many places the wild meat is in high demand as a souvenir that people take back to families and friends.

Yet, most of the surveyed Vietnamese visitors expressed no intention to eat wild meat during the trip to CTNP. Besides feelings of disgust and cruelty, they cited ethical consideration, conservation issues and legal constraints. The belief in the unity of humans/nature, is also paradoxically, invoked in sentiments against such consumption. Emphasising the equality of all organisms in the world, one informant stated that: "In terms of spiritual aspect, I feel that a creature has its right to live" (C5V504f). This awareness very much related to a dimension from Buddhist ideology that called for restraint in the killing of animals. In sum, the experiences of Vietnamese visitors in protected areas informed by traditional ideologies that disproportionately affected urban residents, as well as by more recent Western influences through education and elsewhere and tend to comply with ethical perspectives that also pervade and typify Western perceptions of ecotourism (Fennell, 2001).

The idiosyncracies of Vietnamese culture curtail the degree to which the anthropocentric tendencies of its emerging ecotourist market can be extrapolated to other East Asian and Southeast Asian societies that share similar religious and philosophical traditions. It is also useful to note that culture is not fixed but evolves (Runciman, 2005), and that Vietnamese culture has been changing rapidly over the past few decades. The anthropocentrism revealed in this study is therefore likely to change even more as Vietnamese culture is integrated more broadly and deeply into the global community.

Discussion and conclusion

The results indicate that the domestic visitors to a protected area of Viet Nam that provided the sample for this study are indeed 'ecotourists' as defined in the Western sense through the three core criteria of nature-based attractions, learning, and sustainability (Blamey, 1997). Nevertheless, they do display different behaviour due to the idiosyncratic cultural and social context. In particular, a salient anthropocentrism dimension has been identified for the first time in the literature as the factor that most clearly differentiates the Vietnamese and Western ecotourist segments. This newly identified anthropocentrism, in conjunction with the harder-to-softer spectrum, has been critical for explaining the phenomenon of the domestic Vietnamese ecotourist. It can be seen that anthropocentrism contains many cultural implications, and consequently operationalises earlier discussions of cultural contexts that shape distinctive patterns of Asian ecotourism. For example, human manipulation of the environment that other authors (Lee et al., 2013; Sofield & Li, 2007; Weaver, 2002a) pointed out from observations in Asian national parks is arguably not solely for aesthetic appeal but also a response to the fear of emptiness and danger in wilderness areas. For this and other reasons, anthropocentrism also entails an embedded collectivism that underpins high crowding thresholds among Vietnamese ecotourists (Cochrane, 2006; Weaver, 2002a).

Moreover, despite such conformities to the broader East Asian cultural context, idiosyncracies have been identified in Vietnamese domestic ecotourism that are not consistent with the otherwise ubiquitous 'blossom and waterfall' ecotourism model of the Buddhist/Confucianist cultural realm identified by Weaver and Lawton (2002). This includes a paucity of introduced cultural artefacts, such as temples, cemeteries and gardens, in Vietnamese protected areas, and a preference for interacting with charismatic megafauna rather than with micro-fauna or micro-flora. The extent to which the Vietnamese model can be extrapolated to other East Asian contexts, therefore, is a matter for further investigation.

In terms of practical implication, a new regime of learning facilitation for domestic visitors should take into account the broader regional impulses of collectivism (i.e., cooperative group learning) as well as the proclivities to touch and otherwise interact closely with desirable flora and fauna. The complex learning context of Vietnamese visitors, moreover, should accommodate opportunities, perhaps through sensory botanical gardens and wildlife rehabilitation facilities, to satisfy those who seek sensory impressions (Ballantyne et al., 2011). This could foster a model of what might be described as 'complementary' or 'green' anthropocentrism that seeks deliberately and simultaneously to satisfy and enrich visitors without compromising parallel mandates to protect and restore the environment. It is perhaps even possible that interactions that attract feelings of awe and delight may further stimulate the sensitivities that are apparent in some of the Vietnamese visitors through 'enhancement sustainability', and inspire their participation in different kinds of site enhancement activities. In the longer term, the same effect may pertain to non-ecotourist nature-based domestic visitor segments exposed to the same opportunities (Coghlan, Buckley, & Weaver, 2012; Weaver, 2013). It may be suggested further that Western visitors are also exposed to these opportunities in order to achieve a better understanding of the Vietnamese culture and to meet Vietnamese people, thereby potentially enhancing the visitor experience for both groups. These managerial implications suggest for the question about how Vietnamese strictly protected areas be managed to best accommodate both domestic and Western ecotourists whilst achieving optimal benefits for the natural environment.

Notes

1 C6 shows the number of cluster. V65 means this visitor was Vietnamese and her survey questionnaire was coded number 65. f means female.
2 Pilot means this informant was selected from the Stage I interviews.

References

Acott, T. G., La Trobe, H. L., & Howard, S. H. (1998). An evaluation of deep ecotourism and shallow ecotourism. *Journal of Sustainable Tourism, 6*(3), 238–253.

Anh, M. (2013). Việt Nam bắt đầu có ý thức bảo vệ môi trường [Vietnam starts to have environmental protection awareness]. *RFA*. Retrieved from http://www.viet.rfi.fr/viet-nam/20130929- viet-nam-bat-dau-co-y-thuc-bao-ve-moi-truong.

Ballantine, J. L., & Eagles, P. F. L. (1994). Defining Canadian ecotourists. *Journal of Sustainable Tourism, 2*(4), 210–214.

Ballantyne, R., Packer, J., & Sutherland, L. A. (2011). Visitors' memories of wildlife tourism: Implications for the design of powerful interpretive experiences. *Tourism Management, 32*(4), 770–779. doi:http://dx.doi.org/10.1016/j.tourman.2010.06.012.

Berg, N. (2012). U.S. Urban Population Is Up ... but What Does 'Urban' Really Mean? *The Atlantic Cities Place Matters*. Retrieved from http://www.theatlanticcities.com/neighborhoods/2012/03/us-urbanpopulation-what-does-urban-really-mean/1589/.

Blamey, R. (1997). Ecotourism: The search for an operational definition. *Journal of Sustainable Tourism, 5*(2), 109–130.

Blamey, R., & Hatch, D. (1998). *Profiles and Motivations of Nature-Based Tourists Visiting Australia, Occasional paper No. 25*. Canberra: Bureau of Tourism Research.

Brunn, O., & Kalland, A. (1995). Images of nature: An introduction. In O. Bruun, & A. Kalland (Eds.), *Asian perceptions of nature: A critical approach* (pp. 173–188). Surrey: Curzon Press.

Brunner, J. (2013). Protected Area Management: Vietnam vs. India. *IUCN*. Retrieved from http://www.iucn.org/vi/vietnam/?11096/1/.

Ceballos-Lascuráin, H. (1996). *Tourism, Ecotourism and Protected Areas: The State of Naturebased Tourism and Around the World and Guidelines for its Development*. Gland, Switzerland and Cambridge, UK: IUCN.

Cochrane, J. (2006). Indonesian national parks: Understanding leisure users. *Annals of Tourism Research, 33*(4), 979–997.

Coghlan, A., Buckley, R., & Weaver, D. (2012). A framework for analysing awe in tourism experiences. *Annals of Tourism Research, 39*(3), 1710–1714. doi:http://dx.doi.org/10.1016/j.annals.2012.03.007.

Dao, H. (2013). *Vietnamese Folk Paintings*. Artbook.

Diamantis, D. (1999). The characteristics of UK's ecotourists. *Tourism Recreation Research, 24*(2), 99–102.

Duff, L. (1993). Ecotourism in national parks: Impacts and benefits. *National Parks Journal, June*.

Eagles, P. (1992). The travel motivations of Canadian ecotourists. *Journal of Travel Research, 31*(2), 3–7. doi:10.1177/004728759203100201.

Eagles, P., & Higgins, B. (1998). Ecotourism market and industry structure. In K. Lindberg, & D. E. Hawkins (Eds.), *Ecotourism: A guide for planners and managers* (pp. 11–43). North Bennington: The Ecotourism Society.

Fennell, D. (2001). A content analysis of ecotourism definitions. *Current Issues in Tourism, 4*(5), 403–421. doi:10.1080/13683500108667896.

Gadgil, M., & Vartak, V. D. (1976). The sacred groves of western Ghats in India. *Economic Botany, 30*(2), 152–160. doi:10.2307/4253716.

Goeldner, C. R., & Richie, J. R. B. (2003). *Tourism: Principles, Practices, Philosophies* (9th ed.). Hoboken: John Wiley & Sons.

Goodwin, H. (1996). In pursuit of ecotourism. *Biodiversity and Conservation, 5*, 277.

Hair, J. F., Black, W. C., Babin, B. J., & Anderson, R. E. (2010). *Multivariate Data Analysis*. Upper Saddle River: Pearson.

Hall, C. M., & Weiler, B. (1992). What's special about special interest tourism? In B. Weiler, & C. M. Hall (Eds.), *Special interest tourism* (pp. 1–14). London: Wiley.

Hvenegaard, G., & Dearden, P. (1998). Ecotourism versus tourism in a Thai national park. *Annals of Tourism Research, 25*(3), 700–720.

Kerstetter, D. L., Hou, J., & Lin, C. (2004). Profiling Taiwanese ecotourists using a behavioral approach. *Tourism Management, 25*, 491–498.

Kim, S. S., Lee, C, & Klenosky, D. B. (2003). The influence of push and pull factors at Korean national parks. *Tourism Management, 24*(2), 169–180. doi:http://dx.doi.org/10.1016/S0261-5177(02)00059-6.

King, V. (2008). The middle class in Southeast Asia: Diversities, identities, comparisons and the Vietnamese case. *International Journal of Asian Pacific Studies, 4*(2), 73–109.

Laarman, J. G., & Durst, P. B. (1987). Nature travel in the tropics. *Journal of Forestry, 85*(5), 43–46.

Lee, Y., Lawton, L. J., & Weaver, D. B. (2013). Evidence for a South Korean model of ecotourism. *Journal of Travel Research, 52*(4), 520–533. doi:10.1177/0047287512467703.

Lindberg, K. (1991). *Policies for Maximizing Nature Tourism's Ecological and Economic Benefits* (p. 37). World Resources Institute. http://citeseerx.ist.psu.edu/viewdoc/download?doi=10.1.1.536.5828&rep=rep1&type

Lück, M. (2002). Large-scale ecotourism - A contradiction in itself? *Current Issues in Tourism*, 5(3&4), 361–370.

Masuda, T., Gonzalez, R., Kwan, L., & Nisbett, R. E. (2008). Culture and aesthetic preference: Comparing the attention to context of East Asians and Americans. *Personality and Social Psychology Bulletin, 34*(9), 1260–1275. doi:10.1177/0146167208320555.

McClung, M. R., Seddon, P. J., Massaro, M., & Setiawan, A. N. (2004). Nature-based tourism impacts on yellow-eyed penguins Megadyptes antipodes: Does unregulated visitor access affect fledging weight and juvenile survival? *Biological Conservation, 119*(2), 279–285.

Meric, H. J., & Hunt, J. (1998). Ecotourists' motivational and demographic characteristics: A case of North Carolina travelers. *Journal of Travel Research, 36*(Spring), 57–61.

Müllner, A., Eduard, L. K., & Wikelski, M. (2004). Exposure to ecotourism reduces survival and affects stress response in hoatzin chicks (Opisthocomus hoazin). *Biological Conservation, 118*(4), 549–558.

Nguyen, N., & Nguyen, H. (2012). Lời nguyền của rừng Cấm Dìn Chin [Curses of fobidden forest Cam Din Chin]. *Phu nu today*. Retrieved from http://vietnamnet.vn/vn/kinh-te/123217/luat-phutram-va-loi-nguyen-rung-thieng.html.

Orams, M. B. (2002). Feeding wildlife as a tourism attraction: A review of issues and impacts. *Tourism Management, 23*, 281–293.

Peng, W. (2007). From ghost-belief of Baiyue to Buddhism and Taoism of Han Nationality on the historical change of folk religion in South Fujian. *Journal of Fujian Normal University, 6*.

Petersen, Y. Y. (1995). The Chinese landscape as a tourist attraction: Image and reality. In A. A. Lew, & L. Yu (Eds.), *Tourism in China* (pp. 141–154). Colorado: Westview Press.

Rittichainuwat, B. (2011). Ghosts: A travel barrier to tourism recovery. *Annals of Tourism Research, 38*(2), 437–459. doi:http://dx.doi.org/10.1016/j.annals.2010.10.001.

Runciman, W. G. (2005). Culture does evolve. *History and Theory, 44*(1), 1–13. doi:10.2307/3590778.

Ryan, C., Hughes, K., & Chirgwin, S. (2000). The gaze, spectacle and ecotourism. *Annals of Tourism Research, 27*(1), 148–163.

Sasaki, K., Sasaki, Y., & Fox, S. F. (2010). Endangered traditional beliefs in Japan: Influences on snake conservation. *Herpetological Conservation and Biology, 5*(3), 10.

Saso, M. R. (1972). *Taoism and the Rite of Cosmic Renewal*. Pullman: Washington State University Press.

Singh, T., Slotkin, M. H., & Vamosi, A. R. (2007). Attitude towards ecotourism and environmental advocacy: Profiling the dimensions of sustainability. *Journal of Vacation Marketing, 13*(2), 119–134.

Smith, A. C. (2012). Viet Nam Folk Paintings. Ha Noi: Lao Dong Publishing House.

Sofield, T., & Li, F. M. S. (2007). China: Ecotourism and cultural tourism, harmony or dissonance? In J. Higham (Ed.), *Critical issues in ecotourism* (pp. 368–385). Oxford: Butterworth-Heinemann.

Sterling, E. J., Maud, M., & Le, D. M. (2006). *Vietnam: A Nature History*. New Haven, London: Yale University Press.

Stevenson, D. (2003). *Cities and Urban Cultures*. Maidenhead, Philadelphia: Open University Press.

Suntherland, M., & Britton, D. (1980). *National Parks of Japan*. New York: Kodansha.

Tao, C. (Teresa), Eagles, P. F. J., & Smith, S. L. J. (2004). Profiling Taiwanese ecotourists using a self-definition Approach *Journal of Ecotourism, 12*(2).

Thanh, D. (2012). Bí ẩn khu lăng mộ của các gru giữa rừng già Yok Đôn [Secret gru's tombs in the middle of ancient forest Yok Don]. *Tien Phong Online*.https://baogialai.com.vn/channel/8210/201211/bi-an-lang-mo-cua-cac-gru-giua-rung-gia-yok-don-2201414/

Thien Nhien. (2013). Chinese Game of Collecting Poisonous Weed, Mushrooms. *Vietnamnet Bridge*. Retrieved from http://english.vietnamnet.vn/fms/environment/83610/chinese-game-of-ollectingpoisonous-weed--mushrooms.html.

Thomas, M. (2002). *Moving Landscapes: NATIONAL PARKS & the Vietnamese Experience*. Australia: NSW National Parks and Wildlife Service and Pluto Press Australia.

Valentine, P. S. (1993). Ecotourism and nature conservation: A definition with some recent developments in Micronesia. *Tourism Management, 14*(2), 107–115.

Vietnam Forest Creatures. (2012). Retrieved December 8 2012 from https://vncreatures.net/

Wallace, G. N. (1993). Visitor management: Lessons from Galápagos National Park. In K. Lindberg, & D. E. Hawkins (Eds.), *Ecotourism: A guide for planners & managers* (Vol. 1, pp. 55–81). North Bennington: The Ecotourism Society.

Wallace, G. N., & Pierce, S. M. (1996). An evaluation of ecotourism in Amazonas, Brazil. *Annals of Tourism Research, 23*(4), 843–873.

Waylen, K. A., McGowan, P. J. K., & Milner-Gulland, E. J. (2009). Ecotourism positively affects awareness and attitudes but not conservation behaviours: A case study at Grande Riviere, Trinidad. *Oryx, 43*(03), 343–351. doi:10.1017/S0030605309000064.

Wearing, S., & Neil, J. (1999). *Ecotourism: Impacts, Potentials and Possibilities*. Oxford: Butterworth-Heinemann.

Weaver, D. (2002a). Asian ecotourism: Patterns and themes. *Tourism Geographies, 4*(2), 153–172.

Weaver, D. (2002b). Hard-core ecotourits in Lamington national park, Australia. *Journal of Ecotourism, 1*(1), 19–35.

Weaver, D. (2013). Protected area visitor willingness to participate in site enhancement activities. *Journal of Travel Research, 52*(3), 377–391.

Weaver, D., & Lawton, L. (2002). Overnight ecotourist market segmentation in the Gold Coast hinterland of Australia. *Journal of Travel Research, 40*, 270–280.

Weiler, B., & Richins, H. (1995). Extreme, extravagant and elite: A profile of ecotourists on earthwatch expeditions. *Journal of Tourism Recreation Research, 20*(1), 29–36.

Weller, R. P. (2006). *Discovering Nature: Globalisation and Environmental Culture in China and Taiwan*. Cambridge: Cambridge University Press.

Wen, Y., & Ximing, X. (2008). The differences in ecotourism between China and the West. *Current Issues in Tourism, 11*(6), 567.

Wight, P. A. (1993). Ecotourism: Ethics or eco-sell. *Journal of Travel Research, Winter*, 3–7.

Woods, B., & Moscardo, G. (1998). Understanding Australian, Japanese, and Taiwanese ecotourists in the Pacific Rim region. *Pacific Tourism Review, 1*, 329–339.

World Bank. (2011). *Vietnam Urbanisation Review*: Technical Assistance Report. . World Bank, Ha Noi. https://openknowledge.worldbank.org/handle/10986/2826 License: CC BY 3.0 IGO.

Zukin, S. (1998). Urban lifestyles: Diversity and standardisation in spaces of consumption. *Urban Studies, 35*(5-6), 825–839. doi:http://usj.sagepub.com/archive/.

21

ECOTOURISM AS FORM OF LUXURY CONSUMPTION

Serena Volo and David D'Acunto

Introduction

The goal of this chapter is to explain the nature of luxury consumption in the context of ecotourism, particularly its relationships with traditional forms of conspicuous consumption. There is a growing attention towards this form of tourism with a small but lucrative market segment of luxury ecotourists and an eager set of tourism entrepreneurs and large multinationals interested in harvesting its potential. A scrutiny of the available literature reveals that scholars' positions vary in the understanding of the phenomena. This chapter presents an initial contribution to analyzing different luxury consumption perspectives in ecotourism for the purpose of stimulating more research in this area.

This chapter first introduces and synthesises the literature related to various aspects of luxury consumption in tourism and hospitality. Then, it reviews research focusing on luxury aspects of ecotourism, discussing several aspects, namely: (a) luxury within the frame of soft and hard forms of ecotourism consumption, (b) sustainable and ethical issues in luxury ecotourism, and (c) the role of ecotourists and ecotourism providers. The chapter then discusses luxury ecotourism in light of the recent interest in the tourism literature indicating directions for future research.

Luxury in tourism

Luxury in tourism evokes exclusivity, high social status, personalised experience, and high levels of comfort and convenience (Chen & Peng, 2014; Kurtz, 2004) and tourists looking for luxury when travelling generally seek the most complete spectrum of services and best-quality products (Ikkos, 2003; Yang & Mattila, 2014, 2017). Luxury represents an essential market segment for the tourism industry with luxury tourist expenditures accounting for 25% of the overall international travel market (Park, Reisinger, & Noh, 2010). Luxury tourists have a daily average expenditure eight times higher compared to other tourists (ILTM, 2011). Therefore, the luxury segment represents an interesting subset of the tourism market in which several tourism stakeholders invest in the hope to capture a share of this lucrative business.

Luxury is however subjective and affected by customers' social context (Nueno & Quelch, 1998), therefore it is not easy to provide a universally recognized definition of luxury, given

DOI: 10.4324/9781003001768-21

that social contexts are intrinsically fluid, dynamic, evolutionary, and very culture sensitive (Mortelmans, 2005; Yeoman, 2011). Moreover, luxury depends on the experience and individual consumers' needs (Hennigs, Wiedmann, Klarmann, & Behrens, 2015). Nevertheless, some common traits are identifiable among consumers seeking luxury in services, in that luxury consumers are mainly driven by the positive emotions deriving from the consumption of exclusive experiences (Kapferer, 2015). Some of the reasons behind these behaviours have been identified in the willingness to reach self-actualisation, to create a more idealised image of the self (Danziger, 2006) and to impress others (Mason, 1981). Other scholars attempted to link luxury to the concepts of authenticity, customisation, limited editions, acute personalisation (e.g., Yeoman, 2012), with the expected levels of luxury differing across tourists' income levels (Riley, 1995; Yeoman & McMahon-Beattie, 2006). However, given the intangible nature of luxury in tourism, several dimensions have been studied with regards to this construct. For instance, prestige-seeking behaviours and status purposes (Correia & Moital, 2009), tourists' behavior at the destination (Riley, 1995), or even intimacy and privacy (Park et al., 2010; Correia, Kozak, & Reis, 2016) have been investigated with regards to their relationship with luxury in tourism, while time, wellness, and cultural enrichment have been considered as antecedents of luxury in tourism (Yeoman, 2012).

Luxury in tourism is therefore a social phenomenon enhancing status and prestige, driving tourists towards outrageous consuming behaviours patterns, and a set of value dimensions deriving from luxury consumption has been identified, namely: social acceptance, emotional attachment, and quality assurance (Vigneron & Johnson, 1999). According to past research (Berry, 1994; Swanson, 2004; Allsopp, 2005), tourists seek luxury in different service areas or products, such as sustenance (e.g., caviar, champagne), shelter (e.g., accommodation with spa), clothing and appeal (e.g., branded clothes, perfumes, jewelry), and leisure activities (e.g., entertainment and sporting services), luxury holiday accommodation and home furnishings, as well as fine arts or handmade crafts and antiques. In tourism, luxury has evolved over time, moving from the initial idea of traditional luxury "product" (e.g., five-star resorts and hotels) to a more comprehensive service offer, including unique dining, six-star spa resort, and exclusive and custom-designed experiential tours (Yesawich & Russel, 2004; Bakker, 2005; Park et al., 2010). As such, presently the attractiveness of services has overcome goods in the luxury segment of tourism, with the consumption of luxury experiences being superior to luxury products (Tynan, McKechnie, & Chhuon, 2010). Thus, luxury is not simply associated with the consumption of products or services in luxurious spaces and contexts, but it encompasses experiences of time, space, authenticity, community, individuality, and wellbeing (Yeoman & McMahon-Beattie, 2018).

Contemporary luxury travelers prefer real, unique, authentic experiences in unspoiled destinations, which stimulate them physically and intellectually, adding novelty to their lives (Yeoman, 2008). The growing need of consumers for luxury and authenticity fostered the development of luxury subsegments in tourism markets not originally targeting on such travelers (e.g., ecotourism). Moreover, new generations of consumers are approaching luxury with different behaviour patterns compared to the previous ones. Baby boomers with high disposable income and available leisure time (Park et al., 2010) represent nowadays the biggest share of luxury tourists and they are experienced, informed, and adventurous travelers (Silverstein, Fiske, & Butman, 2003) seeking unique experiences. In the quest for uniqueness, ecotourism destinations and attractions might play an important role. Luxury travel, fine dining, and pampering services (Kim, 2018) are today among the most desired experiences, thus re-shaping the luxury market habits.

Luxury in hospitality

The luxury segment in hospitality has been growing over the course of the last decade, registering one of the highest occupancy rates (Yang & Mattila, 2017; Yang & Mattila, 2014). Although just 3% of travelers seek luxury in hospitality services, the segment represents itself 20% of the total tourism expenditure (Chen & Peng, 2014). It is therefore a crucial market segment for service providers and, given its growing importance, hospitality scholars have started to focus their attention on the luxury consumption segment. Nevertheless, the concept of luxury with specific reference to the hospitality service consumption received relatively less attention compared to luxury goods consumption (Yang & Mattila, 2017), and still little is known about the differences between luxury experiential purchases and material purchases.

Luxury value in hospitality generally involves the three different subdimensions of experiential, symbolic, and functional value. While functional value refers to products' core benefits and quality (Wiedmann, Hennigs, & Siebels, 2009), the experiential one evokes fantasies, fun, and feelings essential to the luxury consumption, while symbolic value reflects the owner's wealth and status (Han, Nunes, & Drèze, 2010). Extant research on luxury consumption has mainly focused on goods rather than on services (Peng, Chen, & Hung, 2020). This is unfortunate since luxury experiences, such as dining or vacations, strongly affect positive emotions in customers compared to material possession of luxury goods (Van Boven & Gilovich, 2003), and experiences are more closely connected to the self than material possessions (Carter & Gilovich, 2012). Indeed, according to the so-called "experience recommendation" phenomenon, consumers prioritise experiences over tangible with hospitality service covering a major component of life experiences (Carter & Gilovich, 2010; Yang & Mattila, 2014).

The perception of luxury value at hotels affects tourists' attitudes and, in turn, their staying behaviours (Chen & Peng, 2014). In this relation, symbolic and experiential values have a significant impact on luxury hotel consumers, directly influencing guests' staying behaviour (Chen & Peng, 2014), while functional value does not elicit the same effect. Yang and Mattila (2016) investigated four different value dimensions, namely: functional, financial, hedonic, and symbolic and found, partly in contrast with Chen and Peng (2014), that only three value categories (functional, financial, hedonic) positively and directly affect consumers' purchase intentions. These findings indicate the less predominant role of symbolic value, which is more likely to be sought by consumers in luxury goods rather than in luxury services, which might be due to the intangible and invisible characteristics of services and their reliance on service quality and atmosphere (Yang & Mattila, 2016).

Luxury with regards to restaurant experience has also been explored by several hospitality scholars. Luxury restaurants are mainly identified as those providing full table-service, premium price menu, and high-quality environment (Kim, Lee, & Yoo, 2006; Ryu & Jang, 2007; Hwang & Hyun, 2013; Han & Hyun, 2013; Chen, Peng, & Hung, 2015; Yang & Mattila, 2016). Customers seeking luxury dining experiences have high expectations with regards to both service elements, the tangible ones (e.g., food quality, food taste, and consistency of menu items), and intangibles (e.g., employee competence, servers' knowledge, restaurant atmosphere, and ambience quality) (Weiss, Feinstein, & Dalbor, 2005; Cullen, 2005; Kwun & Oh, 2007; Njite, Dunn, & Hyunjung Kim, 2008; Kim, Bergman, & Raab, 2010; Wang & Chen, 2012; Yang & Mattila, 2016). The concept of luxury in hospitality has therefore a broader scope since luxury service providers (e.g., hotels and restaurants) do not only sell superior tangible products (e.g., luxury hotel room and amenities, premium menu items, and superb food presentations) but have also to deliver superlative service and highly sophisticated consumption environments (Chen & Hu, 2010; Lee & Hwang, 2011; Dortyol, Varinli, & Kitapci, 2014; Yang & Mattila, 2016).

The luxury dimension in ecotourism

Background

Most of the forms of sustainable travel, including ecotourism, have their origins with the environmental movement of the 1970s, although ecotourism itself became prevalent as a travel concept in the late 1980s (Briney, 2020). Until the end of the '90s, ecotourists were not particularly attracted by the attribute of luxury when choosing accommodation (Pearce & Wilson, 1995) with 56% of them preferring middle-range levels of luxury and only 6% more willing to choose a luxury type of accommodation (Kwan, Eagles, & Gebhardt, 2008). In 1997, Wight surveyed both ecotourists and general consumers in the attempt to cluster these different cohorts according to their accommodation preferences and luxury levels expectations. The author found that ecotourists were basically not interested in luxury accommodations. Wight also pointed out the lower desire of experienced ecotourists seeking luxury compared to general tourists and highlighted the relevance that ecotourists assign to activities at destination as recognized to strongly affect experiences, leaving no relevance to accommodation choices. Indeed, ecotourists are generally more attracted by intimate, adventure-type accommodation and by environmentally sensitive operations. Nevertheless, in the same study the author reported the paradoxical market trend of raising larger, more international and luxury accommodations or resorts in the wilderness, mainly operated by major global chains, thus noting the rise of a small niche but potentially lucrative market for the luxury future ecolodge industry (Higgins-Desbiolles, 2011).

Recent developments: Soft versus hard ecotourism

As predicted by Wight in 1997 in its segmentation of ecotourists according to their different approaches and needs, the luxury ecotourism market niche developed only in the last two decades. Soon after this prediction, Myles (2003) pointed out the paradox of ecotourism derived by different patterns in ecotourists' behavior, by differentiating those seeking luxury experiences (i.e., people preferring to stay in luxury accommodations surrounded by modern conveniences and spending a very modest budget at destinations due to prepaid packages) and low-impact ecotravellers (i.e., people saving money on accommodation but spending more at the destinations and keeping minimal environmental impacts). Ecolodge participants increased their appreciation for softer and more luxurious ecotourism, leading to the development of the new category of "structured ecotourists" (Weaver & Lawton, 2007).

The developing segment of luxury ecotourism and ecolodge accommodations originate from tourists' desire to seek luxury experiences as a temporary escape from their everyday realities (Low, 2010) and from the governments' ability to build and run luxury tourism facilities at cheaper prices in developing countries (Moscardo & Benckendorff, 2010). New forms of accommodation emerged, such as ecolodges, defined as any "nature-dependent tourist lodge that meets the philosophy of ecotourism" by The International Ecotourism Society (TIES) (Higgins-Desbiolles, 2011). By offering small-scale tourism based around wildlife and pristine natural environments, these new ecotourism accommodations are mainly based in remote areas and some offer luxury experiences (Ryan & Stewart, 2009). Nevertheless, access to such destinations requires more time and higher disposable income, thus representing a viable tourism option for a few categories of travelers, on average wealthier and with few time constraints (Moscardo & Benckendorff, 2010) thus resulting in very selective. These "soft" ecotourists seek higher levels of services and facilities to mediate encounters between venues and potentially large numbers of visitors more casually engaged with the natural environment (Weaver & Lawton, 2007), compared to their "hard" counterparts. As a result, different types of ecotourism accommodations are available, with five-star luxury lodges also being admitted to the list of ecolodges (Page & Dowling, 2002). Softer-path ecotourists enjoy staying in more luxurious forms of accommodation (Hunter & Shaw, 2005) with the result of a larger ecological footprint due to their higher resource demands.

Thus, the scope of luxury ecotourism experiences is growing, with more sophisticated offerings (e.g., wildlife observation helicopter rides, luxury cruise lines) yet authors point out the vulnerability of the resources involved in such type of activities and the need for local luxury ecotourism business to import goods not available for the international luxury standards (Fennell, 2015, 2020). Nevertheless, consumers seem to be able to exert a good level of discrimination between alternative types of ecotourism experiences. Indeed, when investigating the relationship between general interest in ecotourism and holiday preferences, Perkins and Grace (2009) found that people seeking ecotourism, wildlife tourism, and volunteer ecotourism tend to negatively relate to luxury resort holidays interest, thus confirming the different views of ecotourism in such market segments. This evidence of understanding of the subtleties offers some perspective into the valence that sustainable and ethical issues still have on the ecotourism market.

Luxury in ecotourism: Sustainable and ethical issues

Luxury in the ecotourism realm also involves ethical issues. Most ecotourism destinations are in developing countries where, in the eyes of the local population, travelling for recreational purpose is itself a luxury (Melubo, 2020). Ecotourism is not for everybody and most travellers to ecotourism destinations belong to the mid-upper class of developed countries, with about 60% coming from the United States, 30% from Western Europe, and the remaining 10% from Australia or Asia (Kwan et al., 2008; Smith, 2019). For instance, by focusing on Canadian ecotourists travelling to Costa Rica, Fennell and Smale (1992) found how ecotourists are on average better educated and more affluent than the general Canadian population. For some, this might be perceived as unethical since the luxury activity takes place within developing countries serving upper classes. Nevertheless, there is evidence that tourists with the highest income levels tend to be interested in natural attractions (Bieger & Laesser, 2002). Paradoxically, when choosing a developing country as destination, luxury ecotourists tend to believe that their expenditures will have a positive impact on the country's ability to prosper, even though this is not always the case (Hanna, 2017). Ecotourists might see going for a luxury trip as a way to distinguish from others expecting also to reduce their negative impacts on the destination, both from a social and environmental viewpoint. Nevertheless, commercial forms of luxury ecotourism are increasingly threatening local and Indigenous communities and leading to the degradation of natural areas while providing few benefits to local communities (Honey, 2008; Stronza & Gordillo, 2008; Regmi & Walter, 2017; Walter, Regmi, & Khanal, 2018). Many luxury resorts are established adjacent to natural reserve boundaries and promote themselves as 'eco-lodges' (Puri, Karanth, & Thapa, 2019), however across many protected areas, most principles of ecotourism are not followed (Banerjee, 2010). This has led to distinguishing between authentic and "pseudo" ecotourism in ecotourism research (Donohoe & Needham, 2008). Despite their claims of authentic ecotourism, such luxury facilities, and the need of infrastructures to welcome luxury ecotourists, have important implications in terms of the ecological footprint on the environment (Gössling, Hansson, Hörstmeier, & Saggel, 2002; Ryan & Stewart, 2009).

Wheeller (2005) points out the need for more attention on luxury ecolodges since, while claiming ecological sensitivity and sustainability, often result in hasty development in the pursuit of quick economic gains, foster elitist consumption by wealthy ecotourists, and might result in ecotourism development characterised by political corruption. Although the foundations of ecotourism lay in sustainability and the environmentally responsible enjoyment of protected areas, the industry is often accused of "greenwashing" and the luxury segment is driving this negative reputation. The current proliferation of new luxury ecotourism resorts claiming themselves as profit-driven companies playing a key role in local community development, disrupts the needs of customer for ethical travel and high-end luxury experience, which involves community participation and development. In the extant literature, community-based ecotourism is generally associated with low-budget travelers and basic accommodation, while new luxury ecolodges aim at blending community-based conservation initiatives with high-end facilities (Duffy, 2008). More research is needed in this direction to examine the extent to which luxury ecotourism holds an ecotourism-oriented value chain in the creation of its tourism experiences.

The role of ecotourists and of ecotourism providers

Ecotourists themselves play a critical role in shaping the process of sustainability in ecotourism. There is evidence of a growing trend of customers seeking highly luxurious and exclusive access to pristine nature in order to enhance their social status. These ecotourists have been defined as "egotourists" (Mowforth & Munt, 2015). Other emerging market segments of travelers looking for luxury ecotourism experience and not primarily characterised in the extant ecotourism literature are the corporate visitors travelling for business and the romantic couples seeking luxury and discretion (Ryan & Stewart, 2009). This trend shows the evolution of the market and the existence of new clusters of customers are gradually approaching luxury ecotourism. Furthermore, new generations are approaching ecotourism and the role of millennials consumers and their relationship with luxury consumption in the ecotourism domain is also under recent investigation (Costa, Abreu, Gestão, & Barbosa, 2019). As "mass ecotourism" emerges as phenomenon, it threatens ecotourism integrity and principles, and thus pushes some ecotourists segments towards higher standards and premium prices, leading to a new subsegment of luxury ecotourists who seek the exclusiveness of a true wilderness experience (Myles, 2003).

Given the developing and mutating ecotourism market, reassuring customers on the impacts that luxury services and practices have on the environment is nowadays of paramount importance for ecolodges. Properly adopting and communicating the implementation of solar energy systems, the organic composting, and wastewater treatment systems may significantly contribute to reduce environmental repercussions (Higgins-Desbiolles, 2011). The role of luxury for ecotourism players has been investigated recently by Buckley and Mossaz (2018) in their study on the marketing tools adopted by ecotourism providers in Africa. The authors found that, after wildlife viewing opportunities which remains as the prime feature for marketing communication and strategies, luxury and exclusiveness are the second most relevant attributes conveyed by tourism providers to customers. These findings show that ecotourism operators believe in luxury facilities to boost sales and are exploiting the interest of most of their clients who are motivated by wildlife viewing opportunities and luxury, whereas only a small subset of customers is actually driven by contributing to conservation.

Conclusion

Ryan and Stewart (2009) argue how ecotourism is not inconsistent with luxury and the two concepts can coexist when environmental conservation is served through an organization's vision that pursues resource regeneration and considers restrictions on visitor penetration of natural spaces. Some claim that having a broader understanding of ecotourism might include the consumption of luxury experiences, Fennell (2015) argues for more care in the understanding of the intrusiveness of some forms of luxury ecotourism consumptions and he still remains open to widening the boundaries of the definition of ecotourism preferably by engaging all the ecotourism constituencies (industry, residents, governments, academia) and aiming at unifying their "vastly different trajectories" (Fennell, 2015, p. 273).

Ecotourism as a form of luxury consumption is an under-researched area; however, there is growing interest in researchers towards the concept and this has been also recently pointed out by Wondirad (2019) in a meta-analysis of ecotourism. Olearnik and Barwicka (2019) included, among the new ecotourism features, luxurious tourist destinations that follow rules of sustainability. Koninx (2019, p. 11) also discusses the centrality of luxury in new ecotourism by defining it as 'a new haven of luxury and indulgence' in some ecotourism destinations. These considerations are particularly interesting when compared with Wight's (1997) findings, as they show how ecotourists' expectations for luxury have radically evolved over time. However, and in contrast with the general luxury vacation literature, the social value perceived from luxury ecotourism experiences still does not seem to predict purchasing intentions for ecotourism products (Jamrozy & Lawonk, 2017), meaning ecotourists, behaviours remain mainly driven by other values. The strength of these values might be a discriminating factor for the definition of luxury ecotourism contributing to providing a better understanding of the "process for a better tourism" in an uncertain post-pandemic landscape.

Future research should include comprehensive and accurate evaluation of this tourism segment in terms of consumers' behaviour at the destination and tourism providers value chains. The ethical aspects of luxury ecotourism consumption ought to be explored to ascertain the outcomes that this form of tourism produces at tourism community level. At large, the tourism political agenda should be directed towards forms of ecotourism that can encompass the consumption of the more affluent yet in line with the "luxury of values" around which ecotourism is built. Finally, tourism scholars research agenda should also include investigations on the experiential dimension of ecotourism in luxury settings and encourage collaborations with the industry to foster novel forms of educational ecotourism.

References

Allsopp, J. (2005). Additional practice papers: Premium pricing: Understanding the value of premium. *Journal of Revenue and Pricing Management*, *4*(2), 185–194.

Bakker, M. (2005). Luxury and tailor-made holidays. *Travel & Tourism Analyst*, (20), 1–47.

Banerjee, A. (2010). Tourism in protected areas: Worsening prospects for tigers? *Economic and Political Weekly*, 27–29.

Berry, C. J. (1994). *The Idea of Luxury: A Conceptual and Historical Investigation* (Vol. 30). UK: Cambridge University Press.

Bieger, T., & Laesser, C. (2002). Market segmentation by motivation: The case of Switzerland. *Journal of Travel research*, *41*(1), 68–76.

Briney, A. (2020). An Introduction to Ecotourism. *ThoughtCo*. Retrieved from http://thoughtco.com/what-is-ecotourism-1435185 (accessed on 11.02.2020).

Buckley, R., & Mossaz, A. (2018). Private conservation funding from wildlife tourism enterprises in sub-Saharan Africa: Conservation marketing beliefs and practices. *Biological Conservation, 218*, 57–63.
Carter, T. J., & Gilovich, T. (2010). The relative relativity of material and experiential purchases. *Journal of Personality and Social Psychology, 98*, 146–159.
Carter, T. J., & Gilovich, T. (2012). I am what I do, not what I have: The differential centrality of experiential and material purchases to the self. *Journal of Personality and Social Psychology, 102*(6), 1304–1317.
Chen, P. T., & Hu, H. H. (2010). How determinant attributes of service quality influence customer-perceived value. *International Journal of Contemporary Hospitality Management, 22*(4), 535–551.
Chen, A., & Peng, N. (2014). Examining Chinese consumers' luxury hotel staying behavior. *International Journal of Hospitality Management, 39*, 53–56.
Chen, A., Peng, N., & Hung, K. P. (2015). The effects of luxury restaurant environments on diners' emotions and loyalty. *International Journal of Contemporary Hospitality Management, 27*(2), 236–260.
Correia, A., Kozak, M., & Reis, H. (2016). Conspicuous consumption of the elite: Social and self-congruity in tourism choices. *Journal of Travel Research, 55*(6), 738–750.
Correia, A., & Moital, M. (2009). Antecedents and consequences of prestige motivation in tourism: An expectancy-value motivation. *Handbook of Tourist Behaviour: Theory and Practice*, 16–34.
Costa, A., Abreu, M., Gestão, A., & Barbosa, B. (2019). Millennials' trends in luxury marketing: The ecoturism. *2019 14th Iberian Conference on Information Systems and Technologies (CISTI)*, 1–6.
Cullen, F. (2005). Factors influencing restaurant selection in Dublin. *Journal of Foodservice Business Research*, 7(2), 53–85.
Danziger, P. (2006). *Shopping: Why We Love it and How Retailers Can Create the Ultimate Customer Experience*. Chicago: Kaplan Publishing.
Donohoe, H. M., & Needham, R. D. (2008). Internet-based ecotourism marketing: Evaluating Canadian sensitivity to ecotourism tenets. *Journal of Ecotourism*, 7(1), 15–43.
Dortyol, I. T., Varinli, I., & Kitapci, O. (2014). How do international tourists perceive hotel quality? *International Journal of Contemporary Hospitality Management, 26*(3), 470–495.
Duffy, R. (2008). Neoliberalising nature: Global networks and ecotourism development in Madagasgar. *Journal of Sustainable Tourism, 16*(3), 327–344.
Fennell, D. A. (2001). A content analysis of ecotourism definitions. *Current Issues in Tourism, 4*(5), 403–421.
Fennell, D. A. (2015). *Ecotourism*. Routledge: New York.
Fennell, D. A. (2020). *Ecotourism*. Routledge: New York.
Fennell, D. A., & Smale, B. J. (1992). Ecotourism and natural resource protection: Implications of an alternative form of tourism for host nations. *Tourism Recreation Research, 17*(1), 21–32.
Gössling, S., Hansson, C. B., Hörstmeier, O., & Saggel, S. (2002). Ecological footprint analysis as a tool to assess tourism sustainability. *Ecological Economics, 43*(2/3), 199–211.
Han, H., & Hyun, S. S. (2013). Image congruence and relationship quality in predicting switching intention: Conspicuousness of product use as a moderator variable. *Journal of Hospitality & Tourism Research, 37*(3), 303–329.
Han, Y. J., Nunes, J. C., & Drèze, X. (2010). Signaling status with luxury goods: The role of brand prominence. *Journal of Marketing, 74*(4), 15–30.
Hanna, P. (2017). Is It Ethical to Take a Luxury Holiday in a 'Developing' Country? *Fair Observer*. Retrieved from www.fairobserver.com/culture/luxury-travel-tourism-industry-culture-news41210/.
Hennigs, N., Wiedmann, K. P., Klarmann, C., & Behrens, S. (2015). The complexity of value in the luxury industry. *International Journal of Retail & Distribution Management, 43*(10/11), 922–939.
Higgins-Desbiolles, F. (2011). Death by a thousand cuts: Governance and environmental trade-offs in ecotourism development at Kangaroo Island, South Australia. *Journal of Sustainable Tourism, 19*(4/5), 553–570.
Honey, M. (2008). *Ecotourism and Sustainable Development: Who Owns Paradise?* Washington, D.C.:Island Press.
Hunter, C., & Shaw, J. (2005). Applying the ecological footprint to ecotourism scenarios. *Environmental Conservation*, 294–304.
Hwang, J., & Hyun, S. S. (2013). The impact of nostalgia triggers on emotional responses and revisit intentions in luxury restaurants: The moderating role of hiatus. *International Journal of Hospitality Management, 33*, 250–262.

Ikkos, A. (2003). Luxury Tourism: A Matter for All, Not Just Hotels. *JBR Hellas Ltd*. Retrieved from http://www.gbrconsulting.gr/articles/Luxury%20Tourism.pdf.

ILTM. (2011). The Future of Luxury Travel. Retrieved from http://www.iltm.net/files/the_future_of_luxury_travel_report.pdf.

Jamrozy, U., & Lawonk, K. (2017). The multiple dimensions of consumption values in ecotourism. *International Journal of Culture, Tourism and Hospitality Research*, *1*(11), 18–34.

Kapferer, J. N. (2015). *Kapferer on Luxury: How Luxury Brands Can Grow Yet Remain Rare*. Philadelphia, PA: Kogan Page Publishers.

Kim, W. G., Lee, Y. K., & Yoo, Y. J. (2006). Predictors of relationship quality and relationship outcomes in luxury restaurants. *Journal of Hospitality & Tourism Research*, *30*(2), 143–169.

Kim, Y. (2018). Power moderates the impact of desire for exclusivity on luxury experiential consumption. *Psychology & Marketing*, *35*(4), 283–293.

Kim, Y. S., Bergman, C., & Raab, C. (2010). Factors that impact mature customer dining choices in Las Vegas. *Journal of Foodservice Business Research*, *13*(3), 178–192.

Koninx, F. (2019). Ecotourism and rewilding: The case of Swedish Lapland. *Journal of Ecotourism*, *18*(4), 332–347.

Kurtz, R. (2004). Marketing travel to the affluent. In *The complete 21st century travel and hospitality marketing handbook* (pp. 525–536). Upper Saddle River, NJ: Prentice Hall.

Kwan, P., Eagles, P. F., & Gebhardt, A. (2008). A comparison of ecolodge patrons' characteristics and motivations based on price levels: A case study of Belize. *Journal of Sustainable Tourism*, *16*(6), 698–718.

Kwun, D. J. W., & Oh, H. (2007). Past experience and self-image in fine dining intentions. *Journal of Foodservice Business Research*, *9*(4), 3–23.

Lee, J. H., & Hwang, J. (2011). Luxury marketing: The influences of psychological and demographic characteristics on attitudes toward luxury restaurants. *International Journal of Hospitality Management*, *30*(3), 658–669.

Low, T. (2010, June). Sustainable luxury: A case of strange bedfellows. In Proceedings of the Tourism and Hospitality Research in Ireland Conference, Shannon, Ireland (pp. 15–16).

Mason, R. S. (1981). *Conspicuous Consumption: A Study of Exceptional Behaviour*. UK: Gower.

Melubo, K. (2020). Is there room for domestic tourism in Africa? The case of Tanzania. *Journal of Ecotourism*, *19*(3), 248–265.

Mortelmans, D. (2005). Sign values in processes of distinction: The concept of luxury. *Semiotica*, *2005*(157), 497–520.

Moscardo, G., & Benckendorff, P. (2010, November). Sustainable luxury: Oxymoron or comfortable bedfellows. In Proceedings of the 2010 international tourism conference on global sustainable tourism, Mbombela, Nelspruit, South Africa (pp. 15–19).

Mowforth, M., & Munt, I. (2015). *Tourism and Sustainability: Development, Globalisation and New Tourism in the Third World*. London: Routledge.

Myles, P. B. (2003). Contribution of wilderness to survival of the adventure travel and ecotourism markets. In Science and Stewardship to Project and Sustain Wilderness Values: Seventh World Wilderness Congress Symposium, 2001 November 2–8, Port Elizabeth, South Africa (p. 185). US Department of Agriculture, Forest Service, Rocky Mountain Research Station.

Njite, D., Dunn, G., & Hyunjung Kim, L. (2008). Beyond good food: What other attributes influence consumer preference and selection of fine dining restaurants? *Journal of Foodservice Business Research*, *11*(2), 237–266.

Nueno, J. L., & Quelch, J. A. (1998). The mass marketing of luxury. *Business Horizons*, *41*(6), 61–61.

Olearnik, J., & Barwicka, K. (2019). Chumbe Island Coral Park (Tanzania) as a model of an exemplary ecotourism enterprise. *Journal of Ecotourism*, 1–15.

Page, S. J., & Dowling, R. K. (2002). *Ecotourism*. Harlow: Pearson.

Park, K. S., Reisinger, Y., & Noh, E. H. (2010). Luxury shopping in tourism. *International Journal of Tourism Research*, *12*(2), 164–178.

Pearce, D. G., & Wilson, P. M. (1995). Wildlife-viewing tourists in New Zealand. *Journal of Travel Research*, *34*(2), 19–26.

Peng, N., Chen, A., & Hung, K. P. (2020). Dining at luxury restaurants when traveling abroad: Incorporating destination attitude into a luxury consumption value model. *Journal of Travel & Tourism Marketing*, *37*(5), 562–576.

Perkins, H., & Grace, D. A. (2009). Ecotourism: Supply of nature or tourist demand? *Journal of Ecotourism*, *8*(3), 223–236.

Puri, M., Karanth, K. K., & Thapa, B. (2019). Trends and pathways for ecotourism research in India. *Journal of Ecotourism, 18*(2), 122–141.

Regmi, K. D., & Walter, P. (2017). Modernisation theory, ecotourism policy, and sustainable development for poor countries of the global South: Perspectives from Nepal. *International Journal of Sustainable Development & World Ecology, 24*(1), 1–14.

Riley, R. W. (1995). Prestige-worthy tourism behavior. *Annals of Tourism Research, 22*(3), 630–649.

Ryan, C., & Stewart, M. (2009). Eco-tourism and luxury–the case of Al Maha, Dubai. *Journal of Sustainable Tourism, 17*(3), 287–301.

Ryu, K., & Jang, S. S. (2007). The effect of environmental perceptions on behavioral intentions through emotions: The case of upscale restaurants. *Journal of Hospitality & Tourism Research, 31*(1), 56–72.

Silverstein, M., Fiske, N., & Butman, J. (2003). *Trading Up: The New American Luxury*. New York: Portfolio.

Smith III, J. B. (2019). *Policy and Market Strategies of the Ecotourism Industry in Developing Countries*(Doctoral Dissertation, The University of Texas at Austin).

Stronza, A., & Gordillo, J. (2008). Community views of ecotourism. *Annals of Tourism Research, 35*(2), 448–468.

Swanson, K. K. (2004). Tourists' and retailers' perceptions of souvenirs. *Journal of Vacation Marketing, 10*(4), 363–377.

Tynan, C., McKechnie, S., & Chhuon, C. (2010). Co-creating value for luxury brands. *Journal of Business Research, 63*(11), 1156–1163.

Van Boven, L., & Gilovich, T. (2003). To do or to have? That is the question. *Journal of Personality and Social Psychology, 85*(6), 1193.

Vigneron, F., & Johnson, L. W. (1999). A review and a conceptual framework of prestige-seeking consumer behavior. *Academy of Marketing Science Review, 1*(1), 1–15.

Walter, P., Regmi, K. D., & Khanal, P. R. (2018). Host learning in community-based ecotourism in Nepal: The case of Sirubari and Ghalegaun homestays. *Tourism Management Perspectives, 26*, 49–58.

Wang, C. H., & Chen, S. C. (2012). The relationship of full-service restaurant attributes, evaluative factors and behavioral intention. *International Journal of Organizational Innovation, 5*(2), 248–262.

Weaver, D. B., & Lawton, L. J. (2007). Twenty years on: The state of contemporary ecotourism research. *Tourism Management, 28*(5), 1168–1179.

Weiss, R., Feinstein, A. H., & Dalbor, M. (2005). Customer satisfaction of theme restaurant attributes and their influence on return intent. *Journal of Foodservice Business Research, 7*(1), 23–41.

Wheeller, B. (2005). Nature-based tourism in peripheral areas: Development or disaster. In *Ecotourism/egotourism and development* (pp. 263–272). UK: Channel View Publications.

Wiedmann, K. P., Hennigs, N., & Siebels, A. (2009). Value-based segmentation of luxury consumption behavior. *Psychology & Marketing, 26*(7), 625–651.

Wight, P. A. (1997). Ecotourism accommodation spectrum: Does supply match the demand? *Tourism Management, 18*(4), 209–220.

Wondirad, A. (2019). Does ecotourism contribute to sustainable destination development, or is it just a marketing hoax? Analyzing twenty-five years contested journey of ecotourism through a meta-analysis of tourism journal publications. *Asia Pacific Journal of Tourism Research, 24*(11), 1047–1065.

Yang, W., & Mattila, A. S. (2014). Do affluent customers care when luxury brands go mass? *International Journal of Contemporary Hospitality Management, 26*(4), 526–543.

Yang, W., & Mattila, A. S. (2016). Why do we buy luxury experiences? *International Journal of Contemporary Hospitality Management, 28*(9), 1848–1867.

Yang, W., & Mattila, A. S. (2017). The impact of status seeking on consumers' word of mouth and product preference—A comparison between luxury hospitality services and luxury goods. *Journal of Hospitality & Tourism Research, 41*(1), 3–22.

Yeoman, I. (2008). *Tomorrows Tourist*. Oxford: Elsevier.

Yeoman, I. (2011). The changing behaviours of luxury consumption. *Journal of Revenue and Pricing Management, 10*(1), 47–50.

Yeoman, I. (2012). *2050-Tomorrow's Tourism* (Vol. 55). UK: Channel View Publications.

Yeoman, I., & McMahon-Beattie, U. (2006). Luxury markets and premium pricing. *Journal of Revenue and Pricing Management, 4*(4), 319–328.

Yeoman, I., & McMahon-Beattie, U. (2018). The future of luxury: Mega drivers, new faces and scenarios. *Journal of Revenue and Pricing Management, 17*(4), 204–217.

Yesawich, P. B., & Russel. (2004). *Portrait of Affluent Travelers*. Orlando, FL: YPB&R.

THEME 4

Environment and learning

22
ECOTOURISM AND THEORIES OF LEARNING/EDUCATION

Manuel Ramón González-Herrera and Silvia Giralt-Escobar

Introduction

The teaching performance of an ecotourism professor is more professional when it is based on science (Ferreiro, 2006). That is why it is increasingly necessary to promote the strengthening and links between ecotourism, education and research (Robledano et al., 2018), especially in the globalised world in which we live, where communicative globalisation has transformed educational potentials (Marginson & Dang, 2017). In this sense, it is considered that alternative tourism linked to education and culture can contribute significantly to human development from a socio-economic, political, cultural, and spiritual point of view, in harmony with itself, with nature, and with other human beings (Torres, Zaldívar, & Enríquez, 2013). Based on this, new professionals will be able to transfer scientific-technological knowledge and socialisation of ecotourism principles to local communities (Cujía, Pérez, & Maestre, 2017), and to the tourists with whom they interact.

Evidently, in tourism practice, educational goodwill is not so coherently associated to the quality of the services offered. Recent research reveals that some companies pay little attention to the educational components of ecotours, and concludes that, while almost all tourist guides minimise their educational role or practice, most tourists prioritise education and new learning in their ecotourism experiences. That is why they expect guides to improve their educational role, recognising that this constitutes a contradiction between the expectations of tourists and the understanding of the guides of their role, with significant implications for management and practice of ecotourism (Duong et al., 2019). This study shows two important educational axes for ecotourism; the first relates to the work of tourist guides (and tourist experiences in general) in relation to their educational role and the second regards the teaching of this discipline.

Such a contradiction reflects one of the challenges facing ecotourism education, so a change from the traditional teacher-centred and classroom-based educational practices, to innovative student-centred approaches is necessary (Ramírez & Santana, 2019), with emphasis on strengthening the binomial formed by ecotourism and sustainability education (Piñar, García, & García, 2012). Given this opportunity, some destinations have opted to increase the number of international and national visitors through human resources education, and thus, have a trained and qualified workforce, especially in ecotourism (Thapa, 2019).

DOI: 10.4324/9781003001768-22

Ecotourism and education

Ecotourism is a growing international tourism trend, with demands on natural, cultural, and human resources (Bustam, Buta, & Stein, 2012). Hector Ceballos-Lascuráin is credited with coining the term *ecotourism* and its preliminary definition in July 1983, when he stated that ecotourism involves travelling to relatively undisturbed or uncontaminated natural areas with the specific object of "studying", admiring, and enjoying the scenery and its wild plants and animals, as well as any existing cultural aspects (Planeta.com, 2007). Since the emergence of this concept in the late twentieth century, the influence of "knowledge and education" on nature tourists' behavior has been explored in many studies (Moghimehfar, Halpenny, & Ziaee, 2014). Therein, it is recognised that "visitor learning" is the central aim in almost all definitions of ecotourism, which involves a "learning experience" (Walter, 2013; Mondino & Beery, 2019).

Ecotourism has several characteristics. It takes place in natural areas; it sustains local communities; includes wildlife encounters and recreational experiences; and may take on aspects of adventure tourism, community-based ecotourism, volunteer tourism, and outdoor education. According to TIES's definition and principles of ecotourism, this is conceptualised as responsible travel to natural areas that conserves the environment and improves the wellbeing of local people (TIES Overview, 1990). This definition recognises the scope of "responsible travel", which supposes the need for an educational perspective. In this chapter, ecotourism is understood as "Travel with a primary interest in the natural history of a destination. It is a non-invasive and participatory form of nature-based tourism that is built around learning, sustainability (conservation and local participation/benefits), and ethical planning, development and management" (Fennel, 2020, p. 20). This definition emphasises the importance of "learning/education" as a central theme of ecotourism.

Based on the previous definitions, it is important to mention that links between ecotourism and education are closely related to the world of tourism and travel, based on which the need for training of different stakeholders arises, either from a professional perspective or from that of general culture. Camargo and Sánchez (2016), argue that good tourism development starts with education. In this order, these authors coincide with the criteria of the Tourism Education Futures Initiative (TEFI), when they affirm that collaborative education and co-creation of knowledge are the basis for creating sustainable tourism (Camargo & Sánchez, 2016), one of the main principles of ecotourism education.

In response to this demand, ecotourism education is offered through different university models and tourism training technicians (Cervera & Ruiz, 2008; Boluk & Corey, 2016), as well as through tourism education and training promoted by UNWTO in response to the needs of employers and tourism professionals (Sancho Pérez, 1995; Fossati, Marín, Pedro, & Sancho, 2003). In the same way, the general formation of the tourist culture of the host community is particularly favored (Alonso, Gallego, & Honey, 1995, 2006), and the experiential learning during tourist trips is inspired (García-Allen, 2019). For this reason, tourism becomes a cognitive and affective-motivational phenomenon for different audiences, which is why this discipline needs to be studied by the pedagogy and education sciences.

Pedagogy is one of the social sciences that studies consciously organised and goal-oriented education (Urías, 2013), as well as unconscious and unintended experiences grouped in the informal education category (Elías, 2016). That is to say, that it studies the formation of the students in all its aspects, in function of which the educational institution as a main factor, the family, and the social organisations intervene. In the opinion of different authors, like Elías (2016), "Pedagogy has a dubious status as a scientific discipline, in conflict, voices inside and outside its domains debating its nature. In this way, it has incorporated theoretical constructs, methodological focuses and instrumental resources from sciences with a greater degree of consolidation – and recognition – (anthropology, sociology, psychology, among others); at the same time that it has developed its own theoretical and methodological body" (Elías, 2016, pp. 33–34). With these approaches, ecotourism education represents the integrated set of knowledge that allows the guidance of the educational process of educational institutions and their environments of influence, as well as of other training institutions.

Didactics is the branch of pedagogy that studies the teaching and learning processes, through which instruction, education, and student development are given (Álvarez de Zaya, 1999; Urías, 2013). This subdiscipline has a dynamic, complex, and multifactorial object of study, at the same time that it involves the active participation of the learner (*Ibid.*). Hernández, Hernández, Capote, and García (2004) consider that "Education is a very complex sociocultural phenomenon and, for its complete study and analysis, the participation and collaboration of multiple disciplines that explain its different dimensions and contexts are necessary. [So] … nobody doubts that without such concurrence of disciplinary perspectives, the analysis of education would be partial and incomplete" (Hernández et al., 2004, p. 40).

Learning and teaching are conceptual categories corresponding to different disciplinary fields, although connected; so, they assume different objects of study. On the one hand, learning is the process of acquiring knowledge, skills, abilities, attitudes, and values, which are generally obtained through observation, study, teaching, experience, or practice. Due to its complexity, there are various theoretical positions and conceptual meanings regarding its definition, methods, and applications. Depending on these, different learning paradigms and theories have been developed regarding the act of learning, which are mainly linked to the disciplinary field of psychology. On the other hand, teaching is the action and effect of teaching or instructing through the transmission of ideas, principles, beliefs, knowledge, experiences, skills, and habits to another person who does not have them. During this activity, teachers or facilitators interact with their students in a given educational context, through which learning or knowledge acquisition is facilitated. To achieve this goal, several paradigms, models, and approaches related to the act of teaching have been developed, mainly linked to the disciplinary field of pedagogy and education sciences.

When pedagogy assumes tourism in particular as an educational study object, a specific branch of this known as tourism pedagogy or leisure pedagogy is created (Colton, 1987; Galles, Graves, & Sexton, 2018; Zavydivska, Zavydivska, & Khanikiants, 2019). Consequently, the didactics of tourism has as a specific study object the teaching and learning processes of tourism and its subdisciplines (like ecotourism), as well as ecotourism experiences (for example, the practice of a tour guide). It therefore covers the study of formal, non-formal, and informal educational events, such as those that result from the daily experience of visitors who come to a destination. In this way, it includes attention to the components of the personal teaching and learning processes -professor and student- and non-personal components such as objectives, contents, methods, means of teaching, and evaluation, among others.

In the opinion of Colom and Brown (1993), the pedagogy of tourism is focused on two main areas that correspond to training for tourism (formal and non-formal) and the educational content of tourism (passing from informality to non-formal education). These perspectives offer the possibility of application in specific areas, such as free time education, education for international and intercultural understanding, among others (Colom & Brown, 1993). Today, tourism academics and educators increasingly accept that education must meet the needs of the industry, and they have begun to wonder what can be done to help students think and learn more broadly and critically (Mair & Sumner, 2017), which contributes to the elevation of quality in tourism training and the improvement of tourism pedagogy.

Different research addresses the pedagogical study of ecotourism incorporating innovative topics such as the pedagogy of climate change, which potentiates the role and responsibility of visitors as key actors in the face of climate change, or the potential use of these sites to promote environmental learning. In addition, social and political actions on climate change are promoted "in situ", all of which facilitates experimental learning, responsibility, and civic action towards conservation (Jamal & Smith, 2017). In the same way, different authors have incorporated pedagogical research into their daily work with the purpose of examining how sustainable the tourism education of future professionals is, recognising that curricular programs with traditionalist approaches still persist (Cole, 2019), which represents a challenge for tourism pedagogy.

On the other hand, the pedagogy of leisure is considered a specialised branch of the pedagogy that is responsible for the study of leisure or free time of people, in order to promote teaching and learning oriented to the educational or productive use of it (Colton, 1987; Colom & Brown 1993; Teplicancova, Almasiova, Krska, & Sedlacek, 2017; Hjalmarsson, 2018). In particular, this pedagogical dimension pays attention to the free time of the whole society, since we all have this space of time in different measures, taking advantage of the free time budget due to educational opportunities. In this order, it facilitates from the pedagogical point of view instruction and education in its cognitive-instrumental, affective-motivational, and developmental dimensions, by integrating the components of the teaching and learning processes. Leisure pedagogy is related to school and out-of-school activities through the links that are generated by acting on the same subject. Thus, the first includes the pedagogical projection of the actions carried out in school institutions, while the extracurricular pedagogy includes education of free time in the family, the community, and other institutions.

Conceptualised from this perspective, the pedagogy of leisure focuses its attention on the categories of rest, fun, and development (Miranda, 2006; Ávila, 2017). In such a way, the forms of occupation of free time are becoming an object of study, which has a favourable impact on human behaviours, generating a model for the use of free time based on education, and preparing us for a better socio-environmental performance. It guides us pedagogically to assume the planning, development, and control processes of different ecotourism activities, promoting appropriate didactic approaches to achieve a better efficiency and effectiveness of learning in different contexts of action.

Under this approach, the pedagogy of leisure promotes the fulfillment of basic educational, sociocultural, and recreational functions that allow the personal and intellectual development of those who participate. The fundamental areas of leisure education concern school, cultural institutions, and other leisure-time institutions of an educational nature that develop excursions, urban visits, parties, and evenings out and community activities linked to local popular traditions, folk crafts, gastronomy, painting, photography, yoga, among many others. It also includes activities in parks and open spaces for leisure time, in which there is usually no face-to-face educational intervention, but which incorporate pedagogical criteria as part of the activities that are promoted. For example, the case of summer camps and other types of holiday stays in natural environments, clubs, and other similar spaces where free time activities take place.

There is a substantive base of information on ecotourism guide training (Black, Weiler, & Chen, 2019). This is due to guides having important roles to perform in the ecotourism experience, such as ensuring the safety of the visitor, the interpretation of sites, and modelling appropriate environmental and cultural behavior. However, to be able to perform these roles, guides need proper training that provides them with the necessary knowledge and skills (Black, Ham, & Weiler, 2001). Tour guide training is an adult education activity, but much training is competency-based with an emphasis on knowledge transmission and skill acquisition; therefore, good training should lead to change, not only in terms of knowledge and skills, but also in attitudes and behaviour (Christie & Mason, 2003). This process makes it possible to develop abilities which could enable guides to communicate and interpret the environment, promote minimal impact practices, ensure the sustainability, and motivate visitors to appreciate the quality of the tourist destination (Skanavis & Giannoulis, 2009). In this way, practical teaching of ecotour guides has a positive impact on their roles and responsibilities (Ballantyne & Hughes, 2001).

On the other hand, host learning and education give them the capacity for local self-determination and control of ecotourism development and management (Regmi, Dev, and Pierre 2016), especially when there is a lack of proper environmental education (Mondino & Beery, 2019). Training local people to be interpretive guides helps achieve not only environmental sustainability, but also economic sustainability; once trained, they may encourage conservation action amongst both tourists and the local community (Skanavis & Giannoulis, 2009).

Ecotourism education process

According to Jafari (2005), tourism is a scientific or academic discipline (Jafari, 2005, pp. 46–55). In this sense, it can be considered that ecotourism is an academic subdiscipline, as this activity becomes an objective of teaching and learning. For this reason, it is necessary to know how to teach and communicate knowledge to students, and how students learn during this education process. The ecotourism education category is understood in this chapter as the teaching and learning processes that take place in different institutions, either formal, not formal or informal, in school or out of school. It is also recognised that there is a dialectical union between instruction and education, through which the student assimilates the content of the teaching, while producing and developing their personality traits, which influences the feelings, development, emotions, values, and so on (Álvarez de Zaya, 1999). It is worth noting that in informal education there are processes that can hardly be referred to as "teaching", however there are lessons learned. All these educational processes have the characteristics of a theoretical system composed of concepts, categories, laws, and a particular structure of its components, which determine an internal logic, in which different external conditions to the object itself intervene (*Ibid.*).

For a long time, traditionalist teaching and learning practises have been characterised by the leading role of the professor, who assumes the function of transmitting information as a part of a rote learning process by reception. Under this approach, the professor explains to the students what they should learn in a unidirectional way, being in charge of the diagnosis of the student's learning needs, of the organisation and presentation of the content of a particular discipline, and of the reproductive evaluation of the students. On the other hand, the student assumes a role of passive receiver of information, who memorises and repeats the contents to face the evaluation administered by the professor. Under this conceptualisation, learning occurs individually and reproductively, so that everything is practically reduced to memory.

Such educational practises do not fully conform to the educational needs of the times in which we live, which is why it is necessary to promote alternative educational models and teaching strategies for ecotourism education and training. Based on these, the ecotourism educational process could contribute to the formation and development of cognitive-instrumental, affective-motivational, and axiological knowledge, and in this way, to assume a positive attitude and consequently responsible behaviours in each tourist destination. In this regard, it is recommended that the following categories be integrated into ecotourism education:

- Knowing: cognitive dimension.
- Knowing how to do: instrumental dimension.
- Wanting to do: motivational affective dimension.
- Knowing how to be: motivational affective dimension.
- Be willing to do: attitudinal dimension.
- Doing: behavioural dimension.
- Making to know: communicational dimension.

The aforementioned categories are related to the four pillars of education proposed by UNESCO: learning to know, learning to do, learning to be, and learning to live together. It is therefore important to strengthen ecotourism education by incorporating the 10 commandments of learning raised by Pozo (2008), which means the development of teaching and learning processes considering:

1. Interests and motives of the students.
2. Previous knowledge of the students.
3. Adequate dosage of the amount of new information presented in each activity.
4. Suitable appropriation of the basic knowledge that will be necessary for future learning.
5. Diversification of tasks and learning scenarios for the same content.
6. Design of learning situations based on the contexts and tasks in which the learners must recover what they have learned.
7. Organisation and connection of each learning activity with the other one, so that the students perceive the explicit relationships between them.
8. Incentive among students to reflect on their knowledge, helping them to generate and resolve cognitive conflicts that arise.
9. Assignment of learning problems and/or open tasks, and the promotion of the cooperation among students for their resolution.
10. Training of the students to plan and organise their own work.

Theoretical and methodological bases of the teaching and learning processes

The theoretical and methodological bases of the teaching and learning processes have been configured according to the specific conditions and educational proposal of each historical period, so they are the result of the transmission of the accumulated knowledge by humanity, and the assimilation of these for new generations (Labarrere & Valdivia, 1988, p. 164). Regarding each of these historical moments, different scientific psycho-pedagogical theories have been formulated that have promoted the design of models, theoretical frameworks and didactic strategies (methods, techniques, activities), among others, which offer professors the premises to conduct the teaching and learning processes. Learning theories make it possible to understand and develop the processes by which students acquire knowledge, while teaching theories allows scientific guidance of the instructional and educational processes of each academic discipline. Both groups of theories complement each other as part of ecotourism education, based on what the construction and appropriation of disciplinary knowledge of this specific field of knowledge materialises.

Different studies at the international level have assumed the research and operationalisation of psycho-pedagogical theories as a basis for improving the educational and transformative practice that ecotourism educators carry out (Benckendorff & Zehrer, 2017). Kay and Kibble (2016) recognise that educators should understand learning theories and be able to apply them in the classroom (Kay & Kibble, 2016). As a basis for the systematisation of collected information in this chapter, the classification elaborated by Elías (2016) is used, in which the author relates the theoretical proposals corresponding to the psychoeducational and pedagogical dimensions of teaching practice as related disciplinary fields (Table 22.1). Based on this classification, the author identifies theoretical proposals that obey different objects of study, that is, on the one hand, the disciplinary field of educational psychology (learning theories), and on the other hand, the disciplinary field of pedagogy (teaching approaches) (Elías, 2016).

Relevant theories of knowledge and psychoeducational paradigms applicable to ecotourism

The study of the psychoeducational dimension of ecotourism is carried out through two constructs, that of learning theories and that of psychoeducational paradigms as a derived concept (Elías, 2016).

- **Cognitive Learning Theory**

  The background of cognitive theories is related to different academic disciplines. This paradigmatic approach is focused on cognition as a way for the search, acquisition, organisation, and use of knowledge. Therefore, it has as its purpose the description and explanation of mental representation as a model of information representation, based on which it incorporates the categories of the cognitive (attention, perception, memory, intelligence, language, thought, among others) for the understanding of the complex mental phenomena that lead to learning (Vega, 1984; Gardner, 2006). In this sense, it takes as an approach the question of how information is processed, and presents learning as the result of the acquisition of knowledge, while considering the possibility of developing the cognitive potential of the subject that learns as a condition for the solution of problems. The proposal of projects and programs "to teach to think" and "to learn to learn" correspond to this paradigm.

  Under this paradigm, the professor of ecotourism is responsible for promoting the development of mental representations in their students with the purpose of ordering the information obtained through these representations, and thus promote the formation of learning skills; for this, different paradigmatic positions are assumed as the meaningful learning approach. For their part, the ecotourism student is not a simple subject that responds to external stimuli, but instead becomes an active processor of information, capable of processing representations (models, schemes) that serve as the basis for interpreting and transforming reality.

  The ecotourism professor and ecoguide must consider knowledge as symbolic representations in the students' minds, so it occurs through "experiences" that are stored and retrieved in memory or cognitive structure. For this reason, the educator must be a stimulator of experiences and contribute to the development of the ability to think and reflect. One example of cognitive learning activity in ecotourism education includes (Grade Power Learning, 2020) who asks students to reflect on their experience on ecotourism trails and identify negative environmental impacts. The professor could encourage them to find new solutions to impacts and promote discussions about the status of the

Table 22.1 Psychoeducational and pedagogical dimensions of teaching practice

PSYCHOEDUCATIONAL DIMENSION			*PEDAGOGICAL DIMENSION*		
Learning Theories		*Psychoeducational Paradigms*	*Educational Paradigms*	*Exercise of Power*	*Teaching Approach*
Environmentalists	Connectionism Behaviorism Contiguous conditioning Operant conditioning	Behaviorist	**Focused on teaching**	Coercive	Executive Approach
Cognitivist	Gestalt theories Information Processing Theories	Humanist	**Focused on learning**	Persuasive Cognitive	Cultivator Approach
Interactionists	Psychogenetic Sociocultural	Psychogenetic			Classic Liberating Approach
		Sociocultural		Collective	Critical Liberating Approach

Source: Based on the multi-referential vision of teaching practice (Elías, 2016, p. 87).

environment based on these negative impacts. The professor should also help students explore and understand how ideas are connected, and to ask students to justify and explain their thinking. Finally, the professor uses visualisations to improve students' understanding and recall.

Practical ecotourism learning is a form of meaningful experiential learning in which students interact with real-life experiences through direct observation and interaction with the environment, both personally, interpersonally, and in teams. For this focus, academic study practices play an important role while recognising that experiential learning is very useful for business careers in programs such as business accounting and finance (McCarthy, 2016; Cea, Jorge, Burgos, & Filgueira, 2018). Under this paradigm, professors develop the ability of their students to learn from their own experiences based on a conceptual theoretical framework and well-planned objectives, leading to the development of skills that allow meaningful learning and possibilities of making decisions in new situations. On the other hand, students must assume an active role during the understanding of the setting in which they learn, reflect on the experience, conceptualise the experience, and apply their experiential knowledge to the solution of new problems. Community-based ecotourism (CBET) is a site of experiential learning which may encourage transformative learning for visitors (Walter, 2016).

- **Constructivism Learning Theory**

Although it is based on several preceding paradigms, derived from the classical theories of learning (sociocultural psychology of L. Vygotsky and the psychogenetic theory of J. Piaget) and pedagogical theories (critical pedagogy of P. Freire, and action-oriented teaching based on the theory of activity formulated by L. Vygotsky and A. Leontiev), this psychological current has a student-centred approach, and is oriented to the question of how knowledge is constructed, that is, how reality is known and how to learn, which reflects a transition from traditional practices of information exposure towards the construction of knowledge and personality. Researchers and academics from different disciplines intervene in this area of study with the purpose of achieving meaningful and cooperative group learning, based on which the professor must create learning situations that allow their development.

The ecotourism educator must be supported by the presentation of problems and conflict scenarios, which favour the appropriation of working methods for their solution. Under this paradigm, the professor is not a simple transmitter of knowledge, but a facilitator or mediator of the most favourable conditions for the learning of the students. Therefore, the main function is to guide the process in such a way that their students learn; in this sense, the educator must relate the contents of the teaching with the needs, interests, and previous experiences of their students, and promote the conditions for the full enjoyment of this process. On their part, the student is not a passive recipient of information, but an active subject, and at the same time is responsible for discovering and building their own learning, as well as providing meaning to that knowledge in order to transfer it into new learning situations. Some recent studies corroborate that learning is an active experience, so, the ideas that students have on the subject and the subject taught will be part of their learning experience, as learning is socially and culturally rooted (Sithara & Marikar, 2017).

There are many examples of constructivist activities, which can be used in ecotourism education. Among these are ecotourism debates based on background readings, experimentation tasks that students develop and then bring to class to discuss the results in classroom, discussion on ecotourism business management issues with ecotourism owners, field trips to different ecotourism destinations, interactive business case studies, and elaboration of monitoring plans for their proposed business. A didactic strategy to put in use during ecotourism classes could be to present a learning situation through which the students must answer what they know, such as when introducing the study of the Triple Bottom Line Model. In this learning context, students are encouraged to carry out a collaborative teaching assignment for the construction of the model; after some time they expose the obtained product in classroom, and lastly the professor proposes a new task to put into practice the knowledge acquired.

In particular, research projects are very useful because students research a topic and can present their findings to the class (project-based learning). This focus allows students to appropriate knowledge, skills, and competencies through the interpretation and investigation of objects, phenomena, and processes of objective reality, in a way that encourages the relationship of their learning with a situation of tourism (challenge) to which they must find a solution. The basic components of the ecotouristic project correspond to the establishment of an idea or topic which is relevant to the students; the elaboration of appropriate evaluation criteria; the approach of a guiding question or challenge; the presentation of learning activities to be addressed during the development of the project; the final product to deliver; and the audience for the presentation (diffusion) of it (Gobierno de Canaria, 2012).

- **Sociocultural Learning Theory**

  This paradigmatic trend recognises the contribution of society to the development of the individual, and highlights the role of the interactions that are established between each subject and the culture of the context in which it operates, which is why cognitive development is carried out collaboratively. Marginson and Dang (2017) point out that Vygotsky's sociocultural theory is widely used in educational research, and relate this potential to the proposal made by this author regarding the four "genetic domains" to investigate higher cognitive processes, that is, phylogenetic (humans in natural evolution), the historical cultural (social activity of humans), the ontogenetic (individual life expectancy), and the microgenetic (immediate events).

The main focus of sociocultural theory (L. Vygotsky) is sociocultural learning, which recognises that cognitive development cannot be isolated from human development, from the social, cultural, political, and economic conditions in which it is carried out (development unit), highlighting the important function of education for this purpose. The proposal of "Zone of Proximal Development" (ZPD) corresponds to the sociocultural paradigm; this represents the distance between the level of real or effective development of the student, in which it is possible to solve learning problems faced by themselves (comfort zone), and the level of potential development in which they would require guidance or assistance for the resolution of these problems, since they are not yet able to perform them independently. This conception explains that in the learning process, there is always a zone of near development towards which the student's knowledge will advance under the guidance of the professor, and that such learning can be stimulated and directed from the educational institutions.

Under this paradigm, ecotourism learning is recognized as a social and interactive construction, in which a bi-directional and dynamic relationship between the subject (ecotourist) and the object (ecotourism destination) adjusted to historical-cultural conditions, is established. For this, the main function of the professor is to teach interactively and promote the progressive advancement of the students from their comfort zone (autonomous) to the area of near development (not autonomous), so that the knowledge and skills required are acquired to gain autonomy and self-regulation. At the same time, the students assume an active and leading role in the construction and reconstruction of their ecotourism knowledge, so they play an important role in the elaboration of procedures that allow the solution of problems. In this way, they are capable of making the ecotourism knowledge their own through a process of progressive internalization from an external plane to an internal one.

Taking into consideration the role of the interactions established among students or visitors and the ecotourism destination, this approach could be applied through for example, environmental interpretation activities. For this purpose, the focus of learning in situ or situated learning could be integrated, which represents a form of cognitive learning that promotes learning in real scenarios, and also the focus of learning by doing, which is represented by a constructivist methodology that assumes learning based on the approach of practical doing as a way to reach knowledge. It is very important to recognise that environmental interpretation is considered an important vehicle for sustainable tourism by creating pro-environmental attitudes and responsible behaviours (Poudel & Nyaupane, 2013); consequently, it is an indispensable tool for achieving the goals of ecotourism (Weiler & Ham, 2000, 2001, 2002; Ham & Weiler, 2002), and a reason why it is recommended to strengthen education oriented towards this goal.

Most of the postulates previously explained by different learning theories, psychoeducational paradigms, educational paradigms, models, and teaching approaches related to the learning and teaching acts (Table 22.1) could be integrated during the ecotourism education processes. The following example (Table 22.2) shows three learning models and teaching approaches that can be integrated through a teaching assignment developed by means of an academic study trip, in which students learn collaboratively, by discovery, and in a meaningful way at the same time.

Table 22.2 Integration of approaches through an ecotourism teaching assignment.

Collaborative learning by significant discovery through Academic Study Trips		
Discovery Learning	**Collaborative Learning**	**Significant Learning**
(Piaget & Bruner)	(L. Vygotsky & Bruner)	(D. Ausubel)
Students must be able to discover the **learning content in a meaningful way**, and solve the practical problems they face in their surroundings, so that they can transfer the learning to new situations and make decisions dependently.	Students **learn collaboratively in small structured groups**, in which two or more members work with the same task or goal to be achieved. The learning is guided, and has a teacher-oriented purpose for the students to learn; participation is promoted through different activities.	Students are able to relate their previous knowledge and the experiences **lived with the new learning content** through a process of readjustment and reconstruction of knowledge, for which they must be motivated to learn by solving problems, and carrying out practical and laboratory activities.
Teaching assignment "Ecotourism Planning" activity type: Academic study trip to understand the geospatial approach of an ecotourism destination through the interpretation of the production process of the ecotourist space, as well as the recognition of the model of territorial implantation, the inventory and evaluation of tourism resources. These activities allow students to understand and justify the contribution of the practical learning process to the formation of ecotourism knowledge, incorporating academic study trips to the development of collaborative learning by significant discovery as a strategy and practice of sustainability.		

Conclusion

During the last decades, more and more attention has been paid to the theoretical and methodological bases of ecotourism teaching and learning processes. Based on that, the study of ecotourism as a scientific discipline, and the teaching and learning ecotourism education process has been promoted. Such development has become a major challenge for professors, ecotour guides, and other educators who have had to find creative answers to questions such as how to teach and communicate knowledge to students and how students learn during this education process. In the search for answers to these questions it is possible to understand a clear synergy between educationally oriented ecotourism, conservation of environmental resources, and sustainable development (Sander, 2010; Coles, Poland, & Clifton, 2014). In function of these, it is recommended to project clearly the educational aims, curriculum, role of the instructor, teaching assignments, and learning outcomes to optimise the educational use of natural and sociocultural environments during the ecotouristic activities.

Reaching this target, educators should study different learning theories, psychoeducational paradigms, educational paradigms, models, and teaching approaches applicable to ecotourism education. In correspondence with each of these paradigms, educational proposals should be formulated based on their own theoretical-conceptual frameworks. Each of these theories has its own limitations and inadequacies, but their study and understanding serve as a guide to incorporate alternative forms of ecotourism learning during the educational processes, which makes the paradigmatic, conceptual, methodological, and instrumental understanding of each proposal necessary.

References

Alonso, C., Gallego, D., & Honey, P. (1995). *Learning styles. Diagnostic Procedures and Improvement.* Bilbao, Spain: Mensajero Editorial.

Alonso, C., Gallego, D., & Honey, P. (2006). *Learning styles. Diagnostic Procedures and Improvement.* Bilbao, Spain: Mensajero Editorial. Retrieved from https://www.researchgate.net/publication/311452891_Los_Estilos_de_Aprendizaje_Procedimientos_de_diagnostico_y_mejora.

Álvarez de Zaya, C. (1999). *Didactics. The School in Life.* Havana, Cuba: Education and Development Editorial.

Ávila, S. (2017). Las tres D en el tiempo libre: descanso, diversión y desarrollo. *Comunicae.es.* Retrieved from https://www.comunicae.es/nota/las-tres-d-en-el-tiempo-libre-descanso-1192288/.

Ballantyne, R., & Hughes, K. (2001). Interpretation in ecotourism settings: Investigating tour guides' perceptions of their role, responsibilities and training needs. *The Journal of Tourism Studies, 12*(2), 2–9. Retrieved from https://www.researchgate.net/publication/285640604_Interpretation_in_Ecotourism_Settings_Investigating_tour_guides'_perceptions_of_their_role_responsibility_and_training_needs.

Benckendorff, P., & Zehrer, A. (Ed.). (2017). *Handbook of Teaching and Learning in Tourism.* Cheltenham, United Kingdom: Edward Elgar Publishing Journals. doi:https://doi.org/10.4337/9781784714802

Black, R., Ham, S., & Weiler, B. (2001). Ecotour guide training in less developed countries: Some preliminary research findings. *Journal of Sustainable Tourism, 9*(2), 147–156. Retrieved from https://www.tandfonline.com/doi/abs/10.1080/09669580108667395.

Black, R., Weiler, B., & Chen, H. (2019). Exploring theoretical engagement in empirical tour guiding research and scholarship 1980–2016: A critical review. *Scandinavian Journal of Hospitality and Tourism, 19*(1), 95–113. doi:https://doi.org/10.1080/15022250.2018.1493396.

Boluk, K., & Corey, J. (2016). Tourism Education Futures Initiative (TEFI Tourism). Retrieved from http://digitalcommons.library.tru.ca/tefi/tefi9/day2/14.

Bustam, T. D., Buta, N., & Stein, T. V. (2012). The role of certification in international ecotourism operators' internet promotion of education. *Journal of Ecotourism, 11*(2), 85–101. Retrieved from https://www.tandfonline.com/doi/abs/10.1080/14724049.2012.683005?scroll=top&needAccess=true&journalCode=reco20.

Camargo, B., & Sánchez, I. (June, 2016). Educators' perceptions and incorporation of critical perspectives in tourism programs: The case of Mexico. In *9th Tourism Education Futures Initiative Conference (TEFI9)*. Thompson Rivers University. Retrieved from https://digitalcommons.library.tru.ca/tefi/?utm_source=digitalcommons.library.tru.ca%2Ftefi%2Ftefi9%2Fday1%2F11&utm_medium=PDF&utm_campaign=PDFCoverPages.

Cea, R., Jorge, S., Burgos, H., & Filgueira, M. E. (2018). Tipos psicológicos y estilos de aprendizaje de estudiantes de una facultad de ciencias económicas y administrativas en Chile. *RAN: Revista Academia & Negocios*, *1*(4), 65–80. Retrieved from https://dialnet.unirioja.es/servlet/articulo?codigo=6599118.

Cervera, A., & Ruiz, M. E. (2008). Tourism education: A strategic analysis model. *Journal of Hospitality, Leisure, Sports and Tourism Education* (Pre-2012), 7(2), 59–70. Retrieved from https://www.researchgate.net/publication/26579093_Tourism_education_a_strategic_analysis_model.

Christie, M. F., & Mason, P. A. (2003). Transformative tour guiding: Training tour guides to be critically reflective practitioners. *Journal of Ecotourism*, *2*(1), 1–16. Retrieved from https://www.tandfonline.com/doi/abs/10.1080/14724040308668130.

Cole, E. (2019). *Sustainable Tourism Pedagogy in Swedish Tourism Master's Programs: Higher Education Curriculum Approach Towards Sustainable Pedagogy*. Independent thesis Advanced level (professional degree). Dalarna University, School of Technology and Business Studies, Tourism Studies. Retrieved from http://du.diva-portal.org/smash/record.jsf?pid=diva2%3A1302555&dswid=-1383.

Coles, T., Poland, R., & Clifton, J. (2014). Ecotourism in an educational context: Promoting learning opportunities through travel. *Journal of Biological Education*, *49*(2), 213–217. doi:http://dx.doi.org/10.1080/00219266.2014.956484.

Colom, C. A., & Brown, G. G. (1993). Turismo y Educación (Bases para una Pedagogía del Turismo). *Revista Española de Pedagogía*, año Ll, 194(enero – abril), 57–75. Retrieved from https://revistadepedagogia.org/wp-content/uploads/2018/03/3-Turismo-y-Educaci%C3%B3n.-Bases-para-una-Pedagog%C3%ADa-del-Turismo.pdf.

Colton, C. (1987). Leisure, recreation, tourism: A symbolic interactionism view. *Annals of Tourism Research*, *14*(3), 345–360. Retrieved from https://www.sciencedirect.com/science/article/abs/pii/0160738387901071.

Cujía, G. E., Pérez, R. S., & Maestre, C. D. (2017). Ecotourism, education, science and technology as factors of sustainable development. Case, Guajira, Colombia. *Educación y Humanismo*, *19*(32), 174–189. doi:https://doi.org/10.17081/eduhum.19.32.2540.

Duong, L., Fountain, J., Stewart, E., Espiner, S., & Shone, M. (2019). Exploring the educative role of ecotourism guides in Vietnam. In *CAUTHE 2019, sustainability of tourism, hospitality & events in a disruptive digital age: Proceedings of the 29th Annual Conference*. Cairns, QLD: Central Queensland University, Australia. Retrieved from https://search.informit.com.au/documentSummary;dn=413005512754589;res=IELBUS.

Elías, J. A. (2016). *Compromisos e implicaciones que subyacen a la práctica docente en Educación Media Superior Una mirada desde la multirreferencialidad*. Maestría en Investigación Educativa Aplicada. México: ICSA, UACJ.

Ferreiro, G. R. (2006). *Estrategias didácticas del aprendizaje cooperativo. El constructivismo social: una nueva forma de enseñar y aprender*. México: Editorial Trillas.

Fennell. D. A. (2020). *Ecotourism*. London: Routledge.

Fossati, R., Marín, A., Pedro, A., & Sancho, A. (2003). *Educando Educadores en Turismo*. Madrid, España: Organización Mundial del Turismo.

Galles, G. M., Graves, P. E., & Sexton, R. L. (2018). Leisure and the production possibility frontier: A two-step pedagogy. *The American Economist*, *64*(1), 123–130. doi:https://doi.org/10.1177/0569434518810425.

García-Allen, J. (2019). Psicología y Mente. Retrieved from https://psicologiaymente.com/desarrollo/estilos-de-aprendizaje.

Gardner, H. (2006). *La nueva ciencia de la mente: Historia de la revolución cognitiva*. Buenos Aires, Argentina: Paidós. Retrieved from http://www.sidalc.net/cgi-bin/wxis.exe/?IsisScript=zamocat.xis&method=post&formato=2&cantidad=1&expresion=mfn=022460.

Gobierno de Canaria. (2012). *Kit de Pedagogía y TIC*. España: Consejería de Educación y Universidades. Retrieved from http://www3.gobiernodecanarias.org/medusa/ecoescuela/pedagotic/aprendizaje-basado-proyectos/.

Grade Power Learning. (2020). The Cognitive Learning Approach. Retrieved from https://gradepowerlearning.com/cognitive-learning-theory/.

Ham, S. H., & Weiler, B. (2002). Interpretation as the centrepiece of sustainable wildlife tourism. In R. Harris, T. Griffin, & P. Williams (Eds.), *Sustainable tourism: A global perspective* (pp. 5–44). London, UK: Butterworth - Heinneman. Retrieved from https://www.researchgate.net/publication/288911466_Interpretation_as_the_centrepiece_of_sustainable_wildlife_tourism.

Hernández, P., Hernández, C., Capote, C., & García, J. (April, 2004). La percepción del alumnado de las teorías psicoeducativas de los mejores y peores maestros. In *IV Congreso Internacional de Psicología y Educación "Calidad Educativa"*. España: Almería. Retrieved from http://www.tafor.net/images/Comunica001.pdf.

Hjalmarsson, M. (2018). The presence of pedagogy and care in leisure-time centres' local documents. *Leisure-Time Teachers' Documented Reflections, 43*(4), 57–63. doi:https://doi.org/10.23965/AJEC.43.4.07.

Jafari, J. (2005). El turismo como disciplina científica. *Política y Sociedad, 42*(1), 39–56. Retrieved from https://revistas.ucm.es/index.php/POSO/article/view/24139.

Jamal, T., & Smith, B. (2017). Tourism pedagogy and visitor responsibilities in destinations of local-global significance: Climate change and social-political action. *Sustainability, 9*(6), 1–27. Retrieved from https://www.mdpi.com/2071-1050/9/6/1082.

Kay, D., & Kibble, J. (2016). Learning theories 101: Application to everyday teaching and scholarship. *Advances in Physiology Education*, 40, 17–25. Retrieved from https://www.ncbi.nlm.nih.gov/pubmed/26847253.

Labarrere, G., & Valdivia, G. (1988). *Pedagogía* (Segunda edición). La Habana, Cuba: Editorial Pueblo y Educación.

Mair, H., & Sumner, J. (2017). Critical tourism pedagogies: Exploring the potential through food. *Journal of Hospitality, Leisure, Sport & Tourism Education, 21*(B), 195–203. Retrieved from https://www.sciencedirect.com/science/article/abs/pii/S1473837617301314?via%3Dihub.

Marginson, S., & Dang, T.-K.-A. (2017). Vygotsky's sociocultural theory in the context of globalization. *Asia Pacific Journal of Education, 37*(1), 116–129. Retrieved from https://www.tandfonline.com/doi/citedby/10.1080/02188791.2016.1216827?scroll=top&needAccess=true.

McCarthy, M. (2016). Experiential learning theory: From theory to practice. *Journal of Business & Economics Research, 14*(3), 91–100. Retrieved from https://doi.org/10.19030/jber.v14i3.9749.

Miranda, G. (2006). El tiempo libre y de ocio reivindicado por los trabajadores. *Revista Pasos, 4*(3), 301–326. Retrieved from http://www.pasosonline.org/Publicados/4306/PASOS09.pdf#page=25

Moghimehfar, F., Halpenny, E. A., & Ziaee, M. (2014). How big is the gap? Comparing the behaviours and knowledge of mountain hikers with ecotourism ideals: A case study of Iran. *Journal of Ecotourism, 13*(1), 1–15. Retrieved from https://www.tandfonline.com/doi/abs/10.1080/14724049.2014.925466.

Mondino, E., & Beery, T. (2019). Ecotourism as a learning tool for sustainable development. The case of Monviso Transboundary Biosphere Reserve, Italy. *Journal of Ecotourism, 18*(2), 107–121. Retrieved from https://www.tandfonline.com/doi/full/10.1080/14724049.2018.1462371.

Piñar, Á. Á., García, S. M. D., & García, C. H. (2012). Ecoturismo y Educación Ambiental para la Sustentabilidad en la Reserva de la Biosfera de Los Tuxtlas (México). *TuryDes, 5*(12). Retrieved from https://pdfs.semanticscholar.org/860a/8d77b5100e84dc1e763b15694572269efdba.pdf.

Planeta.com. (2007). Ecoturismo México. Retrieved from https://planeta.com/hector-ceballos-lascurain/.

Poudel, S., & Nyaupane, G. P. (2013). The role of interpretative tour guiding in sustainable destination management: A comparison between guided and nonguided tourists. *Journal of Travel Research, 52*(5), 659–672. doi:https://doi.org/10.1177/0047287513478496.

Pozo, J. I. (2008). *Aprendices y maestros: la psicología cognitiva del aprendizaje*. España: Alianza Editorial.

Ramírez, F., & Santana, J. (2019). Education and ecotourism. In F. Ramírez, & J. Santana (Eds.), *Environmental education and ecotourism* (pp. 21–25). Cham: Springer. Retrieved from https://link.springer.com/book/10.1007%2F978-3-030-01968-6.

Regmi, K., Dev, W., & Pierre. G. (2016). Conceptualising host learning in community-based ecotourism homestays. *Journal of Ecotourism, 15*(1), 51–63. Retrieved from https://www.tandfonline.com/doi/abs/10.1080/14724049.2015.1118108?journalCode=reco20.

Robledano, F., Esteve, M. A., Calvo, J. F., Martínez-Paz, J. M., Farinós, P., Carreño, M. F., ... & Zamorad, A. (2018). Multi-criteria assessment of a proposed ecotourism, environmental education and research infrastructure in a unique lagoon ecosystem: The Encañizadas del Mar Menor (Murcia, SE Spain). *Journal for Nature Conservation, 43*(June 2018), 201–210. Retrieved from https://www.sciencedirect.com/science/article/abs/pii/S1617138117301218#!.

Sancho Pérez, A. 1995. *Educando educadores en turismo*. Madrid, España: Organización Mundial del Turismo, Instituto de Turismo, Empresa y Sociedad, Universidad Politécnica de Valencia.

Sander, B. (2010). The importance of education in ecotourism ventures. *Substantial Research Paper American University*, May 2010, 1–61. Retrieved from https://www.american.edu/sis/gep/upload/education-ecotourism_srp_ben_sander-2.pdf.

Sithara, F., & Marikar, F. (2017). Constructivist teaching/learning theory and participatory teaching methods. *Journal of Curriculum and Teaching*, *6*(1), 110–122. Retrieved from https://eric.ed.gov/?id=EJ1157438.

Skanavis, C., & Giannoulis, C. (2009). Improving quality of ecotourism through advancing education and training for eco-tourism guides. *Tourismos*, *5*(2), 49–68. Retrieved from https://mpra.ub.uni-muenchen.de/id/eprint/25314.

Teplicancova, M., Almasiova, A., Krska, P., & Sedlacek, J. (2017). Social environment selected aspects determination by school children leisure time movement activities. *Gymnasium*, *XVIII*(2), 187–194. Retrieved from https://www.researchgate.net/publication/322268659_Social_Environment_Selected_Aspects_Determination_by_School_Children_Leisure_Time_Movement_Activities.

Thapa, B. (2019). Ecotourism education and development in Kazakhstan. *Journal of Hospitality & Tourism Education*, *31*(2), 119–124. Retrieved from https://doi.org/10.1080/10963758.2018.1485499.

TIES Overview. (1990). The International Ecotourism Society (TIES). Retrieved from https://ecotourism.org/ties-overview/.

Torres, F. César., Zaldívar, M. Pablo., & Enríquez, G. F. (2013). Turismo Alternativo y Educación. Una propuesta para contribuir al desarrollo humano. *El Periplo Sustentable*, *24* (2013), 125–154. Retrieved from https://dialnet.unirioja.es/servlet/articulo?codigo=4195434.

Urías, G. (2013). *La Pedagogía como fundamento del autodesarrollo comunitario*. Doctorado en Desarrollo Comunitario. UCLV. (unpublished).

Vega, M. (1984). *Introducción a la Psicología Cognitiva*. España, Madrid: Alianza Ed.

Walter, P. (2013). Theorising visitor learning in ecotourism. *Journal of Ecotourism*, *12*(1), 15–32. Retrieved from https://www.tandfonline.com/doi/abs/10.1080/14724049.2012.742093.

Walter, P. G. (2016). Catalysts for transformative learning in community-based ecotourism. *Current Issues in Tourism*, *19*(13), 1356–1371. Retrieved from https://www.tandfonline.com/doi/abs/10.1080/13683500.2013.850063?journalCode=rcit20.

Weiler, B., & Ham, S. H. (2000). Tour guides and interpretation in ecotourism. In D. Weaver (Ed.), *The encyclopedia of ecotourism* (pp. 549–564). Wallingford, UK: CABI Publishing.

Weiler, B., & Ham, S. H. (2001). Tour guide training: Lessons for Malaysia about what works and what's needed. In C. Nyland, W. Smith, R. Smyth, & M. Vicziany (Eds.), *Malaysia business in the new era* (pp. 149–161). Cheltenham, UK: Edward Elgar.

Weiler, B., & Ham, S. H. (2002). Tour guide training: A model for sustainable capacity building in developing countries. *Journal of Sustainable Tourism*, *10*(1), 52–69. Retrieved from https://www.tandfonline.com/doi/abs/10.1080/09669580208667152.

Zavydivska, N., Zavydivska, O., & Khanikiants, O. (2019). Features of free time pedagogy in the conditions of health preserving study of students. *ВИПУСК*, *13*(2019), 15–22. Retrieved from http://repository.ldufk.edu.ua/bitstream/34606048/22447/1/N.%20Zavydivska%2C%20O.%20Zavydivska%2C%20O.%20Khanikiants.pdf.

23

A CRITICAL ANALYSIS OF SUSTAINABLE DESTINATION GOVERNANCE FROM ENVIRONMENTAL PERSPECTIVE

A systematic review

Kadir Çakar

Introduction

Over the last few decades, the tourism industry has faced a enormous growth across the globe impacting specific locations (Hatipoglu, Alvarez, & Ertuna, 2016). This unexpected growth is often linked to and called as overtourism that underscore the consequences resulting from the global mobility and the necessity of discovering of new locations, as well as improved access to more established unpopular destinations as alternative (Cheer, Milano, & Novelli, 2019). One of the basic reasons behind the emergence of this issue is rapid urbanisation (Lalicic, 2019) that derived from the globalisation process and negative impacts of mass tourism practices (Milano, Novelli, & Cheer, 2019) that engender pollutions due to hosting a huge amount of tourists at destinations leading to environmental devastations by threatening sustainability (Singh, 2018). This fact underscores negative and positive impacts discussion of tourism referring to three pillars (e.g., economic, natural, and socio-cultural) (Mihalič, Šegota, Knežević Cvelbar, & Kuščer, 2016).

As a result, more attention has been drawn by scholars on developing solutions in response to such emerging societal problems, which entail moving visitors from hotspots destinations to silent locations as priority of destination planners and managers (Jacobsen, Iversen, & Hem, 2019) or rural areas as alternative places to be consumed (Eckert, Zacher, Pechlaner, Namberger, & Schmude, 2019). This creates the presence of ethical dilemmas that enhance untouched landscapes to be consumed whereas possible distortions of natural environments may arise. To this end, today tourists as contemporary consumers quest for and thus tend to travel to destinations by which they are intriguied that can only visited by few visitors (Zerva, 2018). This kind of tourism which accentautes sustainable forms of tourism are often called as eco-tourism (Spenceley, 2008) or responsible tourism (Hall, 2012). The management of such places are problematic which entails decision-making concerning principles, laws, policies, rules, and management of tourism and visitor referring to governance (Leung, Spenceley, Hvenegaard, & Buckley, 2018) to provide sustainability of touristic landscapes as well as achieve desired

 DOI: 10.4324/9781003001768-23

objectives in ecotourism destinations. Based on a critical approach, the objective of the current book chapter is to elicit present gap on the sustainable governance that relates to ecotourism and environmental policy whereas presenting recommendations for future research studies.

Literature review

Sustainability and tourism governance

Tourism is an activity that is marketed through environment to be consumed by visitors or tourists and, therefore, the environment is treated as a key resource, and as a result its protection and management are crucial both for the future of the tourism industry and the society from a holistic perspective (Dimitriou, 2017). Recently, more attention has been directed to the endorsement of sustainability principles in ensuring tourism development, which gained broader attention and thus has occupied a significant space in the agenda of scholars, governments, and industry practitioners (Cornelisse, 2019). In a similar manner, the term has been acknowledged as an integral part of tourism policymaking process by both the public and private actors at all levels of governance (Hall, 2011b).

To provide destinations' sustainability and achieving desired objectives, there are many preconditions such as knowledge, thought, the application of power, resources, and rules including coordination, cooperation, collaboration, and partnerships as being drivers of governance to form necessary policies to be realized them for tourism in destinations (Bramwell, 2011; Graci, 2013; Mihalič et al., 2016). It is, therefore, sustainable governance depends on the presence of numerous parameters through which public, private actors, and community stakeholders can ensure sustaianbility of tourism destinations or touristic landscapes (World Tourism Organization and Griffith University, 2017).

On one hand, effective governance is treated as an integral part to building and sustaining as a serving environment, necessary to the accomplishment of social and economic objectives for any country (Ndivo & Okech, 2020). It is also regarded as an essential tool to reach objectives of sustainable tourism as the participatory action of a wide variety of different tourism actors in decision-making process can increase both democratic processes and ownership (Bramwell & Lane, 2011). On the other hand, the presence of good governance is addressed as another essential prior requirement for sustainable tourism (Nunkoo, 2017) which refers to some vital criteria through which local actors can govern local policies in an effective way (Beaumont & Dredge, 2010).

So far, the concept of governance has been dealt with by scholars from a wide variety of perspectives as a form of public-private initiative through which main tourism actors can intentionally act in terms of reaching common goals (Dredge & Whitford, 2011). In the extant literature on the typology of tourism governance and modes of governance there are numerous research each of which provides the position of each actors, use of power, and distribution of authority in different manner (Kagermeier, Amzil, & Elfasskaoui, 2019; Richins, 2011). For instance, in their research, Wan and Bramwell (2015) proposed a hybrid mode of governance governance that encapsulates both state control and public-private partnerships whereas Hall (2011a) established typology of governance which split into four main areas: hierarchies, markets, networks, and communities. Hierarchies refers to state-led governance approach, whilst markets addresses the power of private economic actors and their associations. On the other hand, networks reflect partnerships that is ensured by public-private partnerships and associations while communities signals governance that is based on the local level with direct public involvement (Boluk, Cavaliere, & Higgins-Desbiolles, 2019).

All of these approaches have in common in terms of spatial application which are limited to urban areas while overlooking as how to manage and conservation in sustaining tourism in environmental or protected areas through governance. However, typology of the International Union for Conservation of Nature (IUCN) is considered as one of the most salient ones in terms of governance for protected areas proposing four types of governance typology that relies upon management approaches by taking degree to which dissemination of power into account of each type of typology (Leung et al., 2018):

- ***Type A-Governance by government***: This type of governance is often conducted by Federal or national ministry/agency, sub-national ministry or agency in charge at several different levels (e.g., regional, provincial, municipal), Government-delegated management (e.g., NGO);
- ***Type B-Shared governance***: This kind of governance usually refers to transboundary governance which may entail formal or informal arrangements among two or more than two countries; collaborative governance (in which more than actors closely work together to reach particular results or outcomes) and joint governance which works through a pluralist board or other multi-party governing body.
- ***Type C-Private governance***: This typology is mostly preferred for conserved areas established that are implemented by landowners, non-profit organizations (e.g., NGO, universities) and for-profit organizations (e.g., corporate landowners);
- ***Type D-Governance by Indigenious peoples and local communities***: In this typology, Indigenous peoples' conserved areas and landscapes are established and operated by Indigenous peoples; community-conserved territories are established and managed by local communities.

On the other hand, Dredge and Jamal (2013) offered 'soft' and 'hard governance' as typology of governance modes each these reflect differences in use. The mode of soft governance approach emphasises the role of market and production systems that omit state control to form tourism development whilst hard governance involves rule-dependent and hierarchical approach which imposes an authoritative order and control mechanisms which enable vertical and horizontal integration of policy and regulation.

Governance in ecotourism and protected areas

The concept of ecotourism, for which there is no commonly held single definition, is concerned with as sub-genre of sustainable tourism (Chiutsi, Mukoroverwa, Karigambe, & Mudzengi, 2011) and is considered as being one of the new forms of tourism that is among others terms to be easily defined (Wang, Cater, & Low, 2016). From this perspective, Fennell (2001) has identified 85 definitions of the term *ecotourism* derived from the analysis of his research. Ecotourism was improved by means of the twofold energetics of both the enhancement of the environmental movement and the integration of the concept of sustainable development (Higgins-Desbiolles, 2011). It is treated as a concept resulting from a globalisation process that is closely related to environmental governance (Duffy, 2006); it is fuelled by generalised principles regarding local livelihoods and conservation of not only natural resources but also cultural environments (Thompson, Gillen, & Friess, 2018). In its multidisciplinary nature, the concept has both supply and demand side perspectives each of which has many different sub-categories (Weaver & Lawton, 2007). On the other hand, tourism plays regarded as a vital in the way non-governmental organisations govern landscapes, particularly in decentralised conservation contexts not only in developing countries (Pellis, Lamers, & Van der Duim, 2015) but also in urban protected areas for each of which the presence governance is received a greater attention (De Leon & Kim, 2017). In a similar manner, the presence of strong local governance systems addresses whole participation of all community members in decision-making processes concerning issues of local residents as well as community-based ecotourism governance (Farrelly, 2011) that is described as a mode of governance which consists of the interaction of rules, institutions, processes, and principles by which the stakeholders in this approach perform and fulfill decisions (Gan, Nair, & Hamzah, 2019). It is also worth noting that in the application of effective management of protected areas in ensuring sustainable of ecotourism the necessary policy and planning process can be accomplished by the presence of governance (Fennell & Dowling, 2003).

It is suggested by prior research that in a value chain structure there is a interrelation between tourism governance and sustainability of regional development areas (Adiyia, Stoffelen, Jennes, Vanneste, & Ahebwa, 2015). In their research, 18 good governance strategies were found by Pasape, Anderson, and Lindi (2015) that support sustainable ecotourism. In another study which was led by Palmer and Chuamuangphan (2018) it was confirmed that within ecotourism context, governance has been established as possessing a crucial impact on the ways in which local participation in tourism is realised. These results derived from previous research confirm that the presence of good governance structure relies upon many different elements in providing sustainability of protected areas and in ecotourism, which entails the participation of diverse range of actors in decision-making such as government, private sector, and civil society (Nordlund, Kloiber, Carter, & Riedmiller, 2013; Saruman, Razman, Zakaria, & Ern, 2017). For instance, the element of cooperation between the stakeholders in embracing the norms of governance was assessed as a supporting tool in affecting the management in the protection and conservation of landscapes (Noh, Shuib, Tai, & Noh, 2018). Similarly, Zeppel (2012) highlighted the significance of collaborative governance by which destinations can achieve conservation objectives as well as preservation of natural areas may play a vital role in planning and policymaking processes at local levels.

On the other hand, in the existing relevant academic literature it is highlighted that there is a close relationship between good governance and adaptive co-management in ensuring sustainability of protected areas and ecotourism areas (Plummer & Fennell, 2009). Good governance is also defined as one of the seven facets of adaptive co-management (Fennell, Plummer, & Marschke, 2008). The adaptive co-management is also predicated on the participation of various actors representing stakeholder groups in decision-making processes and management of landscapes, as well as adaptive learning to increase the governance approach in order to protect and foster natural resources in ecological areas (Islam, Ruhanen, & Ritchie, 2018a). It refers to the social learning which come true by participatory action of stakeholders to achieve sustainable tourism in protected areas through good governance stucture which entails primarily the presence of cooperation and coordination that based on interactions which provides the building of trust, the establishment of formal and informal rules and norms, the sharing of power, and joint decision-making (Islam, Ruhanen, & Ritchie, 2018b). Over the past decade, this newly emerging management approach has been the focal point of local stakeholders to attain desired goals towards sustainable ecotourism at local level by local actors that implies the inclusive governance approach by involving all actors which centring social learning and adaptation in decision-making mechanism (Armitage, Marschke, & Plummer, 2008; Armitage, Berkes, Dale, Kocho-Schellenberg, & Patton, 2011; Islam et al., 2018a).

UNWTO Sustainable Development Goals (SDGs) and governance for sustainable tourism

The issue of sustainable development has been the intense focal topic of several different actors playing key a key role such as tourism policymakers and industry and destination marketing organisations including tourism researchers (Hall, 2019). The year 2015 has been a turning point for global development since governments have collectively acknowledged the 2030 Agenda for Sustainable Development, together with the Sustainable Development Goals (SDGs) (UNWTO, 2020). The year 2017 was declared a year of Sustainable Development by the World Tourism Organisation (WTO) to assemble the tourism industry and key actors to prioritise tourism as a tool to success SDGs (Dube, 2020). These SDGs are consisting of 17 goals and 169 targets each of which address diverse range of areas by considering the sustainability in all spheres of life (Dube & Nhamo, 2020; Laimer, 2017; UN, 2020), which calls for critical thinking on these areas to provide successfully transition process of tourism to sustainability (Boluk et al., 2019).

By referring to the SDG 8, SDG 12, and SDG 14, the tourism industry is considered in three of these goals through its huge amount of contribution to economic growth and creating jobs, sustainable production and consumption, and marine conservation (Perdomo, 2016). On the other hand, SDG 16 and SDG 17 are featured among remaining others as goals and targets of them are clearly refer to the necessary of presence of governance for sustainability of tourism. SDG 16, which refers to peace, justice, and strong institutions, entails the necessity of effective, accountable, and inclusive institutions, whereas SDG 17 reflects partnerships for the goals to be achieved by related actors that signals vital elements of governance to reach desired outcomes and objectives (Boluk, Cavaliere, & Higgins-Desbiolles, 2017). Goals and targets of SDG 17 are closely and directly related to governance for sustainable tourism (Siakwah, Musavengane, & Leonard, 2019) while the presence of SDG 16 partly reflects governance elements that stands as a supporting tool to foster SDG 17 towards sustainability through effective governance.

Methods

Review and selection of articles for analysis

The present chapter has adopted a qualitative systematic literature review as a method used. Keyword search terms were applied to collect data by using the terms of 'ecotourism and governance' (n = 9), 'environmental governance' (n = 7), 'sustainable destination governance' (n = 27), and 'sustainable governance (n = 7) in SCOPUS that is most widely used and considered as being one of the internationally recognised databases for scholars in social sciences. These keywords terms were extracted and formed after extensive review of the relevant literature (Table 23.1).

Table 23.1 Data collection technique and process

Keywords	*Category*	*Results*	*Data collection date*	*Index*
'sustainable destination governance'	Article title	31 documents	23.04.2020	SCOPUS
'ecotourism and governance'	Article title	11 documents	23.04.2020	SCOPUS
'environmental governance'	Article title	10 documents	23.04.2020	SCOPUS
'sustainable tourism governance'	Article title	9 documents	23.04.2020	SCOPUS

In addition, the review process has revealed peer-reviewed articles published by *Journal of Sustainable Tourism* (n = 25) and *Sustainability* journal (n = 6). These journals' articles included as their topic were closely related to sustainable tourism and governance. Thematic analysis was used to analyse the data which consists of many stages. During the data collection process, which was carried out on 23 April 2020, many inclusion and exclusion criteria were implemented (Papamitsiou & Economides, 2014). In this vein, only peer-reviewed articles were selected for analysis while books, book chapters, conference proceedings, research notes, and research letters were excluded. At the initial stage of data collection stage 62 articles were totally identified that were obtained from the database and finally data set was limited to 33 articles after collated items and eliminating duplicates based on the criteria presented. As a result, the data set was generated for analysis from 64 articles in total, including articles of special issues.

Thematic analysis of selected articles

A thematic analysis of the data has provided to identify limitations in the current evolution of the field as well as research gaps in the literature that were regarded as: 1) governance for sustainability in hotspots places and urban areas (e.g., overtourism) has been overlooked and less emphasised; 2) there is limited study in the literature on the topic of governance relating to issue of degrowth from sustainability of tourism landscapes and; 3) there is a lack of research on the topics of SDGs and governance which relate those themes with sustainable management of protected and conversed areas and interrelated fields such as responsible tourism, ecotourism, wildlife tourism, etc.

Given the interdisciplinary nature of governance as a field of research and its multilayered scope, overall, thematic analysis has revealed that numerous researchers attempt to examine the sustainable tourism governance from a diverse range of topics such as ecotourism (Higgins-Desbiolles, 2011), environmental aspect (Erkuş-Öztürk & Eraydın, 2010), climate change (Jamal & Watt, 2011), wildlife (Moore and Rodger, 2010), surf tourism (Mach & Ponting, 2018), political economy (Bramwell, 2011), event tourism (Dredge & Whitford, 2011), mobilities (Dredge & Jamal, 2013), residents' perception (Gajdošík, Gajdošíková, & Stražanová, 2018), smart bike-sharing (Chen & Zhu, 2020), and SDGs (Siakwah et al., 2019).

Spatial dimensions of governance from sustainable tourism insight in articles are varied; for instance, national (Getzner, Vik, Brendehaug, & Lane, 2014), regional (González-Morales, Álvarez-González, Sanfiel-Fumero, & Armas-Cruz, 2016), local (Chen & Lin, 2017), global (Duffy, 2006), and rural (Frost & Laing, 2015). From a methodological point of view, the number of research adopting qualitative research methods have been found as dominant (e.g., Bichler & Lösch, 2019; Noh et al., 2018) as compared to quantitative ones (e.g., Armas-Cruz, Sanfiel-Fumero, & González-Morales, 2017; Fernández-Tabales, Foronda-Robles, Galindo-Pérez-de-Azpillaga, & García-López, 2017). On the other hand, the proliferation of using a mixed methods approach was also prominently noted (e.g., Conceição, Dos Anjos, & Gadotti dos Anjos, 2019; Dinica, 2009); that is followed by conceptual chapter which remain lesser (Borges et al., 2014; Wray, 2015).

Conclusion

The aims of the current chapter were to identify current gaps on the sustainable governance that relates to ecotourism and environmental policy while presenting recommendations for future research studies by reviewing 64 articles. The critical review has identified that there is a huge gap on the topic of governance of ecotourism areas particuarly related to governance of hotspots places referring to issue of overtourism that is essential to provide sustainability of environments that are supplied to be consumed by visitors.

Governance is concerned with the effort among different state actors to describe and promote the adoption of policies and planning mechanisms for sustainable tourism (Bianchi & de Man, 2020). This present book chapter has confirmed that governments are not counted any longer as the most important source of decision-making in managing and conservation of environments; rather, a different and wide variety of actors are also playing a crucial role in decision-making processes through established new mechanisms and forums (Armitage, De Loë, & Plummer, 2012). As such, an adaptive co-management approach may also be undertaken to overcome the weaknesses in governance issues in managing protected areas as well as ecotourism landscapes (Islam et al., 2018a). This management approach involves (social) learning that accentuates an integral part of sustainable destination governance of ecotourism environments (Islam et al., 2018b) and regarded as a normative goal in environmental science and resource management to provide protection and conservation of landscapes and thus has received the great attention it deserves (Armitage et al., 2008). In doing so, local stakeholders or actors perform collaborative actions and share the outcomes of their actions (Fennell et al., 2008). Further, adaptive co-management embeds the co-production of knowledge as a core element in the centre of a decision-making mechanism within an adaptive governance structure (Armitage et al., 2011) referring to a shifting paradigm that signals adaptive co-management encapsulating both collaborative and the effective learning feature of adaptive management (Plummer & Fennell, 2009). From this perspective, to provide sustainability in destinations, the participation of the local actors has been found as vital not only to form long-term strategies but also forming tourism planning (Graci, 2013) that address the needs for effective governance to provide environmental protection and conservation.

Although the agenda is closely affecting tourism policy, it directly underscores and thus mentions a few times tourism at first glance when teasing out the SGD 17 goals and 169 targets; it can be claimed that the role of tourism industry is significantly ascribed to achieve the UNWTO Sustainable Development by 2030 (Hall, 2019). UNWTO SDGs, especially goals of SDG 16 and SDG 17, play crucial roles to provide sustainability of tourism in urban and/or protected areas for which governance should be considered as a driving force for tourism to support the SDGs (Siakwah et al., 2019).

Given the interdisciplinary nature of the field and governance concept, the search process was limited to the SCOPUS database as it provides a broader content in social sciences and constitutes the chief limitation of the current chapter. Future research may attempt to investigate or discuss the possibility of presence in an ideal sustainable governance model within the context of UNWTO SDGs to effectively manage conservation of natural environment and protected areas.

References

Adiyia, B., Stoffelen, A., Jennes, B., Vanneste, D., & Ahebwa, W. M. (2015). Analysing governance in tourism value chains to reshape the tourist bubble in developing countries: The case of cultural tourism in Uganda. *Journal of Ecotourism*, *14*(2/3), 113–129.

Armas-Cruz, Y., Sanfiel-Fumero, A., & González-Morales, O. (2017). Environmental management of the tourist accommodation industry and sustainable governance in a protected area. *Universia Business Review*, (56), 84–105.

Armitage, D., Berkes, F., Dale, A., Kocho-Schellenberg, E., & Patton, E. (2011). Co-management and the co-production of knowledge: Learning to adapt in Canada's Arctic. *Global Environmental Change*, *21*(3), 995–1004.

Armitage, D., De Loë, R., & Plummer, R. (2012). Environmental governance and its implications for conservation practice. *Conservation Letters, 5*(4), 245–255.

Armitage, D., Marschke, M., & Plummer, R. (2008). Adaptive co-management and the paradox of learning. *Global Environmental Change, 18*(1), 86–98.

Beaumont, N., & Dredge, D. (2010). Local tourism governance: A comparison of three network approaches. *Journal of Sustainable Tourism, 18*(1), 7–28.

Bianchi, R. V., & de Man, F. (2020). Tourism, inclusive growth and decent work: A political economy critique. *Journal of Sustainable Tourism*, 1–19.

Bichler, B. F., & Lösch, M. (2019). Collaborative governance in tourism: Empirical insights into a community-oriented destination. *Sustainability, 11*(23), 6673.

Boluk, K., Cavaliere, C. T., & Higgins-Desbiolles, F. (2017). Critical thinking to realize sustainability in tourism systems: Reflecting on the 2030 Sustainable Development Goals. *Journal of Sustainable Tourism, 25*(9), 1201–1204.

Boluk, K. A., Cavaliere, C. T., & Higgins-Desbiolles, F. (2019). A critical framework for interrogating the United Nations Sustainable Development Goals 2030 Agenda in tourism. *Journal of Sustainable Tourism, 27*(7), 847–864.

Borges, M., Eusébio, C., & Carvalho, N. (2014). Governance for sustainable tourism: A review and directions for future research. *European Journal of Tourism Research*, 7(1), 45–56.

Bramwell, B. (2011). Governance, the state and sustainable tourism: A political economy approach. *Journal of Sustainable Tourism, 19*(4/5), 459–477.

Bramwell, B., & Lane, B. (2011). Critical research on the governance of tourism and sustainability. *Journal of Sustainable Tourism, 19*(4/5), 411–421.

Cheer, J. M., Milano, C., & Novelli, M. (2019). Tourism and community resilience in the Anthropocene: Accentuating temporal overtourism. *Journal of Sustainable Tourism*, 27(4), 554–572.

Chen, M., & Lin, Y. (2017). Integrating co-management and the Satoyama Initiative for forest governance: Community-based ecotourism and conservation of Adiri and Labuwan. *Taiwan Journal of Forest Science, 32*(4), 299–316.

Chen, H., & Zhu, T. (2020). Co-governance of smart bike-sharing schemes based on consumers' perspective. *Journal of Cleaner Production*, 120949.

Chiutsi, S., Mukoroverwa, M., Karigambe, P., & Mudzengi, B. K. (2011). The theory and practice of ecotourism in Southern Africa. *Journal of Hospitality Management and Tourism*, 2(2), 14–21.

Conceição, C. C., Dos Anjos, F. A., & Gadotti dos Anjos, S. J. (2019). Power relationship in the governance of regional tourism organizations in Brazil. *Sustainability, 11*(11), 3062.

Cornelisse, M. (2019). Moral claims in sustainable tourism development. *Tourism Planning & Development*, 1–19.

De Leon, R. C., & Kim, S. M. (2017). Stakeholder perceptions and governance challenges in urban protected area management: The case of the Las Piñas–Parañaque Critical Habitat and Ecotourism Area, Philippines. *Land Use Policy, 63*, 470–480.

Dimitriou, C. K. (2017). From theory to practice of ecotourism: Major obstacles that stand in the way and best practices that lead to success. *European Journal of Tourism, Hospitality and Recreation, 8*(1), 26–37.

Dinica, V. (2009). Governance for sustainable tourism: A comparison of international and Dutch visions. *Journal of Sustainable Tourism*, 17(5), 583–603.

Dredge, D., & Whitford, M. (2011). Event tourism governance and the public sphere. *Journal of Sustainable Tourism, 19*(4/5), 479–499.

Dredge, D., & Jamal, T. (2013). Mobilities on the Gold Coast, Australia: Implications for destination governance and sustainable tourism. *Journal of Sustainable Tourism, 21*(4), 557–579.

Dube, K. (2020). Tourism and Sustainable Development Goals in the African context. *International Journal of Economics and Finance Studies, 12*(1), 88–102.

Dube, K., & Nhamo, G. (2020). Sustainable development goals localisation in the tourism sector: Lessons from Grootbos Private Nature Reserve, South Africa. *GeoJournal*, 1–18.

Duffy, R. (2006). Global environmental governance and the politics of ecotourism in Madagascar. *Journal of Ecotourism, 5*(1–2), 128–144.

Eckert, C., Zacher, D., Pechlaner, H., Namberger, P., & Schmude, J. (2019). Strategies and measures directed towards overtourism: A perspective of European DMOs. *International Journal of Tourism Cities, 5*(4), 639–655.

Erkuş-Öztürk, H., & Eraydın, A. (2010). Environmental governance for sustainable tourism development: Collaborative networks and organisation building in the Antalya tourism region. *Tourism Management, 31*(1), 113–124.

Fennell, D. A. (2001). A content analysis of ecotourism definitions. *Current Issues in Tourism, 4*(5), 403–421.

Farrelly, T. A. (2011). Indigenous and democratic decision-making: Issues from community-based ecotourism in the Boumā National Heritage Park, Fiji. *Journal of Sustainable Tourism, 19*(7), 817–835.

Fennell, D., & Dowling, K. R. (Eds.). (2003). Ecotourism policy and planning: Stakeholders, management and governance. In *Ecotourism policy and planning* (pp. 331–344). Wallingford: CABI.

Fennell, D., Plummer, R., & Marschke, M. (2008). Is adaptive co-management ethical? *Journal of Environmental Management, 88*(1), 62–75.

Fernández-Tabales, A., Foronda-Robles, C., Galindo-Pérez-de-Azpillaga, L., & García-López, A. (2017). Developing a system of territorial governance indicators for tourism destinations. *Journal of Sustainable Tourism, 25*(9), 1275–1305.

Frost, W., & Laing, J. (2015). Avoiding burnout: the succession planning, governance and resourcing of rural tourism festivals. *Journal of Sustainable Tourism, 23*(8–9), 1298–1317.

Gajdošík, T., Gajdošíková, Z., & Stražanová, R. (2018). Residents' perception of sustainable tourism destination development: A destination governance issue. *Global Business & Finance Review, 23*(1), 24–35.

Gan, J. E., Nair, V., & Hamzah, A. (2019). The critical role of a lead institution in ecotourism management: A case of dual governance in Belum-Temengor, Malaysia. *Journal of Policy Research in Tourism, Leisure and Events, 11*(2), 257–275.

Getzner, M., Vik, M. L., Brendehaug, E., & Lane, B. (2014). Governance and management strategies in national parks: Implications for sustainable regional development. *International Journal of Sustainable Society, 6*(1–2), 82–101.

González-Morales, O., Álvarez-González, J. A., Sanfiel-Fumero, M. Á., & Armas-Cruz, Y. (2016). Governance, corporate social responsibility and cooperation in sustainable tourist destinations: The case of the Island of Fuerteventura. *Island Studies Journal, 11*(2), 561–584.

Graci, S. (2013). Collaboration and partnership development for sustainable tourism. *Tourism Geographies, 15*(1), 25–42.

Hall, C. M. (2011a). A typology of governance and its implications for tourism policy analysis. *Journal of Sustainable Tourism, 19*(4–5), 437–457.

Hall, C. M. (2011b). Policy learning and policy failure in sustainable tourism governance: From first-and second-order to third-order change? *Journal of Sustainable Tourism, 19*(4–5), 649–671.

Hall, M. (2012). Governance and responsible tourism. In D. Leslie (Ed.), *Responsible tourism. Concepts, theory and practice* (pp. 107–118). Wallingford:CABI.

Hall, C. M. (2019). Constructing sustainable tourism development: The 2030 agenda and the managerial ecology of sustainable tourism. *Journal of Sustainable Tourism, 27*(7), 1044–1060.

Hatipoglu, B., Alvarez, M. D., & Ertuna, B. (2016). Barriers to stakeholder involvement in the planning of sustainable tourism: The case of the Thrace region in Turkey. *Journal of Cleaner Production, 111*, 306–317.

Higgins-Desbiolles, F. (2011). Death by a thousand cuts: Governance and environmental trade-offs in ecotourism development at Kangaroo Island, South Australia. *Journal of Sustainable Tourism, 19*(4–5), 553–570.

Islam, M. W., Ruhanen, L., & Ritchie, B. W. (2018a). Adaptive co-management: A novel approach to tourism destination governance? *Journal of Hospitality and Tourism Management, 37*, 97–106.

Islam, M. W., Ruhanen, L., & Ritchie, B. W. (2018b). Exploring social learning as a contributor to tourism destination governance. *Tourism Recreation Research, 43*(3), 335–345.

Jacobsen, J. K. S., Iversen, N. M., & Hem, L. E. (2019). Hotspot crowding and over-tourism: Antecedents of destination attractiveness. *Annals of Tourism Research, 76*, 53–66.

Jamal, T., & Watt, E. M. (2011). Climate change pedagogy and performative action: Toward community-based destination governance. *Journal of Sustainable Tourism, 19*(4–5), 571–588.

Kagermeier, A., Amzil, L., & Elfasskaoui, B. (2019). The transition of governance approaches to rural tourism in Southern Morocco. *European Journal of Tourism Research, 23*, 40–62.

Laimer, P. (2017). Tourism indicators for monitoring the SDGs. In *Sixth UNWTO International Conference on Tourism Statistics* (pp. 21–24). 21–24 June, Manila, Philippines.

Lalicic, L. (2019). Solastalgia: An application in the overtourism context. *Annals of Tourism Research*, 102766.
Leung, Y. F., Spenceley, A., Hvenegaard, G., & Buckley, R. (Eds.) (2018). Tourism and visitor management in protected areas: Guidelines for sustainability. In *Best practice protected area guidelines series No. 27*. Gland, Switzerland: IUCN.
Mach, L., & Ponting, J. (2018). Governmentality and surf tourism destination governance. *Journal of Sustainable Tourism*, *26*(11), 1845–1862.
Mihalič, T., Šegota, T., Knežević Cvelbar, L., & Kuščer, K. (2016). The influence of the political environment and destination governance on sustainable tourism development: A study of Bled, Slovenia. *Journal of Sustainable Tourism*, *24*(11), 1489–1505.
Milano, C., Novelli, M., & Cheer, J. M. (2019). Overtourism and tourismphobia: A journey through four decades of tourism development, planning and local concerns. *Tourism Planning & Development*, *26*(4), 353–357.
Moore, S. A., & Rodger, K. (2010). Wildlife tourism as a common pool resource issue: Enabling conditions for sustainability governance. *Journal of Sustainable Tourism*, *18*(7), 831–844.
Ndivo, R. M., & Okech, R. N. (2020). Tourism governance in transition period: Restructuring Kenya's tourism administration from centralized to devolved system. *Tourism Planning & Development*, *17*(2), 166–186.
Noh, A. F. M., Shuib, A., Tai, S. Y., & Noh, K. M. (2018). Indicators of governance of marine ecotourism resources: Perception of communities in Pulau Perhentian, Terengganu. *International Journal of Business and Society*, *19*(S1), 17–25.
Nordlund, L. M., Kloiber, U., Carter, E., & Riedmiller, S. (2013). Chumbe Island Coral Park—governance analysis. *Marine Policy*, *41*, 110–117.
Nunkoo, R. (2017). Governance and sustainable tourism: What is the role of trust, power and social capital? *Journal of Destination Marketing & Management*, *6*(4), 277–285.
Palmer, N. J., & Chuamuangphan, N. (2018). Governance and local participation in ecotourism: Community-level ecotourism stakeholders in Chiang Rai province, Thailand. *Journal of Ecotourism*, *17*(3), 320–337.
Papamitsiou, Z., & Economides, A. A. (2014). Learning analytics and educational data mining in practice: A systematic literature review of empirical evidence. *Journal of Educational Technology & Society*, *17*(4), 49–64.
Pasape, L., Anderson, W., & Lindi, G. (2015). Good governance strategies for sustainable ecotourism in Tanzania. *Journal of Ecotourism*, *14*(2–3), 145–165.
Pellis, A., Lamers, M., & Van der Duim, R. (2015). Conservation tourism and landscape governance in Kenya: The interdependency of three conservation NGOs. *Journal of Ecotourism*, *14*(2–3), 130–144.
Perdomo, Y. (2016). Key issues for tourism development–the AM-UNWTO contribution. *Worldwide Hospitality and Tourism Themes*, *8*(6), 625–632.
Plummer, R., & Fennell, D. A. (2009). Managing protected areas for sustainable tourism: Prospects for adaptive co-management. *Journal of Sustainable Tourism*, *17*(2), 149–168.
Richins, H. (2011). Issues and pressures on achieving effective community destination governance: A typology. In E. Laws, H. Richins, J. Agrusa, & N. Scott (Eds.), T*ourist destination governance: Practice, theory and issues* (pp. 51–66). UK: CAB International.
Saruman, M. F. M., Razman, M. R., Zakaria, S. Z. S., & Ern, L. K. (2017). Influence of hierarchy sustainability governance in ecotourism management: Case study in Paya Indah Wetlands. *International Journal of Academic Research in Business and Social Sciences*, 7(3), 514–527.
Siakwah, P., Musavengane, R., & Leonard, L. (2019). Tourism governance and attainment of the Sustainable Development Goals in Africa. *Tourism Planning & Development*, 1–29.
Singh, T. (2018). Is over-tourism the downside of mass tourism? *Tourism Recreation Research*, *43*(4), 415–416.
Spenceley, A. (2008). Requirements for sustainable nature-based tourism in transfrontier conservation areas: A southern African Delphi consultation. *Tourism Geographies*, *10*(3), 285–311.
Thompson, B. S., Gillen, J., & Friess, D. A. (2018). Challenging the principles of ecotourism: Insights from entrepreneurs on environmental and economic sustainability in Langkawi, Malaysia. *Journal of Sustainable Tourism*, *26*(2), 257–276.
UN. (2020). Sustainable Development Golas. Retrieved from, http://www.undp.org/content/undp/en/home/sustainable-development-goals.html (accessed on 27.05.2020).

Wan, Y. K. P., & Bramwell, B. (2015). Political economy and the emergence of a hybrid mode of governance of tourism planning. *Tourism Management, 50*, 316–327.
Wang, C. C., Cater, C., & Low, T. (2016). Political challenges in community-based ecotourism. *Journal of Sustainable Tourism, 24*(11), 1555–1568.
Weaver, D. B., & Lawton, L. J. (2007). Twenty years on: The state of contemporary ecotourism research. *Tourism Management, 28*(5), 1168–1179.
World Tourism Organization and Griffith University. (2017). *Managing Growth and Sustainable Tourism Governance in Asia and the Pacific*. Madrid: UNWTO.
Wray, M. (2015). Drivers of change in regional tourism governance: A case analysis of the influence of the New South Wales Government, Australia, 2007–2013. *Journal of Sustainable Tourism, 23*(7), 990–1010.
UNWTO. (2020). Tourism in the Agenda 2030. Retrieved from https://www.unwto.org/tourism-in-2030-agenda (accessed on 27.05.2020).
Zeppel, H. (2012). Collaborative governance for low-carbon tourism: Climate change initiatives by Australian tourism agencies. *Current Issues in Tourism, 15*(7), 603–626.
Zerva, K. (2018). 'Chance Tourism': Lucky enough to have seen what you will never see. *Tourist Studies, 18*(2), 232–254.

24
WILL WORK FOR FOOD
Positioning animals in ecotourism

Georgette Leah Burns

Introduction

Discourse about the benefits and harm of ecotourism remains polarised, particularly when it involves non-human animals. Ecotourism should at least be ecologically based, but the ideal is a form of tourism that is ecologically responsible and does the right thing—morally, ethically, sustainably—for all stakeholders. Proponents argue that ecotourism should adopt and promote environmentally responsible practices, offer economic benefits to local communities, and support environmental conservation initiatives. Embraced with gusto in the 1990s, ecotourism has become such a popular marketing term that almost any form of tourism has been labelled in this idealised why. Indeed, tourism involving animals, thus infiltrating the tourist experience with something that may be perceived as akin to being 'natural', seems to automatically fulfill the ecologically based criteria. Yet, while many ventures claim the ecotourism label, most do not, perhaps cannot, fulfil its (unrealistic?) ideologies. This is the reality in which animals used in ecotourism find themselves.

By animals, I am referring to non-human animal species; thus, encompassing all living beings except humans and plants. However, because many tourists have a bias towards species they consider more charismatic, ecotourism is inherently speciesist (Kerley, Geach, & Vial, 2003). Most of the examples in this chapter focus on the animals most popular in ecotourism though there is of course scope for all involvement of all species. Vertebrates have long been more popular than invertebrates, for example, though tourism centred on invertebrates such as glow worms and butterflies does exist (Valentine & Birtles, 2004) and Lemelin (2015) writes about the mainstreaming of entomotourism.

A wide range of literature on the use of animals in ecotourism is presented here to provide an understanding of the current state of the field. The question of why animals are an important aspect of some forms of ecotourism is explored, before examining how animals are put to use in this context. Possible benefits and harm are discussed, before turning to some potential scenarios for the future.

Why animals are used in ecotourism

The history of using animals as the prime focus of tourism experiences is long. Birdwatching trips and safaris to observe African mammals were taking place in the 1800s (Adler, 1989) but

DOI: 10.4324/9781003001768-24

well before then animals were transported across the globe to be displayed in zoos and perform for humans. Queen Hatshepsut of Egypt's extensive fifteenth century BC palace menagerie is often described as being the first zoo (Mason, 2000) and the African elephant 'Jumbo', shipped from Ethiopia to London in 1865, one of the most famous examples of an animal used in tourism (Hancocks, 2003).

Not only is the history long, but the number of animals involved is large. With an estimated three million animals kept in zoos and aquariums (Fennell, 2013), the number of individuals used annually across all aspects of the ecotourism industry is likely to reach tens of millions. Both the number of individuals and diversity of species keeps growing, to the extent that intensification of the modern tourism industry has facilitated access and proximity to a vast range of species across the globe (Winter, 2020, p. 1).

Animals are used in ecotourism primarily for the entertainment of people. They are held captive for the purposes of display, used as porters to carry humans and our belongings, considered quarry to be hunted, and as objects used in competition and sport (Fennell, 2013). In all cases, their role is to serve a purpose for humans. This is the simple answer to the question of why animal-based ecotourism exists. Valuing animals extrinsically, or instrumentally, for what they can do for us, and separating 'us' and 'them' in this dichotomy, enables this situation to exist and persist. The animals provide a human audience with entertainment whilst simultaneously providing tourism operators with financial benefit.

Tourism is a human concept and a form of human-centred hedonism (Burns, 2015; Lück & Porter, 2018). It is anthropocentric: devised by and for human. The intention of most tourism experiences is to enable humans to escape from daily activities: the opportunity to do something different. Animal-based ecotourism shares that intention, offering access to species different from the ones most people are familiar with in their everyday lives. Or, if the animals are familiar, then it is a chance to see them do something different; for example, when a working dog rounds up sheep for a display. In this way, animal-based ecotourism is embedded in anthropocentric assumptions about human exceptionalism, prioritising human needs and desires over those of all other animals (Ivanov, 2019). The objective is human enjoyment and, as I will discuss, often comes at the cost of animal suffering (Winter, 2020, p. 17).

How animals are used in ecotourism

Much of the literature considering animals in ecotourism focuses on wildlife, which is an interesting curiosity in itself. For example, of the 18 chapters in Markwell's (2015) edited book on *Animals in Tourism*, the only chapter on domesticated animals is about travelling with them rather than them being the object of the tourism experience (Gretzel & Hardy, 2015). One reason for the popularity of wildlife suggested by Young and Carr (2018) is because domestic animals which are part of the mundane, every-day life, lack the attraction of exotic, strange, or uncommon animals. However, the uneven appeal is perhaps also due to the nature of the label of ecotourism, which is embedded with aims of conservation and preservation—subjects associated more readily with wild animals than domestic ones. It is also perhaps due to the often synonymous treatment of 'ecotourism' and 'nature-based tourism' (e.g., Hidinger, 1996, p. 49), with nature implying tourism with undomesticated animals. A rare exception recognising the role of domestic animals in ecotourism is the book *Domestic animals, humans and leisure: Rights, welfare, and wellbeing* by Young and Carr (2018).

Ecotourism involving animals can be targeted (direct), in which being entertained by animals is the central experience such as in a zoo or wildlife sanctuary. It can also be non-targeted (indirect), in which the animals are secondary to the main experience or even accidental such as sighting a turtle while visiting a beach for the purpose of swimming (Burns, 2017). Ecotourism involving animals occurs within captive or non-captive settings, with Cohen (2019) further dividing these for wildlife into a continuum of four categories: fully natural, semi-natural, semi-contrived, and fully contrived. These different motivational and spatial settings are important because they can have vastly differing consequences for the animals in terms of the way they are used. In targeted and captive settings, the animals are more likely to experience artificial, monotonous environments that restrict their movement and ability to engage in natural behaviours and thus may be harmful to their welfare. In non-targeted and non-captive settings, the animals may be able to engage in more natural, and thus healthier, behaviours.

Ecotourism involving animals may be considered non-consumptive or consumptive, depending on how the animals are used. In non-consumptive activities, the animals may be used for observational encounters in non-captive settings, such as bird watching in a national park. They may also be used for interactive presentations in captive settings, such as dolphin shows at a marine park, where they are displayed and often also touched by tourists as part of the entertainment. In contrast, consumptive activity types are generally considered as those in which the end product for the animals is their deliberate death at the hands of the tourists. Hunting and fishing, for example, are consumptive animal tourism activities. The distinction, however, is not always clear cut. Even where death is not deliberate, it can still be the outcome for the animal. For example, some ecotourism depends on the use of vehicles to access natural areas and, consequently, wildlife inhabiting destinations with large visitation number per year are susceptible to being struck by vehicles (Tablado & D'Amico, 2017). Similarly, studies increasingly show that disturbance from ecotourism activities can affect survival rates of species (e.g., Müllner, Linsenmair, and Wikelski's (2004) study of hoatzin chicks in Amazonian rainforest lakes, and D'Cruze et al. (2018) study of wild animals across Latin America). Further, some authors argue that in the context of ecotourism animals are constructed as commodities for consumption, perhaps not always physically but at least metaphorically in the sense of them being an object that has market value and the experience humans have with them can be bought and sold (Burns, 2015). The line drawn then, that consumptive tourism is only that resulting in deliberate death, seems narrow and it could be inferred that all forms of ecotourism involving animals are consumptive. Nevertheless, the simplistic perception of difference between deliberate versus non-deliberate death and the labels associated with them remains dominant in the literature.

The list of how animals are engaged in ecotourism for, and by, humans is long. Activities range from those that are more natural (for the animal), such as watching seals from the shore in Iceland (Burns, 2015), that enable the animals to continue with uninterrupted natural behaviour, to very artificial encounters, such as orangutans boxing in Thailand (Sellar, 2018), where animals are forced to perform unnatural behaviours to entertain the human audience. Jones (2013), for example, describes an ecotourism attraction in Florida that illegally fed wild alligators to entertain customers. Practices like this encourage unnatural behaviour in wild animals. They learn to approach humans for food, increasing their likelihood of being reported as a nuisance and killed by wildlife control officers. These different engagements result in differing effects on the wellbeing of the animals involved.

Animals as workers in the ecotourism industry

Animals expend physical and emotional labour as they work for humans in ecotourism. Dashper (2019, p. 29) argues that conceptualising animals as workers within tourism importantly highlights their contributions to the tourism product. As workers, animals are engaged in a wide range of activities. Animals are used as transport for both people and equipment; for example, yaks in Nepal (Ning, Oli, Gilani, Joshi, & Bisht, 2016), mules in Morocco (Cousquer & Allison, 2012), camels in Botswana (Seiful, Angassa, & Boitumelo, 2019), horses in Iceland (Helgadóttir & Dashper, 2016), sled dogs in Canada (Fennell & Sheppard, 2011), and elephants in Thailand (Flower, Burns, & Jones, 2021).

Animals work in a range of activities that can be labelled as sport. Being entertained by animals racing against each other is typically based on the labour of horses (Winter & Frew, 2018), greyhounds (Markwell, Firth, & Hing, 2017), and camels (Seiful et al., 2019) but in ecotourism contexts can extend to less expected species such as toads and crabs (Birts, 2017). In the context of sport tourism, animals can be something we gamble on or something we kill. Hunting and fishing return to the category of sport, thus demonstrating that the use of animals in this context can be both consumptive and non-consumptive. Bullfighting, another sport, causes the death of approximately 250,000 bulls annually (Humane Society International, 2020). Meanwhile, alligator wrestling does not purposefully aim to kill the animal (Frank, 2012).

Animals work for humans as trained performers in staged encounters. Circuses involving live animals, a prime example in this category, are not the only venues in which this type of work occurs. At the Samui Monkey Theater in Thailand, chained macaques play musical instruments (Discovery Thailand, 2020). At Sea World in Australia, dolphins are ridden and seals 'kiss' staff (Village Roadshow Theme Parks, 2020). At Sheep World in New Zealand, dogs round up sheep (Sheep World, 2020). At the Kayabukiya Tavern in Japan, monkeys are 'employed' as waiters (Financial Express, 2017).

There are contexts in which animals look less like workers in the ecotourism industry. These are usually non-captive, less managed settings such as whale watching. However, even here the animals are sources of entertainment for us. Although the performance is less likely to be forced, the animals may be coerced into close contact.

A primary objective for many ecotourists is to have a close encounter with the animal (Tully & Carr, 2020). Animals not caged and specifically trained for this purpose may be enticed by feeding, also referred to as provisioning. There are many examples of this around the world, involving many different species. In some cases, the feeding may be systematic and controlled; for example, with wild birds at O'Reilly's Rainforest Retreat in Australia (O'Reilly's, 2020) and with wild macaques in monkey parks in Japan (Knight, 2010). However, feeding may also be unregulated and uncontrolled which can lead to instances of human-animal conflict (e.g., Burns & Howard, 2003). We feed animals as part of a tourism experience. We also eat them.

Eating animals as part of the ecotourism experience may be connected with the use for sport, where the animal killed through hunting or fishing activities is then eaten. However, in most cases eating animal flesh could easily be dismissed as unrelated to ecotourism; for example, if the tourist is consuming meat products in a restaurant, the same as they would at home. Sometimes though the eating is more obviously part of an ecotourism experience, with tourists motivated by a desire to taste the flesh of animals they may not normally have the opportunity to eat (e.g., Burns, Óqvist, Angerbjörn, & Granquist, 2018; Mkono, 2015).

In all of the contexts described in this section, perhaps with the exception of being eaten, the ecotourism is obviously based on the physical labour of the animals. While the physicality might be obvious, more hidden is a type of labour that is emotional (Burns & Benz-Schwarzburg, 2021). While the concept of emotional labour has long been part of an anthropocentric discourse (e.g., Hochschild, 1983/2012), extending the concept to animal workers has come to attention more recently. Dashper's (2019) analysis of trail-riding tourism provides an example of this by recognising the horses as engaged in emotional labour in the service of humans. Burns and Benz-Schwarzburg (2021) note that the requirement of emotional labour from animals is often a crucial component of animal performances. Animals are trained to display behaviour that can be interpreted as expressing a particular emotion; for example, when a sea lion kisses a keeper the audience sees affection. This in turn effects human emotion about the experience. In this way, the animals become staff members trained to display fake emotions for the purpose of enhancing the ecotourism experience for humans (Burns & Benz-Schwarzburg, 2021).

Treatment of animals in ecotourism

Tully and Carr (2020) describe tourism as a facilitator of animal oppression because it interacts with animals through an unequal power relationship dominated by economics and control that establishes animals as objects for our use. This links with the earlier discussion about consumptive tourism and the plethora of literature on animals as tourism objects and commodities (e.g., Burns, 2015; Cohen & Cohen, 2019; Tully & Carr, 2020). Establishing them as such denies them agency and allows for their use in entertainment (Burns, 2015; Burns & Benz-Schwarzburg, 2021; Notzke, 2019). In this context, their sentience is also often denied or ignored (Burns & Benz-Schwarzburg, 2021).

What tourists may not realise, or choose not to realise, is that many ecotourism encounters are made possible because of the power relationship based on abuse, manipulation, cruelty, and control over the animal. The closer the encounter, the more evident this becomes; for example, when we ride an elephant (Schmidt-Burbach, 2017) or pat a tiger (Cohen, 2012, 2013). This type of tourism can create the illusion that high-level predators are akin to toys or pets. Through training of animals and language used with tourists, the illusion is created that the animals are docile, domesticated, and tame (Winter, 2020, p. 18), but this is rarely the case. Instead, particularly for once wild animals, they are often "systematically abused to conform their behaviour to the needs of their human handlers" (Winter, 2020, p. 12) and the desires of the tourists. Winders (2017) describes the language often used in these contexts as part of a 'humane-washing scheme'. Lack of regulation of marketing material means that some ecotourism operators can use names like 'haven', 'refuge', and 'sanctuary' to attract welfare-minded tourists even though the terms themselves are often meaningless (Winders 2017).

The list of how animals are mistreated in the context of ecotourism is lengthy and has already been alluded to. For the purposes of ecotourism animals are hunted and captured in the wild, permanently confined, used as transport, cruelly trained, intentionally killed, and eaten. Training may include the animals being drugged, and control through painful practices. For example, elephants in some ecotourism ventures in Thailand are trained using bull hooks that pierce their skin (Bansiddhi et al., 2018), and bears that perform in Russia's Petersburg Circus are restrained in a standing position, tethered by their necks to a wall, to strengthen their legs (Daly, 2019).

We know that animals suffer by being forced to live in places and under conditions that do not meet their innate needs. They also suffer by being forced to perform unnatural behaviours which cause them harm. Cetaceans confined in small tanks, for example, suffer sensory deprivation (Winter, 2020). Elephants forced to perform handstands suffer pain and long-term physical problems (Barnes, 2006).

Cohen (2013) describes tigers chained on short leads, left in the hot sun, kept in small cages, drugged, and beaten. This treatment occurs so the tigers can be handled by tourists. Gürsoy (2020) describes horses exposed to dehydration, heavy loads, and lack of food that sometimes results in serious accidents and even death. This treatment occurs so the horses can profitably pull tourists in carriages. Bauer (2017) describes lambs being malnourished, subjected to rough handling, not bonding with their mothers, suffering from hypothermia, and in need of urgent veterinary care. This occurs so they can be used as props for tourists' photographs. These are all examples of animal abuse and exploitation in captive ecotourism settings.

Justifying the treatment of animals in ecotourism

Situations like those described above persist because they are tolerated by tourists. Mistreatment of animals may be tolerated because the tourists do not know it exists. It can be hidden in the discourse used for tourists; for example, in the labelling of venues as described previously and in words like 'training' used without explanation of how that training takes place and what it entails. Cruelty may be tolerated when the tourists are uncertain about the impacts of their activities, and thus choose to ignore them. This is linked with the earlier contention that ecotourism is essentially a hedonistic human-focused industry. Some tourists may not care about the quality of life that ecotourism ventures offer for the animals. Others may care but choose not to match their actions with this care in the context of their desire for escapism. As a form of cognitive dissonance, this scenario is referred to frequently in the ecotourism literature (e.g., Burns et al., 2018; Curtin & Wilkes, 2007; Font, Bonilla-Priego, & Kantenbacher, 2019; Moorhouse, D'Cruze, & Macdonald, 2017).

A further alternative is that tourists may justify their actions as a type of care in which they find reasons to legitimise their desires—to be entertained by animals. This can manifest in several ways that focus on the believed benefits for the animals, and for the people. The justification may be financial: believing the money spent at an animal-based attraction contributes to conservation of the species. Justification can also be based on the captive animal being 'sacrificial' for the purpose of education. The captive individual enables people to learn about the species and thus, ideally, fosters a desire to protect it. Therefore, although life might not be ideal for a singular animal, its captivity is warranted to benefit the species as a whole. The Tiger Trek at Taronga Zoo in Australia embodies this education ideal. This immersive exhibit takes visitors on a simulated flight to Sumatra where they disembark to view tigers in an enclosure that appears very natural. On exit, visitors are prompted to make shopping choices based on sustainable palm oil consumption to help conserve tiger habitat (Kelly, 2018).

Yet another, and there are plenty more, justification for tourists to prioritise their own desires can be found in the belief that the animal has a better life as part of the tourist attraction than it would outside it. This is supported by data showing that some animals live longer in captivity and are less stressed because they avoid the fear of being predated upon and not knowing when they will obtain their next meal (Longley, 2011).

Justification can also be explicitly based on benefits to humans, and the connection between human wellbeing and human–animal interactions in the context of tourism is a well-studied phenomenon (e.g., Webb & Drummond, 2001; Weiler, Ham, & Smith, 2011; Yerbury & Boyd, 2018; Yerbury & Weiler, 2020). Curtin (2009), for example, found that encounters with wildlife can positively influence psychological well-being in humans and 'animal-assisted therapy', such as swimming with dolphins, has long been used to treat a range of human illnesses (Williamson, 2008). Thus, the 'feel-good' aspect embodied in an encounter can have positive outcomes for humans perceived to outweigh any negative outcomes for the animals.

In addition to being accepted by some tourists, cruelty to animals in the context of ecotourism may also persist due to lack of regulation and policy to prevent it. The United Nations World Tourism Organisation's (UNWTO) Global Code of Ethics, for example, positions itself strongly in terms of human rights and benefits but recognition of animals here, as in much other relevant legislation, is ignored (Burns, 2015; Fennell, 2014). The more focused Code of Ethics and Animal Welfare adopted by the World Association of Zoos and Aquariums (WAZA, 2003) covers animals in captive ecotourism. However, no global body regulates wildlife tourism attractions (Moorhouse et al., 2017) and thus responsibility for legal control usually lies at national and local levels. Some countries enforce strict regulations concerning the treatment of animals in tourism, for example, the Australian Animal Welfare Standards and Guidelines – [for] Exhibited Animals (Harding & Rivers, 2014); however, others are less strict. The problem is exacerbated because control is often left to codes of conduct (for example with seal [Óqvist, Granquist, Burns, & Angerbjörn, 2017] and whale watching [Garrod & Fennell, 2004]), which are not legally enforceable and lack conformity. Within the remit of tourism policy, Sheppard and Fennell (2019, p. 141) found focus remains on enhancing human welfare and rights through tourism development despite recent consideration of a broader and deeper range of impacts that includes concern for animals. Thus, changes may be ahead, but they are slow in coming. Globally encompassing regulations may not necessarily be the best answer as there is need to take into account local contexts and opportunities for compliance. However, without clear regulation, policy and the ability to enforce these, protection of both animals and people during ecotourism encounters can be compromised.

Cruelty implies intent to cause harm, is largely confined to captive settings and has existed since the beginning of animal-based ecotourism. Evidence is now increasing of negative consequences of ecotourism for animals in non-captive settings. Müllner et al.'s (2004) study of hoatzin chicks in Amazonian rainforest revealed exposure to ecotourism reduced survival and affected stress responses. Stress responses and heightened anxiety were also found amongst Barbary Macaques by Maréchal et al. (2011). More recently, D'Cruze et al.'s (2018, p. 1563) study of close interactions (feeding, swimming and petting) between humans and wild animals across Latin America identified a range of consequences including altered feeding patterns and reproductive behaviour, increased stress and other physiological responses, injury, disease, and death. Not all species are affected equally, however.

Hidinger's (1996) study in Tikal National Park, Guatemala, found that in areas where more tourism was present, some species increased in density, some decreased, and others seemed unaffected. From ecological studies we know that some species adapt better than others to more humanised environments (Chace & Walsh, 2006). Eastern water dragons thrive along the heavily touristed riverbank in Brisbane, Australia, for example, while population numbers of eastern bearded dragons around Brisbane have declined (Garden, McAlpine, Possingham, & Jones, 2007). These studies demonstrate that consequences are species specific.

While ecotourists may not know the details of their impacts on animals, the spread of global knowledge through internet sources means that information is not hard to find should one endeavour to look. It may, however, be easy to ignore. As mentioned earlier, a growing body of literature evokes cognitive dissonance as the answer to why tourists willing engage in activities that they know harm animals (e.g., Curtin & Wilkes, 2007, swimming with dolphins; Burns et al., 2018, eating whale meat; Lück & Porter, 2018, feeding sea birds; Ziegler et al., 2018, feeding whale sharks).

Despite the negative examples of animal abuse and exploitation described previously, ecotourism should be based on positive ideals. Conservation ideology is often at the heart of ecotourism, and this is particularly relevant for many forms of animal-based ecotourism. Ecotourism experiences may foster in humans a greater appreciation of animals. Encountering live animals through ecotourism experiences can increase awareness of species and their conservation status. In turn, this may promote pro-environmental attitudes (Powell & Ham, 2008) and create incentives to help conserve habitat and protect animals. This conservation argument is a key element in the World Association of Zoos and Aquariums' strategy for the future (Barongi, Fisken, Parker, & Gusset, 2015). However, even if the ecotourism ventures using animals originate from good intentions, they can still have negative consequences for individual animals. As discussed in the examples, they can be counterproductive to the goals of ecotourism, threaten the survival of non-captive species, and cause animals mental and physical harm.

A two-way relationship

In the two-way, although unevenly power-based, relationship between humans and animals in ecotourism it is important to also consider possible benefit and harm to humans. Benefits, primarily in terms of human satisfaction and wellbeing, used as justification for the use and treatment of animals in ecotourism, were discussed previously. However, harm can also come to humans in the context of animal-based ecotourism.

Humans have been killed by animals in the context of ecotourism, in both captive and non-captive settings. For example, a child was killed by a dingo on Fraser Island in a non-captive non-targeted encounter in 2001 (Burns & Howard, 2003). A trainer was killed by an orca at SeaWorld in Orlando in 2010 and a child killed by painted dogs in the Pittsburgh Zoo in 2012 (Coyne, 2019); both in a captive, targeted encounters. Regardless of the setting, such events are rare, far rarer than the animal dying as a consequence of the encounter.

Animals can transfer diseases to humans through direct encounters, such as being bitten by a diseased monkey or eating game meat. Reverse zoonoses (zooanthroponoses) also occur and can threaten the health of animals, with pathogen transmission from humans to great apes (Dunay et al., 2018) and mountain gorillas (Hanes, Kalema-Zikusoka, Svensson, & Hill, 2018) becoming a concern for their conservation. Thus, we can have a situation where ecotourists may wish to assist conservation of animals such as mountain gorillas and pay for tours to see them believing this will assist them when in fact the tourists' presence may be a threat to the animal.

Current directions in animal-based ecotourism

Is it possible to achieve the dual benefits of entertaining humans and protecting animals under the label of ecotourism? Given the ideals of ecotourism, this should be possible. Currently, however, it would seem that in much animal-based ecotourism "The objective is tourist enjoyment, but the effect is animal suffering" (Winter, 2020, p. 16). Suffering is a long way from protection, and increased entertainment for humans often comes at the expense of conservation and welfare for the animals (Burns & Benz-Schwarzburg, 2021). It would seem, however, that the "innumerable examples of abuse and cruel treatment" (Winter, 2020, p. 2) are becoming less acceptable. This is evident through widespread publicity (e.g., Daly, 2019) and pressure from social media influencing tourists to vote with their feet and avoid ecotourism venues with poor welfare reputations. In addition, tourism researchers are increasingly raising issues that demand we address the treatment of animals in all forms of tourism from an ethical standpoint (e.g., Burns, 2017; Burns & Benz-Schwarzburg, 2021; Fennell, 2011). Increased concern for the way we use animals in tourism can lead to improvements in their protection and welfare. At the very least, we should strive to bring entertainment and protection together in a more balanced way that better reflects the ideals of ecotourism.

Respect for animals as sentient beings having intrinsic value could be a start to achieving a more balanced relationship (Burns & Benz-Schwarzburg, 2021). Changes in our relationships with all animals, not just in the arena of ecotourism, towards recognition and consideration of intrinsic value, are clearly gaining momentum. Animals gained legal recognition as sentient beings in New Zealand in 2015 (New Zealand Parliament, 2015) and the Norwegian Animal Welfare Act states that animals have intrinsic value independent of their usable value for humans, despite this being difficult to legally enforce (Blattner, 2019).

Evidence is also mounting of shifts toward greater recognition of how to improve the welfare of animals in ecotourism. The evolution of tourism policies to include concern for the welfare of animals, even though economics and human welfare remain dominant (Sheppard & Fennell, 2019), is an example. We also see captive tourism operations such as zoos being rebranded as centres for education and conservation, and semi-captive settings as 'sanctuaries': though we need to be wary of praise here as rebranding without a change in practice is not a solution (Carr & Broom, 2018; Shani, 2012). An increasing number of operations, however, seem to be altering the ecotourism activities they offer to be more considerate of animal welfare. This is demonstrated by the 2019 opening of an elephant sanctuary in Thailand where tourists are not permitted to interact directly with the elephants (World Animal Protection, 2019). The Maesa Elephant Camp, also in Thailand, used the months it was forced to close due to the COVID-19 pandemic in 2020 to remove activities where elephants were chained, ridden, and controlled by a hook (Maesa Elephant Camp, 2020). Save the Bilby Fund in Australia used the same period of time to close its Charleville breeding facility to tourists, thus enabling the nocturnal animals to not be disturbed during the day (Save the Bilby Fund, 2020).

Coupled with rebranding and changing activities, the captivity of at least some species of animals for the purposes of ecotourism is in decline. Marine mammals are perhaps the most obvious of these, with many countries discontinuing displays of them. A Senate Public Bill was passed in Canada in 2019, for example, banning the captivity of whales and dolphins (Parliament of Canada, 2019). In parallel, ecotourism experiences that involve decreased use of real animals are on the rise. Circus Roncalli recently experimented with holographic images of animals to replace real ones (Burns & Benz-Schwarzburg, 2021). Robotic dolphins are being developed to substitute for captive ones (ABC, 2020). Moving even further into virtual reality there is Planet Zoo (Planet Zoo, 2020), an online game which, although it could be argued continues the themes of anthropocentrism and of power and domination over animals for the benefit of humans, offers hedonism and escapism without harming any real animals as the gamer creates the experience they want. Does an argument need to be mounted for these type of experiences still being ecotourism, or are they in fact the ultimate in ecotourism? By not using real animals, the goal of entertainment for humans while still protecting the welfare of animals is achieved.

What is clear is that how we engage with animals in ecotourism contexts is changing. Changes are occurring in not just the type of activities undertaken by tourists and offered by venues, but also in the way we think about our relationships with animals: how we conceptualise these relationships and then act on notions of right and wrong. At their extreme, this leads to the complete removal of encounters with real animals.

Conclusion

Animals are used in across a wide continuum of types of ecotourism activities. At one end we have activities such as birdwatching which are likely to have minimal impact on the animals, especially if conducted on a small scale. However, small-scale encounters could have large impacts on certain species, and expansion of research in this area is increasingly providing us with more information about those potential impacts. At the opposite end of the continuum are activities where captive animals are forced to work for human entertainment. These activities have a very large impact on individual animals, but perhaps less on the total population of a species. Feeding and confinement of animals allows humans to have close proximity to them, but often at considerable cost to the individual animal. That cost, labour, and welfare is increasingly being recognised and questioned. Throughout animal-based ecotourism activities we can see an anthropocentric dominance that denies the moral consideration of animals (Winter, 2020, p. 17), but this seems to be changing.

Global regulation is needed for some ventures such as wildlife attractions and zoos (Moorhouse et al., 2017), but to maximise effectiveness the local context requires consideration. The lack of regulation and policy regarding the use of animals in ecotourism urgently needs addressing. Global legislation is not necessarily the best answer because the local context, the needs of particular species, and the opportunity for compliance all need to be taken into account. Current legislation and guidelines require expansion, development, and implementation to adequately protect both animals and people engaged in ecotourism.

Winter (2020, p. 19) describes an "anthropocentric wall of defence built upon economic imperatives" that drives the use of animals in tourism and provides a barrier to improving conditions for them. The argument that educating tourists can lead to changes in their attitudes, and in turn lead to positive outcomes for animals, has long been discussed. Malchrowicz-Mośko, Munsters, and Korzeniewska-Nowakowska (2020, p. 21), for example, assert that "Sustainable solutions for controversial animal tourism have to be found by raising tourists' awareness by means of information and education". While there appears to be merit in pursuing this approach, the evidence presented above that some tourists know and care, but choose to ignore, suggests that this alone is not enough. Peer pressure through social media is emerging as one way the barrier is breaking down and clearly has potential for greater use.

Above all else, we should not assume that because animals are involved in a tourism activity it fits definitions of ecotourism and confirms to an ideal type of tourism that offers benefits for all its stakeholders. Across the spectrum of ecotourism, some activities are definitely worse than others in terms of animal welfare and we should be mindful of, and questioning of, all.

The history of animal use in ecotourism is long and entails animals working for the pleasure of humans. Treatment of animal workers has not always been kind and has certainly never been equitable. Consideration of animal welfare and tourism responsibility, however, is changing. The outlook for the future, based on current trends of recognition of animal welfare and decreasing use of—particularly captive—animals, promises to be very different.

References

ABC. (2020). Animatronic Dolphin Unveiled by Creators of Hollywood Creatures, Who Hope They'll Replace Captive Animals at Parks. Retrieved from https://www.abc.net.au/news/2020-10-16/robot-animatronic-dolphin-replace-captive-animal-edge-innovation/12774116?fbclid=IwAR24_EvQdIqA6ePVHiujmSeWI2q2RxD0jWd3v3Rvppa5EEVEBtnnhsFHUes (accessed on 12.11.2020).

Adler, J. (1989). Origins of sightseeing. *Annals of Tourism Research*, *16*, 7–29.

Bansiddhi, P., Brown, J. L., Thitaram, C., Punyapornwithaya, V., Somgird, C., Edwards, K. L., & Nganvongpanit, K. (2018). Changing trends in elephant camp management in northern Thailand and implications for welfare. *PeerJ*, *6*(3), e5996. doi:10.7717/peerj.5996.

Barnes, J. (2006). *Elephants at Work*. Canada: Gareth Stevens Publishing.

Bauer, I. (2017). The plight of the cute little lambs: Travel medicine's role in animal welfare. *Journal of Ecotourism, 16*, 95–111.

Birts, P. (2017). A Guide to Australia's Wackiest Races. *Discovery Holiday Parks*. Retrieved from https://travel.nine.com.au/recommended/australias-best-and-strangest-outback-races/ab721467-3f29–4384-8b52–89658ed04a3c (accessed on 10.09.2020).

Blattner, C. E. (2019). The recognition of animal sentience by the law. *Journal of Animal Ethics, 9*, 121–136. doi:10.5406/janimalethics.9.2.0121.

Burns, G. L. (2015). Animals as tourism objects: Ethically refocusing relationships between tourists and wildlife. In K. Markwell (Ed.), *Animals and tourism: Understanding diverse relationships*. Bristol: Channel View Publications.

Burns, G. L. (2017). Ethics and responsibility in wildlife tourism: Lessons from compassionate conservation in the Anthropocene. In R. Green, & I. Lima (Eds.), *Wildlife tourism, environmental learning and ethical encounters*. Cham: Springer.

Burns, G. L., & Benz-Schwarzburg, J. (2021). Representing wild animals to humans: The ethical future of wildlife tourism. In G. Bertella (Ed.), *Wildlife tourism futures*. Bristol: Channel View Publications.

Burns, G. L., & Howard, P. (2003). When wildlife tourism goes wrong: A case study of stakeholder and management issues regarding Dingoes on Fraser Island, Australia. *Tourism Management, 24*, 699–712.

Burns, G. L., Óqvist, E. L., Angerbjörn, A., & Granquist, S. (2018). When the wildlife you watch becomes the food you eat: Exploring moral and ethical dilemmas when consumptive and non-consumptive tourism merge. In C. Kline (Ed.), *Animals as food: Ethical implications for tourism*. London: Routledge.

Carr, N., & Broom, D. M. (2018). *Tourism and Animal Welfare*. London: CABI International.

Chace, J. F., & Walsh, J. J. (2006). Urban effects on native avifauna: A review. *Landscape and Urban Planning, 74*, 46–69.

Cohen, E. (2012). Tiger tourism: From shooting to petting. *Tourism Recreation Research, 37*, 193–204.

Cohen, E. (2013). Buddhist compassion and animal abuse in Thailand's Tiger Temple. *Society and Animals, 21*, 266–283.

Cohen, E. (2019). Crocodile tourism: The emasculation of ferocity. *Tourism, Culture and Communication, 19*, 83–102.

Cohen, S. A., & Cohen, E. (2019). New directions in the sociology of tourism. *Current Issues in Tourism, 22*, 153–172.

Cousquer, G., & Allison, P. (2012). Ethical responsibilities towards expedition pack animals: The mountain guide's and expedition leader's ethical responsibilities towards pack animals on expedition. *Annals of Tourism Research, 39*, 1839–1858.

Coyne, J. (2019). Orcas, tigers and painted dogs, oh my! The need for targeted zoo safety and security regulation. *University of Illinois Law Review, 3*, 801–831.

Curtin, S. (2009). Wildlife tourism: the intangible, psychological benefits of human–wildlife encounters. *Current Issues in Tourism, 12*, 451–474. DOI:10.1080/13683500903042857.

Curtin, S., & Wilkes, K. (2007). Swimming with captive dolphins: Current debates and post-experience dissonanc. *International Journal of Tourism Research, 9*, 131–146. DOI:10.1002/jtr.599.

Daly, N. (2019). Suffering Unseen: The Dark Truth Behind Wildlife Tourism. *National Geographic*. Retrieved from www.nationalgeographic.com/magazine/2019/06/global-wildlife-tourism-social-media-causes-animal-suffering (accessed on 12.11.2020).

Dashper, K. (2019). More-than-human emotions: Multispecies emotional labour in the tourism industry. *Gender, Work & Organization, 27*, 24–40.

D'Cruze, N., Niehaus, C., Balaskas, M., Vieto, R., Carder, G., Richardson, V. A., ... & Macdonald, D. W. (2018). Wildlife tourism in Latin America: Taxonomy and conservation status. *Journal of Sustainable Tourism, 26*, 1562–1576.

Discovery Thailand. (2020). Samui Monkey Theater in Koh Samui. Retrieved from www.discoverythailand.com/Koh_Samui_Samui_Monkey_Theater.asp (accessed on 13.11.2020).

Dunay, E., Apakupakul, K., Leard, S., Palmer, J. L., & Deem, S. L. (2018). Pathogen transmission from humans to great apes is a growing threat to primate conservation. *EcoHealth, 15*, 148–162. doi:https://doi.org/10.1007/s10393-017-1306-1.

Fennell, D. A. (2011). *Tourism and Animal Ethics*. London: Routledge.

Fennell, D. A. (2013). Contesting the zoo as a setting for ecotourism, and the design of a first principle. *Journal of Ecotourism, 12*, 1–14. doi:10.1080/14724049.2012.737796.

Fennell, D. A. (2014). Exploring the boundaries of a new moral order for tourism's global code of ethics: An opinion piece on the position of animals in the tourism industry. *Journal of Sustainable Tourism, 22*, 983–996. doi:10.1080/09669582.2014.918137.

Fennell, D. A., & Sheppard, V. A. (2011). Another legacy for Canada's 2010 Olympic and Paralympic Winter games: Applying an ethical lens to the post-games' sled dog cull. *Journal of Ecotourism, 10*, 197–213.

Financial Express. (2017). Monkeys Serve as Waiter in This Japanese Restaurant! *Financial Express*. Retrieved from https://www.financialexpress.com/world-news/monkeys-serve-as-waiter-in-this-japanese-restaurant-they-get-bananas-as-salary-and-soya-bean-as-tip/802356 (accessed on 10.09.2020).

Flower, E., Burns, G. L., & Jones, D. (2021). How tourist preference and satisfaction can contribute to improved welfare standards at elephant tourism venues in Thailand. *Animals, 11*, 1094.

Font, X., Bonilla-Priego, M., & Kantenbacher, J. (2019). Trade associations as corporate social responsibility actors: An institutional theory analysis of animal welfare in tourism. *Journal of Sustainable Tourism, 27*, 118–138.

Frank, A. K. (2012). Grappling with tradition: The seminoles and the commercialization of alligator wrestling. In F. A. Salamone (Ed.), *The native American identity in sports: Creating and preserving a culture*. Plymouth, UK: Scarecrow Press.

Garden, J. G., McAlpine, C. A., Possingham, H. P., & Jones, D. N. (2007). Habitat structure is more important than vegetation composition for local-level management of native terrestrial reptile and small mammal species living in urban remnants: A case study from Brisbane, Australia. *Austral Ecology, 32*, 669–685.

Garrod, B., & Fennell, D. A. (2004). An analysis of whale watching codes of conduct. *Annals of Tourism Research, 31*, 334–352.

Gretzel, U., & Hardy, A. (2015). Pooches on wheels: Overcoming pet-related travel constraints through RVing. In K. Markwell (Ed.), *Animals and tourism: Understanding diverse relationships*. Bristol: Channel View Publications.

Gürsoy, I. T. (2020). Horse-drawn carriages: sustainability at the nexus of human–animal interaction. *Journal of Sustainable Tourism, 28*, 204–221. doi:10.1080/09669582.2019.1671852.

Hancocks, D. (2003). *A Different Nature: The Paradoxical World of Zoos and Their Uncertain Future*. Berkeley: University of California Press.

Hanes, A. C., Kalema-Zikusoka, G., Svensson, M. S., & Hill, C. M. (2018). Assessment of health risks posed by tourists visiting mountain gorillas in Bwindi Impenetrable National Park, Uganda. *Primate Conservation*, 32.

Harding, T., & Rivers, G. (2014). *Australian Animal Welfare Standards and Guidelines*. Exhibited animals – Consultation regulation impact statement. Orange: NSW Department of Primary Industries.

Helgadóttir, G., & Dashper, K. (2016). "Dear international guests and friends of the Icelandic Horse": Experience, meaning and belonging at a niche sporting event. *Scandinavian Journal of Hospitality and Tourism, 16*, 422–441. doi:10.1080/15022250.2015.1112303.

Hidinger. L. A. (1996). Measuring the Impacts of ecotourism on animal populations: A case study of Tikal National Park, Guatemala. *Yale Forestry and Environment Bulletin*, P49–P59. Retrieved from http://gip.uniovi.es/docume/ecotu.pdf (accessed on 13.08.2020).

Hochschild, A. (1983/2012). *The Managed Heart*. Berkeley: University of California Press.

Humane Society International. (2011). Bullfighting in Europe. Retrieved from www.hsi.org/news-media/bullfighting_europe/ (accessed on 13.11.2020).

Ivanov, S. (2019). Tourism beyond humans – robots, pets and teddy bears. In G. Rafailova, & S. Marinov (Eds.), *Tourism and intercultural communication and innovations*. Newcastle upon Tyne: Cambridge Scholars Publishing.

Jones, B. K. (2013). Marshmallows for alligators: Defining ecotourism in Southwest Florida. *Culture, Agriculture, Food and Environment, 37*, 116–123. doi:10.1111/cuag.12058.

Kelly, A. (2018). *Inspiring Pro-Conservation Behavior through Innovations in Zoo Exhibit and Campaign Design* (Master of Science Thesis, Kansas State University).

Kerley, G. I. H., Geach, B. G. S., & Vial, C. (2003). Jumbos or bust: Do tourists' perceptions lead to an under-appreciation of biodiversity? *South African Journal of Wildlife Research, 33*, 13–21.

Knight, J. (2010). The ready-to-view wild monkey: The convenience principle in Japanese wildlife tourism. *Annals of Tourism Research*, 37, 744–762.

Lemelin, R. H. (2015). From the recreational fringe to mainstream leisure: The evolution and diversification of entomotourism. In K. Markwell (Ed.), *Animals and Tourism: Understanding Diverse Relationships*, Bristol: Channel View Publications.

Longley, L. (2011). A review of ageing studies in captive felids. *International Zoo Yearbook, 45*, 91–98. doi:10.1111/j.1748-1090.2010.00125.x.

Lück, M., & Porter, B. A. (2018). The ethical dilemma of provisioning pelagic birds in exchange for a close encounter. *Journal of Ecotourism, 17*, 401–408.

Maesa Elephant Camp. (2020). The Chang Chiangmai. Retrieved from https://maesaelephantcamp.com (accessed 11.11.2020).

Markwell, K. (2015). *Animals and Tourism: Understanding Diverse Relationships*. Bristol: Channel View Publications.

Markwell, K., Firth, T., & Hing, N. (2017). Blood on the race track: An analysis of ethical concerns regarding animal-based gambling. *Annals of Leisure Research, 20*, 594–609. doi:10.1080/11745398.2016.1251326.

Malchrowicz-Mośko, E., Munsters, W., & Korzeniewska-Nowakowska, P. (2020). Controversial animal tourism considered from a cultural perspective. *Tourism, 30*, 21–30. doi:http://dx.doi.org/10.18778/0867-5856.30.1.

Maréchal, L., Semple, S., Majolo, B., Qarro, M., Heistermann, M., & MacLarnon, A. (2011). Impacts of tourism on anxiety and physiological stress levels in wild Male Barbary Macaques. *Biological Conservation, 144*, 2188–2193.

Mason, P. (2000). Zoo tourism: The need for more research. *Journal of Sustainable Tourism, 8*, 333–339.

Mkono, M. (2015). Eating the animals you come to see: Tourist's meat-eating discourses in online communicative texts. In K. Markwell (Ed.), *Animals and Tourism: Understanding Diverse Relationships*. Bristol: Channel View Publications.

Moorhouse, T., D'Cruze, N. C., & Macdonald, D. W. (2017). Unethical use of wildlife in tourism: What's the problem, who is responsible, and what can be done? *Journal of Sustainable Tourism, 25*, 505–516. Doi:doi.org/10.1080/09669582.2018.1533019.

Müllner, A., Linsenmair, K. E., & Wikelski, M. (2004). Exposure to ecotourism reduces survival and affects stress response in hoatzin chicks (Opisthocomus hoazin). *Biological Conservation, 118*, 549–558.

New Zealand Parliament. (2015). Animal Welfare Amendment Bill. Retrieved from https://www.parliament.nz/en/pb/bills-and-laws/bills-proposedlaws/document/00DBHOH_BILL12118_1/animal-welfare-amendment-bill (accessed on 14.11.2020).

Barongi, R., Fisken, F. A., Parker, M., & Gusset, M. (eds). (2015). *Committing to Conservation: The World Zoo and Aquarium Conservation Strategy* (pp. 69). Gland: WAZA Executive Office.

Ning, W., Oli, K. P., Gilani, H., Joshi, S., & Bisht, N. (2016). Yak raising challenges: Transboundary issues in Far Eastern Nepal. In W. Ning, Y. Shaoliang, S. Joshi, & N. Bisht (Eds.), *Yak on the move: Transboundary challenges and opportunities for Yak raising in a changing HKH Region*. Nepal: International Centre for Integrated Mountain Development.

Notzke, C. (2019). Equestrian tourism: Animal agency observed. *Current Issues in Tourism, 22*, 948–966.

O'Reilly's. (2020). O'Reilly's Activities and Experiences. Retrieved from https://oreillys.com.au/rainforest-retreat-activities (accessed 30.10.2020).

Óqvist, E. L., Granquist, S., Burns, G. L., & Angerbjörn, A. (2017). Seal watching: An investigation of codes of conduct. *Tourism in Marine Environments, 13*, 1–15. doi:https://doi.org/10.3727/154427317X14964473293699.

Parliament of Canada. (2019). Senate Public Bill. Retrieved from https://www.parl.ca/LegisInfo/BillDetails.aspx?Language=e&Mode=1&billId=8063284&View=5. (accessed 14.11.2020).

Planet Zoo. (2020). Planet Zoo. Retrieved from https://www.planetzoogame.com (accessed 11.11.2020).

Powell R. B., & Ham, S. H. (2008). Can ecotourism interpretation really lead to pro-conservation knowledge, attitudes and behaviour? Evidence from the Galapagos Islands. *Journal of Sustainable Tourism, 16*, 467–489.

Save the Bilby Fund. (2020). Charleville Bilby Experience. Retrieved from https://savethebilbyfund.com/faq-re-not-opening-2020-due-to-covid-19/ (accessed 11.11.2020).

Schmidt-Burbach, J. (2017). *Taken for a Ride: The Conditions for Elephants Used in Tourism in Asia*. London: World Animal Protection. Retrieved from https://www.worldanimalprotection.org.au/sites/default/files/au_files/taken_for_a_ride_report.pdf (accessed on 13.11.2020).

Seiful, E., Angassa, A., & Boitumelo, W. S. (2019). Community-based camel ecotourism in Botswana: Current status and future perspectives. *Journal of Camelid Science, 11*, 33–48.
Sellar, J. M. (2018). Animal welfare and tourism: The threat to endangered species. In N. Carr, & D. M. Broom (Eds.), *Tourism and animal welfare*. London: CABI International.
Shani, A. (2012). A quantitative investigation of tourists' ethical attitudes towards animal-based attractions. *Tourism: An International Interdisciplinary Journal, 60*, 139–158.
Sheep World. (2020). New Zealand Sheep and Wool Centre. Retrieved from https://www.sheepworldfarm.co.nz (accessed on 10.09.2020).
Sheppard, V. A., & Fennell, D. A. (2019). Progress in tourism public sector policy: Toward an ethic for non-human animals. *Tourism Management, 73*, 134–142.
Tablado, Z, & D'Amico, M. (2017). Impacts of terrestrial animal tourism. In D. T. Blumstein, B. Geffroy, D. S. M. Samia, & E. Bessa (Eds.), *Ecotourism's promise and peril: A biological evolution*. New York: Springer International Publishing. Doi:10.1007/978-3-319-58331-0_7.
Tully, P. A., & Carr, N. (2020). The oppression of donkeys in seaside tourism. *International Journal of the Sociology of Leisure, 3*, 53–70.
Valentine, P., & Birtles, A. (2004). Wildlife watching. In K. Higginbottom (Ed.), *Wildlife tourism: Impacts, management and planning*. Altona: Common Ground Publishing.
Village Roadshow Theme Parks. (2020). Seal Guardians Presentation. Retrieved from https://seaworld.com.au/attractions/shows-and-presentations/seal-guardians-presentation (accessed on 13.11.2020).
WAZA. (2003). WAZA Code of Ethics and Animal Welfare. Retrieved from www.waza.org/wp-content/uploads/2019/05/WAZA-Code-of-Ethics.pdf (accessed on 14.11.2020).
Webb, N. L., & Drummond, P. D. (2001). The effect of swimming with dolphins on human well-being and anxiety. *Anthrozoös, 14*, 81–85.
Weiler, B., Ham, S., & Smith, L. D. (2011). The impacts of profound wildlife experiences. *Anthrozoös, 24*, 51–64. doi:10.2752/175303711x12923300467366.
Williamson, C. (2008). Dolphin assisted therapy: Can swimming with dolphins be a suitable treatment? *Developmental Medicine & Child Neurology, 50*, 477. doi:10.1111/j.1469-8749.2008.00477.x.
Winders, D. J. (2017). Captive wildlife at a crossroads – Sanctuaries, accreditation, and humane-washing. *Animal Studies Journal, 6*, 161–178.
Winter, C. (2020). A review of animal ethics in tourism: Launching the annals of tourism research curated collection on animal ethics in tourism. *Annals of Tourism Research, 84*, 102989.
Winter, C., & Frew, E. (2018). Dark racing: Backstage at the sport of kings. *Leisure Studies, 37*, 452–465.
World Animal Protection. (2019). Thai Elephant Venue Reopens Without Cruelty. Retrieved from https://www.worldanimalprotection.org.au/news/thai-elephant-venue-reopens-without-cruelty (accessed on 14.11.2020).
Yerbury, R. M., & Boyd, W. E. (2018). Wild dolphins, nature and leisure: Whose wellbeing? In N. Carr, & J. Young (Eds.), *Wild animals and leisure: Rights and wellbeing*. New York: Routledge.
Yerbury, R. M., & Weiler, B. (2020). From human wellbeing to an ecocentric perspective: How nature-connectedness can extend the benefits of marine wildlife experiences. *Anthrozoös, 33*, 461–479. doi:10.1080/08927936.2020.1771054.
Young, J., & Carr, N. (2018). *Domestic Animals, Humans and Leisure: Rights, Welfare, and Wellbeing*. London: Routledge.
Ziegler, J. A., Silberg, J. N., Araujo, G., Labaja, J., Ponzo, A., Rollins, R., & Dearden, P. (2018). A guilty pleasure: Tourist perspectives on the ethics of feeding whale sharks in Oslob, Philippines. *Tourism Management, 68*, 264–274.

25

BIODIVERSITY CONSERVATION THROUGH AN AGROECOTOURISM PROJECT

The case of Ovacık Village, Turkey

Burcin Kalabay Hatipoglu, Fatma Cam Denizci, and Tümay Imamoğlu

Introduction

The tourism sector has the potential to contribute to sustainable development at all levels and help us live in a more sustainable world. Within the sustainable development agenda, tourism has a role in empowering local and indigenous communities, including women and youth, promoting local cultures and support change for more sustainable consumption and production patterns (United Nations, 2016). Many nations started integrating Sustainable Development Goals (SDGs) into their tourism policy frameworks (UNWTO, 2018), and ecotourism and community-based tourism are promoted as tools for achieving some of these goals, like biodiversity conservation (UNDP, 2020). Developing methodological approaches for evaluating the results of tourism interventions and sharing best case examples among stakeholders through open exchanges can accelerate the change towards sustainable development.

The purposes of this chapter are twofold, to broadly examine how a community-based agroecotourism project contributes to biodiversity conservation in a small forest town in Turkey, and to develop an approach for evaluating the benefits of agroecotourism in community-based projects. The literature suggests conducting studies that are reflecting the views of the locals and that are culturally embedded for investigating the benefits of ecotourism (Stronza, Hunt, & Fitzgerald, 2019). Thus, we utilise a participative action research methodology, and we base our evaluation on participant observations, semi-structured interviews, surveys, and group meetings that we conducted before and after project implementation.

Background to the study

In the literature, small-scale ecotourism projects are presented as a means for improving the livelihoods of people living in and around protected areas (Choo & Jamal, 2009; Hunt, Durham, Driscoll, & Honey, 2015; Nyaupane & Poudel, 2011) and biodiversity conservation (Das & Chatterjee, 2015; Kiss, 2004). Buckley (1994) described ecotourism to offer nature-based products and services, sustainably managed to lower negative impacts of tourism, provide

DOI: 10.4324/9781003001768-25

financial support for conservation efforts, and alter the environmental attitudes of the people. More recently, the International Ecotourism Society (TIES) defined ecotourism as "responsible travel to natural areas that conserves the environment, sustains the well-being of the local people, and involves interpretation and education" (TIES, 2015). Overall, the literature implies that ecotourism as a strategy should broadly satisfy four interconnected criteria that are; 1) diversify livelihoods and support socio-economic development and wellbeing of residents living in and around protected areas, 2) contribute to the conservation of the wildlife and protected areas, 3) focus on learning and education of visitors through nature-based activities, and 4) strengthen resource management institutions (Blamey, 2001; Das & Chatterjee, 2015; Fennell, 2012; Hunt et al., 2015; Nyaupane & Poudel, 2011; Stronza et al., 2019). For ecotourism projects to be considered highly successful, they should also affect land-use policies and tourism policies and plans (Hunt et al., 2015).

Ecotourism has been investigated widely within different settings (Duffy, Kline, Swanson, Best, & McKinnon, 2017; Stone, 2015; Weaver & Lawton, 2007; Wondirad, Tolkach, & King, 2020), resulting in diverse conclusions about the benefits (Buckley, 2009). For instance, in Nepal, the economic benefits of ecotourism to the residents living near the Royal Chitwan National Park were found to be limited (Bookbinder, Dinerstein, Rijal, Cauley, & Rajouria, 1998). Conversely, some studies reported more positive outcomes. For instance, in a comparative study in the Amazon region (Peru, Ecuador, and Bolivia), Stronza and Gordillo (2008) showed both economic (e.g., direct employment and income from the sale of products and services) and non-economic gains from ecotourism (e.g., learning opportunities and networking). Hunt et al. (2015) compared the effects of ecotourism as opposed to the other forms of tourism in the Osa Peninsula of Costa Rica. Their study displayed that ecotourism offered the best employment opportunity on the peninsula, and at the same time, ecotourism improved attitudinal support for conservation. Likewise, Choo and Jamal (2009) investigated the organic farms in South Korea and assessed their potential for being involved in ecotourism and environmental conservation. They concluded that tourism in farms provided additional income to the farmers, and the organic management encouraged the use of indigenous seeds and livestock as well as local products.

Discussing the reasons for failure in achieving environmental conservation through ecotourism, Kiss (2004) argued that the incentives for ecotourism are short-lived, involve small areas, and few people with limited economic income. Therefore, it cannot truly contribute to conservation. After an overview of the literature, Das and Chatterjee (2015) listed some of the reasons for failure as negative socio-economic results (e.g., revenue leakages, unequal distribution of income, joblessness, and social problems), negative environmental results (damage to the crop and livestock by wildlife), insensitive visitor behaviours, and limited learning outcomes for the visitors and the locals.

Scholars proposed that there should be co-ownership, and co-management of the projects shared between the operators and the communities to improve local support for conservation (Bookbinder et al., 1998; Coria & Calfucura, 2012). Furthermore, some income gained from ecotourism should be re-invested in conservation efforts (Fennell, 2012; Kiss, 2004). The involvement of non-governmental organisations (NGOs) in ecotourism operations helped achieve community development and nature conservation outcomes (Buckley, 2009), especially when community-based approaches are employed (Romero-Brito, Buckley, & Byrne, 2016; Stronza & Gordillo, 2008). In support of these findings, a recent review of the articles that examined the biodiversity hotspots displayed that ecotourism facilitates conservation when these four criteria are met: (a) a specific forest conservation mechanism is in place; (b) there is a spatial boundary delineating the area governed by the conservation mechanism; (c) local families receive direct economic benefits; and (d) strong community-oriented monitoring and enforcement (Brandt & Buckley, 2018, p. 114).

Despite fruitful discussions on the definition of ecotourism, its role in improving livelihoods and conservation, and the reasons for failure and success, there is a paucity of research in *defining its benefits* and *measuring its outcomes* objectively and reliably (Buckley, 2009). In a review of the academic articles, Agrawal and Redford (2006) noted that many case studies reported on the programs without focusing on the important elements of the context (e.g., population, distance from the markets, how residents value biodiversity) that could have significant explanatory power for the variety of results recorded in different settings. The authors proposed to collect data before and after the interventions, adopt a longitudinal approach to research design, and to conduct comparative studies using objective measures. The need for observing and measuring effects over time has been recorded by other scholars as well (Stronza et al., 2019). Furthermore, reviewing the literature, Stronza et al. (2019) highlighted that there was a need for studies that are reflecting the views of the locals and that are culturally embedded. They advocated that participatory approaches could close this gap. The authors recommended developing indicators of success in collaboration with the participants and to use them together with theoretically developed indicators.

Methods

Study site—Ovacık Village

Ovacık Village is one of the 57 villages of Şile, a large-sized district of Istanbul next to the Black Sea. Şile, unlike many other areas of the city, is underpopulated as a result of its distance from the Istanbul city center (60 km) and its vast forests (80% of the land). In the winter months, the town has under 50,000 residents, and it is mostly visited by university students that reside in the nearby campuses, patrons for the few fish restaurants on the shore, and nature lovers visiting the hikes in the nearby forests. However, in the summer months, the town is flooded by the daily visitors (on average, 1 million on the weekends) from nearby cities for its sandy beaches. Unfortunately, the overtourism in this period places incredible stress on the infrastructure of the town, causing the Municipality and the residents concern. On the other hand, this mass type of tourism does not economically benefit tourism businesses nor the residents (Interview with the Municipality). Thus, in recent years, the Şile Municipality has been highly supportive of alternative tourism types that will change the tourism type of the city towards more sustainable forms. For instance, in 2014, an Earth Market has been set up in the town centre resulting from a collaboration between the Şile Slow Food Convivium and the Municipality. Earth Market is an activity of Slow Food Network and promotes good, clean, and fair food production and consumption (Slow Food Foundation for Biodiversity, 2020).

Ovacık is a small forest village with 165 residents, and it is 10 kilometres away from the Şile town centre. There is a mix of ethnicities in the village as in other parts of Şile (e.g., Bosniaks). Furthermore, there are those people who bought large-sized land and moved to the village after retiring from their professional lives in the Istanbul center. The residents' average age is 50, and there are 10 children and 3 youth in the village (interview with Ovacık Mukhtar). Family income in the village is low, and it is mostly derived from family farming and animal raising (poultry, buffalos, and bees). The surroundings of the village are covered with chestnut trees, and chestnut honey is a unique product of the area. Unfortunately, quarries, that pop up in nearby villages, pose a threat to the forest land and the natural habitat (Demir, 2019). The central government gives permits for these quarries, and the local municipality has little power to stop their operations but to raise awareness and place an appeal to stop the activities.

The farmers in Ovacık village are mostly following traditional forms of agriculture. It is a self-regulated, sustainable form of agriculture. They replant heirloom seeds (heritage seeds), grow a variety of crops, use natural fertilisers, pesticides, and herbicides and utilise outdoor grazing for the livestock. The seed law of Turkey (Law No 5553), which was enacted in 2006, differentiates between the "genetic resources" and "plant varieties" for seed production and sales (Republic of Turkey Ministry of Agriculture and Forestry, 2006). According to the law, farmers cannot commercially sell the genetic varieties (local varieties) that they have saved and improved over generations. In contrast, the plant varieties (e.g., hybrids), that are developed by breeders and scientists can be commercially sold after registration and certification. The law allows the genetic verities to be used by the farmers and only to be exchanged among farmers as long as there is no commercial exchange (Grain, 2007). This new law raises much concern for the small-scale farmers and environmentalists as it risks the local varieties to be lost and makes the farmers dependent on large-sized multinational companies. Since the enactment of the law, environmental NGOs (e.g., Buğday Association), local authorities (e.g., Eskisehir Municipality) and communities have been involved in creating a database of heritage seeds and organised many local seed exchange events and festivals.

Similarly, the residents of Ovacık Village were also concerned about the implications of the new law. In the year 2012, with the leadership of the second author, the women of Ovacık Village founded the Women's Seed Exchange Association. The same year, they organised their first seed exchange activity in Ovacık Village. The event attracted more than 7000 visitors from out of town, and it was considered a milestone for Şile as it was the beginning of other nature-based activities in the district.

Future is in Tourism Program

Future is in Tourism—Sustainable Tourism Support Fund (FIIT) is a national program in Turkey that was founded by a multi-stakeholder partnership. Since 2012, Anadolu Efes Company (alcoholic beverages), United Nations Development Programme (UNDP), and Turkish Ministry of Culture and Tourism collaborate to support small-scale sustainable tourism projects in rural areas (Hatipoglu, Ertuna, & Salman, 2019a). Each year they provide approximately 20,000 €s to three projects that are chosen among many applications (e.g., 400 applications in 2015). Besides providing financial capital for a year, the partnership offers technical assistance, know-how, and public relations activities during and after a project's completion. It is the longest-running sustainable tourism support program in the country (Hatipoglu, Ertuna, & Salman, 2019b). Until the year 2020, including the first rural tourism project in the Eastern Anadolu Region of Turkey, 19 projects have been supported by the FIIT partnership.

Research design

This chapter employs a participatory action research approach (PAR) to investigate a community-based ecotourism project supported by the Future is in Tourism (FIIT) program in Turkey. Researchers that utilise PAR methodology accept that there could be multiple interpretations of the phenomenon being studied and that when they collaborate with practitioners, they can work together to change it for the better (Kindon, Pain, & Kesby, 2007). The data gathered during PAR research can aid in understanding the complexity of applying for community development programmes and bring explanations into behaviors of the examined actors (Bertella, 2019; Stronza et al., 2019). Furthermore, PAR can help close the gap between academia and practice (Moscardo, Konovalov, Murphy, McGehee, & Schurmann, 2017). PAR was previously used by other scholars investigating ecotourism (Gutiérrez-Montes, 2005; Stronza & Gordillo, 2008). It is recommended for studying the link between ecotourism and conservation because it integrates the views of the residents into the research design (Stronza et al., 2019).

Data collection and the sample

Initially, the second (Ovacık Village Seed Association) and third authors approached the first author (Şile Tourism Association) for sharing expertise on sustainable tourism development in rural areas. Previously, the first author had investigated the Slow Food Earth Market in Şile and knew the region well. The graduate students of the first author were involved with the Ovacık community for a term, and as a result, they developed a tourism project proposal in the year 2017. The same year, the two associations, together with Şile Municipality, formed a partnership, and applied for the FIIT program. Their community-based ecotourism project was chosen as one of the three projects to be supported during 2017–2018.

It is recommended to conduct on-site audits, collect data from internal and external documentary sources, and to follow interviews with staff for the rigorous evaluation of an ecotourism project (Buckley, 2009). Previously scholars examined ecotourism and organic operations utilising content analysis on websites (Choo & Jamal, 2009; Fennell & Markwell, 2015), semi-structured (Duffy et al., 2017; Morgan & Winkler, 2020) and in-depth interviews (Choo & Jamal, 2009), and participant observations (Duffy et al., 2017; Morgan & Winkler, 2020). Thus, to increase the trustworthiness, the data collection was triangulated by source, methods, and investigators, and the period included both before and after project implementation phases (2016–2020). In addition to secondary data (e.g., project documents), primary data included informal discussions, participant observations, semi-structured interviews, and open-ended surveys with community members, local authorities, and visitors (Table 25.1). The first author visited the village since 2016 and acted as both an observer and facilitator in the meetings. The data through interviews and open-ended surveys are collected by the second and third authors and audited by the first author.

We gathered the questions around two themes that are 1) the benefits of the tourism project for the community members, their families, and visitors, and 2) attitudes for biodiversity conservation. In this research, biodiversity is taken as the "variability among living organisms from all sources, including terrestrial, marine, and other aquatic ecosystems and the ecological complexes of which they are part; this includes diversity within species, between species, and of ecosystems" (Stronza et al., 2019, p. 231). The questions were structured open-ended to capture how the respondents defined the benefits of ecotourism. Including demographic questions, there were 16 questions for the community members, eight questions for the other residents, the Mukhtar, and the group visitors each.

Data analysis

We analysed data through thematic analysis in the guidance of theory-based indicators. We transferred the transcribed interviews and the surveys into the qualitative software

Table 25.1 Overview of the data collection (2017–2020)

Source	*Type*	*Emphasis*	*Authors*	#
Community members working at the center	Unstructured group meeting	Needs, capabilities, intentions to be involved in tourism activity	1, 2, 3	2
Community members working at the center	Open-ended surveys	Demographics, understanding of biodiversity conservation, and benefits of the project.	2	11
Residents not employed in the center	Open-ended surveys	Benefits of the project for the village	2	2
Ovacık Village Mukhtar* (2017)	Unstructured interview	Intentions to be involved in tourism activity and suitable types of tourism	1, 2,3	1
Ovacık Village Mukhtar** (2020)	Open-ended survey	Benefits of the project for the village	2	1
Residents from nearby villages	Open-ended surveys	Benefits of the project for the village and Şile	2	2
Group visitors	Open-ended surveys	Activity characteristics, reasons for visiting, evaluations of the learning activities.	2	6
Local municipality	Semi-structured interviews before and after	Benefits of the project for Sile as a destination	2, 3	2

Source: Authors.

Notes

* *The elected head of the government in the village.*

** *The mukhtar changed in 2018.Source:* Authors.

program NVivo V12, and we conducted coding and analysis of data by using the program. Based on the themes discussed in the literature (Choo & Jamal, 2009; Nyaupane & Poudel, 2011; Stronza et al., 2019), we used the broad indicators of *biodiversity conservation, diversified livelihoods, environmental interpretations, and strengthened resource institutions* to analyse the results. During the thematic analysis, we also recorded the emerging themes to answer the research questions.

Findings

Needs assessment

Ovacık Seed Association and Sile Tourism Association, in collaboration with Sile Municipality, applied for the FIIT program in 2017. Before the application, they conducted a needs assessment that identified the social and economic needs of the community as alternative sources of income for women, access to short food supply chains, and a way to stop the migration of youth to city centres (Group meeting with women, 2017). During the informal discussions with the women of the village and the Ovacık Village Mukhtar, the women informed the two associations that they would like to participate in a tourism model, which operates one or two days a week, and visitors are to be served at a communal area with no overnight stays.

Project activities

The goals of the "A day in the Ovacık Village" were to provide nature-based tourism services that economically benefited the farmers and facilitated visitors to experience the village life, become aware of sustainable agriculture principles and sustainable living, and understand local cultures through food (Project application document). The one-year program included activities that would build on the local assets and enable the achievement of these goals. We can group the activities as; *infrastructure improvements* (roads in the village, renovation of an old school as a community centre, installation of a kitchen, rainwater collection and bokashi compost bins, and establishment of vegetable and plant gardens based on permaculture principles); *capacity building* (skill-building in food preparation, food service, marketing, and local recipe development, personal development, and embroidery on traditional Şile cloth) and *marketing of the project*. Eleven women from the village participated in the training modules of the program, and later they worked in the community centre. The second author organised all the project activities and took part in managing and delivering of the tourism services.

Description of tourism services

The tourism services started during the project implementation with smaller groups, which enabled the women to learn and practice. The promotion of the events was made via social media, FIIT program partners, and extended network of the founders of the project. The activities were structured in and around the community center and extended to the village and the forest trail nearby. Visitor groups needed to make a reservation ahead, indicating which type of activity they wanted to participate in.

After the project was completed in 2018, the centre continued to accept visitor groups (e.g., schools, Rotary Club groups, special-interest groups, and corporations) mostly on the weekends and occasionally on the weekdays. Walk-in visitors are only accepted if there is availability. The capacity of the centre was set at less than 40 visitors during the winter months, and women worked in groups of four to deliver the services. A typical daily program starts with a breakfast at the community centre, followed by presentations on the zero-waste kitchen, compost making and production of organic fertilizers (e.g., worm castings using kitchen waste). The day continues with a one-hour walk in the forest accompanied by explanations on local plants and species (e.g., mushrooms and bees) and ends with a demonstration session at the centre. The demonstrations include local food making (e.g., pasta, vinegar, pickles, milk jam, and cheese) and eco-friendly products (e.g., toothpaste).

Benefits of the ecotourism project

The benefits of the project are gathered and evaluated by using the views of multiple beneficiaries. We have identified the direct beneficiaries as the women working in the project, residents of the village, mukhtar, local municipality, visitors and workshop organisers, and women from nearby villages that have taken part in workshops at the centre.

The ecotourism project was implemented in a village that had a certain awareness of biodiversity conservation and applied the principles of sustainable agriculture at their farms. Through the introduction of tourism in the village, the community was able to sell their products directly to consumers at a fair price. As a result, they recognised that consumers valued sustainably produced food. The positive valuation of their efforts strengthened their willingness to continue with these good farming practices and ignited their desire to learn more about biodiversity conservation (Table 25.2).

Table 25.2 Benefits of agroecotourism in Ovacık Village.

Dimensions	*Quotes*
1. Biodiversity conservation	Preserve and use heirloom seeds (OWSA & STA) Slow down contamination of soil and water through using natural fertilisers and avoiding the extensive use of agrochemicals (OWSA & STA) Protection of bees (OWSA & STA)
2. Diversified livelihoods	
Alternative income	"Through the center, we sell what we produce, and we gain additional income for our family." (CR) "We sell our produce at a fair price." (CR) "Even if I do not directly work at the center, I gain income from the sales of my products." (OR)
Capacity building	"Through the project activities, we are more informed." (CR) "I have developed my interpersonal skills." (CR) "We can spend our extra income on self-development activities for women." (CR)
Social capital development	"For our students, the walk in the forest and meeting with the local children were valuable as an experience." (VG) "I am happy to work at the center because I can get together with other women of the village." (CR) "I am happy to meet new people living out of our village." (CR) "Through the Ovacık project, we visit festivals and events in Şile and other towns." (CR) "We meet celebrities and media through the center." (CR) "The center provides a place to socialize for the women; I wish we had a similar center at our village." (PR)
Food heritage & culture conservation	"We co-host sessions with Ovacık Workshop that involve bringing local women of different ethnicity, together with from surrounding villages. The women prepare their traditional foods, which we serve and enjoyed together at the Workshop. Food is the catalyst for the conversation we have during the session." (VG). "We visited Ovacık to learn about village life and examine the socio-economic dynamics of life in villages." (VG) "Visitors enjoy learning to cook our traditional recipes with us using local products." (CR)
Community development	"I wished my village to be known by others, and I wanted to meet new people from other cultures as a result of the project." (CR) "I never expected the results to be this positive; the project contributed to our village and us." (CR)
Women empowerment	"We feel psychologically stronger." (CR) "As women, we have proven to ourselves that we can stand on our own feet." (CR) "Women from the nearby villages envy our success in Ovacık; they ask local authorities to coordinate similar projects in other parts of Şile." (MR)

(*Continued*)

Table 25.2 (Continued)

Dimensions	*Quotes*
3. Environmental interpretations	
Environmental ethics in destination communities	"We learned not to waste any organic waste from our kitchens and to use it in our gardens as fertilizers." (CR)
	"I make my toothpaste, vinegar, and soap at home." (CR)
	"During making home-made pasta, we used the branches of the quince tree to give form to pasta; I recognized that many ingredients for food are available in the nearby forest." (CR)
	"I learned that home-made vinegar is better for our health." (CR)
	"I demonstrate to the visitors how we make compost." (CR)
Impact beyond destination	"This project can be a role model to the nearby villages in Şile." (SR)
	"Our group members learned that it is easy to make ecological consumer products at home." (VG)
	"Our group members appreciated the value of cooking traditional food with sustainably grown local produce." (VG)
	"The participants in the event learned they could start composting and better manage waste at their homes." (VG)
4. Strengthened resource institutions and amenities	"We are happy that an idle building (school) is renovated and used by our community." (CR)
	"The permaculture and eco-conscious approach, combined with the traditional values of the village, provides the perfect location for educational and nature immersed experience." (VG)

OWSA = Ovacık Women's Seed Association, STA = Şile Tourism Association, CR = Community Center Respondent, OR = Ovacık Village Respondent, VG = Visitor Group, PR = Participant from the nearby village, SR = Şile Municipality Respondent, MR = Mukhtar of Ovacık Village.

The ecotourism activities had tangible (e.g., alternative income and infrastructure development) and intangible (e.g., capacity building and women empowerment) effects. The alternative income opportunities were not restricted to the 11 women that took part in the project, but the community, in general, benefitted from tourism. Even if only a small group of community members worked at the centre and gained direct income, the other community members could still sell their products through the centre and benefit economically (Table 25.2). This approach aimed to lessen the economic inequalities resulting from the initiative within the community.

An equally significant outcome of the project was the development of social capital. Women got together at the community centre, attended capacity building and self-development workshops, and worked around a common goal (bonding social capital), and they hosted visitors from other towns and even other countries (bridging social capital). The children of the village also connected with the youth from the visiting schools (bridging social capital). The project gained visibility in the social and printed media and also on the national TV. Consequently, the project partners and the community members gained further access to experts, NGOs, and local authorities (linking social capital).

The capacity development activities of the project resulted in the women to change their conservation and waste management behaviours beyond the community centre, at their daily lives (Table 25.2). They also started to appreciate natural resources more than before. For instance, after a workshop, one of the community respondents recognised that she could find materials from the forest to use as tools in her kitchen. The enjoyment and appreciation of nature were also mirrored in the visitor groups. Furthermore, the learning activities that the visitors have participated in increased their awareness of biodiversity conservation, sustainable consumption, and waste management.

Discussion and conclusions

This chapter firstly examined the ways a community-based agroecotourism project contributes to biodiversity conservation in a small forest town in Turkey. Secondly, the chapter developed a participatory approach for evaluating the benefits of ecotourism in community-based projects. The awareness of the community on biodiversity conservation (preserving and using of heirloom seeds, growing a variety of crops and using of natural fertilizers) together with the women's willingness to take part in tourism formed a suitable setting for ecotourism to be implemented as a tool for social wellbeing and biodiversity conservation in Ovacık (Choo & Jamal, 2009; Das & Chatterjee, 2015; Nyaupane & Poudel, 2011).

The project coordinators implemented the ecotourism project in Ovacık by considering the significant contextual elements of the town (Agrawal & Redford, 2006). Given that the coordinators exercised a community-based approach, they considered the views of multiple stakeholders from the start. The gendered expectations of the community effectively shaped where the project would be applied in the village, how it would be managed, and the frequency of the events (Morgan & Winkler, 2020). The co-ownership and co-management of the ecotourism project by the women of the village facilitated local support for the project goals (Bookbinder et al., 1998; Coria & Calfucura, 2012). This is not to say that all community members agreed, as some refrained from taking part in the project.

When planning ecotourism, the project coordinators considered the carrying capacity of their ecosystem, limiting the activities only to small groups (Buckley, 1994). Working with smaller groups enabled the community center to provide individualized service to visitors and to develop personal relationships between the visitors and the community. Furthermore, they charged a fair price for the tourism services and the products, signalling the good production and quality services they provide to the visitors. Indeed, the prices and the quality helped them to differentiate themselves from other food-service providers in Şile, placing them at the top end of the market. Consequently, the fairness in prices motivated the farmers to continue with sustainable agriculture production methods and be interested to learn more. Ecotourism did not replace farming as an economic activity but supported its existence. Thus, confirming the literature, we observed a *positive feedback loop* between sustainable agriculture, ecotourism, and biodiversity conservation in the "A day in the Ovacık Village" project (Stronza et al., 2019).

The ecotourism activities modified the ecological behaviors of the destination community and created awareness in the visitors through learning (Das & Chatterjee, 2015). The learning component combined elements of natural habitat (e.g., a walk in the forest), environmental conservation (e.g., waste management and rainwater collection), and heritage and cultural interest (e.g., local recipes) (Choo & Jamal, 2009; Duffy et al., 2017). The visitors were pleased because they were spending quality time in the village and learning, and at the same time, they were directly contributing to the livelihoods of the community (Coria & Calfucura, 2012).

In this research, we wanted to reflect the views of the locals in evaluating the outcomes of the ecotourism activity rather than imposing our ideas on what social wellbeing and biodiversity conservation should be (Stronza et al., 2019). Accordingly, we asked key stakeholders of the project, whether they found the results of the ecotourism activity successful, without defining what success is. Based on their answers, we gather the benefits of ecotourism in Ovacık Village as 1) provides alternative income to the community, 2) develops social capital, 3) builds capacity and empowers women, 4) helps to conserve heritage and culture, 5) changes the behaviors of community and the visitors through learning, and 6) develops infrastructure and amenities. This definition of the benefits is grounded in empirical data, reflecting the views of multiple stakeholders. Nevertheless, it is context-specific and culturally embedded, and a similar inquiry in another setting may result in different dimensions. Meanwhile, these dimensions are in line with the broader definition of ecotourism found in the recent literature (Das & Chatterjee, 2015; Stronza et al., 2019; TIES, 2015).

There are certain implications of this research for practitioners and policymakers. Ecotourism development is often seen as a policy tool for improving the livelihoods of people living in rural areas. In Turkey, alternative tourism types are also viewed as a tool for rural development. The tourism strategy plan of Turkey 2023 identifies new ecotourism cities and corridors to be developed throughout the country (Ministry of Culture and Tourism, 2007).

Relying on how the stakeholders view the benefits of the project in Ovacık, agroecotourism can be linked to multiple UN SDGs. In particular, the results contribute to SDG 12, which is responsible consumption and production, and SDG 5, which is gender equality. The chapter also displays the essential elements of ecotourism development for it to contribute to sustainable development. First, the continued collaboration between several stakeholders and participation of the community in the planning and implementation stages contributes to the positive outcomes of the project (Wondirad et al., 2020). Second, the existence of powerful grassroots organisations for raising awareness and improving local support for ecotourism development is vital (Buckley, 2009). Moreover, land development and natural area conservation policies of the government must be in alliance with ecotourism development (Hunt et al., 2015; Stone, 2015) and not prioritise immediate gains from other sectors like mining or real estate development. Last, natural resources should not be seen as an asset to be exploited for the sake of tourism development, but tourism should be approached as a tool for conservation and sustainable development.

In the case of Ovacık, the Şile Municipality acknowledged the significance of "A Day in Ovacık Project" for displaying a best example case for other villages around Şile. The Municipality should use this opportunity to prove its cause for terminating quarry permits and ask the central government to back ecotourism development in Şile and the surrounding forest land. In Ovacık, the community centre gained extra income from their tourism activities; however, the women are divided between spending the money for social responsibility, visits to other towns, and capacity development. We can recommend women to use some of the money for biodiversity conservation to maintain the sustainability of the natural environment. These activities could be designed in line with the goals of the project, such as establishing a local heirloom seed bank and extending the educational programs to other farmers, installing new waste management equipment for demonstrations, or networking with other centres that support seed preservation.

Finally, the collaboration with practitioners and PAR methodology applied in this chapter has proven to be instrumental in gathering data for a broader understanding of the phenomenon under study. Furthermore, the approach enabled the recommendations of the chapter to be more specific and provide improved guidance for action.

References

Agrawal, A., & Redford, K. H. (2006). *Poverty, Development, and Biodiversity Conservation: Shooting in the Dark?* New York: Wildlife Conservation Society.

Bertella, G. (2019). Participatory action research and collaboration in CSR initiatives by DMOs. *Journal of Ecotourism, 18*(2), 165–173.

Blamey, R. K. (2001). Principles of ecotourism. In D. B. Weaver (Ed.), *The encyclopedia of ecotourism* (Vol. 2001). Wallingford, UK: Cabi Publishing.

Bookbinder, M. P., Dinerstein, E., Rijal, A., Cauley, H., & Rajouria, A. (1998). Ecotourism's support of biodiversity conservation. *Conservation Biology, 12*(6), 1399–1404.

Brandt, J. S., & Buckley, R. C. (2018). A global systematic review of empirical evidence of ecotourism impacts on forests in biodiversity hotspots. *Current Opinion in Environmental Sustainability, 32,* 112–118.

Buckley, R. (1994). A framework for ecotourism. *Annals of Tourism Research, 21*(3), 661–665.

Buckley, R. (2009). Evaluating the net effects of ecotourism on the environment: A framework, first assessment and future research. *Journal of Sustainable Tourism, 17*(6), 643–672.

Choo, H., & Jamal, T. (2009). Tourism on organic farms in South Korea: A new form of ecotourism. *Journal of Sustainable Tourism, 17*(4), 431–454.

Coria, J., & Calfucura, E. (2012). Ecotourism and the development of indigenous communities: The good, the bad, and the ugly. *Ecological Economics, 73,* 47–55.

Das, M., & Chatterjee, B. (2015). Ecotourism: A panacea or a predicament? *Tourism Management Perspectives, 14,* 3–16.

Demir, S. (2019). Bakanlık devreye girdi Şile ormanları kurtuldu! (Sile forests were saved when the Ministry intervened!). Retrieved from https://www.sozcu.com.tr/2019/gundem/bakanlik-devreye-girdi-sile-ormanlari-kurtuldu-5434859/.

Duffy, L. N., Kline, C., Swanson, J. R., Best, M., & McKinnon, H. (2017). Community development through agroecotourism in Cuba: An application of the community capitals framework. *Journal of Ecotourism, 16*(3), 203–221.

Fennell, D. (2012). Ecotourism. In A. Holden, & D. Fennell (Eds.), *The Routledge handbook of tourism and the environment* (pp. 345–355). Abington: Routledge.

Fennell, D., & Markwell, K. (2015). Ethical and sustainability dimensions of foodservice in Australian ecotourism businesses. *Journal of Ecotourism, 14*(1), 48–63.

Grain. (2007). Turkey's New Seed Law. Retrieved from https://www.grain.org/es/article/entries/623-turkey-s-new-seed-law.

Gutiérrez-Montes, I. A. (2005). *Healthy communities equal healthy ecosystems? Evolution (and breakdown) of a participatory ecological research project towards a community natural resource management process* (PhD Thesis, Iowa State University, Ames, US).

Hatipoglu, B., Ertuna, B., & Salman, D. (2019a). Corporate social responsibility in tourism as a tool for sustainable development. *International Journal of Contemporary Hospitality Management, 23*(3), 339–357.

Hatipoglu, B., Ertuna, B., & Salman, D. (2019b). Evaluation of a Turkish company's progress towards a CSR 2.0 approach to corporate governance. In *Corporate sustainability and responsibility in tourism* (pp. 343–360). Cham: Springer.

Hunt, C. A., Durham, W. H., Driscoll, L., & Honey, M. (2015). Can ecotourism deliver real economic, social, and environmental benefits? A study of the Osa Peninsula, Costa Rica. *Journal of Sustainable Tourism, 23*(3), 339–357.

Kindon, S., Pain, R., & Kesby, M. (2007). *Participatory Action Research Approaches and Methods: Connecting People, Participation and Place* (Vol. 22). London: Routledge.

Kiss, A. (2004). Is community-based ecotourism a good use of biodiversity conservation funds? *Trends in Ecology & Evolution, 19*(5), 232–237.

Ministry of Culture and Tourism. (2007). Tourism Strategy of Turkey-2023. Retrieved from https://www.ktb.gov.tr/Eklenti/43537,turkeytourismstrategy2023pdf.pdf?0&_tag1=796689BB12A54 0BE0672E65E48D10C07D6DAE291.

Morgan, M. S., & Winkler, R. L. (2020). The third shift? Gender and empowerment in a women's ecotourism cooperative. *Rural Sociology, 85*(1), 137–164.

Moscardo, G., Konovalov, E., Murphy, L., McGehee, N. G., & Schurmann, A. (2017). Linking tourism to social capital in destination communities. *Journal of Destination Marketing & Management, 6*(4), 286–295.

Nyaupane, G. P., & Poudel, S. (2011). Linkages among biodiversity, livelihood, and tourism. *Annals of Tourism Research*, *38*(4), 1344–1366.

Republic of Turkey Ministry of Agriculture and Forestry. (2006). Seed Law. Retrieved from https://www.tarimorman.gov.tr/BUGEM/TTSM/Menus/107/Seed-Law.

Romero-Brito, T. P., Buckley, R. C., & Byrne, J. (2016). NGO partnerships in using ecotourism for conservation: Systematic review and meta-analysis. *PloS one*, *11*(11), 1–19.

Slow Food Foundation for Biodiversity. (2020). Earth Markets. Retrieved from https://www.fondazioneslowfood.com/en/what-we-do/earth-markets/.

Stone, M. T. (2015). Community-based ecotourism: A collaborative partnerships perspective. *Journal of Ecotourism*, *14*(2–3), 166–184.

Stronza, A. L., & Gordillo, J. (2008). Community views of ecotourism. *Annals of Tourism Research*, *35*(2), 448–468.

Stronza, A. L., Hunt, C. A., & Fitzgerald, L. A. (2019). Ecotourism for conservation? *Annual Review of Environment and Resources*, *44*, 229–253.

TIES. (2015). What Is Ecotourism. Retrieved from https://ecotourism.org/.

UNDP. (2020). The GEF Small Grants Program: Biodiversity. Retrieved from https://sgp.undp.org/areas-of-work-151/biodiversity-170.html?view=summary.

United Nations. (2016). Sustainable Development Goal 12. Retrieved from https://sustainabledevelopment.un.org/sdg12.

UNWTO. (2018). Tourism and the Sustainable Development Goals –Journey to 2030. Retrieved from https://www.e-unwto.org/doi/book/10.18111/9789284419401.

Weaver, D. B., & Lawton, L. J. (2007). Twenty years on: The state of contemporary ecotourism research. *Tourism Management*, *28*(5), 1168–1179.

Wondirad, A., Tolkach, D., & King, B. (2020). Stakeholder collaboration as a major factor for sustainable ecotourism development in developing countries. *Tourism Management*, *78*, 104024.

26
ECOTOURISM AND REWILDING EUROPE

Nils Lindahl Elliot

In 2016 Rewilding Europe, an organisation based in the Netherlands but with campaigns across the European Union, published a short video on its website that offered a vision of its mission. Amid wide shots and aerial footage of stunning mountains, forests, coasts, rivers and plains, the video focalised various wild animals: bears, lynxes, and golden eagles, but also 'aurochs', European bison, wild boar, and griffon vultures. According to the video, large parts of the countryside in Europe were being abandoned, and this constituted an 'historic opportunity' to make Europe 'a wilder place!' (Rewilding Europe, 2016a, bold lettering in the originals). To this end, Rewilding Europe was working to 'support the comeback' of grazers, as well as their predators and scavengers in areas such as the Laplands and the Central Apennines, western Iberia, and the Danube Delta. The plan was to rewild 1 million hectares, i.e., an area of some 10,000 km^2 or 3900 mi^2, 'where rivers flow freely', 'where wildlife roams again', and 'where nature takes care of itself' (Rewilding Europe, 2016a).

Rewilding Europe is just one—albeit one particularly ambitious—example of the many rewilding initiatives that have emerged across the world over the last two decades. There are significant variations in the meaning of rewilding (see for example Lorimer et al. 2015;Jørgensen, 2015), as well as debate regarding its scientific basis (see for example Rubenstein, Rubenstein, Sherman, & Gavin, 2016; Rubenstein & Rubenstein, 2015; Nogués-Bravo, Simberloff, Rahbek, & Sanders, 2016). Rewilding may nevertheless be broadly defined as a form of conservation that entails restoring ecosystems and landscapes to a state that is considered to be somehow 'closer to nature'. However, 'rewilders' (as Rewilding Europe refer to themselves) are prepared to intervene across trophic levels in order to address what they regard as one of the more serious forms of anthropogenic degradation: the extirpation of large carnivores, and more generally of so-called keystone species, with adverse consequences for the biodiversity of ecosystems.

Many of the earliest scientific proposals for this kind of intervention have centred on the beyond-human aspects of ecological interaction; any attention given to human involvement has typically been a negative one in so far as conservation has focussed on limiting human impact on certain ecosystems (e.g., park-making), or on enabling wild animals to overcome the manifold barriers to mobility/migration placed in their way by human development. However, the Rewilding Europe video, like the organisation's mission statements, suggests a way of approaching rewilding that is premised on a more proactive conception of human involvement.

 DOI: 10.4324/9781003001768-26

In so far as there is across Europe a tendency towards 'rural depopulation', and in so far as 'more and more Europeans want to *enjoy* wild nature and wildlife' (Rewilding Europe, 2016a, bold in the original), this constitutes a *business* opportunity which can and should itself be capitalised. Central to such an opportunity is the development of tourism.

Rewilding Europe is by no means alone in proposing this kind of change. Across the world, there are many examples of initiatives that rely on tourism (and leisure practices more generally) as a means of at once funding, and engaging in what is, in effect if not by express design, a hegemonic turn with which to meet the often significant *opposition* to rewilding (for accounts of opposition and conflict, see for example Nilsen, Milner-Gulland, Schofield, Mysterud, Stenseth & Coulson, 2007; Barkham, 2017; Pellis, 2019).

In this chapter I problematise such proposals, and with them the interrelation between rewilding and ecotourism—an interrelation which, somewhat surprisingly, has thus far attracted comparatively little critical scrutiny (Hall, 2019). In the absence of an extensive corpus on the subject, I will employ the example of Rewilding Europe as an at once convenient, and strategic case study. However *sui generis* Rewilding Europe's proposals may be on one level, the organisation has developed its proposals with reference to broader discourses of rewilding. This being the case, an analysis of Rewilding Europe may well have implications that go beyond the particularities of this organisation and its plans for rewilding certain areas across Europe.

Rewilding Europe's mission as discursive formation

To begin with, Rewilding Europe's initiative may be approached via its mission statements and promotional literatures. These constitute discourses (Foucault, 1971, 1993) that invite subjects to attend to some objects and not others (what I will describe as indexing); to categorise those objects in particular ways (classifying); and in so doing to promote certain forms of practice, subjectivity, and interrelation (framing) (Lindahl Elliot, 2019, p. 216). Of course, such discourses are only a part of a much broader ensemble which includes institutions, fields of interaction (Bourdieu, 1992), as well as techniques and technologies of intervention in what are themselves complex beyond-human settings. As such, rewilding discourses may be regarded as aspects of an assemblage devoted to exercising biopower (Foucault, 1978). As Foucault famously put it,

> [i]f one can apply the term *bio-history* to the pressures through which the movements of life and the processes of history interfere with one another, one would have to speak of *bio-power* to designate what brought life and its mechanisms into the realm of explicit calculations and made knowledge-power an agent of transformation of human life. (Foucault, 1978, p. 143, italics in the original)

I suggest that in the case of rewilding, the more immediate object of transformation is beyond-human life.

Approached from this perspective, Rewilding Europe's mission statements constitute not just organisational expressions of intent, but more or less complex ways of indexing, classifying, and framing an ostensibly natural world for the purposes of a specific kind of intervention. That intervention is articulated by way of a discursive formation (Foucault, 1993) that includes discourses of nature, science, socio-economics, and tourism. With an important exception that I will consider below, the formation in question is premised on what is, philosophically speaking, a realist ontology of nature.

Even a summary analysis of Rewilding Europe's communications reveals that the organisation treats nature as being quite separate from culture, and as being independent of any vicissitudes of representation. What is true for nature in general is particularly true for wildness, which Rewilding Europe, like rewilders more generally, treat as being, ecologically speaking, the ultimate expression of a *nonhuman* nature.

Starting from this very strong classification of the human and the nonhuman, rewilders define an arena of intervention that indexes particular kinds of animals (viz., wild animals) and geographies (wildernesses). These are framed as being inherently deserving of what is, in effect, a sacralising modality of classification: where other animals and geographies might be expendable, wild animals, and wildernesses ought to be preserved, and actively protected. In general, this articulation constitutes one of the more important, if silent ways with which rewilders establish a normative subjectivity vis-a-vis conservation(ist) practice. In so far as certain animals are wild, and certain geographies are said to constitute wildernesses, they should be preserved.

On one level, this kind of discourse suggests a continuity with respect to romantic-sublime discourses of nature that are centuries old. However, rewilders specify the frame further: they index some kinds of wild animals and wildernesses, which they classify as being particularly worthy of conservation; and do so with reference to a scientific discourse about species and their interrelations. That discourse is the one associated with the field of conservation biology, which was established in the 1980s (see for example, Soulé & Wilcox, 1980; Meine, Soulé, & Noss, 2006).

A detailed account of this field, or even of its proposals regarding rewilding, is beyond the scope of this article. Here it will suffice to note that one of the founders of conservation biology was Michael E. Soulé, and in 1998 Soulé, writing with the biologist Reed Noss (Soulé & Noss, 1998), put forward proposals that are now widely regarded as being formative for what has become an increasingly globalised rewilding movement (see also Foreman, 2004). Soulé and Noss characterised rewilding as a form of conservation that goes beyond the 'biodiversity conservation' paradigm, and so beyond the 'representation of vegetation or physical features diversity and the protection of particular biotic features' (Soulé & Noss, 1998, p. 19). Rewilding, they suggested, involves intervention on three levels, or 'the three C's': the preservation of *cores* (large, strictly protected reserves, i.e., 'the wild' or 'wilderness'); the creation and/or preservation of *corridors* between such cores, across anthropogenically developed areas (thereby ensuring connectivity amongst the reserves); and the reintroduction where necessary of large *carnivores* (or more generally, of keystone species) (Soulé & Noss, 1998, p. 22).

As this account begins to explain, Soulé and Noss were particularly keen to index large carnivores, which they argued were responsible for initiating what they framed as 'top-down' ecological interactions, known as *trophic cascades* (Paine, 1980). In the absence of large carnivores capable of controlling, say, medium to large herbivores, but also smaller predators, an ecosystem might eventually lose its biodiversity thanks to overgrazing, or to predation by an exploding population of the smaller predators, with cascading effects across several trophic levels (see for example Terborgh et al., 1999). A classic example involves the extirpation of wolves from the Yellowstone Park in the 1920s, and then their reintroduction in 1995–1996 (see for example, Singer, Mark, & Cates, 1994; Beschta & Ripple, 2007).

Since the publication of Soulé and Noss's proposals, rewilders have recontextualised aspects of those proposals in ways that have reflected the specificities of their own fields of interaction and the ecologies in which they wished to intervene. The proposals of Rewilding Europe are no exception. In some of the earlier policy statements, Rewilding Europe have been particularly keen on reintroducing large herbivores, or what one document classified more generally as 'Large Apex *Consumer* Species' (Wild10, 2015, p. 9, italics added to the original. See also the signature 'aurochs' programme, Goderie, Helmer, & Kerkdijk-Otten, 2013). This has generated controversy in so far as some ecologists have regarded the emphasis on herbivores as a kind of ecological exceptionalism that reflects Dutch rewilders' debt to the discourse of '*Natuurontwikkeling*' (roughly translated as 'nature development') (see Fisher, 2018, 2019). Perhaps in response to this kind of criticism, later documents produced by Rewilding Europe have gone out of their way to recognise the importance of both leaving nature to 'do what it may', and of reintroducing large carnivores: 'Rewilding Europe recognizes the *crucially important ecological role of large carnivores* (Rewilding Europe, 2019, p. 3, italics in the original). This emphasis to one side, Rewilding Europe recognised from an early stage that the absence of large predators *and* herbivores might 'generate extensive cascading of detrimental effects in marine, terrestrial and freshwater ecosystems. This "trophic downgrading" affects functions and resilience of ecosystems and has negative impacts on biodiversity as well as the spread of infectious diseases, carbon sequestration, invasive species, and biochemical cycles' (Wild10, 2015, p. 9).

Thus far, I've highlighted the role of discourses of science and nature. However, to understand how Rewilding Europe links tourism to its rewilding initiative, it is necessary to consider the role of socio-economic discourse, and with it, a discourse of tourism. Let's begin with the former. In its earliest promotional literature, Rewilding Europe explained that 'The backdrop for Rewilding Europe is an ailing agricultural economy in rural Europe, propped up by inefficient EU subsidies and heading towards a period of rapid change as these subsidies are replaced and restructured' (Rewilding Europe, 2012, p. 20). But whereas 'nature conservation has often been seen as an influence which seeks to slow or indeed halt economic activity', Rewilding Europe would aim to recognise 'the vital role of business, investment and job opportunities for the success of conservation' (Rewilding Europe, 2012, p. 20). This being so, Rewilding Europe would 'seek to exploit rapidly evolving new markets – for example, nature-based tourism', which it noted was growing 'at three times the rate of conventional tourism globally … with the right investments in tourism facilities and promotion, several areas in Europe have the potential to become world-class wildlife tourism attractions' (Rewilding Europe, 2012, p. 20).

This, and numerous other invocations of the need to embrace 'business' suggest that the Rewilding Europe initiative has developed what is arguably a neoliberal discourse on rewilding (for an account of neoliberalism, and neoliberalisation, see Harvey, 2005; England & Ward, 2016, respectively). This is particularly evident in one of the economic discourses that is invoked by Rewilding Europe's mission statements—what may be described as a discourse of 'natural capital' (after the influential concept proposed by Constanza & Daly, 1992 in an article for *Conservation Biology*). Natural capital, and the later 'natural capitalism' (see Hawken, Lovins, & Lovins, 1999), are at once responses to, but in the end themselves arguably forms of the neoliberalisation of nature (Castree, 2008). Simplifying somewhat, those who conceptualise natural capital seek to 'economise' ecological relations so as to reveal the ecological costs of economic development. However, this strategy is integrally linked to a much broader social process that involves dynamics of privatisation and marketisation, deregulation and reregulation, market proxies in the residual public sector, and the construction of flanking mechanisms in the civil society (Castree, 2008, p. 142). Rewilding Europe is, as an organisation, arguably a good example of the last of these aspects.

One especially prominent version of the neoliberal discourse may be found in the work of the biologist Gretchen C. Daily. In the late 1990s, Daily suggested that if 'goods and services flowing from natural ecosystems are greatly undervalued by society', this was at least partly because 'the benefits those ecosystems provide are not traded in formal markets and do not send price signals of changes in their supply or condition' (Daily, 1997, p. 2). The solution was to develop a scheme with which to render economically accountable not just the ecological 'goods' that have long been employed by economies—e.g., seafood, forage, timber, biomass fuels, and many pharmaceuticals—but also the 'ecological services', viz., the 'actual life-support functions, such as cleansing, recycling, and renewal' as well as the 'intangible aesthetic and cultural benefits' of ecosystems (Daily, 1997, p. 3). Amongst the latter, Daily included the provision of 'the aesthetic beauty and the intellectual stimulation that lift the human spirit' (Daily, 1997, p. 4).

During the 2000s, this discourse came to echoed by a growing number of scientific and conservation(ist) institutions—amongst them, the United Nation's Millennial Ecosystem Assessment (see for example, Millennial Ecosystem Assessment, 2005). The concept also began to be used in discussions about nature tourism, and ecotourism (see for example de Groot, Wilson, & Baumans, 2002). Its influence is evident across the promotional literature of Rewilding Europe. For example, *A Vision for a Wilder Europe* (henceforth *Vision*), a document produced by Rewilding Europe and other participants in the tenth World Wildlife Conference (Wild10, held in Salamanca in 2013), spoke of the 'the need for wild land to provide *ecosystem services* like clean water and air, as base-line scientific reference areas, [and] *for recreation and tourism* … and indeed, to refresh our human spirit and wellbeing' (Wild10, 2015, p. 3, italics added). According to *Vision,* by 2025 the signatories of the proposals hoped 'to give more people a closer relationship with nature in contrast to our highly technological world, increase resilience to the effects of climate change, and generate new economic opportunities and better *services* for society' (Wild10, 2015, p. 7, italics added). An identical modality of framing is evident in later documents, including Rewilding Europe's current Three-Year Strategic Plan (2019–2021), which suggests that 'We have, as a society, begun to recognise the need for wild land, to provide ecosystem services like clean water and air, as base-line scientific reference areas, for recreation and economic development, and [once again], to refresh our human spirit and wellbeing' (Rewilding Europe, 2019, p. 8).

To be clear, not all rewilders embrace neoliberalisation, and indeed some actively oppose it (see for example, Monbiot, 2019). In the case of Rewilding Europe, the proximity to neoliberalism may be explained with reference to various factors, including the corporate backgrounds of several of its organisational members and backers; the importance of a neoliberal ethos for environmental policy frameworks in the early twenty-first-century European Union; but also, and as I have just explained, the ironic rise of a neoliberal discourse even amongst conservation biologists who might otherwise be highly critical of untrammelled economic development.

Now if Rewilding Europe has been keen to develop certain socio-economic relations as part of its initiative, the statements quoted above reveal the importance that a discourse of *tourism* plays in this process. One of the 'tools' established by Rewilding Europe was the European Safari Company, which it launched in late 2016. According to its website, the European Safari Company specialises in 'experiential nature-based travel across Europe':

> The European Safari Company offers unique adventures that directly support nature, wildlife and local cultures in unique places across our continent. Places where nature is still thriving or bouncing back due to rural depopulation, legal protection and re-wilding efforts. Places that have the opportunity to build a new future based on their unique setting, landscapes, wildlife, local culture and people. (European Safari Company, 2016; see also Rewilding Europe, 2016b)

As these statements begin to suggest, the kind of tourism that the Rewilding Europe initiative seeks to promote entails a combination of wildlife tourism and adventure tourism. However, both are framed in ways that suggest a commitment to ecotourism—albeit, what might be characterised as 'soft' ecotourism (Weaver, 2001). There are of course numerous ways of defining ecotourism (Fennell, 2015), but for the purposes of this chapter it will suffice to employ a definition likely to be endorsed by Rewilding Europe: the one proposed by The International Ecotourism Society (TIES), 'responsible travel to natural areas that conserves the environment, sustains the well-being of the local people, and involves interpretation and education' (TIES, 2015). Soft ecotourism is easily mistaken for 'nature-based tourism', i.e., the kind of mass tourism whose members are attracted to sites and/or events deemed to be of outstanding ecological significance, but who require as a condition for travel the kinds of luxuries associated with modern tourism—for example, easy access, a certain standard of accommodation, visitor programmes, etc.

Should such tourism still be called *eco*tourism? A number of scholars have noted that the 'eco' in ecotourism may be subject to greenwashing (see for example, Honey, 2008; Fennell, 2015). Soft ecotourism may itself be a sign of the neoliberalisation of alternative forms of tourism. To be sure, in the context of an increasingly catastrophic climate change, a case must be made that *no* form of tourism that relies on greenhouse gas-producing infrastructure may be regarded as being 'ecological'. In recognition of these issues, I use the expression (eco)tourism to underscore the contested nature of the term. The last points notwithstanding, and as the earlier quotes show, Rewilding Europe does at least *profess* a commitment to sustainable forms of development, and with it, tourism with a strong element of localism. While practices on the ground may end up contradicting any such intent, it seems clear that, if rewilding is to involve any tourism, it can, discursively at least, only really 'go' with some form of ecotourism; to do otherwise would flatly contradict, even on the level of official policies and mission statements, the very aims of rewilding.

We can say then that, even as rewilding might restore wilderness areas and their wild animals, according to Rewilding Europe those same areas can and should in turn provide tourists with an 'ecological service'—a *cultural* ecological service—which travellers might access via entities such as the European Safari Company. But doing so should be part of a broader business strategy designed to ensure that rewilding projects might still obtain funding despite the privatisation of environmental initiatives—itself one of the many consequences of neoliberalisation.

As I began to suggest at the start of this chapter, the discursive formation that articulates the relation between rewilding and (eco)tourism in the case of Rewilding Europe is based on a realist ontology. There is, however, an important exception to this orientation: what has to be characterised as an overtly *idealist* turn at the core not just of Rewilding Europe's vision, but of rewilding initiatives more generally. The turn in question takes us back to the earlier mention of a romantic-sublime discourse of nature.

Across some of the quotes presented thus far, the reader will have noticed several references to an '*imperative* for wilderness' (Wild10, 2015, p. 3), and a need for wild land 'to refresh our human *spirit* and wellbeing' (Rewilding Europe, 2019, p. 8). Daily also refers to 'the aesthetic beauty and the intellectual stimulation that lift the *human spirit*' (Daily, 1997, p. 4) (italics added to the originals). Of course, there can be no scientific evidence for 'refreshment', let alone the existence of a human 'spirit'. The logic of this frame would thus appear to be a version of the following: if wild places and animals have essential attributes (as per the realist ontology), then those essences may somehow be transferred to, or reactivated in the 'human spirit' if individuals are able to travel to, or otherwise 'commune' with the wild, with wilderness. From this perspective, it seems that 'going back to nature' is the *human* equivalent of rewilding.

At least in the thought of some prominent rewilders, what might be characterised as *auto-rewilding* is not as far-fetched as might be assumed. *Vision*, the Wild10 conference statement mentioned earlier, included a long epigraphic quotation from George Monbiot's (2013)*Feral: Searching for Enchantment on the Frontiers of Rewilding*. Monbiot and *Feral* are credited with having played an important role in popularising rewilding across the English-speaking world (see for example Jørgensen, 2015). In *Feral*, Monbiot employs the representation of his own rather extreme leisure pursuits to illustrate what he himself describes as the 'rewilding of human life' (Monbiot, 2013, p. 10). According to Monbiot, this kind of rewilding should go hand in hand with the rewilding of beyond-human species and ecosystems. While Monbiot is careful not to oppose human rewilding to 'civilisation', he also makes it clear that it entails something like *becoming hominid*. In one example in the third chapter of *Feral*, Monbiot explains that, after finding a dead deer while '[f]oraging for herbs and fungi in a wood in southern England', he 'gathered up the [deer's] ankles and heaved it onto [his] shoulders', and the 'effect was remarkable. As soon as I felt its warmth on my back, I wanted to roar … This, my body told me, was why I was here. This was what I was for' (Monbiot, 2013, p. 33). According to Monbiot, he believed, though he recognised he could not prove it, that he was 'experiencing a genetic memory … These genetic memories—these unconsidered urges—are printed onto our chromosomes, an irreducible component of our identity' (Monbiot, 2013, pp. 33–34).

From this kind of perspective, auto-rewilding via adventure tourism and wildlife tourism may be regarded as a way of returning to something like one's 'Pleistocene roots'—an idea that suggests a sociobiological discourse. It is arguably no coincidence that Monbiot, who is a zoologist by training and is well versed in rewilding conservation theory, adopts such a frame vis-a-vis his own practices. One of the main theories used by Soulé and Noss to justify their version of rewilding is island biogeography theory, which was developed in part by Edward O. Wilson (see MacArthur & Wilson, 1967). As well as being a famous conservation activist in his own right (see for example Wilson 2016), Wilson is the renowned founder of sociobiology and evolutionary psychology (Wilson, 1975, 1978, 1998), and the proponent of the biophilia hypothesis (Wilson, 1984; Kellert & Wilson, 1993). Sociobiology, evolutionary psychology, and biophilia embrace adaptationism, and so diminish *a priori* the importance of cultural-historical contingency in any account of human natures (see Orzack & Forber, 2017).

From discursive formation to hybrid geography

Researchers in both the physical and critical social sciences have raised numerous questions regarding the claims made by evolutionary psychologists in general (see for example, Gould, 1979; Lewontin, Rose, & Kamin, 1984; Rose & Rose, 2001), and those who advocate the biophilia hypothesis in particular (see for example Joye & Blocke, 2011). Many have done so from discursive perspectives that reflect a *nominalist* ontology. A discussion of the different varieties of nominalism is beyond the scope of this chapter (see for example Rodríguez-Pereyra, 2016). Here it will suffice to note that the kind of nominalism that is generally used to critique evolutionary psychology underscores the discontinuities between general categories such as nature, human nature, the wild, or wilderness, and the particulars that they refer to. Where the subjects of realist, and positivist discourses tend to adopt a 'nature-endorsing' stance (Soper, 1995, p. 4), the advocates of nominalist and what are generally constructivist discourses adopt a 'nature-sceptical' stance (Soper, 1995, p. 4). The latter is the kind of perspective adopted not only by Foucault himself, but also by poststructuralist scholars concerned with discourses of nature, including Raymond Williams (1983), William Cronon (1996), and in a less markedly culturalist manner, Donna Haraway (1989, 1991).

The proposals of these scholars have been widely discussed, and in some cases hotly disputed (for a particularly pertinent example, see Soulé, 1995). This being so, here I will consider a more recent set of proposals. As suggested earlier, rewilding initiatives may be regarded as a form of *bio*power. However, in Foucaultian scholarship, as in poststructuralist theory more generally, analyses of biopower tend to be short on the 'bio', and long on the 'power' (for a critique of this tendency, see Lindahl Elliot, 2019, pp. 235–237). Discourse analysis also runs the risk of logocentrism. These problems can be at least partly addressed by turning to research that is itself thoroughly nominalist (to echo the expression used by Barbara Herrnstein Smith, 2014, p. 492), but which does at least engage with beyond-human actors. I refer initially to the work of the social theorist and philosopher Bruno Latour, one of the founders of Actor-Network-Theory (Latour, 2005); but also to that of Sarah Whatmore, the leading scholar in a closely related subfield of human geography known as hybrid geography (Whatmore, 2002).

Simplifying greatly, ANT researchers argue that the meanings and roles of things are entirely contingent on their place in networks. However, the networks are themselves regarded as a matter of trajectories and inter-relations that are always in a state of flux. The networks can thus never be fully represented; a nominalist frame is evident in so far as it is assumed that there is, indeed there *must* be, a chasm between the necessarily generalising representations produced by scientists (or any other actors, for that matter), and the individuals (objects, processes, etc.) that such representations refer to. The challenge is thereby to develop a form of inquiry that remains very close to an explanatory degree zero vis-a-vis particulars. Any researcher's attempt to develop a metalanguage (a vocabulary of theoretical concepts, more or less technical explanations, etc.) with which to explain a certain state of affairs is likely to be misleading in so far as it changes the practices that actors engage in: when that happens, sociology stops being empirical and becomes "vampirical"' (Latour, 2005, p. 50). As Latour puts it, 'The project of ANT is simply to extend the list and modify the shapes and figures of those assembled as participants and to design a way to make them act as a durable whole' (Latour, 2005, p. 72). The list in question includes not only humans and their institutions, but all manner of *things* (Latour, 2005, p. 46)—not just 'human', but also 'nonhuman' things, and indeed those that entail mixtures of nature and culture, e.g., hybrids.

In the context of this chapter, perhaps the best way of engaging with this kind of approach is via the research developed by Sarah Whatmore (2002), a human geographer who has built on ANT and other fields to develop a series of a proposals vis-a-vis the study of beyond-human animals and geographies. In particular, I would like to highlight three aspects of Whatmore's work which seem particularly pertinent to a problematisation of rewilding: its emphasis on the *hybridity* of ostensibly wild geographies; its rejection of a wilderness conceived as an *outside*; and its focus on *topological*, as opposed to metric space.

The first of these aspects is what gives the name to the geographic subfield: hybrid geographies rejects the culture-nature dualism that is so integral to rewilding. Instead, it seeks to recognise the hybridity of modern geographies. As concieved by Whatmore, this requires 'an upheaval in the binary terms in which the question of nature has been posed and a recognition of the intimate, sensible and hectic bonds through which people and plants; devices and creatures; documents and elements take and hold their shape in relation to each other in the fabric-ations of everyday life' (Whatmore, 2002, p. 3). Hybrid geographies oppose 'the purifying impulse to fragment living fabrics of association and designate the proper places of "nature" and "society"', and seek instead to countenance the world 'as an always already inhabited achievement of heterogeneous social encounters where, as Donna Haraway puts it, "all of the actors are not human and all of the humans are not "us" however defined"' (Whatmore, 2002, p. 3).

Rewilding Europe would doubtless enthusiastically embrace the idea that not all of the actors are human; its managers might suggest that the *whole point* of the initiative is to privilege nonhuman actors. Note, however, that central to this endeavour there is a 'purifying impulse' in the form of the indexing the wild and of wilderness, and with it the tacit opposition/classification of nature and culture. From this kind of perspective, wilderness is not only symbolically constructed as being wholly separate, but as being 'outside'.

According to Whatmore, this *a priori* of a nature that is 'outside' must itself be refused. In as much as the wild is defined as that which lies beyond of human civilisation's historical and geographical reach, it 'renders the creatures that live "there" inanimate figures in unpeopled landscapes, removing humans to the "here" of a society from which all trace of animality has been expunged' (Whatmore & Thorne, 1998, pp. 435–436). Instead of starting from an exterior world or an 'original nature', Whatmore would like to start from the premise that '"wild" animals have been, and continue to be, routinely imaged and organized within multiple social orderings in different times and places' (Whatmore, 2002, p. 14). Such orderings 'confound the moral geographies of wilderness which presuppose an easy coincidence between the species and spaces of a pristine nature, confining their place to the margins and interstices of the social world' (Whatmore, 2002, pp. 9–10). The problem is thereby not to start with the wild or with wildlife as 'the outside', but to approach it from an 'inside' where 'the everyday worlds of people, plants and animals are already in the process of being mixed up' (Whatmore & Thorne, 1998, p. 437).

Two aspects of this critique seem particularly relevant to Rewilding Europe's initiative. First, however much Rewilding Europe and figures such as Monbiot have emphasised the importance of leaving nature to be what it will (both Monbiot and some of the later Rewilding Europe mission statements acknowledge the impossibility of a return to a historic nature [see Monbiot, 2013, pp. 9–10, and Rewilding Europe, 2019, p. 6]), there is a contradiction to the extent that rewilders can and must, according to their own discourse, *intervene*. As Monbiot has put it,'Rewilding, to me, is about resisting the urge to control nature and allowing it to find its own way'. Alas, he adds in almost the same expository breath that '[rewilding] involves reintroducing absent plants and animals (and in a few cases culling exotic species which cannot be contained by native wildlife), pulling down fences, blocking the drainage ditches, but otherwise stepping back' (Monbiot, 2013, pp. 9–10). In this double movement – claiming on the one hand that nature must find its own way, but on the other hand suggesting what steps should be taken to ensure that this happens—we find what is, discursively speaking, a key ambivalence in the rewilding movement, as well as the clearest evidence that a certain biopower must be involved.

A second implication is for (eco)tourism. Far from simply constituting the kind of 'return' of people implied by Rewilding Europe's use of concepts such as 'rural depopulation' or 'land abandonment' (see for example Wild10, 2015, p. 27), tourism constitutes but one instance of an ongoing, and almost always ancient entanglement between both human and beyond-human actors. In this context, to speak of 'rural depopulation' arguably constitutes a convenient case of what critical discourse analysts would describe as a non-transactive, and nominalised statement (Kress & Hodge, 1993, p. 20). It is convenient in so far as it symbolically empties rural areas, and thereby prepares them for rewilding—but this without having to engage with the politics behind any real or imagined exodus. From a hybrid geography perspective, far from being the purely restorative process that is envisioned by many rewilding initiatives, and to which people may be 'added' as part of a general shift towards something like, say, an 'experience economy' (Pine & Gilmore, 1998, 2013), rewilding tourism constitutes the latest chapter in nature-culture entanglement, however *sui generis* (eco)tourism might be from a cultural-historical point of view.

Taking this point further, it may be argued that it is precisely the entanglement which makes rewilding possible—this not just thanks to the hybridity of the geographies in question, but because in many cases rewilding initiatives do quite literally generate, as Rewilding Europe itself puts it, 'business'. From this perspective, if the many who oppose rewilding on the grounds of agribusiness have a vested interest in maintaining a certain rural order, so do those who seek to generate wildernesses mainly if not expressly for the purposes of tourism. A case in point, the rewilding game reserves in South Africa (see for example Hoogendoorn, Meintjes, Kelso, & Fitchett, 2018). In such contexts, it might even be argued that there is a particularly convenient 'goodness of fit' between the taxa indexed by many rewilding initiatives, and tourism: the kinds of animals that many travellers would most like to see are often precisely the large, and 'charismatic' carnivores and herbivores.

Rewilders would doubtless object that the choice of species/taxa has, or should have, nothing to do with charisma and everything to do with trophic cascades—themselves dynamics objectively observed and well-documented by science. If, however, some recent research is anything to go by, then such claims may not rest on as solid a foundation as might be expected. At the very least, it is likely that there is a great deal more complexity than is allowed for by accounts such as Soulé and Noss's (see for example Alston et al., 2019; see also Marshall, Thompson Hobbs, & Cooper, 2013; Nogués-Bravo et al., 2016).

The last aspect I will consider with respect to hybrid geographies involves the shift from metric to topological space, for which Whatmore draws on the work of the philosopher Michel Serres (Serres & Latour, 1995). Serres explains the difference between the two kinds of space via a remarkable analogy: when one takes a handkerchief and spreads it out as if to iron it, one can see in it certain fixed distances and proximities. If, for example, one sketches a circle in one area, one can mark out nearby points and measure far-off distances. This is, arguably, precisely the kind of spatial logic that is used by rewilding initiatives to delimit wildernesses, however much they then seek to interconnect those wildernesses via biotic corridors. If, however, one takes that same handkerchief and one crumples it by putting it in one's pocket, then suddenly two hitherto distant points will be close, or even superimposed. Or if one tears the handkerchief in certain places, the opposite happens: two points that were close now become distant. Topology is the science of nearness and rifts, while metrical geometry is the science of stable and well-defined distances (Serres in Serres & Latour, 1995, p. 60).

Whatmore approaches hybrid geographies and their creatures in ways that acknowledge a certain 'crumpling' and tearing of space—in the context of rewilding, precisely the kind of phenomenon that occurs when humans remove specimens from one place, and reintroduce them in areas where they had previously become locally extinct; or indeed, what happens when humans reintroduce *themselves*, as tourists but also as rewilders, into habitats that nevertheless continue to be conceived and objectified along the lines of a metric conception of space, and on the basis of the traditional conception of wilderness: as an 'out there', a non-human space.

If scholars such as Latour and Whatmore offer a valid perspective—and certainly some substantial objections may be raised (see for example Lindahl Elliot, 2019, pp. 359–364)—then the advocates of discourses of rewilding and (eco)tourism should put far more emphasis on the hybridity of both practices. By this kind of account, (eco)tourism in rewilding areas involves not so much going back to a restored nature, but the latest in a long line of interventions, or perhaps one should say in-terventions: interventions 'from inside'—inside, that is, an at once topologically and metrically conceived space. From this kind of perspective, the challenge is to recognise and explain the hybrid contingencies of encounter, as opposed to simply celebrating, in the way that the Rewilding Europe video does, the opportunity to go to the wilderness and to see wild animals.

References

Alston, J. M., Maitland, B. M., Brito, B. T., Esmaeli, S., Ford, A. T., Hays, B., ... & Goheen, J. R. (2019). Reciprocity in restoration ecology: When might large carnivore reintroduction restore ecosystems? *Biological Conservation, 234*(June), 82–89. doi:doi.org/10.1016/j.biocon.2019.03.021.

Barkham, P. (2017). It Is Strange to See the British Struggling with the Beaver: Why Is Rewilding So Controversial? *The Observer.* Retrieved from https://www.theguardian.com/environment/2017/jul/01/rewilding-conservation-ecology-national-trust (accessed on 10.03.2020).

Beschta, R. L., & Ripple, W. J. (2007). Increased willow heights along northern Yellowstone's Blacktail Deer Creek following wolf reintroduction. *Western North American Naturalist, 67*(4), 613–617. doi:-doi.org/10.3398/1527-0904(2007)67[613:IWHANY]2.0.CO;2.

Bourdieu, P. (1992). The purpose of reflexive sociology (The Chicago Workshop)'. In P. Bourdieu, & L. J. D. Wacquant (Eds.), *An invitation to reflexive sociology* (pp. 61–215). Cambridge: Polity & Blackwell.

Castree, N. (2008). Neoliberalising nature: Processes, effects, and evaluations. *Environment and Planning A: Economy and Space,* 40(1), 153–173.

Coffey, B. (2016). Unpacking the politics of natural capital and economic metaphors in environmental policy discourse. *Environmental Politics, 25*(2), 203–222. doi:doi.org/10.1080/09644016.2015.1090370.

Constanza, R., & Daly, H. E. (1992). Natural capital and sustainable development. *Conservation Biology, 6*(1), 37–46. doi:doi.org/10.1046/j.1523-1739.1992.610037.x.

Cronon, W. (1996). The trouble with wilderness; or, getting back to the wrong nature. In W. Cronon (Ed.), *Uncommon ground: Rethinking the human place in nature* (pp. 69–90). New York: W.W. Norton.

Daily, G. (1997) Introduction: What are ecosystem services? In G. Daily (Ed.), *Nature's services: Societal dependence on natural ecosystems* (pp. 1–11). Washington, D.C.: Island Press.

de Groot, R. S., Wilson, M. A., & Baumans, R. M. J. (2002). A typology for the classification, description and valuation of ecosystem functions, goods and services. *Ecological Economics, 41,* 393–408. doi:doi.org/10.1016/S0921-8009(02)00089-7.

England, K., & Ward, K (2016). Theorizing neoliberalization. In S. Springer, K. Birch, & J. MacLeavy (Eds.), *The handbook of neoliberalism* (pp. 50–60). London: Routledge.

European Safari Company. (2016). About Us. Retrieved from https://www.europeansafaricompany.com/about-us/ (accessed on 10.03.2020).

Fennell, D. (2015). *Ecotourism* (4th ed.). London: Routledge.

Fisher, M. (2018). Comment on P. Jepson, The Story of a Recoverable Earth. Originally published in *Resurgence & The Ecologist,* 311 (Nov/Dec), republished in *Rewilding Earth* blog. Retrieved from https://rewilding.org/the-story-of-a-recoverable-earth/ (accessed on 10.03.2020).

Fisher, M. (2019). Drifting from rewilding. *Rewilding Earth.* Retrieved from https://rewilding.org/drifting-from-rewilding/ (accessed on 10.03.2020).

Foreman, D. (2004). *Rewilding North America: A Vision for Conservation in the 21st Century.* Washington D.C.: Island Press.

Foucault, M. (1971). The orders of discourse. *Social Science Information, 10*(2), 7–30. doi:doi.org/10.1177/053901847101000201.

Foucault, M. (1978). *History of Sexuality, Volume I: An introduction,* translated by R. Hurley. New York: Pantheon Books.

Foucault, M. (1993)[1969]. *The Archeology of Knowledge and the Discourse on Language,* translated by A. M. Sheridan Smith. New York: Barnes & Noble.

Goderie, R., Helmer, W., & Kerkdijk-Otten, H. (2013). *The Aurochs - Born To Be Wild.* Zutphen, NL: Roodbont Publishers.

Gould, S. J. (1979). *Ever since Darwin: Reflections in Natural History.* New York: W.W. Norton.

Hall, C. M. (2019). Tourism and rewilding: An introduction – definition, issues and review. *Journal of Ecotourism, 18*(4), 297–308. doi:doi.org/10.1080/14724049.2019.1689988.

Haraway, D. (1989). *Primate Visions: Gender, Race, and Nature in the World of Modern Science.* London: Routledge.

Haraway, D. (1991). *Simians, Cyborgs and Women: The Reinvention of Nature.* London: Free Association Books.

Harvey, D. (2005). *A Brief History of Neoliberalism.* Oxford: Oxford University Press.

Hawken, P., Lovins, A., & Lovins, L. H. (1999). *Natural Capitalism: Creating the Next Industrial Revolution.* London: Little, Brown and Co.

Honey, M. (2008). *Ecotourism and Sustainable Development: Who Owns Paradise?* (2nd ed.). Washington D.C.: Island Press.

Hoogendoorn, G., Meintjes, D., Kelso, C., & Fitchett, J. (2018). Tourism as an incentive for rewilding: The conversion from cattle to game farms in Limpopo province, South Africa. *Journal of Ecotourism, 18*(4), 309–315. doi:doi.org/10.1080/14724049.2018.1502297.

Jørgensen, D. (2015). Rethinking rewilding. *Geoforum; Journal of Physical, Human, and Regional Geosciences, 65*, 482–488. doi:doi.org/10.1016/j.geoforum.2014.11.016.

Joye, Y., & Blocke, A. D. (2011). Nature and I are two: A critical examination of the biophilia hypothesis. *Environmental Values, 20*(2), 189–215. doi:doi.org/10.3197/096327111X12997574391724.

Kellert, S. R., & Wilson, E. O. (Eds.). (1993). *The Biophilia Hypothesis*. Washington D.C.: Island Press.

Kress, G., & Hodge, R. (1993). *Language as Ideology* (2nd ed.). London: Routledge.

Latour, B. (2005). *Reassembling the Social: An Introduction to Actor-Network-Theory*. Oxford: Oxford University Press.

Lewontin, R. C., Rose, S., & Kamin, L. J. (1984). *Not in Our Genes: Biology, Ideology and Human Nature*. Harmondsworth: Penguin.

Lindahl Elliot, N. (2019). *Observing Wildlife in Tropical Forests, Vol. 1: A Geosemeiotic Approach*, rev. ed. Bristol: Delome.

Lorimer, J., Sandom, C., Jepson, P., Doughty, C., Barua, M, & Kirby, K. J. (2015). Rewilding: Science, practice, and politics. *Annual Review of Environmental Resourses, 40*, 39–62. doi:doi.org/10.1146/annurev-environ-102014-021406.

MacArthur, R. H., & Wilson, E. O. (1967). *The Theory of Island Biogeography*. Princeton, NJ: Princeton University Press.

Marshall, K. N., Thompson Hobbs, N., & Cooper, D. (2013). Stream hydrology limits recovery of riparian ecosystems after wolf reintroduction. *Proceedings of the Royal Society B 280*, 20122977. doi:doi.org/10.1098/rspb.2012.2977.

Meine, C., Soulé, M., & Noss, R. F. (2006). A mission-driven discipline: The growth of conservation biology. *Conservation Biology, 20*(3), 631–651. doi:doi.org/10.1111/j.1523-1739.2006.00449.x.

Millennial Ecosystem Assessment. (2005). *Ecosystems and Human Wellbeing: Synthesis*. Washington D.C.: Island Press.

Monbiot, G. (2013). *Feral: Searching for Enchantment on the Frontiers of Rewilding*. London: Penguin Books.

Monbiot, G. (2019). Neoliberalism Promised Freedom – Instead It Delivers Stifling Control. *Guardian*. Retrieved from https://www.theguardian.com/commentisfree/2019/apr/10/neoliberalism-freedom-control-privatisation-state (accessed on 10.03.2019).

Nilsen, E. B., Milner-Gulland, E. J., Schofield, L. Mysterud, A., Stenseth, N. C., & Coulson, T. (2007). Wolf reintroduction to Scotland: Public attitudes and consequences for red deer management. *Proceedings of the Royal Society B, 274*(1612), 995–1003. doi:doi.org/10.1098/rspb.2006.0369.

Nogués-Bravo, D., Simberloff, D., Rahbek, K., & Sanders, N. J. (2016). Rewilding is the new Pandora's box in conservation. *Current Biology, 26*, R83–R101. doi:doi.org/10.1016/j.cub.2015.12.044.

Orzack, S., & Forber, P. Adaptationism. In E. N. Zalta (Ed.), *The Stanford encyclopedia of philosophy*. Retrieved from https://plato.stanford.edu/archives/spr2017/entries/adaptationism [accessed on 10.03.2020).

Paine, R. T. (1980). Food webs: Linkage, interaction strength and community infrastructure. *Journal of Animal Ecology, 49*(3), 666–685.

Pellis, A. (2019). Reality effects of conflict avoidance in rewilding and ecotourism practices – the case of Western Iberia. *Journal of Ecotourism, 18*(4), 316–331. doi:doi.org/10.1080/14724049.2019.1579824.

Pine, B. J., & Gilmore, J. H. (1998). Welcome to the experience economy. *Harvard Business Review, July–August*, 97–105.

Pine, B. J., & Gilmore, J. H. (2013). The experience economy: Past, present and future. In J. Sundbo, & F. Sørensen (Eds.), *Handbook on the experience economy* (pp. 21–44). Northampton, MA: Edward Elgar.

Rewilding Europe. (2012). *Rewilding Europe: Making Europe a Wilder Place*. Nijmegen: Rewilding Europe.

Rewilding Europe. (2016a). Video: Making Europe a Wilder Place. Retrieved from https://www.youtube.com/watch?v=Xyir6xQsQr0&feature=youtu.be (accessed on 10.03.2020).

Rewilding Europe. (2016b). A New Way of Travelling - the European Safari Company Launches Today. Retrieved from https://rewildingeurope.com/news/a-new-way-of-travelling-the-european-safari-company-launches-today/ (accessed on 10.03.2020).

Rewilding Europe. (2019). *Three-Year Strategic Plan 2019-2021*. Amsterdam: Rewilding Europe.

Rewilding Europe. (2020). About Rewilding Europe. Retrieved from https://rewildingeurope.com/our-story/ (accessed on 10.03.2020).

Rodríguez-Pereyra, G. (2016). Nominalism in metaphysics. In E. N. Zalta (Ed.), *The Stanford encyclopedia of philosophy*. Retrieved from https://plato.stanford.edu/archives/win2016/entries/nominalism-metaphysics/ (accessed on 10.03.2020).

Rose, H., & Rose, S. (Eds.). (2001). *Alas, Poor Darwin: Arguments Against Evolutionary Psychology*. London: Vintage.

Rubenstein, D. R., Rubenstein, D. I., Sherman, P. W., & Gavin, T. A. Pleistocene park: Does re-wilding North America represent sound conservation for the 21st century? *Biological Conservation 132*, 232–238. doi:doi.org/10.1016/j.biocon.2006.04.003.

Rubenstein, D. R., & Rubenstein, D. I. (2015). From Pleistocene to trophic rewilding: A wolf in sheep's clothing. *PNAS, 113*(1). doi:doi.org/10.1073/pnas.1521757113.

Serres, M., & Latour, B. (1995). *Conversations on Science, Culture, and Time*, translated by R. Lapidus. Ann Arbor: University of Michigan Press.

Singer, F. J., Mark, L. C., & Cates, R. C. (1994). Ungulate herbivory of willows on Yellowstone's northern winter range. *Journal of Range Management. 47*, 435–443.

Smith, B. H. (2014). Review of the book *An inquiry into modes of existence: An anthropology of the Moderns*, by Bruno Latour. *Common Knowledge, 20*(3), 491–493.

Soper, K. (1995). *What Is Nature? Culture, Politics and the Non-Human*. Oxford: Blackwell.

Soulé, M. E. (1995). The social siege of nature. In M. E. Soulé, & G. Lease (Eds.), *Reinventing nature: Responses to postmodern deconstruction* (pp. 137–170). Washington D.C.: Island Press.

Soulé, M., & Noss, R. (1998). Rewilding and biodiversity: Complementary goals for continental conservation. *Wild Earth, 8*, 19–28.

Soulé, M. E., & Wilcox, B.A. (1980). *Conservation Biology: An Evolutionary-Ecological Perspective*. Sunderland, MA: Sinauer Associates.

Terborgh, J. Estes, J. A., Paquet, P. Ralls, K., Boyd-Herger, D., Miller, B. J., Noss, R. F. (1999). The role of top carnivores in regulating terrestrial ecosystems. In M. E. Soule, & J. Terborgh (Eds.), *Continental Conservation: Scientific Foundations of Regional Reserve Networks* (pp. 39-64). Washington, D.C., USA:The Wildlands Project. Island Press.

TIES. (2015). Press Release: TIES Announces Ecotourism Principles Revision. Retrieved from http://www.ecotourism.org/news/ties-announces-ecotourism-principles-revision (accessed on 10.03.2020).

Weaver, D. B. (2001). Ecotourism as mass tourism: Contradiction or reality? *Cornell Hotel and Restaurant Administration Quarterly, 42*, 104–112.

Whatmore, S. (2002). *Hybrid Geographies: Natures Cultures Spaces*. London: Sage.

Whatmore, S., & Thorne, L. (1998). Wild(er)ness: Reconfiguring the geographies of wildlife. *Transactions of the Institute of British Geographers, 23*(4), 435–454. doi:doi.org/10.1111/j.0020-2754.1998.00435.x.

Williams, R. (1983). Nature. *Keywords: A Vocabulary of Culture and Society* (pp. 219–224). London: Fontana.

Wild10. (2015). *A Vision for a Wilder Europe* (2nd ed.). Salamanca: World Wildlife Congress.

Wilson, E. O. (1975). *Sociobiology: The New Synthesis*. Cambridge, MA: Harvard University Press.

Wilson, E. O. (1978). *On Human Nature*. Cambridge, MA: Harvard University Press.

Wilson, E. O. (1984). *Biophilia: The Human Bond with Other Species*. Cambridge, MA: Harvard University Press.

Wilson, E. O. (1998). *Consilience: The Unity of Knowledge*. London: Abacus.

Wilson, E. O. (2016). *Half Earth: Our Planet's Fight for Live*. London: Liveright.

27

THE ROLE OF ECOTOURISM IN NATURE NEEDS HALF VISION

Helen Kopnina

Introduction: Ecotourism and Half-Earth vision

In recent years, tourism has grown to one of the top industries (Ingles, 2005), increasing accessibility to the most remote areas for anyone with the time and money to travel (Stronza, 2001). The number of tourist arrivals worldwide has increased from 527 million in 1995 to over 2000 million in 2018 and is expected to grow with 3.5% by 2030, providing over 9% global economic activity (UNWTO, 2019). Simultaneously, many varieties of tourism have emerged, including adventure-, eco-, green-, or socially responsible tourism.

The International Ecotourism Society defines ecotourism as "responsible travel to natural areas that conserves the environment, sustains the well-being of local people and involves interpretation and education". Ecotourism is normally associated with the protection of local flora and fauna and the economic and social development of local communities (Kiper, 2013). Ecotourists are stimulated by both "push" and "pull" factors, as they are "pushed" to flee "overcrowded, unpleasant conditions" at home (Honey, 2008, p. 12), and "pulled" by a quest for self-realisation, discovery, authenticity, connection with nature and with local people, serving as a bridge between different cultural backgrounds (Cohen & Kennedy, 2007; Teunissen, 2016). In this context, ecotourism is seen as stimulating local people to invest in and preserve their rituals, the process which oscillates between reification and invention (Picard, 1999, p. 16).

Where some forms of ecotourism offer viable economic alternatives to mass tourism, local communities have been known to move away from cultivated land, allowing nature areas to regenerate (Stem, Lassoie, Lee, Deshler, & Schelhas, 2003). Indirect tourism benefits create education opportunities and stronger associations with conservation attitudes and behaviors (Stronza, 2001; Stem et al., 2003). While ecotourism can be effective as a component of a broader conservation strategy (Stem et al., 2003), the presence of (eco)tourists has also in some cases caused over-development and over-building in protected areas (Leung, Spenceley, Hvenegaard, & Buckley, 2018). The model of ecotourism synthesized by Fennell and Weaver (2015) involves global but differentiated forms of different forms of ecotourism, or what they call ecotouria, along social and ecological lines.

In a broader sense, (eco)tourism has laid bare larger questions related to the human relationship to other cultures, to the quest for economic development, and environmental and

DOI: 10.4324/9781003001768-27

social change (Stronza, 2001). In this context, tourists' relationship with the Nature Needs Half (otherwise known and hereby referred to as Half-Earth) vision deserves more attention. It has been established that to maintain viable populations of most of the Earth's remaining species, around 50% of landscapes and seascapes need protection from intensive human economic activity. The Half-Earth vision, committed to pathways to resolution of anticipated new relational arrangements and potential conflicts, is championed by conservation biologists and social scientists, including Noss and Cooperrider (1994), Noss, Quigley, Hornocker, Merrill, and Paquet (1996), Terborgh et al. (1999), Olson et al. (2001), Svancara et al. (2005), Wilson (2016), Kopnina (2016), Cafaro et al. (2017), Dinerstein et al. (2017), Kopnina, Washington, Gray, and Taylor (2018), and Schleicher et al. (2019).

However, there is little research on the social and economic costs and benefits of Half-Earth (Fletcher & Büscher, 2016). The Half-Earth has been "ambiguous about the exact forms and locations of the new conserved areas being called for" (Schleicher et al., 2019). This chapter poses that some forms of ecotourism can potentially resolve the ambiguity of the Half-Earth vision as beneficial to both communities and the environment. The model of ecotourism that encourages the development of an international network of protected areas, starting with biodiversity "hotspots" (Cincotta, Wisnewski, & Engelman, 2000) and explicitly meant to promote positive socio-economic change fits well within this vision (Fennell & Weaver, 2015). By examining the Half-Earth proposal, the possible strategies of creating synergies between human and nonhuman interests on the ground, focusing on the case of ecotourism, are explored. This is hereby illustrated by two case studies from Mondulkiri, Cambodia and Vlieland, the Netherlands. Both case studies involved ethnographic fieldwork employing the methodology of participant observation, conducted by the author in August 2019 in Cambodia and December 2019 in the Netherlands. As will be outlined in this chapter, domestic ecotourism, including ethnic minority groups, offer a unique and distinct opportunity for a more socially and ecologically responsible tourism.

Wild areas and local communities

Critical social scientists (see an overview of their views in Kopnina et al., 2018; Piccolo, Washington, Kopnina, & Taylor, 2018; Washington et al., 2018) have pointed out that ecotourism is informed by a particular 'ecotourist gaze', a subject to fashion, taste, and the caprice of the affluent (Honey, 2008; Fletcher, 2015). This gaze, according to critics, is naïve in its romanticism at best, and at worst, it represents neocolonial, elitist, Western practices that drive displacement (Fletcher, 2015). The non-consumptive use of wildlife does not always pay the environmental dividends expected, for example, trophy hunting might (Nowak et al., 2019), the local people are not always major beneficiaries of ecotourism (Honey, 2008). It was also noted that ecotourists embark on transcontinental flights to move closer to nature, in effect negating the "sustainability" element of their activity (Saletta, 2014). Ecotourism has been criticised for the rebound effects, both in terms of unsustainable flights (Walters, 2002) and social and environmental costs. Tourism in nature areas necessitated transport and accommodation that can hardly be seen as "green" (Walters, 2002). Nature has become commodified as 'resources' or 'services' (Kopnina, 2017). The motivation behind ecotourism, whether it be big game viewing safaris in Tanzania or whale watching in Iceland, is to make the wildlife 'pay their way' (Blewitt, 2012). If the wildlife provides more income sold on the black market or consumed as bushmeat (Crist & Cafaro, 2012; Peterson, 2012), the motivation to preserve this "commodity" by local people becomes less pronounced, with practices like trophy hunting sometimes presenting themselves as part of ecotourism (McGranahan, 2011).

Critics have stressed that not only does ecotourism often hurt the environment, but it also neglects the poor, vulnerable, and minority groups (Lee & Jamal, 2008). Therefore, environmental (social) justice (Lee & Jamal, 2008) is undermined unless the local communities have an equitable benefit in ecotourism and unless nonhuman species and habitats are protected. Pointing out that tourism has become a multi-billion-dollar industry, profiting from 'ecosystem services' and 'natural resources', and nature spectacles for the benefit of wealthy elites, critical anthropologists and social geographers have problematized the practice of tourism (Stronza, 2001; Ingles, 2005), be it branded mainstream, eco-, green, or socially-responsible tourism. Fletcher (2015) adds that ecotourism is not as innocent as it appears, but a product of "profoundly Eurocentric", "white, upper-middle-classes in postindustrial, predominantly western societies" (p. 341). As a panacea for this, Fletcher (2015) proposed the following:

> Hence, while critics worry that the type of aestheticized, virtual nature experiences found in video games and television documentaries may diminish people's concern for more mundane 'real' nature [...] in point of fact, the opposite may be true: encounters with spectacular, hyperreal, virtual environments may enhance support for conservation more than sustained contact with a 'real' nature that is far more messy, dirty, and inconvenient. (Ibid, 345)

Fletcher (2015) might be right that in some cases development of environmental values can occur through video games and not through immersion in nature. After all, people who grow up next to or within a forest do not necessarily support its protection, while some urban-born city-dwellers might (Kopnina, 2015). Also, as wilderness is rapidly disappearing, there are less and less natural places where people can be "immersed" in nature. However, it is hard to see how video games are not part of the neoliberal economy and an expression of Western consumerism.

Fletcher's critique of the ecotourist gaze as elitist (even though he is white and highly educated himself), tends to polarize Western and non-western values. There is evidence that biophilia (Kellert & Wilson, 1993; Kopnina, 2015), ecocentrism (Naess, 1973) and concern about future generations are not the inventions of Western "elites", but a cross-cultural phenomenon (Piccolo et al., 2018; Washington et al., 2018). Animal rights and welfare have been part of traditional ecocentric, holistic animist religions (Harvey, 2005; Sponsel, 2013). By contrast, human rights discourse and preoccupation with economic benefits are part and parcel of Western, European Enlightenment perspective that has been internalized elsewhere through colonisation.

Also, the formation of new social classes of relatively young, caring individuals who attempt to connect beyond their immediate social network circles offers hope of contributing to a larger societal and environmental change. The negative portrayal undermines the positive potential of ecotourism to actively improve the living conditions of local communities and provide incentives for cultural preservation and nature protection (Stem et al., 2003; Kiper, 2013; Teunissen, 2016). The critics of ecotourism disregard the intention of care in situations where no easy choices can be made. These situations have to do with demographic and economic growth that push both wild habitats and local communities to the margins of industrialised development landscapes. Furthermore, the critics of "western practices" rarely consider issues of population growth (e.g., Olson et al., 2001) and displacement caused by industrial and agricultural development, with conservation and biodiversity serving as a scapegoat. To quote Eileen Crist (2015, p. 93):

> The literature challenging traditional conservation strategies as locking people out, and as locking away sources of human livelihood, rarely tackles either the broader distribution of poverty or its root social causes; rather, strictly protected areas are scapegoated, and wild nature, once again, is targeted to take the fall for the purported betterment of people, while domination and exploitation of nature remain unchallenged.

In this context, the "elitist" argument is used to silence those who speak for nonhumans and their habitats. What is more helpful, is distinguishing between a range of activities within both conservation and ecotourism activities that have the potential to contribute—or not—to environmental protection and positive socio-economic development. Fennell and Weaver (2015) helpfully distinguish between "hard" (smaller-scale, less invasive, more environmentally strict with FITs—free and independent travelers) and "soft" (larger, often superficially labeled "eco-") types of tourism, with a comprehensive system of accreditation that could help determine more or less invasive or indeed "positive" forms of activity, are observed and which localities.

Human relationship to the environment

The human relationship with the environment and the human place in the environment has been a subject of scientific as well as philosophical debates. There is some archeological evidence that even in pre-industrial and probably pre-agricultural times, humans were already acting as key ecosystem engineers, playing an important role in shaping their surroundings (Turner, 1993; Barnosky, Koch, Feranec, & Wing, 2004).

In the idealised framing, today's forest-dependent people have seen the keepers of traditional ecological wisdom of their ancestors, and their practices, such as traditional swidden farming, are not harming but sometimes even enhance biodiversity (McSweeney, 2005; Sponsel, 2013), suggesting the early humans or present indigenous communities did not hurt biodiversity. Following this, it is assumed that local beliefs and practices that influence the use of biodiversity are essential for understanding sustainable use and conservation policies (Van Vliet et al., 2018).

In turn, this belief is contested by evidence of early human-caused extinctions, such as in the cases from Madagascar to Northern Europe (Diamond, 1989). Also, in today's monetised, industrialized world, ecological harm has been inflicted by different groups due to the expansion of the human population and extractive activities (Holt, Bilsborrow, & Oña, 2004) and modern hunting weapons (Nunez-Iturri, Olsson, & Howe, 2008). The romantic idea of the pre-industrial peoples living in harmony with their environment, protecting nature against the encroachment of extractive industries (McSweeney, 2005) has been challenged (Shoreman-Ouimet & Kopnina, 2016). There is also evidence that a global pattern of human arrival to new areas was followed by faunal collapse and other ecological changes, without known exception due to overhunting, fire use, and clearing activities (Burney & Flannery, 2005). Even traditional swidden farming is hardly congruent with nature conservation (Henley, 2011).

As for traditional hunting practices, its "sustainability" depends on the number of hunters. Simply, eight billion relatively large apex predators cannot be "balanced" with the availability of wildlife, derogatorily known as "bushmeat" (Peterson 2012) without intensifying agricultural production. While hunting for bushmeat in biodiversity-endangered or depleted areas has a different impact on the environment than, for example, production of consumer goods in Western countries, these actions collectively threaten environmental systems, in the former case on the local, in the latter case on the global scale (Crist & Cafaro, 2012). Approximately 67% of terrestrial mammals are agricultural animals, 30% are humans, and only 3% are wild terrestrial vertebrate animals (Hamilton, 2017).

Half-Earth vision

Half-Earth vision aims to safeguard nature, halting mass extinction while providing long-term livelihood perspectives for local communities (Wilson, 2016). In addition to affording robust natural solutions to the ecological exigencies, the Half-Earth initiative charts a course toward a sustainable and equitable human coexistence alongside other living beings.

Pragmatically, it is doubtful that human survival and flourishing are possible with the rate of environmental degradation (Steffen et al., 2015; IPBES, 2019; IPCC, n.d.). In anthropocentric terms, it is doubtful whether a viable and flourishing human population of the present size can be sustained on biologically severely degraded earth, even in the short term. There is clear scientific evidence of human-caused climate change (IPCC, n.d.), and biodiversity loss (IPBES, 2019). Biodiversity loss is a problem for both present and future generations, e.g., people who depend on bushmeat and those who derive mental and physical health benefits from being in contact with wild nature.

Ecocentrically speaking, extinction reduces the evolutionary potential of the Earth's living beings, as: "Death is one thing—an end to birth is something else" (Soulé & Wilcox, 1980). Also, the question of "resources" exposes not only to the social and economic impacts of the Half-Earth proposal on a single species (*Homo sapiens*) but also the impact of consumption on survival and flourishing of billions of individuals and collectives of other species (Piccolo et al., 2018; Washington et al., 2018). While Schleicher et al. (2019) take the impact on livelihoods of communities as a normative moral "good", they leave out of consideration of the collective (ecosystem- or species-based) eco-, bio-, or geo-centric ethics, deep ecology (Naess, 1973), ecodemocracy (Gray & Curry, 2020), animal welfare and/or rights (Singer, 1977; Regan, 1986; Bisgould, 2014; Borràs, 2016), and the rights of nature (Chapron, Epstein, & López-Bao, 2019).

Negative impacts on communities (economic benefits or giving up a certain practice, "traditional" or "modern") can at times be unavoidable and necessary but we have to weigh them up as they can hamper longer-term conservation efforts. The critics ignore the fact that the highest concentrations of biodiversity are found in areas with the highest rates of population growth (Cincotta et al., 2000; Crist, Mora, & Engelman, 2017). The aggregate problem is that the human population growth rate in biodiversity "hotspots" is substantially higher than the population growth rate (Cincotta et al., 2000). This suggests that human-induced environmental changes are "likely to continue in the hotspots and that demographic change remains an important factor in global biodiversity conservation" (Cincotta et al., 2000).

Ways forward: How ecotourism fits into Half-Earth

A more promising strategy would be balance impacts on species and future humans, stressing the likelihood that civilization cannot long sustain the loss of species that much less ambitious conservation strategies would likely cause. This does not negate the importance of the species themselves but provides a less strident and more sensitive framework. For example, regarding population dynamics and sustainability, there is the need not only to provide equal economic opportunities but also to consider family planning, support of women's reproductive rights, and elimination of child bride marriages and unwanted pregnancies (Crist et al., 2017).

While some negative impacts on the economy or certain practices, whether "traditional" or "modern" can be necessary, we have to weigh them up against longer-term conservation efforts. While not all traditional practices can be allowed when habitats and species are critically endangered, the level of protection proposed in Half-Earth vision will certainly permit practices that are concerned with non-consumptive nature consumption and will bar most corporate ventures, such as mining, logging, industrial agriculture, and also some types of "extractive tourism" from profiteering at the expense of the natural world and local people.

Solutions rest in potentially win-win scenarios that have to do with reducing human population pressure through the exercise of human rights, women's reproductive rights through family planning (Crist et al., 2017), and greater engagement of scientists, particularly conservationists, with local communities. In this context, Schleicher et al. (2019) are right to point out that we need to take into account the social and economic impacts of any conservation proposals to assess "their acceptability and feasibility". Pragmatically, conservation without engaging and convincing people about the benefits of nature protection can ultimately backfire, and ecotourists are well-positioned to engage in a dialogue with local people. The "ways forward" could include, as suggested by Nowak et al. (2019, p. 434):

> Locally adjusted and bottom-up management practices, granting communities land titles, conservation-compatible agriculture, and coexistence approaches can also benefit communities and conservation more than trophy hunting. Also, tourism reforms could invigorate domestic tourism, minimize leakage of tourism income to foreign investors, and reduce the footprint of wildlife-viewing tourism through green development investment. Diversified nature-based tourism beyond photographing and viewing wildlife could incorporate survival skills/bushcraft training and agritourism, emphasizing local knowledge and cultural exchange.

Case study 1 elephant project: Mondulkiri

An example of ecotourism initiated by local communities and involving Indigenous people in the Mondulkiri project in Cambodia. The rubber plantation Varanasi in the Bousraa has erased much of the forest that used to be used by the indigenous tribe Bunong, who used to practice subsistence and more recently, swidden agriculture (Van der Eynden, 2011). The plantation has dispossessed the Bunong of their land and livelihood, leading to protests, and sabotage actions protesting economic development (Van der Eynden, 2011), with the clearing of Bunong's ancestral land continuing (O'Byrne, 2017). Chrouet Kloeurt, a Bunong representative, stated in the interview: "If the company takes all the land, my children and grandchildren won't have land. They'll take all the land and cut all the trees" (O'Byrne, 2017).

By applying the moral economy concept to the Bunong, their traditional relationship to nature can be seen as "total", while commercial activities of more powerful development agents tap the resources for commodity production for profit (Van der Eynden, 2011). However, simultaneously, the Bunong used to exploit elephants for agricultural work and later for entertainment in the tourist industry (Ma, 2013). While elephant rides generate a large income for indigenous peoples, which are increasingly becoming dependent on the global monetary economy, in many cases the "elephants are not receiving proper care" (Highwood in Ma, 2013; Figure 27.1).

According to one of the tour guides (see Figure 27.2), which has given an introductory presentation to a group of 25 tourists on August 19, 2019, getting new elephants for the sanctuary is a problem. Aside from aiming to keep the forest as a sanctuary for elephants that the organisation can buy from the mahouts or farmers, or circus operators, the NGO also aims to

Figure 27.1 "Mr. Tree story"

start a breeding program. Most of the five elephants are presently too old (above forty) except for one female, and the NGO tries to rent male elephants to ensure that the younger female can produce offspring. Getting new and younger elephants is a problem as they become increasingly expensive (an elephant used to cost around 40.000 dollars, now more)—the animal parts, sold to Vietnam and China, are more valuable than the whole elephant. Also, some circuses and elephant rides are still interested in exploiting elephants for entertainment, with the number of wild elephants declining rapidly. According to the Cambodian guide, the Bunong used elephants for heavy farm and building work, sometimes physically abusing them. The elephants were not allowed to mate as pregnant females were seen as a liability, so new baby elephants were taken from the forest every time the new supply was needed. This was seen as an ancestor-held tradition. Some members of the Bunong communities, through employment at the Mondulkiri project, are starting to change their minds. The guide also reflected that local Cambodian operators of Mondulkiri project explain to the indigenous communities that a dead elephant or its parts give you money now, but ecotourism will do so for foreseeable future and will feed their children and grandchildren. There is the hope they can lend a young male elephant "to make a baby"—but first, an "elephant wedding" will have to be performed, including a relatively expensive money injection for a "party", a sacrifice of multiple other animals, such as sheep, goats, or dogs (Figure 27.3).

Figure 27.2 Mondulkiri project's guides (from Mondulkiri project's website)

Figure 27.3 Elephants bathing with tourists (by author)

Case study 2 (domestic) ecotourism: Vlieland

The Netherlands, where this author resides, is a densely populated country dominated by agricultural and urban development where a few natural areas find themselves squeezed between the farmlands and cities. The Netherlands has committed itself to nature objectives stated in the EU Biodiversity Strategy and thus indirectly to those in the Convention on Biological Diversity (CBD, 2019). A constant struggle between economic development and nature protection policies takes place (Kopnina, Leadbeater, & Cryer, 2019). In 2019, there were plans disband protected status of natural areas in the Netherlands due to other economic interests (further expansion of agriculture and industry) and to avoid government regulation of nitrate pollution (Staatsbosbeheer, 2019). The reasoning goes that if there are no strictly protected areas, and Dutch farmers and builders do not have to "suffer" under limitations on nitrate emissions, there is no need for regulation (Leijten & Rutten, 2019; Winterman, 2019). While at the time of writing this article the plans to lift the protected status of natural areas are off the table, the Netherlands already has less protected nature areas than most other European countries (Winterman, 2019).

Simultaneously, urban areas in the Netherlands struggle with increasing numbers of tourists, attracted by the liberal drug and prostitution policies and associated attractions such as coffee shops and the Red-Light District, as well as "high culture" (Boffey, 2019). The case of Vlieland's tourism discussed here shows how domestic tourism to the nature-dominated island can provide alternative income to residents and yet keep the natural areas intact.

The nature island of Vlieland in the Wadden area (Sijtsma, Broersma, Daams, Hoekstra, & Werner, 2015) offers as one of the "attractions" its gray seals, whose population has been relatively stable in the last few years (Brasseur et al., 2010). Vlieland is mostly visited by domestic tourists and a few German tourists. "Animal watching" including birdwatching, is marketed as one of the key attractions of the island (www.holland.com/global/tourism/destinations/regions/wadden-islands/vlieland.htm), feeding the local economy and helping motivate local villagers to preserve fragile nature (Figure 27.4).

One of the issues is that during seal watching some tourists approach the seals during the time they have and nurse pups (in December and January) too closely, and will at times disturb and scare away the mothers, and in some cases even cause them to abandon cubs altogether. However, despite disturbances, the Vlieland authority has reported a steady rise in the seal population. Local tourism opportunities in Vlieland also attract a wide range of local populations from the mainland, including, increasingly, minority groups, although "ethnic Dutch" nature tourism is more widespread (Loon & Berkers, 2008; Figure 27.5).

Discussion

While both Cambodia and the Netherlands are not directly known for their ecotourism, both countries demonstrate that certain types of activities that contribute to both habitat protection and animal welfare do attract tourists and contribute to local economies. Be it forest for the elephants in Cambodia, or the beaches for seals in the Netherlands, the habitats that sustain both local communities and wildlife offer an example of how places can be shared for mutual benefit, perhaps even as a blueprint for Half-Earth. While the ancient temple complexes of Cambodia and the marijuana coffee-shops of the Netherlands are unlikely to cease attracting massive tourism, alternative ecotourism presents an opportunity to incorporate both environmentally and socially responsible alternatives.

Figure 27.4 Gray seal (by Engelbert Fellinger)

Figure 27.5 Gray seal (by author)

The complexity of the relationship between eco-tourists and local communities, as well as "traditional" practices and culture preservation, comes to the fore. As the operators of Mondulkiri project suggested, creative engagement with, but also respect for local traditions, such as a compromise of an "elephant wedding" can be a good starting point. Tackling traditional beliefs about the elephants being the indigenous people's pride, their heritage, part of their beautiful and unique land, or spirits of their ancestors, and, in a more utilitarian perspective, their chance of long-term financial security, may be key to mutually beneficial co-existence.

While in the idealised terms both ecotourism and Half-Earth vision are about sharing the planet and enriching social and environmental relationships, "win-win" scenarios are not always possible. Some high protection nature areas, for example, those with a high concentration of unique or endangered species, might need prohibition of any economic activities. As discussed previously, not all cultural practices can be preserved in the age of mass extinction. Also, while some human groups (typically, those that maintain high-consumption lifestyle) have a higher impact on the environment that poorer ones, or the ones that still maintain a "traditional" lifestyle (Sponsel, 2013), one may speak of collective culpability for climate change and biodiversity loss (Washington et al., 2018). While hunting for bushmeat in biodiversity-endangered or depleted areas has a different impact on the environment than, for example, production of consumer goods in Western countries, these actions collectively threaten environmental systems, in the former case on the local, in the latter case on the global scale (Crist & Cafaro, 2012). While the idea of Half-Earth supports sharing multi-species spaces, not diving the earth literally in half, support needs to be generated for protection *outside* the reserve system. Ecotourist regions, as proposed by Fennell and Weaver (2015) can be helpful in this regard. The strategies for integrating various regions with sensitivity to local conditions but also, simultaneously, a commitment to sustainability objectives, can occur in different localities. These localities include agricultural lands (for example, through regenerative agriculture, retaining at least some biodiversity hotspots and corridors next to growing crops), roads (protecting roadside vegetation), or cities, which could accommodate more biodiversity through cradle-to-cradle designs, nature is not meant to displace or disadvantage local communities (Schleicher et al., 2019), but to co-exist with them.

Conclusions

Considering the discussion, ecotourists represent a very diverse group of people, not to be reduced to a generalized group of "elites". While some of the ecotourists staying at the expensive resorts might be wealthy, others are young, idealistic backpackers, caring about local culture, and respecting nature (Cohen, 2003; Butcher, 2008; Teunissen, 2016). Some of these travellers include young people who are vegetarian or vegan (Delgado 2003), increasingly aware of the history of oppression both in terms of human and animal rights (Fennell, 2012; Fennell & Markwell, 2015), and attempting to 'make a difference' to the countries visited (Butcher, 2008). Some individuals even choose to remain childless to reduce population pressure (Fleming, 2018). It might be hypothesised that ecotourists, at least as a broad generalised category, are inspired by the ideals of contributing to a better future for human and nonhuman inhabitants of this planet (Cohen, 2003; Teunissen, 2016). Half-Earth suggests that environmental protection requires some economic costs, specifically banning certain types of larger commercial activities (thus restricting "economic elites" if the term is still needed).

There are significant differences between the newly emerging social classes, and the polarising focus on division between "rich" and "poor" countries can obscure the reality of sociocultural, political, and economic changes. Examples of such change are the growth of middle classes in developing countries (Crist & Cafaro, 2012), or the shifting of consumption patterns, for example, from vegetarian/vegan lifestyle to more meat-eating diets in traditional Hindu countries (Delgado, 2003). Ecotourism can no longer be seen as unidirectional—from developed (rich) to developing (poor) countries—but as a multidimensional complex process. Further research needs to determine what types of ecotourism and under which local conditions can be seen to contribute to these ideals.

The context of ecotourism allows for a broader discussion of how the needs of the planet with all her inhabitants can be balanced with increasing demographic and economic pressures. The Half-Earth vision underlines conservation strategies that consider the needs of non-human species with that of the survival and health of future human beings. As discussed in this chapter, it is doubtful whether a viable and flourishing human population of the present size can be sustained on biologically severely degraded Earth, even in the short term. Proponents of the Half-Earth vision have suggested that conservationists, scientists, and policymakers should work in concert with local populations, not dividing the Earth in two, but living next to other beings without degrading them or their habitats. Promoting domestic ecotourism and protected area "ecotouriums" (Fennell & Weaver, 2015) are some of the examples of how this ambitious vision can be implemented. This chapter has focused on the cases of Cambodian and Dutch ecotourism that demonstrate the potential to successfully combine conservation, local economy, and animal welfare.

References

Barnosky, A. D., Koch, P., Feranec, R., & Wing, S. (2004). Assessing the causes of late Pleistocene extinctions on the continents. *Science*, *306*, 70–75.

Bisgould, L. (2014). It's Time to Re-evaluate Our Relationship with Animals [Video file]. Retrieved from https://www.youtube.com/watch?v=Fr26scqsIwk.

Blewitt, J. (2012). *Understanding Sustainable Development*. New York: Routledge.

Boffey, D. (2019). 'We Must Act Now': Netherlands Tries to Control Tourism Boom. *The Guardian*. Retrieved from https://www.theguardian.com/world/2019/may/06/we-must-act-now-netherlands-tries-to-control-tourism-boom.

Borràs, S. (2016). New transitions from human rights to the environment to the rights of nature. *Transnational Environmental Law*, *5*(1):113–143.

Brasseur, S. M. J. M., van Polanen Petel, T., Aarts, G. M., Meesters, H. W. G., Dijkman, E. M., & Reijnders, P. J. H. (2010). *Grey Seals (Halichoerus grypus) in the Dutch North Sea: Population Ecology and Effects of Wind Farms* (No. C137/10). Wageningen: IMARES. https://research.wur.nl/en/publications/grey-seals-halichoerus-grypus-in-the-dutch-north-sea-population-e

Burney, D. A., & Flannery, T. F. (2005). Fifty millennia of catastrophic extinctions after human contact. *Trends in Ecology & Evolution*, *20*(7), 395–401.

Butcher, J. (2008). Ecotourism as life politics. *Journal of Sustainable Tourism*, *16*(3), 315–326.

Cafaro, P., Butler, T., Crist, E., Cryer, P., Dinerstein, E., Kopnina, H., ... & Washington, H. (2017). If we want a whole earth, nature needs half. A reply to Half-Earth or Whole Earth? Radical ideas for conservation, and their implications. *Oryx—The International Journal of Conservation*, *53*(1), 400.

CBD. (2019). 6TH National Report for the Convention on Biological Diversity. Retrieved from https://zoek.officielebekendmakingen.nl/blg-893827.pdf.

Chapron, G., Epstein, Y., & López-Bao, J. V. (2019). A rights revolution for nature. *Science*, *363*(6434), 1392–1393.

Cincotta, R. P., Wisnewski, J., & Engelman, R. (2000). Human population in the biodiversity hotspots. *Nature*, *404*(6781), 990.

Cohen, E. (2003). Backpacking: Diversity and change. *Journal of Tourism and Cultural Change, 1*(2), 95–110.

Cohen, R., & Kennedy, P. (2007). *Global Sociology*. New York: Palgrave.

Crist, E. (2015). I walk in the world to love it. In G. Wuerthner, E. Crist, & T. Butler (Eds.), *Protecting the wild: Parks and wilderness, the foundation for conservation* (pp. 82–95). Washington: The Island Press.

Crist, E., & Cafaro, P. (2012). Human population growth as if the rest of life mattered. In P. Cafaro, & E. Crist (Eds.), *Life on the brink: Environmentalists confront overpopulation* (pp. 3–15). Athens: University of Georgia Press.

Crist, E., Mora, C., & Engelman, R. (2017). The interaction of human population, food production, and biodiversity protection. *Science, 356*(6335), 260–264.

Delgado, C. L. (2003). Rising consumption of meat and milk in developing countries has created a new food revolution. *The Journal of Nutrition, 133*(11), 3907S–3910S.

Diamond, J. M. (1989). The present, past and future of human-caused extinctions. *Philosophical Transactions of the Royal Society of London. B, Biological Sciences, 325*(1228), 469–477.

Dinerstein, E., Olson, D., Joshi, A., Vynne, C., Burgess, N. D., Wikramanayake, E., ... & Saleem, M. (2017). An ecoregion-based approach to protecting half the terrestrial realm. *BioScience, 67*(6), 534–545.

Fennell, D. A. (2012). Tourism and animal rights. *Tourism Recreation Research, 37*(2), 157–166.

Fennell, D., & Markwell, K. (2015). Ethical and sustainability dimensions of foodservice in Australian ecotourism businesses. *Journal of Ecotourism, 14*(1), 48–63.

Fennell, D., & Weaver, D. (2015). The ecotourium concept and tourism-conservation symbiosis. *Journal of Sustainable Tourism, 13*(4), 373–390.

Fleming, A. (2018). Would You Give Up Having Children to Save the Planet? Meet the Couples Who Have. *The Guardian*. Retrieved from https://www.theguardian.com/world/2018/jun/20/give-up-having-children-couples-save-planet-climate-crisis.

Fletcher, R. (2015). Nature is a nice place to save but I wouldn't want to live there: Environmental education and the ecotourist gaze. *Environmental Education Research, 21*(3), 338–350.

Fletcher, R., & Büscher, B. (2016). Why E. O. Wilson Is Wrong About How to Save the Earth. Retrieved from https://aeon.co/opinions/why-e-o-wilson-is-wrong-about-how-to-save-the-earth.

Gray, J., & Curry, P. (2020). Representation for nature: Ecodemocratic decision-making as a practical means of integrating ecological and social justice. In H. Kopnina, & H. Washington (Eds.), *Conservation: Integrating social and ecological justice* (pp. 155–166). Heidelberg: Springer.

Hamilton, C. (2017). *Defiant Earth: The Fate of Humans in the Anthropocene*. Cambridge, UK: Polity Press.

Harvey, G. (2005). *Animism: Respecting the Living World*. Kent Town, Australia: Wakefield Press.

Henley, D. (2011). Swidden farming as an agent of environmental change: Ecological myth and historical reality in Indonesia. *Environment and History, 17*, 525–554.

Holt, F. L., Bilsborrow, R. E., & Oña, A. I. (2004). Demography, household economics, and land and resource use of five indigenous populations in the Northern Ecuadorian Amazon: A summary of ethnographic research. *Occasional Paper, Carolina Population Center*. Chapel Hill, NC: University of North Carolina.

Honey, M. (2008). *Ecotourism and Sustainable Development: Who Owns Paradise?* (2nd ed.). New York: Island Press.

Ingles, P. (2005). More than nature: Anthropologist as interpreters of culture for nature-based tours' in American Anthropological Association (red.). *NAPA Bulletin, 23*, 219–223.

IPBES. (2019). Nature's Dangerous Decline, Intergovernmental Science-Policy Platform on Biodiversity and Ecosystem Services. Retrieved from https://www.ipbes.net/news/Media-Release-Global-Assessment.

IPCC. (n.d.). Global Warming of 1.5 °C. Retrieved from https://www.ipcc.ch/sr15/.

Kellert, S. R., & Wilson, E. O. 1993. *The Biophilia Hypothesis*. Washington, D.C.: Island Press.

Kiper, T. (2013). Role of Ecotourism in Sustainable Development. *InTech*. Retrieved from https://tamug-ir.tdl.org/bitstream/handle/1969.3/28978/InTechRole_of_ecotourism_in_sustainable_development_[1].pdf?sequence=1.

Kopnina, H. (2015). Revisiting the Lorax complex: Deep ecology and biophilia from a cross-cultural perspective. *Environmental Sociology, 43*(4), 315–324.

Kopnina, H. (2016). Half the earth for people (or more)? Addressing ethical questions in conservation'. *Biological Conservation, 203*(2016), 176–185.

Kopnina, H. (2017). Commodification of natural resources and Forest Ecosystem Services: Examining implications for forest protection. *Environmental Conservation, 44*(1), 24–33.

Kopnina, H., Leadbeater, S., & Cryer, P. (2019). Learning to rewild: Examining the failed case of the Dutch "new wilderness" oostvaardersplassen. *The International Journal of Wilderness, 25*(3). Retrieved from https://ijw.org/learning-to-rewild/.

Kopnina, H., Washington, H., Gray, J., & Taylor, B. (2018). The 'future of conservation' debate: Defending ecocentrism and nature needs half movement. *Biological Conservation, 217*(2018), 140–148.

Lee, S., & Jamal, T. (2008). Environmental justice and environmental equity in tourism: Missing links to sustainability. *Journal of Ecotourism*, 7(1), 44–67.

Leijten, J., & Rutten, R. (2019). Met Stikstof Beleeft Rutte 'Heftige Crisis'. *NRC*. Retrieved from https://www.nrc.nl/nieuws/2019/11/13/met-stikstof-beleeft-rutte-heftige-crisis-a3980311.

Leung, Y. F., Spenceley, A., Hvenegaard, G., & Buckley, R. (2018). *Tourism and Visitor Management in Protected Areas: Guidelines for Sustainability*. Gland: IUCN.

Loon, M. V., & Berkers, R. (2008). De toekomst van toerisme, recreatie en vrije tijd. *Kennisdocument voor de Strategische Dialoog Recreatie. Den Haag: Kenniscentrum Recreatie*. Retrieved from https://www.kennisbanksportenbewegen.nl/?file=1972&m=1422883000&action=file.download.

Ma, L. (2013). Elephants at centre of tourism tiff. *Phnom Penh Post*. Retrieved from https://www.phnompenhpost.com/business/elephants-centre-tourism-tiff.

Maltby, M., & Bourchier, G. (2011). Current status of Asian elephants in Cambodia. *Gajah, 35*, 36–42.

McGranahan, D. A. (2011). Identifying ecological sustainability assessment factors for ecotourism and trophy hunting operations on private rangeland in Namibia. *Journal of Sustainable Tourism, 19*(1), 115–131.

McSweeney, K. (2005). Indigenous population growth in the lowland Neotropics: Social science insights for biodiversity conservation. *Conservation Biology, 19*(5), 1375–1384.

Naess, A. (1973). The shallow and the deep: Long-range ecology movement. A summary. *Inquiry*, 16, 95–99.

Noss, R., & Cooperrider, A. (1994). *Saving nature's legacy: Protecting and restoring biodiversity*. Washington, DC: Island Press.

Noss, R. F., Quigley, H. B., Hornocker, M. G., Merrill, T., & Paquet, P. (1996). Conservation biology and carnivore conservation. *Conservation Biology, 10*, 949–963.

Nowak, K., Lee, P. C., Marino, J., Mkono, M., Mumby, H., Dobson, A., … & Sillero-Zubiri, C. (2019). Trophy hunting: Bans create an opening for change. *Science, 366*(6464), 434.

Nunez-Iturri, G., Olsson, O., & Howe, H. F. (2008). Hunting reduces the recruitment of primate-dispersed trees in Amazonian Peru. *Biological Conservation, 141*(6), 1536–1546.

O'Byrne, B. (2017). As company clears land, bunong villagers fight for ancestral home. *The Cambodian Daily*. Retrieved from https://english.cambodiadaily.com/news/as-company-clears-land-bunong-villagers-fight-for-ancestral-home-132609/.

Olson, D. M., Dinerstein, E., Wikramanayake, E. D., Burgess, N. D., Powell, G. V. N., Underwood, E. C., & Kassem, K. R. (2001). Terrestrial ecoregions of the world: A new map of life on earth: A new global map of terrestrial ecoregions provides an innovative tool for conserving biodiversity. *BioScience, 51*(11), 933–938.

Peterson, D. (2012). Talking about bushmeat. In, M. Bekoff (Ed.), *Ignoring nature no more: The case for compassionate conservation* (pp. 64–76). Chicago: Chicago University Press.

Picard, M. (1999). The discourse of Kebalian: Transcultural constructions of balinese identity. In R. Rubinstein, & L. H. Connor (Eds.), *Staying local in the global village: Bali in the twentieth century* (pp. 15–49). Honolulu: University of Hawaii Press.

Piccolo, J., Washington, H., Kopnina, H., & Taylor, B. (2018). Back to the future: Why conservation biologists should re-embrace their ecocentric roots. *Conservation Biology, 32*(4), 959–961.

Regan, T. (1986). A case for animal rights. In M. W. Fox, & L. D. Mickley (Eds.), *Advances in animal welfare science* (pp. 179–189). Washington, DC: The Humane Society of the United States.

Saletta, M. (2014). Can You Be a Sustainable Tourist Without Giving Up Flying? *The Conversation*. Retrieved from https://theconversation.com/can-you-be-a-sustainable-tourist-without-giving-up-flying-33099.

Schleicher, J., Zaehringer, J. G., Fastré, C., Vira, B., Visconti, P., & Sandbrook, C. (2019). Protecting half of the planet could directly affect over one billion people. *Nature Sustainability*, 1–3.

Shoreman-Ouimet, E., & Kopnina, H. (2016). *Culture and conservation: beyond anthropocentrism*. New York: Routledge.

Sijtsma, F. J., Broersma, L., Daams, M. N., Hoekstra, H., & Werner, G. (2015). Tourism development in the Dutch Wadden area: Spatial-temporal characteristics and monitoring needs. *Environmental Management and Sustainable Development*, *4*(2), 217–241.

Singer, P. (1977). *Animal Liberation: A New Ethics for our Treatment of Animals*. New York: Random House.

Soulé, M. E., & Wilcox, B. A. (1980). Conservation biology: Its scope and its challenge. In M. Soulé, & B. Wilcox (Eds.), *Conservation biology: An evolutionary-ecological perspective* (pp. 1–8). Sunderland (MA): Sinauer.

Sponsel, L. E. (2013). Human impact on biodiversity, overview. *Encyclopedia of Biodiversity*, *4*, 137–152.

Staatsbosbeheer. (2019). Dossier Natura 2000. Retrieved from https://www.staatsbosbeheer.nl/over-staatsbosbeheer/dossiers/natura-2000/programma-aanpak-stikstof/veelgestelde-vragen-pas.

Steffen, W., Richardson, K., Rockström, J., Cornell, S. E., Fetzer, I., Bennett, E. M., ... & Folke, C. (2015). Planetary boundaries: Guiding human development on a changing planet. *Science*, *347*(6223), 1259855.

Stem, C. J., Lassoie, J. P., Lee, D. R., Deshler, D. D., & Schelhas, J. W. (2003). Community participation in ecotourism benefits: The link to conservation practices and perspectives. *Society &Natural Resources*, *16*(5), 387–413.

Stronza, A. (2001). Anthropology of tourism; Forging new ground for ecotourism and other tourism alternatives. *Annual Review of Anthropology*, *30*, 261–283.

Svancara, L. K., Brannon, R., Scott, J. M., Groves, C. R., Noss, R. F., & Pressey, R. L. (2005). Policy-driven vs. evidence-based conservation: a review of political targets and biological needs. *Biological Sciences*, *55*, 989–995.

Terborgh, J., Estes, J. A., Paquet, P., Ralls, K., Boyd-Heger, D., Miller, B. J., & Noss, R. F. (1999). The role of top carnivores in regulating terrestrial ecosystems. In M. E. Soulé, & J. Terborgh (Eds.), *Continental conservation: Scientific foundations of regional reserve networks*(pp. 39–64). Washington, D.C.: Island Press.

Teunissen, N. (2016). Adapting new knowledge or reviving old values? The influence of sustainable tourism on hosts' understanding of environmental sustainability (Master's thesis, Cultural Anthropology and Development Sociology Faculty, Leiden University).

Turner, T. (1993). The role of indigenous peoples in the environmental crisis: The example of the Kayapo of the Brazilian Amazon. *Perspectives in Biology and Medicine*, *36*(3), 526–545.

UNWTO. (2019). International Tourist Highlights. Retrieved from https://www.unwto.org/publication/unwto-tourism-highlights-2018.

Van der Eynden, A. (2011). Rubber and soul: Moral economy, development and resistance in the bousraa villages, Mondulkiri, Cambodia (Master's thesis, Norges teknisk-naturvitenskapelige universitet, Fakultet for samfunnsvitenskap og teknologiledelse, Sosialantropologisk institutt).

Van Vliet, N., L'Haridon, L., Gomez, J., Vanegas, L., Sandrin, F., & Nasi, R. (2018). The use of traditional ecological knowledge in the context of participatory wildlife management: Examples from indigenous communities in Puerto Nariño, Amazonas-Colombia. In R. Alves (Ed.), *Ethnozoology: Animals in our lives* (pp. 497–512). Academic Press, Elsevier.

Walters, J. (2002). Save the Planet... Stay on the Ground. *The Guardian*. Retrieved from https://www.theguardian.com/travel/2002/may/12/ecotourism.observerescapesection.

Washington, H., Piccolo, J., Chapron, G., Gray, J., Kopnina, H., & Curry, P. (2018). Foregrounding ecojustice in conservation. *Biological Conservation*, *228*, 367–374.

Wilson, E. O. (2016). *Half-earth: Our Planet's Fight for Life*. New York: WW Norton & Company.

Winterman, P. (2019). Coalitie verdeeld over stikstofmaatregelen. *Het Parool*. Retrieved from https://www.parool.nl/nederland/coalitie-verdeeld-over-stikstofmaatregelen~b6d39715/.

28

ECOTOURISM FOR CONSERVATION?

Amanda L. Stronza, Carter A. Hunt, and Lee A. Fitzgerald

Introduction

Conservation and tourism have worked in tandem since the early twentieth century (Sellars, 1997). Indeed, the first U.S. National Parks were created with both in mind. The architects of parks such as Yosemite, Yellowstone, Grand Canyon, and Sequoia envisioned setting aside public lands to "conserve the scenery and the natural and historic objects therein," and to ensure people of all backgrounds, in melting pot fashion, could enjoy the natural wonders of their (newly united) nation while also keeping such places "unimpaired" for "present and future generations" (National Park Service (NPS), 1916). Tourism and recreation in the parks were meant to serve as engines for nation building and economic development (Machlis & Field, 2000). Innovations included expanded railway lines, visitor centers, hiking trails, campgrounds, and scenic overlooks (Runte, 1979). Through national parks, conservation and tourism have always been connected (Budowski, 1976).

Ecotourism is both an expansion and a refinement of the connection between tourism and conservation. It builds on the idea of using tourism to reinforce conservation and vice versa, while deepening the criteria for sustainability. It emerged in the late 1980s, in the dawn of sustainable development. The early planners saw it as a form of tourism that could and should be designed and managed proactively with concern for channeling revenues to conservation and community development. It was meant to take place in parks, in keeping with the older ideas about tourism from the first national parks, but also to extend beyond parks, to enhance the livelihoods of people in local communities, and to protect not just recreation opportunities or the scenery, but also to meet more contemporary priorities of protecting biodiversity and maintaining ecosystem integrity (Gössling, 1999).

Ecotourism is designed to ensure a positive feedback loop between tourism and conservation—not simply that they can work together, but that they must. Explicit in all definitions of ecotourism is the hypothesis that tourism, when designed and practiced as ecotourism, can benefit wildlife and biodiversity, create incentives to protect landscapes, and support local communities (Krüger, 2005). In this way, ecotourism is a specific kind of tourism, distinguished from nature tourism and outdoor recreation by its conservation and development goals. Although there are many definitions of ecotourism, all adhere at least to a principle of making tourism support an array of social and environmental goals. The International

 DOI: 10.4324/9781003001768-28

Ecotourism Society offers the following—widely cited—definition: "responsible travel to natural areas that conserves the environment, sustains the well-being of the local people, and involves interpretation and education" (The International Ecotourism Society (TIES), 2018). The deepened focus on sustainability includes the concepts of "responsibility," "the well-being of local people," and "education."

Since the late 1980s, scholars and conservationists have questioned the feasibility, significance, and true value of ecotourism (Belsky, 1999; Cater, 2006; Goodwin, 1996; Kiss, 2004; Lindberg, Enriquez, & Sproule, 1996; Orams, 1995; Wallace & Pierce, 1996; Wheeller, 1994; Wight, 1993; Weaver, 1993). Others have challenged the fundamental, neoliberal philosophy of marketing communities and ecosystems, cultural traditions and endemic species, and "consuming" them to "conserve" them (Büscher & Davidov, 2013; Butcher, 2007; Duffy, 2002; Fletcher, 2009; Fletcher & Neves, 2012; Meletis & Campbell, 2007; Sharpley, 2006; West & Carrier, 2004).

Recently, scholars in ecology and conservation biology have begun to take harder aim at ecotourism (Geffroy, Samia, Bessa, & Blumstein, 2015), arguing it is not only helpful to conservation, but in fact, may be harmful to wildlife. Much of the work is conducted by biologists, basing their perspectives on theories related to risks of predation or physiological measures related to stress (e.g., Frid & Dill, 2002). Authors contributing to the recent literature state ecotourism habituates animals to human presence, increases the likelihood of being preyed upon by both other animals and humans, and decreases a population's overall fitness for survival (Frid & Dill, 2002; Fernández-Juricic, Venier, Renison, & Blumstein, 2005; Thomas, Kvitek, & Bretz, 2003; Steidl & Anthony, 2000; Goss-Custard, Triplet, Sueur, & West, 2006; Beale & Monaghan, 2004; Kerbiriou et al., 2009). A counterargument questions the plausibility of habituation transferred to a suite of wild predator species and suggests, instead, that an "ecotourism shield" can serve to protect entire wildlife populations over vast areas with human-wildlife interactions occurring in a few small locations (Fitzgerald & Stronza, 2016).

As Weaver and Lawton (2007) noted, "Despite the essential nature of this research to the management of the ecotourism experience, almost none of the empirical studies have been undertaken by tourism specialists or found in specialized tourism journals. Rather, just one scientific journal, *Biological Conservation*, appears to account for most of them" (but see Krüger, 2005; Kiss, 2004). Although there is evidence of the biologists' findings being overreported (e.g., Bateman & Fleming, 2017), the recent critiques have tended to conflate ecotourism with other kinds of tourism (i.e., the more conventional ideas of what people do in parks and visitor centers, hiking trails, and campgrounds), missing, misunderstanding, or misstating how and why ecotourism is or ever was heralded or established in later decades as a tool for conservation (e.g., Blumstein, Geffroy, Samia, & Bessa, 2017, and references therein).

Assuming all tourism that occurs outdoors or somehow involves nature is "ecotourism," and then arguing such activities fail to achieve conservation, is problematic. All research depends on careful definition and measurement of terms. The hypothesis that ecotourism is beneficial to conservation and development cannot be rigorously tested when assessments are biased by inclusion of data from activities that were not designed with the goals of ecotourism. As behavioral scientists Paul Ferraro and Merlin Hanuaer (Ferraro & Hanauer, 2011, 2014a, 2014b) describe, many conservation programs have depended on intuition and anecdote to guide both the design of conservation programs and the evaluation of their impacts (Ferraro & Pattanayak, 2006). Generalising critiques of tourism can undermine support for ecotourism and potentially thwart efforts that would otherwise build incentives for conservation, sustain protected areas, or facilitate community development (Fitzgerald & Stronza, 2016; Buckley, 2009, 2010, 2011).

Our intent is to provide an overview of ecotourism research, building clarity and cohesion from the literature to summarize how and under what conditions ecotourism works for conservation. We are not reporting a new, empirical analysis of ecotourism in a specific place or time, but rather offering a synthesis. We first provide a history of ecotourism, with definitions and aims, and we give attention to the rise and fall of the idea, mirrored by greenwashing in marketing and analysis. We distinguish ecotourism from other kinds of nature-based tourism, noting how ecotourism is a specific concept with specific ideas and principles for implementation to achieve conservation. In doing so, we also acknowledge the real and potential benefits of other forms of tourism, and we provide a table for comparison (Table 28.1).

Table 28.1 Types of tourism associated with conservation, categorised by their predicted impact on biodiversity conservation

Type of tourism		*Conservation impact*			
Term	*Description*	*PA*	*IL*		*SI*
Outdoor recreation	"Experiences that result from recreational activities occurring in natural environments" (Moore & Driver, 2005, p. 11)	+/−	−	−	−
Wildlife tourism	"The viewing of, and non-consumptive encounters with, wildlife solely in natural areas" (Newsome, Moore, & Dowling, 2013, p. 23)	+/−	−	−	−
Nature-based tourism	"Any form of tourism which uses natural resources in a wild or undeveloped form" (Fennell, 2008, p. 25)	+/−	−	−	−
Pro-poor tourism	"Tourism that generates net benefits for the poor. Benefits may be economic, but they may also be social, environmental or cultural" (Ashley, Roe, & Goodwin, , p. 2)	−	+	−	−
Responsible tourism	Widely considered a pre-cursor for ecotourism: "(Sellars, 1997) minimum environmental impact; (National Park Service (NPS), 1916) minimum impact on—and maximum respect for---host cultures; (Machlis & Field, 2000) maximum economic benefits to the host country 'grassroots'; and (Runte, 1979) maximum 'recreational' satisfaction to participating tourists" (Epler Wood, Gatz, & Lindberg, 1991)	−	−	−	−
Sustainable tourism	"Tourism that takes full account of its current and future economic, social and environmental impacts, addressing the needs of visitors, the industry, the environment and host communities" (United Nations World Tourism Organization (UNWTO), 2005, p. 12)	+/−	+	+	−

(Continued)

Table 28.1 (Continued)

Type of tourism		*Conservation impact*			
Term	*Description*	*PA*	*IL*		*SI*
Geotourism	"A form of tourism that specifically focuses on geology and landscape. It promotes tourism to geo-sites and the conservation of geodiversity and an understanding of earth sciences through appreciation and learning" (Newsome et al., 2013, p. 25)	+	–	–	–
The International Ecotourism Society	"Responsible travel to natural areas that conserves the environment, sustains the well-being of local people, and involves interpretation and education" (The International Ecotourism Society (TIES), 2018)	+	+	+	+
Ecotourism (academic)	"Sustainable, non-invasive form of nature-based tourism that focuses primarily on learning about nature first-hand, and which is ethically managed to be low impact, non-consumptive, and locally oriented (control, benefits and scale). It typically occurs in natural areas, and should contribute to the conservation of such areas" (Fennell, 2008, p. 24)	+	+	+	+
Conservation tourism	"Commercial tourism that makes an ecologically significant net positive contribution to the effective conservation of biological diversity" (Buckley, 2010, p. 2)	+	+	+	+

Abbreviations: ED, environmental interpretation and ethics; IL, diversified livelihoods; PA, support for wildlife and protected areas; SI, strengthened resource management institution

Second, we provide an overview of the economic, ecological, and social benefits that have resulted from committed application of ecotourism principles. We summarize the ecotourism literature over the past 30 years, citing a range of studies from the social and natural sciences, including some of our own, and cataloging ways ecotourism has supported conservation either directly or indirectly. Finally, we offer a research agenda for the future and a framework for conducting rigorous analyses of ecotourism. We include six research design principles for assessing the net positive conservation benefits over time and place.

Ecotourism: Rise and fall?

In the mid-twentieth century, with the rise of international development, governments and newly formed aid agencies promoted tourism as a tool for advancing traditional or underdeveloped societies (Foster, 1973; Rostow, 1960). Market integration through tourism was meant to catalyse a transition to new societies (Escobar, 1991). Economies were perceived as following "stages to modernization," and tourism was an explicit indicator of national progress

(Stronza & Hunt, 2012; Mowforth & Munt, 2015). Large-scale tourism, in particular, with high-rise hotels and transportation networks, was heralded enthusiastically and often uncritically as fuel for development. The concept of comparative advantage resulted in entire island nations and coastal areas of the world marketing themselves as paradises, promising sun, sand, sea, and sex as they lured foreign and multilateral investors with tax breaks, fee exemptions, and devalued local currencies (Ferraro & Hanauer, 2011).

By the late 1980s, development specialists began to reject these modernizing, top-down approaches. They questioned the value and impacts of economic growth and challenged the idea that tourism could provide countries with a "passport to development" (de Kadt, 1979). They favored more democratic and holistic concerns for people and nature—a new paradigm of "sustainable development," best summarized in the 1987 WCED Report "Our Common Future" (aka the Brundtland Report), which drew strong attention to social and environmental dimensions of development (World Commission on Environment and Development (WECD), 1987).

In the realm of conservation, new thinking around sustainable development led to community-centered conservation strategies aimed at improving human welfare while simultaneously protecting the environment (Brandon & Wells, 1992). Sustainability challenged growth as the ultimate goal of development, and new forms of alternative tourism came to be viewed as a "green passport" to development (Smith & Eadington, 1992). The fresh coining of ecotourism led development specialists and conservationists in public, private, and NGO sectors to promote ecotourism as a "win-win" for both communities and ecosystems (Brandon & Wells, 1992; Ziffer, 1989; Boo, 1990) (Figure 28.1). The expectations for ecotourism were high. It was

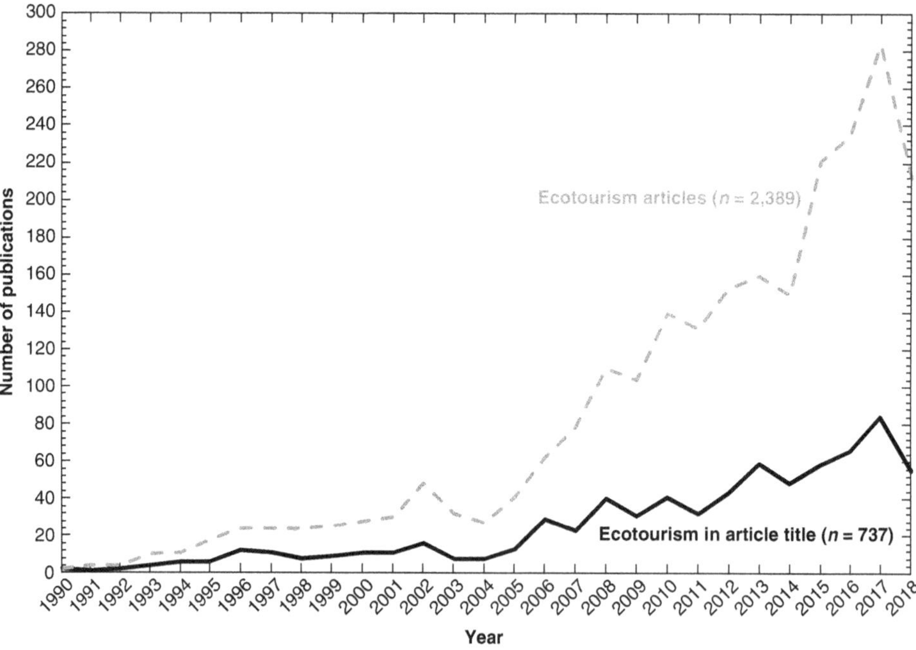

Figure 28.1 The literature referring to ecotourism has increased substantially since 1990. A search on Web of Science recovered 737 journal articles with "ecotourism" in the title. A similar search with "ecotourism" as a broad topic recovered 2389 articles. Many of these articles were studies of several types of tourism, ranging from ecotourism, to park visitation, to recreational activities with ecotourism mentioned in the article

meant to provide sustainable economic development (Wight, 1993; Tobias & Mendelsohn, 1991), effective mechanisms for biodiversity conservation (Weaver, 1993; Ferraro & Hanauer, 2014a, 2014b; Boo, 1990; Fennell & Eagles, 1990; Lindberg, 1991), strategies for empowering marginalised peoples (Scheyvens, 1999; Butler & Hinch, 2007), ethical practices for reversing colonial legacies of social and environmental injustice (Gardner, 2016), and better cross-cultural understanding (Stronza, 2001).

As the goals and standards for tourism shifted to ecotourism, stakeholders throughout the industry took on new possibilities and roles. Local communities partnered with tour companies and NGOs, hoping to channel outside attention on their lands, traditions, and resources to positive changes for their communities (Stronza & Gordillo, 2008). Regional and national governments adopted discourses of using ecotourism to protect biodiversity and alleviate poverty. Tourists were encouraged to gaze more respectfully, listen more closely, ask where their money goes, and change their worldview. NGOs increasingly served as mediators among stakeholders collaborating in new partnerships, lobbying for policies favorable to tourism, and promoting the idea of environmental responsibility in tourism.

By the early 2000s, several scholars began to publish critiques of ecotourism, demonstrating empirically that the practice does not always live up to the ideals. For example, Weaver (Weaver, 2001) and Kontogeorgopoulos (Kontogeorgopoulos, 2004) described ecotourism as a vanguard activity that is likely to create a foothold in culturally and biologically sensitive areas that are later exploited through mass tourism development. Kiss (Kiss, 2004) questioned whether resources used to develop community-based ecotourism would not in fact be better spent on direct, fortress-style conservation over large areas, whereas others characterised ecotourism as a Western construct that privileges tourists' pleasure at the expense of local communities and environments (e.g., Cater, 2006; West & Carrier, 2004).

Several environmental anthropologists and geographers brought critical social theory to ecotourism (see Duffy, 2002, 2008; Fletcher, 2009, 2011, 2014; Fletcher & Neves, 2012; Cater, 2006; Vivanco, 2001; West & Carrier, 2004), analysing its meaning and effects as a Western phenomenon in relatively poor countries. Their work has interrogated ecotourism as an expression and manifestation of Western values about nature and its inhabitants, including humans. They argue ecotourism is inseparable from a political-economic context of neoliberalism (West & Carrier, 2004).

As examples, in his ethnography, *Romancing the Wild* (Fletcher, 2014), Fletcher showed how ecotourism is an organised set of ideas, practises, and values that does not simply represent—but rather shapes—places and peoples to cohere to Western values and market forces. In her analysis of ecotourism in Madagascar, Duffy (Duffy, 2008) argued that ecotourism is popular among a range of powerful interest groups, including the World Bank and global donors "precisely because a commitment to ecotourism by national governments, NGOs and local communities does not challenge the wider policy framework of liberalizing and diversifying economies, and in fact, relies on opening them up to the global market through the neoliberalisation of nature" (p. 341). Neoliberalisation of nature is a process in which nonhuman phenomena are subject to market-based systems of management and development. Indeed, Duffy (Duffy, 2008) contends, ecotourism seems to address numerous agendas: capitalist development, community development, poverty alleviation, wildlife conservation, and environmental protection.

The enthusiasm and extensive promotion of ecotourism is partly the reason so many kinds of tourism have been relabeled as ecotourism, while lacking accountability to the core principles of the idea (Harrison, 1997). Honey (2008) called this greenwashing. Tourism ventures mislabeled as ecotourism range from those that promise minimal impacts on the environment without tangible support for conservation to those that include no more than visiting a natural area with no connection to conservation actions or policies (e.g., nature-based tourism, wildlife tourism, adventure tourism, and outdoor recreation) (Higham, 2007). Ceballos-Lascurain (1996, p. 2) noted "a lingering problem in any discussion on ecotourism is that the concept of ecotourism is not well understood, and therefore, it is often confused with other types of tourism development." With so much greenwashing, some have argued ecotourism is so overapplied that it is meaningless (Chirgwin & Hughes, 1997; Bjork, 2000). Recent literature in ecology and conservation biology takes an iconoclastic view of tourism and argues that "ecotourism" harms wildlife and ecosystems. Such critiques may be the upside down of greenwashing—rather than calling everything ecotourism and lauding the positive results, they are calling everything ecotourism and decrying the negative.

Greenwashing and its opposite are problematic both in marketing and research. By failing to measure or distinguish tourism and ecotourism carefully, scholars risk dismissing or missing altogether the specifically defined conservation purposes of ecotourism. Also, conflating ecotourism with all forms of nature tourism creates an apples and oranges problem in research, making rigorous understanding difficult (Buckley, 2011). Compounding the challenge of mixed terms is a lack of time series data in many studies. This can preclude understanding of how or under what conditions ecotourism affects local conservation practices, levels of biodiversity, ecosystem integrity, governance of resources, or any other social or ecological indicator over time. Ultimately, poorly designed impact studies of ecotourism can thwart conservation efforts on the ground. For a counterpoint to this, see the work of Ferraro and Hanauer (2011, 2014a, 2014b), which draws on the ability to infer causality from nonexperimental data, estimating the effects of a range of conservation programs on social and environmental outcomes.

The biological literature on the effects of ecotourism on animal populations has lacked consideration of scale. Mismatches of observation and conclusions seem to arise from a combination of case studies from a mix of tourist and recreational activities conducted at relatively small scales. In many cases, physiological and behavioral studies have been focused on a sample of animals in contact with people, then they have been discussed in terms of much broader effects, such as on entire wildlife populations or communities (see Bateman & Fleming, 2017). This sort of sampling bias can lead to conclusions that the population is different than it actually is, and it masks the degree to which patterns may or may not scale up to the level of populations across entire landscapes. Ironically, most studies of the effects of tourists on animals take place in areas that are protected by and for tourism, and are subject to strict permits and protocols.

Shannon, Larson, Reed, Crooks, and Angeloni (2017) reviewed population and community-level effects of ecotourism. Unfortunately, they relied heavily on a meta-analysis (Larson, Reed, Merenlender, & Crooks, 2016) that included all sorts of interactions among people and wildlife resulting from recreational activities that ranged from winter sports to boating to dog-walking. Some of the impacts on biodiversity cited were also clearly not related to tourism of any kind (e.g., feral animals, invasive weeds, zebra mussels). Although ecotourists certainly engage in recreational activities such as hiking on trails and viewing wildlife from boats and platforms, ecotourism programs regulate these activities that take place in a relatively small area compared to the lands protected by ecotourism (e.g., guided visitation, restricting hiking to specific trails, etc.). Likewise, comparisons of samples from animal populations in areas with and without tourists only show the degree to which the two samples differ, and may not account for alternative hypotheses.

Using the nontourist area as a baseline assumes that effects of tourism have not already spread throughout a panmictic population. Thus, it is not possible to parse out effects of ecotourism from meta-analyses of recreational encounters with wildlife that are studied only at one scale. Discussions of population level effects of ecotourism on wildlife populations are highly speculative, and it remains a tall order to rigorously assess how wildlife interactions with people in ecotourism areas might affect population-level parameters, such as survivorship, reproduction, dispersal, and population growth. To address issues, scholars need to work at multiple scales using similar methods.

All kinds of tourism, including ecotourism, have positive, neutral, or negative effects at the scale where tourists view and interact with plants and animals along trails and in accessible areas (Buckley, 2011). However, the ecotourism shield in many instances covers an area vastly larger than the spaces where tourist interactions occur. In general, tourists are restricted to certain zones and trails and are accompanied by guides. Even in widely accessible parks, the majority of visitors do not venture into the back-country. Indeed, the survival of many threatened species would not be possible without the direct conservation benefits of ecotourism activities (Buckley, Morrison, & Castley, 2016; Steven, Castley, & Buckley, 2013; Buckley, Castley, de Vasconcellos Pegas, Mossaz, & Steven, 2012).

As examples, protected areas were established for penguin colonies in Patagonia Argentina, New Zealand, Australia, and South Africa (Fowler, 1999; Lewis, Turpie, & Ryan, 2012). Each area receives thousands of visitors annually, which helps justify their existence (Lewis et al., 2012). Across the board, researchers make explicit recommendations aimed at minimizing human disturbance in the colonies (Fowler, 1999; Ellenberg, Mattern, Seddon, & Jorquera, 2006). In another example, coati mundis (*Nasua nasua*) and tegu lizards (*Salvator merianae*) are habituated around the viewing areas of Iguazu Falls, a World Heritage site with national parks in both Argentina and Brazil (UNESCO, 2019). But these species range throughout these largely inaccessible parks, which protect 240,000 hectares of uninhabited interior Atlantic Forest. These examples can be transferred by analogy to most places where tourists interact with wildlife. Exceptions could be found in instances such as when tourists are allowed to view and interact with gorillas and other great apes, where it is feasible the tourists could affect a significant portion of the population through transmission of diseases (Woodford, Butynski, & Karesh, 2002; Muehlenbein et al., 2010). However, ecotourism is regulated in these instances and has provided a shield of protection for these vulnerable species (Sandbrook, 2010). Such a shield can alter movements of animals at landscape scales in some instances. In Grand Teton National Park, calving moose (*Alces alces*) aggregated close to roads to avoid brown bears in more remote areas. Non-calving females and males did not show this response. Brown bears are recent colonists to the park, but over time the attractiveness of roadsides may fade with increasing presence of bears "as a landscape of fear envelopes the entire ecosystem" (Berger, 2007). In summary, the generalisation is that tourism is regulated in all these places and mismanagement, when it occurs, is generally at a small scale. In exceptional cases, behavioral change can occur at landscape scales, but the changes have not been shown to be associated with detriment to populations. As Buckley (2012) noted, "for over half of the red-listed mammal species with available data, at least five per cent of all wild individuals rely on tourism revenue to survive. For one in five species, including rhinos, lions and elephants, that rises to at least 15% of individuals. Yes, that's risky, because tourism is fickle: but take it away and animals are killed by hunters. It happens every single day, every time patrols stop or hungry locals lose conservation incentives. Simply put, if tourism money is cut abruptly, poaching will increase" (p. 29). At large scales, ecotourism can protect landscapes and entire wildlife populations.

The conservation benefits of ecotourism

Ecotourism addresses both social and environmental goals, and it can benefit biodiversity conservation in four direct and indirect ways. As summarised in Table 28.1, these are

(a) support for wildlife and protected areas, (b) diversified livelihoods, (c) environmental interpretation and ethics, and (d) strengthened resource management institutions.

Support for wildlife and protected areas

One documented conservation benefit of ecotourism is the protection of endangered species. Early writings on ecotourism emphasised the impacts on individual species, often those serving as the main attraction in particular destinations and projects. For instance, scholars assessed ecotourism based on numerous flagship species such as sea turtles (Jacobson & Lopez, 1994; Campbell, 2002; de Vasconcellos Pegas, Coghlan, Stronza, & Rocha, 2013; Hunt & Vargas, 2018), howler monkeys in Belize (Belsky, 1999; Alexander, 2000), cetaceans (Walker & Hawkins, 2013), macaws (Munn; 1992), polar bears (Lemelin, Fennell, & Smale, 2008), lemurs (Buckley, 2010), African wild dogs (Lindsey, Alexander, du Toit, & Mills, 2005), Komodo dragons (Walpole & Goodwin, 2000; Walpole & Leader-Williams, 2002), and coral reefs (Diedrich, 2007; Walters & Samways, 2001; Spalding et al., 2017). Although the conservation value of ecotourism may not always offset the perils of extractive industries or less responsible forms of tourism for wildlife, these studies show evidence of increased capacity for conservation within protected areas and increased support for conservation among local populations.

In other recent studies, Ralf Buckley and colleagues used a population accounting approach to measure ecotourism's contribution to conserving IUCN Red-List mammal, bird, and amphibian species (Buckley et al., 2012, 2016; Steven et al., 2013). Their results showed that in the majority of situations, ecotourism provided conservation benefits that outweighed its impacts by increasing survivorship of highly threatened species, including lions, tigers, elephants, wolves, rhinos, and other large species (Buckley et al., 2016). Although much effort is needed on the ground to protect threatened individual animals, in the face of larger commercial and industrial threats, the data suggest a positive influence of ecotourism on endangered species conservation (e.g., Buckley, 2009; Buckley et al., 2012, 2016; Steven et al., 2013; Kirby et al., 2011; Hunt, Durham, Driscoll, & Honey, 2015).

Writings on the conservation benefits of ecotourism include impacts not just on species but also across larger regions. In exploring landscape-level conservation across protected areas, researchers have documented ecotourism's (mostly) positive impacts in Tanzania's Ngorongoro Crater Conservation Area (Charnley, 2005); Peru's Tambopata National Reserve (Kirby et al., 2011; Kirkby, Giudice-Granados, Day, Turner, & Velarde-Andrade, 2010); and Ecuador's Galapagos Islands National Park (Durham, 2008; Powell & Ham, 2008). Although these studies highlight the institutional challenges to implementing conservation across landscape scales, they reinforce the value of ecotourism for conservation in comparison to other competing uses of natural resources, as well as the contributions to local communities. Assessing land use changes attributed to ecotourism using more sophisticated computational analyses, researchers have demonstrated how ecotourism in Costa Rica contributes not only to a reduction in land degradation but also to net reforestation in several independent cases (Hunt et al., 2015; Broadbent et al., 2012; Almeyda, Broadbent, Wyman, & Durham, 2010; Zambrano, Broadbent, & Durham, 2010); parallel ethnographic research in the same regions has confirmed increased economic earning potential and support for protected areas among local populations (Hunt et al., 2015).

A recent global assessment in biodiversity hotspots found that ecotourism supports conservation when the following four criteria are met: (a) a specific forest conservation mechanism is in place, such as a protected area, payment for ecosystem service program, or other conservation pledge; (b) there is a spatial boundary delineating the area governed by the conservation mechanism; (c) local families receive direct economic benefits; and (d) community-oriented monitoring and enforcement are strong (Brandt & Buckley, 2018). These criteria are concordant with the tenets of ecotourism. Other forms of nature-based tourism that do not adhere to these criteria did not lead to similar outcomes. The study provides evidence that tourism works best for conservation when it manifests as ecotourism—that is, when it increases the capacity for conservation in protected areas and in local communities.

Diversified livelihoods

A documented contribution of ecotourism is diversifying the livelihoods of people who live in and near protected areas (Ferraro & Hanauer, 2011, 2014a, 2014b). By combining conservation and development, ecotourism is a classic approach to sustainable development just as it is to other paradigms of sustainable use, integrated conservation development, or community-based natural resource management. Defenders and critics of ecotourism alike tend to describe ecotourism as "a promising route for generating benefits for those living close to tropical biodiversity without undermining its existence" (Agrawal & Redford, 2006, p. 20).

Some have described the connection between ecotourism, livelihoods, and conservation through an "alternative income hypothesis" (Brown & Decker, 2005). This is the notion that local residents who are dependent on wildlife and ecosystem services for their livelihoods will lessen their reliance on natural resources when they switch to work in ecotourism. Langholz (Langholz, 1999), for example, assessed how ecotourism income caused people to reduce their reliance on commercial agriculture, hunting, logging, cattle ranching, and gold mining. Wunder (Wunder, 2000) identified income and employment from ecotourism in the Cuyabeno Wildlife Reserve of Ecuador as influential in building local engagement in conservation. In Costa Rica, Troëng and Drews (2004) found that economic benefits from ecotourism around Tortuguero National Park became incentives for residents to protect sea turtles. In this way, people in host communities can become a first line of defense in the "ecotourism shield" (Fitzgerald & Stronza, 2016).

The alternative income hypothesis is tied with an understanding that working in ecotourism is more sustainable than working in mining, logging, uncontrolled hunting, or farming. The logic further holds that more employment and income from ecotourism can encourage more conservation, and conversely, the loss of benefits may signal degradation (Stronza & Pêgas, 2008). The hypothesis has not always proven true. In Nepal, Bookbinder, Dinerstein, Rijal, Cauley, and Rajouria (1998) found ecotourism benefits were insufficient to provide incentives for local residents to conserve wildlife. In Mexico, Barkin (Barkin, 2003) found ecotourism employment opportunities from the Monarch Butterfly Reserve were not enough to curb logging of the forest. Lindberg et al. (1996) reported similar results in Belize, where tourism activities failed to generate financial support for protected area management. Belsky (1999) explained that decreased local livelihood security associated with ecotourism in Belize actually triggered a "violent backlash against conservation" (Belsky, 1999). In Mexico, Young (1999) found that economic revenues from gray whale watching did not reduce external pressures on inshore fisheries. In the Peruvian Amazon, Stronza (2007) measured the effects of ecotourism benefits among the same households before and after a community lodge opened, and between households with varying levels of participation. She found the economic benefits from ecotourism were ambiguous for conservation—employment in ecotourism led to a general decline in farming and hunting, whereas new income enabled greater market consumption and expansion of agriculture. Taken together, these studies indicate promise from ecotourism and potential scaling limits of ecotourism enterprises. There is a clear need for further analysis of the conditions under which economic benefits can work effectively for conservation.

Although specific conservation outcomes like resource use and habitat protection are often the focus of research on ecotourism impacts, outcomes related to community development have effects for conservation as well. At the scale of entire communities, ecotourism has been associated with communities setting aside tracts of land and vital habitats, with rules assigned to protect resources and species (Wunder, 2000; Mbaiwa & Stronza, 2010; Stronza, 2008, 2010; Borman, 2008; Hoole, 2009). This suggests it is in the social, cultural, and political spheres where ecotourism continues to hold promise for improving local living conditions in ways that reduce pressure on natural resources and biodiversity. In such contexts, ecotourism has been shown to contribute directly to a sense of cultural pride as well as the opportunity to showcase and support local arts and, in some cases, revitalise ethnic traditions, customs, shared identities, and even languages, many of which are tied to intact ecosystems and iconic, endemic wildlife species (Butler & Hinch, 2007; Zeppel, 2006; Stronza, 2008; Coria & Calfucura, 2012).

Environmental interpretation and ethics

Ecotourism's indirect benefits to conservation extend beyond the communities and regions where it occurs by influencing the behavior of ecotourists. Despite early doubts about the potential to convert tourists to "greenies" (e.g., Orams, 1997), more recent research has shed light on the ways interpretation, guiding, and messaging during ecotourism experiences can be leveraged for conservation behaviors in destinations and in tourists' places of origin (Powell & Ham, 2008; Ballantyne & Packer, 2013). For instance, Ham (2011) assessed ecotourists' experiences during trips with National Geographic/Lindblad Expeditions in the Galapagos Islands. Beyond the support the Galapagos National Park received from visitors' entry fees, Ham's informational strategy led to a philanthropic campaign that secured up to $400,000/year in additional donations to the Charles Darwin Foundation. This has inspired other tour operators to explore similar conservation philanthropy opportunities with their clients (Ardoin, Wheaton, Hunt, Schuh, & Durham, 2016).

Ecotourism experiences can also lead to new attitudes, knowledge, and behaviors once visitors return home (Ardoin, Wheaton, Bowers, Hunt, & Durham, 2015). Scholars have explored how free-choice science learning during guided, interpretive experiences in ecotourism settings can be developed in accordance with informal science education theory (e.g., Ballantyne & Packer, 2013; Falk & Staus, 2013). There is emerging evidence that such experiences lead to promotion of parks and conservation messages via social media, as well as increased support for local parks in tourists' places of origin (Wheaton et al., 2016). One path for promoting conservation, or "proenvironmental," behaviors among tourists when they return home is to use postvisit action resources that connect the new knowledge and experiences gained in ecotourism settings to opportunities for conservation action at home (Hughes, Packer, & Ballantyne, 2011; Wu, Huang, Liu, & Law, 2013), especially reducing consumption (Chieh-Wen, Shen, & Chen, 2008; Hall, 2011).

Another indirect benefit of ecotourism is new or newly deepened feelings of stewardship and environmental ethics among host destination communities. Heyman and Stronza (2011) found that cultural interactions between locals and outsiders in ecotourism destinations helped build awareness of local resource scarcity, a concept that gained new meaning for people as they discussed or witnessed habitat degradation or species declines outside of their own communities. Other researchers have highlighted positive changes in the environmental ethic of both local resident hosts (e.g., Hunt et al., 2015; Wunder, 2000; Hunt & Stronza, 2011) and their visiting guests (Ballantyne & Packer, 2013). In Nicaragua, Hunt and Stronza (2011) described how ecotourism employees acquired new environmental concern and stewardship ethics, so much so that they became critical of their own employer's environmental policies (see also Stem et al., 2003).

Strengthened resource management institutions

An indirect but powerful way ecotourism can work for conservation is by strengthening local institutions. Species, landscapes, communities, habitats, and places at the heart of ecotourism (and tourism) operations are often common pool resources. When common pool resources, such as wildlife and forests, are commodified as "attractions" and "destinations," the ways in which they are used and perceived, and by whom, shift, requiring strong institutions to ensure they are governed and managed sustainably (Stronza, 2010). Two basic challenges of managing common pool resources are exclusion and subtraction. The challenge of exclusion is controlling access to potential users (e.g., too many tourists "ruining" a "pristine" habitat); the challenge of subtraction is keeping single users from diminishing or degrading the resource for all others (i.e., hunting or harassing wildlife makes it scarce and skittish) (Ostrom, 1990, 2008). Tourism—or ecotourism—development can compound the problem of exclusion by opening habitats to commercial operators, tourists, and other outsiders, and by expanding the numbers of users, revenues, and technologies that can accelerate subtraction (Campbell, 2002; Young, 1999; Kellert, Mehta, Ebbin, & Lichtenfeld, 2000; Moreno, 2005). Strong local institutions are essential for overcoming these challenges.

Ecotourism, with its emphasis on engagement with local communities and participatory approaches to development, can provide the incentives and social capital to strengthen institutions (Jones, 2005; Marcinek & Hunt, 2015; Snyman, 2013). The quality and stability of local institutions influence how people in local communities are able to monitor wildlife and other resources, establish rules for use and conservation, and sanction rule breakers (Ostrom, Burger, Field, Norgaard, & Policansky, 1999; Pretty & Smith, 2004). Community-based ecotourism operations that help strengthen local institutions have had clearer success in conservation (Stronza & Pêgas, 2008; Stronza, 2010; Romero-Brito, Buckley, & Byrne, 2016). Conversely, ecotourism operations with little attention to local governance have had less success in conservation (Stronza, 2008).

A framework for evaluation

Can ecotourism work for conservation? In this section, we point to studies that provide the way forward for conducting rigorous, empirical research to evaluate the conservation effects of ecotourism. These include comparative approaches designed to test the fundamental predictions of ecotourism, summarised in Table 28.2. Ferraro and Pattanayak (2006) have argued scholars of conservation policy must adopt "state-of-the-art" evaluation methods to determine what works and when. This includes evaluating effects of ecotourism on both environmental and social outcomes, emphasizing quality of research design, and exercising care in measurement and analysis.

Define ecotourism

A first step toward a more rigorous analysis is conceding that scholars have been measuring and judging a wide variety of things and labeling all of it "ecotourism." Muddled definitions

Table 28.2 Framework for rigorous analysis of ecotourism

Research principle	*How?*	*Why?*
Define ecotourism	Adhere to accepted definitions	Avoid false equivalency and definition fallacies ("apples and oranges")
Gather longitudinal data	Panel data; long-term assessment of biodiversity	Understand changes over time on the same criteria with baseline data
Address scale	Test questions at multiple scales using the same methodology, define scale and units of analysis explicitly	Avoid scaling mismatches and identify scaling limits, the fundamental consequences for the interpretations and conclusions drawn from analysis
Measure noneconomic benefits	Shift emphasis from biology and tourist studies to social science in local communities	Noneconomic factors have tremendous influence on conservation institutions, values, and behaviors
Conduct participatory evaluations	Ethnographic research emphasizing emic data, empowering participatory action research approaches	Deepens and expands range of possible variables that will have impact on conservation; enables local monitoring by engaging local residents a priori rather than after the fact
See the larger context	Incorporate broader socio-ecological and political ecological systems-level analysis into the study of ecotourism	Avoid "throwing the baby out with bath water"

of ecotourism make it difficult to assess or compare conservation impacts across sites. In research, this is the proverbial problem of "apples and oranges," or false equivalence, describing a situation where there is a logical and apparent equivalence, for example, between outdoor recreation and ecotourism or between conventional tourism and ecotourism, when, in fact, there is none. The phenomena may share some common characteristics, but they have important differences that are overlooked, often for the purposes of the argument (Dann, Nash, & Pearce, 1988). Problematically, this approach allows cherry-picking cases to prove a point, i.e., "ecotourism is harmful to wildlife," rather than conducting rigorous analysis. Ferraro and Hanauer (2014a, 2014b) have noted that evaluators often ignore the implications of measurement error in their treatment variable, in their outcome variable, and in their control variables. Recent research has demonstrated, however, that these errors are often not random, and ignoring them can lead to serious bias.

Despite the multiple definitions of ecotourism, it is possible to make rigorous and thoughtful comparisons of ecotourism impacts across sites. The key is providing clarity in measurement. No two communities or ecosystems or ecotourism destinations are the same, and establishing controls as one would in a laboratory setting is impossible. Nonetheless, one can identify average effects of treatments across sites and populations. This requires careful measurement or operationalisation of the phenomenon studied—ecotourism—as a causal variable (Bernard, 2013). Without providing clarity in how ecotourism is defined, operationalised, or measured, researchers risk further confusing and confounding different activities and impacts.

Clarity in measurement will ensure more rigorous assessments of ecotourism, a necessary endeavor given ecotourism remains a major conservation strategy environmentalists are busy promoting and implementing around the world (Fitzgerald & Stronza, 2016; Buckley, 2010).

Although one 2001 content analysis outlined as many as 85 different definitions of ecotourism (Fennell, 2001), a number that has almost certainly grown in the intervening years, that study made it clear that despite the large proliferation of definitions, several key variables are common to the vast majority of ecotourism definitions: (a) reference to where ecotourism occurs, for example, in natural areas; (b) ecotourism's net benefits to conservation; (c) ecotourism's respect for local culture; (d) direct benefits of ecotourism for local communities; and (e) ecotourism's educational value for both travelers and local residents. Perhaps the most thorough definition comes from Fennel (2008): "sustainable, non-invasive form of nature-based tourism that focuses primarily on learning about nature first-hand, and which is ethically managed to be low impact, non-consumptive, and locally oriented (control, benefits, and scale). It typically occurs in natural areas, and should contribute to the conservation of such areas" (p. 24).

In Table 28.1, we considered how the definitions of nine different forms of tourism that have some connection to nature, sustainability, or conservation and that are often conflated in the literature with ecotourism compare to these two definitions of ecotourism. Among them, ecotourism is the one activity specifically designed with proactive concern and intent for channeling revenues from visitors to conservation activities and to enhancing the welfare of local people.

Gather longitudinal data

A second principle for conducting rigorous research on the impacts of ecotourism is evaluating changes over time. This entails collecting data on indicators before and after the program (Ferraro & Hanauer, 2011). Long-term conservation is an implicit goal of ecotourism, and longitudinal studies are needed to identify patterns and processes related to the presence of ecotourism, such as rebounding of wildlife populations, resilience of ecotourism ventures, and how negative and positive changes accumulate over time. Indicators of direct and indirect effects of ecotourism, either good or bad for conservation, can be measured only with understanding of the same indicators across sites, and also with panel data over time, such as in longitudinal case studies. Such controls allow researchers to evaluate impacts on species, populations, or communities in ecotourism destinations as well as on what happens to visitors' behaviors during and after travel.

Examples in the literature include long-term research in Tambopata, Peru, by social scientists and biologists (Stronza & Gordillo, 2008; Stronza, 2000, 2007, 2008, 2010; Brightsmith, Stronza, & Holle, 2008), anthropologists, and other social scientists in Roatan, Honduras (Stonich, 1998; Stonich, 2000), the Okavango Delta of Botswana (Mbaiwa, 2003, 2008, 2015; Mbaiwa, Thakadu, & Darkoh, 2008), Madagascar (Gezon, 2014), and in both Guanacaste (Campbell, 1999, 2002; Gray & Campbell, 2007) and the Osa Peninsula regions of Costa Rica (Hunt et al., 2015; Zambrano et al., 2010; Ardoin et al., 2015). These studies provide greater context for understanding how ecotourism plays out against other economic activities and how ecotourism reverberates within local communities, changing how people think about, use, harvest, protect, or interact with wildlife and other natural resources. Such changes are often not discernable in one "field season" or through a single set of observations or single application of a survey instrument. In longitudinal research in the Peruvian Amazon, Stronza (Stronza, 2000, 2007, 2010), for example, showed how economic benefits from ecotourism that were distributed across a community with secure land tenure fostered participation in management and decision making, generating local support for wildlife and forest conservation.

Address scale

A third principle of rigorous research on ecotourism is attention to scale. Ecotourism bears consequences for conservation across multiple scales, ranging from individual tourists' encounters with individual animals, to broader reductions in hunting pressure and opportunities for news skills, benefits, and development for individuals, households, communities, and national governments. In the same way ecologists have recognised for decades their studies are influenced by the scale of observation (Wiens, 1989; Levin, 1992), the scale at which ecotourism is viewed will influence conclusions about its value (Hunt & Stronza, 2009). Ecological research on effects of ecotourism on biodiversity will benefit from explicit definition of the scale at which studies on flora and fauna are conducted, and careful consideration when extrapolating results, positive or negative, to larger scales. Although it is important to document effects of people's actions on biodiversity at any scale, it is also important to frame research questions, and their answers, at the appropriate scale if one is evaluating ecotourism as a conservation endeavor. If the goal of ecotourism is to conserve biodiversity and enhance the wellbeing of people, then a meaningful overarching question is "What are the impacts of ecotourism at the scales that matter to biodiversity conservation, and to local communities?"

How does ecotourism scale in terms of overall benefits? The conservation benefits of ecotourism thus extend from the scale of an individual local guide to an entire community, and they bear a strong influence on national policy aimed at conservation (Hunt & Stronza, 2009). The umbrella of protection provided by ecotourism, which depends not only on land sparing but just as importantly on sustaining incentives for people to conserve biodiversity, can bring a net benefit to conservation of biodiversity at landscape and regional scales, provide revenue to support habitat conservation over large areas for decades, and influence major conservation and development policies (Buckley, 2009, 2010; Hunt & Stronza, 2009). For example, communities set aside tracts of forest surrounding ecolodges, and the positive cumulative effects of individual lodges in a region may be more than additive in terms of lands protected and positive development outcomes. In this way, multiple community-based ecotourism projects can support conservation over large areas (Hunt et al., 2015; Zambrano et al., 2010; Salafsky et al., 2001). Multiscale studies can identify thresholds where ecotourism is more or less impactful, as well as the governance regimes required to sustain them. Testing for and describing the scaling functions of multiple ecotourism ventures and how they interact would be a step forward in understanding its broader role in conservation. Also, understanding how far conservation incentives from ecotourism can reach, depending on markets, location, and ecosystems is a rich area for integrative research (Woodward, Stronza, Shapiro-Garza, & Fitzgerald, 2014).

Measure noneconomic benefits

Measuring the conservation impacts of ecotourism often entails gathering data on numbers of visitors, rooms occupied, and expenditures, as well as calculating revenues, number of jobs, volume of local commerce, and other economic indices (Taylor, Dyer, Stewart, Yunez-Naude, & Ardila, 2003; Wilson & Tisdell, 2003). Income and employment opportunities sometimes appear in studies as indicators of successful ecotourism projects (Gössling, 1999; Bookbinder et al., 1998). However, direct monetary benefits are not sufficient to ensure social and environmental objectives of ecotourism are achieved. In the absence of equitable distribution of economic benefits, secure land tenure for local residents, and social impacts in line with existing social and cultural aspirations, ecotourism is unlikely to result in conservation (Belsky, 1999; Charnley, 2005; Bookbinder et al., 1998; Zeppel, 2006). Scholars must look beyond economic measures of employment and income to other social, cultural, ecological, and political factors to understand the full value of ecotourism.

The next step in proper valuation of ecotourism is recognising that economic benefits are "necessary but not sufficient" for ensuring conservation (Stronza, 2007; Hunt & Stronza, 2011). Aside from providing employment and revenue (Campbell, 1999), community-based ecotourism can help build stewardship of natural resources and strengthen local institutions for managing wildlife, forests, and other common pool resources (Stronza, 2010). Therefore, measuring impacts of ecotourism requires seeing and evaluating nonmonetary indicators—things like social capital (Pretty & Smith, 2004; Pretty & Ward, 2001), feelings of well-being (Scheyvens, 1999; Stronza & Gordillo, 2008), and capacity to work collectively (Stronza & Pêgas, 2008; Zeppel, 2006; van Riper et al., 2017). Adding such social science indicators can provide greater understanding of how ecotourism helps protect wildlife and ecosystems beyond protected areas (Buckley et al., 2016; Hunt et al., 2015).

Conduct participatory evaluations

Relatively few studies of ecotourism are conducted at the local level (Stone & Wall, 2004). Even fewer assessments have emerged from the experiences and perceptions of local residents themselves. A more thorough analysis of ecotourism must include evaluatory criteria derived from local residents. A participatory approach implies gathering and interpreting data in ways that differ from those in studies directed solely by scholars. In participatory analyses, indicators of success are determined by emic (i.e., subjective and culturally embedded views) as well as etic ones (i.e., those defined by scholars, NGOs, conservationists, or other external actors). In cultural anthropology, an emic account of behavior is one that is couched in terms meaningful to the actor; an etic account is one that is given in terms that can be applied to other groups. Emic is culturally specific, whereas etic is culturally neutral.

Various scholars in the social sciences have taken this approach. Ross and Wall (1999a, 1999b) developed an evaluative framework, which they used to compare ecotourism in three protected areas in Indonesia, evaluating field observations and interview responses with indicators of success. Similarly, Weinberg, Bellows, and Ekster (2002) compared ecotourism projects in New Zealand and Costa Rica, using interviews to solicit perceptions of ecotourism's failures and successes along specific criteria. Stronza and Gordillo (2008) conducted a year of ethnographic research, gathering local narratives, insights, and experiences in ecotourism, combined with south-south peer assessments of ecolodges in three indigenous communities in Ecuador, Peru, and Bolivia (Stronza, 2008; Heyman & Stronza, 2011).

The participatory approach entails asking people not just to respond to questions, but also to help determine which questions are most relevant to ask, help gather the data, and then help interpret and present the results. This approach takes evaluation out of solely academic realms and puts it back into communities for applied learning and action. Although others have written about the role of participation in ecotourism planning and management (Garrod, 2010; Guevara, 1986), this framework carries participation to the latter phases of evaluation. Participation in evaluation can be empowering, as people in local communities represent and express their own experiences with ecotourism, in their own languages, both literal and metaphorical.

See the larger context

A productive way to assess connections between ecotourism and conservation is to evaluate impacts in relation to other livelihood strategies and economic activities. The Union of

Concerned Scientists has outlined the primary drivers of deforestation and forest degradation as emerging primarily from the soybean, beef, palm oil, timber, fuelwood, and small-scale farming sectors (Boucher et al., 2011). Each of these agents of degradation represents a land use that often competes directly with ecotourism, particularly in high biodiversity regions of the tropics. The conservation value of ecotourism in such contexts, where it competes with other economic activities that are more likely to lead to deforestation, endangered species loss, environmental degradation, and reductions in biodiversity (Weaver & Lawton, 2007; Mowforth & Munt, 2015; Higham, 2007; Boucher et al., 2011), is particularly high. However, few, if any, studies make such direct comparisons, and instead only compare ecotourism's impacts on wildlife to the absence of human activity. This fails to account for ecotourism's role as an alternative to other economic activities and forms of tourism.

One of the intentions of ecotourism is to provide alternatives to activities that are more likely to lead to environmental degradation and to reduce the perverse incentives that draw marginalised residents into less sustainable livelihood activities and forms of development that create greater damage to wildlife and ecosystems (Epler Wood et al., 1991; Honey, 2008; Higham, 2007). Thus, the relevant questions are not "What are ecotourism's impacts?" or "Is ecotourism better than a national park?" but rather "What are ecotourism's impacts relative to industrial logging?". What are its impacts relative to land conversion for commercial agriculture such as soybean or African oil palm production? What are its impacts relative to mining, fishing, or illegal hunting? What is the role of ecotourism in stemming overexploitation of biodiversity, such as bushmeat hunting or fuelwood gathering? How do the impacts of ecotourism differ from those of other forms of tourism? After all, these are the things ecotourism was invented to combat.

A fruitful line of research with particular relevance to conservation policy is measuring the potential impacts of ecotourism relative to other economic activities, and modeling or predicting impacts across different spatial and temporal scales. Although such research remains scarce, it is needed to demonstrate ecotourism's value as an alternative for rural communities. Although it may be a foregone conclusion that ecotourism's monetary benefits cannot offset oil and gas development, mining, and industrial agriculture (Büscher & Davidov, 2013; Mowforth & Munt, 2015), studies that address scale, are participatory, and consider nonmonetary valuations will serve to identify better the place of ecotourism in an array of conservation strategies and ideas. Moving forward, researchers should delineate ecotourism's impacts in relation to the activities that would be most likely to occur in its absence. This echoes the recent call for more counterfactual modeling of ecotourism's conservation impacts, a methodological approach that would better delineate ecotourism's impacts from those of other economic sectors, competing land uses, and forms of tourism (Brandt & Buckley, 2018).

Conclusion

Earth has entered the Great Acceleration of the Anthropocene, an era of unprecedented environmental change and species loss resulting from human activity (Redmore, Stronza, Songhurst, & McCulloch., 2017; Steffen, Crutzen, & McNeill, 2009; Steffen, Broadgate, Deutsch, Gaffney, & Ludwig, 2015). It is more critical than ever that scholars and practitioners gain better understanding of how human activities can be managed to support the survival of species—including our own—on the planet. Ecotourism remains a major conservation strategy, and increased clarity about its net positive benefits is necessary if we are to leverage opportunities provided by the world's largest industry for further protection of global biodiversity.

Ecotourism is no more a panacea than any other conservation strategy. It is subject to scaling issues and there is variance around its overall effect. Despite recent claims, ecotourism can still hold promise among an array of strategies for justifying large protected areas and building local stewardship, support, and institutional capacity for managing wildlife. As with many conservation programs, the evaluation of ecotourism impacts has lacked rigor (Ferraro & Hanauer, 2014a, 2014b; Ferraro & Pattanayak, 2006). Defining and measuring ecotourism carefully and writing about its impacts—both positive and negative, social, and ecological—is critical also for subjecting all forms of tourism operations to scrutiny. Added rigor in evaluation can help distinguish greenwashing from legitimate and effective forms of ecotourism.

We have provided an overview of the economic, environmental, and social benefits that can result from committed application of ecotourism principles. We identified a trend in the literature, which suggests ecotourism holds more peril than promise, and we identified problematic fallacies and mismatches in the research program. We arrived at a set of research principles that, if embraced, can lead to more rigorous empirical research that will better account for the net benefits ecotourism can offer for people, wildlife, and ecosystems over time.

Summary points

1. Ecotourism was designed by conservationists in the 1980s, at the dawn of sustainable development, to channel tourism revenues into support for conservation and local development.
2. Ecotourism has many definitions, but one clear set of principles. It is an alternative to other forms of tourism development, designed to ensure a positive feedback loop between tourism and conservation.
3. Explicit in all definitions of ecotourism is the hypothesis that tourism, when designed and practiced as ecotourism, can benefit wildlife and biodiversity, create incentives to protect landscapes, and support local communities.
4. Despite research over 30 years on the economic, environmental, and social benefits of ecotourism, it has been dismissed and critiqued as ineffective.
5. Although ecotourism efforts are not always successful, much of the lack of success noted in the scholarship is associated with flaws in research design.
6. Many critiques are a result of evaluating conventional tourism and outdoor recreation and calling it ecotourism. These activities are not synonymous with ecotourism, but rather are the activities to which ecotourism is designed to provide the alternative.
7. The conflation can preclude rigorous analysis of ecotourism, create a misleading implication that ecotourism is worse for conservation than the forms of resource use likely to occur in its absence, and thus impede efforts to make ecotourism work effectively as a strategy for meeting human needs while protecting the environment.
8. We provide a history of ecotourism and a review of the documented impacts. Can ecotourism work for conservation? We point to ways for conducting rigorous research to evaluate the effects and net social and ecological benefits at different scales. These include comparative and longitudinal approaches to testing the fundamental predictions of ecotourism.

Future issues

1. Because ecotourism does not occur in a void, researchers need to place greater attention on the contexts in which ecotourism is occurring so that the environmental impacts of competing uses of natural resources are compared with the impacts of ecotourism activities.

2. In addition to species-level assessments, greater emphasis on landscape and/or ecosystem-level outcomes is needed.
3. Further attention to social outcomes related to environmental ethics, shifting attitudes toward conservation, and changing social relations of power and capacity, particularly in longitudinal studies, will better account for the overall conservation effects of ecotourism.

Disclosure statement

The authors are not aware of any affiliations, memberships, funding, or financial holdings that might be perceived as affecting the objectivity of this review.

Key terms

Biodiversity: the variability among living organisms from all sources, including terrestrial, marine, and other aquatic ecosystems and the ecological complexes of which they are part; this includes diversity within species, between species, and of ecosystems

Ecotourism: responsible travel to natural areas that conserves the environment, sustains the wellbeing of the local people, and involves interpretation and education

Institutions: formal rules, informal norms, or shared understandings that structure political, economic, and social interactions

Livelihoods: means of making a living; encompass people's capabilities, assets, income, and activities required to secure the necessities of life

Sustainable development: development that meets the needs of the present without compromising the ability of future generations to meet their own needs

Acknowledgments

This work is the result of fruitful discussions with many colleagues in many fields and places over many years. We thank in particular the residents of local communities and ecotourism destinations who have shared their insights and first-hand experiences.

Article reproduced with permission from the *Annual Review of Environment and Resources*, Volume 44 © 2019 by Annual Reviews, http://www.annualreviews.org.

References

Almeyda, A. M., Broadbent, E. N., Wyman, M. S., Durham, W. H. (2010). Ecotourism impacts in the Nicoya Peninsula, Costa Rica. *International Journal of Tourism Research, 12*(6), 803–819.

Alexander, S. E. (2000). Resident attitudes towards conservation and black howler monkeys in Belize: The Community Baboon Sanctuary. *Environmental Conservation, 27*(04), 341–350.

Ardoin, N. M., Wheaton, M., Bowers, A. W., Hunt, C. A., & Durham, W. H. (2015). Nature-based tourism's impact on environmental knowledge, attitudes, and behavior: A review and analysis of the literature and potential future research. *Journal of Sustainable Tourism, 23*(6), 838–858.

Ardoin, N. M., Wheaton, M., Hunt, C. A., Schuh, J. S., & Durham, W. H. (2016). Post-trip philanthropic intentions of nature-based tourists in Galapagos. *Journal of Ecotourism, 15*(1), 21–35.

Ballantyne, R., & Packer, J. (2013). *International Handbook on Ecotourism*. Northampton, UK: Edward Elgar Publishing.

Bateman, P. W., & Fleming, P. A. (2017). Are negative effects of tourist activities on wildlife over-reported? A review of assessment methods and empirical results. *Biological Conservation, 211*, 10–19.

Beale, C. M., & Monaghan, P. (2004). Human disturbance: People as predation-free predators? *Journal of Applied Ecology, 41*(2), 335–343.

Belsky, J. M. (1999). Misrepresenting communities: The politics of community-based rural ecotourism in Gales Point Manatee, Belize. *Rural Sociology, 64*(4), 641–666.

Berger, J. (2007). Fear, human shields and the redistribution of prey and predators in protected areas. *Biology Letters, 3*, 620–623.

Bernard, H. R. (2013). *Social Research Methods: Qualitative and Quantitative Approaches*. Los Angeles: Sage.

Bjork, P. (2000). Ecotourism from a conceptual perspective, an extended definition of a unique tourism form. *International Journal of Tourism Research, 2*(3), 189–202.

Blumstein, D. T., Geffroy, B., Samia, D. S. M., & Bessa E., (2017). *Ecotourism's Promise and Peril*. Springer International Publishing doi:10.1007/978-3-319-58331-0.

Boo, E. (1990). *Ecotourism: The Potentials and Pitfalls* (Vol. 1). Washington, DC: World Wildlife Fund.

Bookbinder, M. P., Dinerstein, E., Rijal, A., Cauley, H., & Rajouria, A. (1998). Ecotourism's support of biodiversity conservation. *Conservation Biology, 12*(6), 1399–1404.

Borman, R. (2008). Ecotourism and conservation: The Cofan experience. In A. Stronza, & W. H. Durham (Eds.), *Ecotourism and conservation in the Americas* (pp. 21–29). Wallingford: CABI.

Boucher, D., Elias, P., Lininger, K., May-Tobin, C., Roquemore, S., & Saxon, E. (2011). *The Root of the Problem: What's Driving Tropical Deforestation Today?* Cambridge, UK: Union of Concerned Scientists.

Brandon, K. E., & Wells, M. (1992). Planning for people and parks: Design dilemmas. *World Development 20*(4), 557–570.

Brandt, J. S., & Buckley, R. C. (2018). A global systematic review of empirical evidence of ecotourism impacts on forests in biodiversity hotspots. *Current Opinion in Environmental Sustainability, 32*, 112–118.

Brightsmith, D. J., Stronza, A., & Holle, K. (2008). Ecotourism, conservation biology, and volunteer tourism: A mutually beneficial triumvirate. *Biological Conservation, 141*(11), 2832–2842.

Broadbent, E. N., Almeyda Zambrano, A. M., Dirzo, R., Durham, W. H., Driscoll, L., et al. (2012). The effect of land use change and ecotourism on biodiversity: A case study of Manuel Antonio, Costa Rica, from 1985 to 2008. *Landscape Ecology, 27*(5), 731–744.

Brown, T. L., & Decker, D. J. (2005). Research needs to support community-based wildlife management: Global perspectives. *Human Dimensions of Wildlife, 10*(2), 137–140.

Buckley, R. (2009). Evaluating the net effects of ecotourism on the environment: A framework, first assessment and future research. *Journal of Sustainable Tourism, 17*(6), 643–672.

Buckley, R. (2010a). *Conservation Tourism*. Wallingford, UK: CABI.

Buckley, R. (2010b). Protecting lemurs: Ecotourism. *Science, 344*(6182), 358.

Buckley, R. (2011). Tourism and environment. *Annual Review of Environment and Resources, 36*(1), 397–416.

Buckley, R. (2012). Endangered animals caught in the tourist trap. *New Scientist*, (October 2012), 28–29.

Buckley, R. C., Castley, J. G., de Vasconcellos Pegas, F., Mossaz, A. C., & Steven, R. (2012). A population accounting approach to assess tourism contributions to conservation of IUCN-Redlisted mammal species. *PLOS ONE*, 7(9), e44134.

Buckley, R. C., Morrison, C., & Castley, J. G. (2016). Net effects of ecotourism on threatened species survival. *PLOS ONE, 11*(2), e0147988.

Budowski, G. (1976). Tourism and environmental conservation: Conflict, coexistence, or symbiosis? *Environmental Conservation, 3*(01), 27.

Bul̀ˆscher, B., & Davidov, V. (2013). *The Ecotourism-Extraction Nexus: Political Economies and Rural Realities of (Un)Comfortable Bedfellows*. Florence, Italy: Routledge.

Butcher, J. (2007). *Ecotourism, NGOs and Development: A Critical Analysis*. New York: Routledge.

Butler, R., & Hinch, T. (2007). *Tourism and Indigenous Peoples: Issues and Implications*. Oxford: Butterworth-Heinemann.

Campbell, L. M. (1999). Ecotourism in rural developing communities. *Annals of Tourism Research, 26*(3), 534–553.

Campbell, L. M. (2002). Conservation narratives in Costa Rica: Conflict and co-existence. *Development and Change, 33*(1), 29–56.

Cater, E. (2006). Ecotourism as a western construct. *Journal of Ecotourism, 5*(1–2), 23–39.

Ceballos-Lascurain, H. (1996). *Tourism, Ecotourism, and Protected Areas: The State of Nature-Based Tourism Around the World and Guidelines for its Development.* Cambridge, UK: IUCN.
Charnley, S. (2005). From nature tourism to ecotourism? The case of the Ngorongoro Conservation Area, Tanzania. *Human Organization, 64*(1), 75–88.
Chieh-Wen, S., Shen, M., & Chen, M. (2008). Special interest tour preferences and voluntary simplicity lifestyle. *International Journal of Culture, Tourism, and Hospitality Research, 2*(4), 389–409.
Chirgwin, S., & Hughes, K. (1997). Ecotourism: The participants' perceptions. *J. Tour. Stud., 8*(2), 2–7.
Coria, J., & Calfucura, E. (2012). Ecotourism and the development of indigenous communities: The good, the bad, and the ugly. *Ecological Economics, 73*, 47–55.
Dann, G., Nash, D., & Pearce, P. (1988). Methodology in tourism research. *Annals of Tourism Research, 15*(1), 1–28.
de Kadt, E. (1979). *Tourism: Passport to Development?* Oxford: Oxford University Press.
de Vasconcellos Pegas, F., Coghlan, A., Stronza, A., & Rocha, V. (2013). For love or for money? Investigating the impact of an ecotourism programme on local residents' assigned values towards sea turtles. *Journal of Ecotourism, 12*(2), 90–106.
Diedrich, A. (2007). The impacts of tourism on coral reef conservation awareness and support in coastal communities in Belize. *Coral Reefs, 26*(4), 985–996.
Duffy, R. (2002). *A Trip Too Far: Ecotourism, Politics, and Exploitation.* London: Earthscan.
Duffy, R. (2008). Neoliberalising nature: Global networks and ecotourism development in Madagasgar. *Journal of Sustainable Tourism, 16*(3), 327–344.
Durham, W. H. (2008). The challenge ahead: Reversing vicious cycles through ecotourism. In A. Stronza, & W. H. Durham (Eds.), *Ecotourism and conservation in the Americas* (pp. 265–271). Wallingford, UK: CABI.
Ellenberg, U., Mattern, T., Seddon, P. J., & Jorquera, G. L. (2006). Physiological and reproductive consequences of human disturbance in Humboldt penguins: The need for species-specific visitor management. *Biological Conservation, 133*(1), 95–106.
Epler Wood, M., Gatz, F., & Lindberg, K. (1991). The ecotourism society: An action agenda. Washington, DC: The Ecotourism Society.
Escobar, A. (1991). Anthropology and the development encounter: The making and marketing of development anthropology. *American Ethnologist* , *18*(4), 658–682.
Falk, J., & Staus, N. L. (2013). Free-choice learning and ecotourism. In R. Ballantyne, & J. Packer (Eds.), *International handbook of ecotourism* (pp. 155–168). Northampton, UK: Edward Elgar Publishing.
Fennell, D. A. (2001). A content analysis of ecotourism definitions. *Current Issues in Tourism, 4*(5), 403–421.
Fennell, D. A. (2008). *Ecotourism.* New York: Routledge.
Fennell, D. A., & Eagles, P. F. J. (1990). Ecotourism in Costa Rica: A conceptual framework. *Journal of Park and Recreation Administration, 8*(1), 23–34.
FernÃ¡ndez-Juricic, E., Venier, M. P., Renison, D., & Blumstein, D. T. (2005). Sensitivity of wildlife to spatial patterns of recreationist behavior: A critical assessment of minimum approaching distances and buffer areas for grassland birds. *Biological Conservation, 125*(2), 225–235.
Ferraro, P. J., & Hanauer, M. (2014a). Advances in measuring the environmental and social impacts of environmental programs. *Annual Review of Environment and Resources, 39*, 495–517.
Ferraro, P. J., & Hanauer, M. (2011). Protecting ecosystems and alleviating poverty with parks and reserves: "win-win" or tradeoffs? *Environmental and Resource Economics, 48*(2), 269–286.
Ferraro, P. J., & Hanauer, M. M. (2014b). Quantifying causal mechanisms to determine how protected areas affect poverty through changes in ecosystem services and infrastructure. *PNAS 111*(11), 4332–4337.
Fitzgerald, L. A., & Stronza, A. L. (2016). In defense of the ecotourism shield: A response to Geffroy et al. *Trends in Ecology & Evolution, 31*(2), 94–95.
Fletcher, R. (2009). Ecotourism discourse: Challenging the stakeholders theory. *Journal of Ecotourism, 8*(3), 269–285.
Fletcher, R. (2011). Sustaining tourism, sustaining capitalism? The tourism industry's role in global capitalist expansion. *Tourism Geographies, 13*(3), 443–461.
Fletcher, R. (2014). *Romancing the Wild: Cultural Dimensions of Ecotourism.* Durham, NC: Duke University Press.
Fletcher, R., & Neves, K. (2012). Contradictions in tourism: The promise and pitfalls of ecotourism as a manifold capitalist fix. *Environmental Sociology, 3*(1), 60–77.

Foster, G. M. (1973). *Traditional Societies and Technological Change*. Manhattan: Harper & Row.
Fowler, G. S. (1999). Behavioral and hormonal responses of Magellanic penguins (Spheniscus magellanicus) to tourism and nest site visitation. *Biological Conservation, 90*, 143–149.
Frid, A., & Dill, L. M. (2002). Human-caused disturbance stimuli as a form of predation risk. *Conservation Ecology, 6*(1), 11.
Gardner, B. (2016). *Selling the Serengeti: The Cultural Politics of Safari Tourism*. Athens, GA: University Georgia Press.
Garrod, B. (2010). Local participation in the planning and management of ecotourism: A revised model approach. *Journal of Ecotourism, 2*(1), 33–53.
Geffroy, B., Samia, D. S. M., Bessa, E., & Blumstein, D. T. (2015). How nature-based tourism might increase prey vulnerability to predators. *Trends in Ecology & Evolution, 30*(12), 755–765.
Gezon, L. L. (2014). Who wins and who loses? Unpacking the "local people" concept in ecotourism: A longitudinal study of community equity in Ankarana, Madagascar. *Journal of Sustainable Tourism, 22*(5), 821–838.
Goodwin, H. (1996). In pursuit of ecotourism. *Biodiversity Conservation, 5*(3), 277–291.
Goss-Custard, J. D., Triplet, P., Sueur, F., & West, A. D. (2006). Critical thresholds of disturbance by people and raptors in foraging wading birds. *Biological Conservation, 127*(1), 88–97.
Gössling, S. (1999). Ecotourism: A means to safeguard biodiversity and ecosystem functions? *Ecological Economics, 29*(2), 303–320.
Gray, N. J., & Campbell, L. M. (2007). A decommodified experience? Exploring aesthetic, economic and ethical values for volunteer ecotourism in Costa Rica. *Journal of Sustainable Tourism, 15*(5), 463–482.
Guevara, J. R. (1986). Learning through participatory action research for community ecotourism planning. *Convergence, 29*(3), 24–40.
Hall, C. M. (2011). Consumerism, tourism and voluntary simplicity: We all have to consume, but do we really have to travel so much to be happy? *Tourism Recreation Research, 36*(3), 298–303.
Ham, S. (2011). The ask-or is it the offer? In M. Honey (Ed.), *Travelers' philanthropy handbook* (pp. 141–149). Washington, DC: Center for Responsible Travel.
Harrison, D. (1997). Ecotourism in the South Pacific: The case of Fiji. In D. Bruni (Ed.), *World Ecotour '97 Abstracts Volume* (p. 75). Rio de Janeiro, Brazil: BIOSFERA.
Heyman, W., & Stronza, A. (2011). South-South exchanges enhance resource management and biodiversity conservation at various scales. *Conservation Society, 9*(2), 146.
Higham, J. (2007). Ecotourism: Competing and conflicting schools of thought. In J. Higham (Ed.), *Critical issues in ecotourism: Understanding a complex phenomenon* (pp. 1–19). Oxford: Elsevier.
Honey, M. (2008). *Ecotourism and Sustainable Development: Who Owns Paradise?* Washington, DC: Island Press.
Hoole, A. F. (2009). Place-power-prognosis: community-based conservation, partnerships, and ecotourism enterprises in Namibia. *International Journal of the Commons, 4*(1), 78.
Hughes, K., Packer, J., & Ballantyne, R. (2011). Using post-visit action resources to support family conservation learning following a wildlife tourism experience. *Environmental Education Research, 17*(3), 307–328.
Hunt, C. A., & Stronza, A. (2009). Bringing ecotourism into focus: Applying a hierarchical perspective to ecotourism research. *Journal of Ecotourism, 8*(1), 1–17.
Hunt, C. A., & Vargas, E. (2018). Turtles, Ticos, and tourists: Protected areas and marine turtle conservation in Costa Rica. *Journal of Park and Recreation Administration, 36*. doi:10.18666/JPRA-2018-V36-I3-8820.
Hunt, C. A., Durham, W. H., Driscoll, L., & Honey, M. (2015). Can ecotourism deliver real economic, social, and environmental benefits? A study of the Osa Peninsula, Costa Rica. *Journal of Sustainable Tourism, 23*(3), 339–357.
Hunt, C., & Stronza, A. (2011). Missing the forest for the trees?: Incongruous local perspectives on ecotourism in Nicaragua converge on ethical issues. *Human Organization, 70*(4), 376–386.
Jacobson, S. K., & Lopez, A. F. (1994). Biological impacts of ecotourism: Tourists and nesting turtles in Tortuguero National Park, Costa Rica. *Wildlife Society Bulletin, 22*, 414–419.
Jones, S. (2005). Community-based ecotourism: The significance of social capital. *Annals of Tourism Research, 32*(2), 303–324.
Kellert, S. R., Mehta, J. N., Ebbin, S. A., & Lichtenfeld, L. L. (2000). Community natural resource management: Promise, rhetoric, and reality. *Society & Natural Resources, 13*(8), 705–715.

Kerbiriou, C., Le Viol, E., Robert, A., Porcher, E., Gourmelon, F., & Julliard, R. (2009). Tourism in protected areas can threaten wild populations: From individual response to population viability of the chough Pyrrhocorax pyrrhocorax. *Journal of Applied Ecology, 46*(3), 657–665.
Kirby, C. A., Giudice, R., Day, B., Turner, K., Silvera Soares-Filho, B., et al. (2011). Closing the ecotourism-conservation loop in the Peruvian Amazon. *Environmental Conservation, 38*(01), 6–17.
Kirkby, C. A., Giudice-Granados, R., Day, B., Turner, K., & Velarde-Andrade, L. M. (2010). The market triumph of ecotourism: An economic investigation of the private and social benefits of competing land uses in the Peruvian Amazon. *PLOS ONE, 5*(9), e13015.
Kiss, A. (2004). Is community-based ecotourism a good use of biodiversity conservation funds? *Trends in Ecology & Evolution 19*(5), 232–237.
Kontogeorgopoulos, N. (2004). Conventional tourism and ecotourism in Phuket, Thailand: Conflicting paradigms or symbiotic partners? *Journal of Ecotourism, 3*(2), 87–108.
Krüger, O. (2005). The role of ecotourism in conservation: Panacea or Pandora's box? *Biodiversity and Conservation, 14*(3), 579–600.
Langholz, J. (1999). Exploring the effects of alternative income opportunities on rainforest use: Insights from Guatemala's Maya Biosphere Reserve. *Society & Natural Resources, 12*(2), 139–149.
Larson, C. L., Reed, S. E., Merenlender, A. M., & Crooks, K. R. (2016). Effects of recreation on animals revealed as widespread through a global systematic review. *PLOS ONE, 11*(12), e0167259.
Lemelin, R. H., Fennell, D., & Smale, B. (2008). Polar bear viewers as deep ecotourists: How specialised are they? *Journal of Sustainable Tourism,, 16*(1), 42–62.
Levin, S. A. (1992). The problem of pattern and scale in ecology: The Robert H. MacArthur Award Lecture. *Ecology, 73*(6), 1943–1967.
Lewis, S., Turpie, J., & Ryan, P. (2012). Are African penguins worth saving? The ecotourism value of the Boulders Beach colony. *African Journal of Marine Science, 34*(4), 497–504.
Lindberg, K. (1991). *Policies for Maximizing Nature Tourism's Ecological and Economic Benefits.* Washington, DC: World Resources Institute.
Lindberg, K., Enriquez, J., & Sproule, K. (1996). Ecotourism questioned: Case studies from Belize. *Annals of Tourism Research, 23*(3), 543–562.
Lindsey, P. A., Alexander, R. R., du Toit, J. T., & Mills, M. G. L. (2005). The potential contribution of ecotourism to African wild dog Lycaon pictus conservation in South Africa. *Biological Conservation, 123*(3), 339–348.
Machlis, G., & Field, D. (2000). *National Parks and Rural Development: Practice and Policy in the United States.* Washington, DC: Island Press.
Marcinek, A. A., & Hunt, C. A. (2015). Social capital, ecotourism, and empowerment in Shiripuno, Ecuador. *International Journal of Tourism Anthropology, 4*(4), 327.
Mbaiwa, J. (2008). The realities of ecotourism development in Botswana. In A. Spenceley (Ed.), *Responsible tourism: Critical issues for conservation and development* (pp. 205–224). New York: Earthscan.
Mbaiwa, J. E. (2003). The socio-economic and environmental impacts of tourism development on the Okavango Delta, north-western Botswana. *Journal of Arid Environments, 54*, 447–467.
Mbaiwa, J. E. (2015). Ecotourism in Botswana: 30 years later. *Journal of Ecotourism, 14*(2–3), 204–222.
Mbaiwa, J. E., & Stronza, A. L. (2010). The effects of tourism development on rural livelihoods in the Okavango Delta, Botswana. *Journal of Sustainable Tourism, 18*(5), 635–656.
Mbaiwa, J. E., Thakadu, O. T., & Darkoh, M. B. K. (2008). Indigenous knowledge and ecotourism-based livelihoods in the Okavango Delta in Botswana. *Botswana Notes & Records, 39*, 62–74.
Meletis, Z. A., & Campbell, L. M. (2007). Call it consumption! Re-conceptualizing ecotourism as consumption and consumptive. *Geography Compass, 1*(4), 850–870.
Moore, R. L., & Driver, B. L. (2005). *Introduction to Outdoor Recreation: Providing and Managing Natural Resource Based Opportunities.* State College, PA: Venture Publishing.
Moreno, P. S. (2005). Ecotourism along the meso-American Caribbean reef: The impacts of foreign investment. *Human Ecology, 33*(2), 217–244.
Mowforth, M., & Munt, I. (2015). *Tourism and Sustainability: Development, Globalisation and New Tourism in the Third World* (4th ed.). New York: Routledge.
Muehlenbein, M. P., Martinez, L. A., Lemke, A. A., Ambu, L., Nathan, S., et al. (2010). Unhealthy travelers present challenges to sustainable primate ecotourism. *Travel Medicine and Infectious Disease, 8*, 169–175.
Munn, C. A. (1992). Macaw biology and ecotourism or 'When a bird in the bush is worth two in hand.' In S. R. Beissinger, & N. F. R. Snyder (Eds.), *New world parrots in crisis: Solutions from conservation biology*, (pp. 47–72). Washington, DC: Smithsonian Inst. Press.

National Park Service. (NPS). (1916). *National Park Service Organic Act (16 U.S.C. l 2 3, and 4)*. Washington, DC: NPS. Retrieved from https://www.nps.gov/parkhistory/online_books/fhpl/nps_organic_act.pdf .

Newsome, D., Moore, S. A., & Dowling, R. K. (2013). *Natural Area Tourism: Ecology, Impacts and Management*. Bristol, UK: Channel View Publishing.

Orams, M. B. (1995). Towards a more desirable form of ecotourism. *Tourism Management, 16*(1), 3–8.

Orams, M. B. (1997). The effectiveness of environmental education: Can we turn tourists into "greenies"? *Progress in Tourism and Hospitality Research, 3*, 295–306.

Ostrom, E. (1990). *Governing the Commons: The Evolution of Institutions for Collective Action*. Cambridge, UK: Cambridge University Press.

Ostrom, E. (2008). The challenge of common-pool resources. *Environment: Science and Policy for Sustainable Development, 50*(4), 8–21.

Ostrom, E., Burger, J., Field, C. B., Norgaard, R. B., & Policansky, D. (1999). Revisiting the commons: Local lessons, global challenges. *Science, 284*(5412), 278–282.

Powell, R. B., & Ham, S. H. (2008). Can ecotourism interpretation really lead to pro-conservation knowledge, attitudes and behaviour? Evidence from the Galapagos Islands. *Journal of Sustainable Tourism, 16*(4), 467–489.

Pretty, J., & Smith, D. (2004). Social capital in biodiversity conservation and management. *Conservation Biology, 18*(3), 631–638.

Pretty, J., & Ward, H. (2001). Social capital and the environment. *World Development, 29*(2), 209–227.

Redmore, L., Stronza, A., Songhurst, A., & McCulloch, G. (2017). Which way forward? Past and new perspectives on community-based conservation in the Anthropocene. *Encyclopedia Anthropology, 3*, 453–460.

Romero-Brito, T. P., Buckley, R. C., & Byrne, J. (2016). NGO partnerships in using ecotourism for conservation: Systematic review and meta-analysis. *PLOS ONE, 11*(11), 1–19.

Ross, S., & Wall, G. (1999a). Evaluating ecotourism: The case of North Sulawesi, Indonesia. *Tourism Management, 20*(6), 673–682.

Ross, S., & Wall, G. (1999b). Ecotourism: Towards congruence between theory and practice. *Tourism Management, 20*(1), 123–132.

Rostow, W. W. (1960). *The Stages of Economic Growth: A Non-Communist Manifesto*. Cambridge, UK: Cambridge University Press.

Runte, A. (1979). *National Parks: The American Experience*. Lincoln: University of Nebraska Press.

Salafsky, N., Cauley, H., Balachander, G., Cordes, B., Parks, J., et al. (2001). A systematic test of an enterprise strategy for community-based biodiversity conservation. *Conservation Biology, 15*(6), 1585–1595.

Sandbrook, C. G. (2010). Local economic impact of different forms of nature-based tourism. *Conservation Letters, 3*(1), 21–28.

Scheyvens, R. (1999). Ecotourism and the empowerment of local communities. *Tourism Management, 20*(2), 245–249.

Sellars, R. W. (1997). *Preserving Nature in the National Parks: A History*. New Haven, CT: Yale University Press.

Shannon, G., Larson, C. L., Reed, S. E., Crooks, K. R., & Angeloni, L. M. (2017). Ecological consequences of ecotourism for wildlife populations and communities. In D. T. Blumstein, B. Geffroy, D. M. S. Samia, & E. Bessa (Eds.), *Ecotourism's promise and peril* (pp. 29–46). Cham: Springer Int. Publ.

Sharpley, R. (2006). Ecotourism: A consumption perspective. *Journal of Ecotourism, 5*(1–2), 7–22.

Smith, V. L., & Eadington, W. R. (1992). *Tourism Alternatives: Potentials and Problems in the Development of Tourism*. Philadelphia: Univ. PA Press.

Snyman, S. (2013). Household spending patterns and flow of ecotourism income into communities around Liwonde National Park, Malawi. *Development Southern Africa, 30*(4–05), 640–658.

Spalding, M., Burke, L., Wood, S. A., Ashpole, J., Hutchinson, J., & zu Ermgassen, P. (2017). Mapping the global value and distribution of coral reef tourism. *Marine Policy, 82*(May), 104–113.

Steffen, W., Broadgate, W., Deutsch, L., Gaffney, O., & Ludwig, C. (2015). The trajectory of the Anthropocene: The great acceleration. *Anthropology Review, 2*(1), 81–98.

Steffen, W., Crutzen, P. J., & McNeill, J. R. (2009). The Anthropocene: Are humans now overwhelming the great forces of nature. *Ambio, 36*, 614–621.

Steidl, R. J., & Anthony, R. G. (2000). Experimental effects of human activity on breeding bald eagles. *Ecological Applications, 10*(1), 258–268.

Stem, C. J., Lassoie, J. P., Lee, D. R., Deshler, D. D., & Schelhas, J. W. (2003). Community participation in ecotourism benefits: The link to conservation practices and perspectives. *Society & Natural Resources, 16*(5), 387–413.
Steven, R., Castley, J. G., & Buckley, R. (2013). Tourism revenue as a conservation tool for threatened birds in protected areas. *PLOS ONE, 8*(5), e62598.
Stone, M., & Wall, G. (2004). Ecotourism and community development: case studies from Hainan, China. *Environmental Management, 33*(1), 12–24.
Stonich, S. C. (1998). The political ecology of tourism. *Annals of Tourism Research, 25*, 25–54.
Stonich, S. C. (2000). *The Other Side of Paradise: Tourism, Conservation and Development in the Bay Islands.* Elmsford, UK: Cognizant Comm. Corp.
Stronza, A. (2000). *Because it is ours: Community-based ecotourism in the Peruvian Amazon* (PhD Thesis, University of Florida, Gainesville). Retrieved from https://elibrary.ru/item.asp?id=5309930.
Stronza, A. (2001). Anthropology of tourism: Forging new ground for ecotourism and other alternatives. *Annual Review of Anthropology, 30*(1), 261–283.
Stronza, A. (2007). The economic promise of ecotourism for conservation. *Journal of Ecotourism, 6*(3), 210–230.
Stronza, A. (2008). Partnerships for tourism development. In G. Moscardo (Ed.), *Building community capacity for tourism development* (pp. 101--115). Wallingford, UK: CABI.
Stronza, A. (2008). Through a new mirror: Reflections on tourism and identity in the Amazon. *Human Organization, 67*(3), 244–257.
Stronza, A. L. (2010). Commons management and ecotourism: Ethnographic evidence from the Amazon. *International Journal of Commons, 4*(1), 56.
Stronza, A., & Gordillo, J. (2008). Community views of ecotourism. *Annals of Tourism Research, 35*(2), 448–468.
Stronza, A., & Hunt, C. (2012). Visions of tourism: From modernization to sustainability. *Practice of Anthropology, 34*(3), 19–22.
Stronza, A., & Pêgas, F. (2008). Ecotourism and conservation: Two cases from Brazil and Peru. *Human Dimensions of Wildlife, 13*(4), 263–279.
Taylor, J. E., Dyer, G. A., Stewart, M., Yunez-Naude, A., & Ardila, S. (2003). The economics of ecotourism: A Galápagos Islands economy-wide perspective. *Economic Development and Cultural Change, 51*(4), 977–997.
The International Ecotourism Society (TIES). (2018). *What Is Ecotourism?* Washington, DC: TIES. Retrieved from http://www.ecotourism.org/what-is-ecotourism.
Thomas, K., Kvitek, R. G., & Bretz, C. (2003). Effects of human activity on the foraging behavior of sanderlings Calidris alba. *Biological Conservation, 109*(1), 67–71.
Tobias, D., & Mendelsohn, R. (1991). Valuing ecotourism in a tropical rain-forest reserve. *Ambio, 20*, 91–93.
TroÃ«ng, S., & Drews, C. (2004). Money Talks: Economic Aspects of Marine Turtle Use and Conservation. *WWF International.* Retrieved from http://wwf.panda.org/wwf_news/?153802/wwwpandaorglacmarineturtlespublications .
UNESCO. (2019). Iguazu National Park - UNESCO World Heritage Centre. United Nations Educational, Scientific, and Cultural Organization (UNESCO). Retrieved from https://whc.unesco.org/en/list/303/.
United Nations World Tourism Organization (UNWTO). (2005). *Making Tourism More Sustainable: A Guide for Policy Makers.* Madrid: UNWTO.
van Riper, C. J., Landon, A. C., Kidd, S., Bitterman, P., Fitzgerald, L. A., et al. (2017). Incorporating sociocultural phenomena into ecosystem-service valuation: The importance of critical pluralism. *Bioscience, 67*(3), 233–244.
Vivanco, L. A. (2001). Spectacular quetzals, ecotourism, and environmental futures in Monte Verde, Costa Rica. *Ethnology, 40*(2), 79–92.
Walker, K., & Hawkins, E. (2013). Watching and swimming with marine mammals: International scope, management and best practice in cetacean ecotourism. In R. Ballantyne, & J. Packer (Eds.), *International handbook on ecotourism* (pp. 365–381). Northampton, UK: Edward Elgar Publ.
Wallace, G. N., & Pierce, S. M. (1996). An evaluation of ecotourism in Amazonas, Brazil. *Annals of Tourism Research, 23*(4), 843–873.
Walpole, M. J., & Goodwin, H. J. (2000). Local economic impacts of dragon tourism in Indonesia. *Annals of Tourism Research, 27*(3), 559–576.

Walpole, M. J., & Leader-Williams, N. (2002). Tourism and flagship species in conservation. *Biodiversity Conservation, 11*, 543–547.

Walters, R. D. M., & Samways, M. J. (2001). Sustainable dive ecotourism on a South African coral reef. *Biodiversity Conservation, 10*(12), 2167–2179.

Weaver, D. B. (1993). Ecotourism in the small island Caribbean. *GeoJournal, 31*(4), 457–465.

Weaver, D. B. (2001). Ecotourism as mass tourism: Contradiction or reality? *Cornell Hotel and Restaurant Administration Quarterly, 42*(2), 104–112.

Weaver, D. B., & Lawton, L. J. (2007). Twenty years on: The state of contemporary ecotourism research. *Tourism Management, 28*(5), 1168–1179.

Weinberg, A., Bellows, S., & Ekster, D. (2002). Sustaining ecotourism: Insights and implications from two successful case studies. *Society & Natural Resources, 15*(4), 371–380.

West, P., & Carrier, J. (2004). Ecotourism and authenticity: Getting away from it all? *Current Anthropology 45*(4), 483–498.

Wheaton, M., Ardoin, N., Hunt, C. A., Schuh, J., Kresse, M., et al. (2016). Using web and mobile technology to motivate pro-environmental action after a nature-based tourism experience. *Journal of Sustainable Tourism, 24*(4), 594–615.

Wheeller, B. (1994). Ecotourism: A ruse by any other name. *Prog. Tour. Recreat. Hosp. Manag. 6*, 3–11

Wiens, J. A. (1989). Spatial scaling in ecology. *Functional Ecology, 3*(4), 385.

Wight, P. A. (1993). Sustainable ecotourism: Balancing economic, environmental and social goals within an ethical framework. *J. Tour. Stud., 4*(2), 54–66.

Wilson, C., & Tisdell, C. (2003). Conservation and economic benefits of wildlife-based marine tourism: Sea turtles and whales as case studies. *Human Dimensions of Wildlife, 8*(1), 49–58.

Woodford, M. H., Butynski, T. M., & Karesh, W. B. (2002). Habituating the great apes: The disease risks. *Oryx, 36*(2), 153–160.

Woodward, R. T., Stronza, A., Shapiro-Garza, E., & Fitzgerald, L. A. (2014). Market-based conservation: Aligning static theory with dynamic systems. *Natural Resources Forum, 38*(4), 235–247.

World Commission on Environment and Development (WECD). (1987). *Our Common Future (The Brundtland Report)*. New York: Oxford Univ. Press.

Wu, J., Huang, D., Liu, J., & Law, R. (2013). Which factors help visitors convert their short-term pro-environmental intentions to long-term behaviors? *International Journal of Tourism Sciences, 13*(2), 33–56.

Wunder, S. (2000). Ecotourism and economic incentives–an empirical approach. *Ecological Economics., 32*(3), 465–479.

Young, E. H. (1999). Balancing conservation with development in small-scale fisheries: Is ecotourism an empty promise? *Human Ecology, 27*(4), 581–620.

Zambrano, A. M. A., Broadbent, E. N., & Durham, W. H. (2010). Social and environmental effects of ecotourism in the Osa Peninsula of Costa Rica: The Lapa Rios case. *Journal of Ecotourism, 9*(1), 62–83.

Zeppel, H. (2006). *Indigenous Ecotourism: Sustainable Development and Management*. Wallingford, UK: CABI.

Ziffer, K. (1989). *Ecotourism: The Uneasy Alliance*. Washington, DC: Conserv. Int.

CONCLUSION

David A. Fennell

The two main objectives of this book were to first, provide researchers, graduate students, and upper-level undergraduates with a state-of-the-art and provocative overview of the essential nature of ecotourism; and second, provide a platform for scholars in both developed and developing country contexts, to have their voices heard. There is good balance in this latter objective as the authors provide a rich tapestry of different themes and issues that will have value for the intended audience. And while I lamented in the Introduction over whether ecotourism's best days were behind it, this notion was quickly rejected according to statistics from the *Journal of Ecotourism,* where the volume of submitted manuscripts based on 2020 statistics has never been higher. Ecotourism is alive and well.

In this Conclusion my purpose is to capture the richness in these chapters that preserves past traditions in ecotourism, but also opens up new avenues for future studies. I do this in two ways. The first is to extract key takeaway concepts for each chapter that characterise the essence of each section. The second aim is to provide a rudimentary overview of the degree to which the core principles of ecotourism are supported through the chapters and sections of the book (Table 29.1). I adopt the stance that ecotourism should be defined according to four main criteria: sustainability from the perspective of conservation (biodiversity and heritage) and community development; a nature-based focus; learning and/or education; and ethics/morality (Fennell, 2020). There is general support that ecotourism is defined by at least the first three of these core criteria and mounting support for the fourth (Nowaczek & Smale, 2010). Manifest content analysis was used to count the number of times that these criteria were mentioned in each of the chapters, with sums generated for each of the four sections of the book (unweighted).

Key concepts in chapters and sections

The chapters included in this book provided a range of different perspectives on ecotourism that both overlap and deviate with past ecotourism research. Using a content analysis approach, I identified a list of key terms that characterise what I thought to represent the essence of each chapter and section (Figure 29.1). Such a list is thought to be useful in helping to underscore where we've been in ecotourism research, and perhaps a pathway for the future.

DOI: 10.4324/9781003001768-102

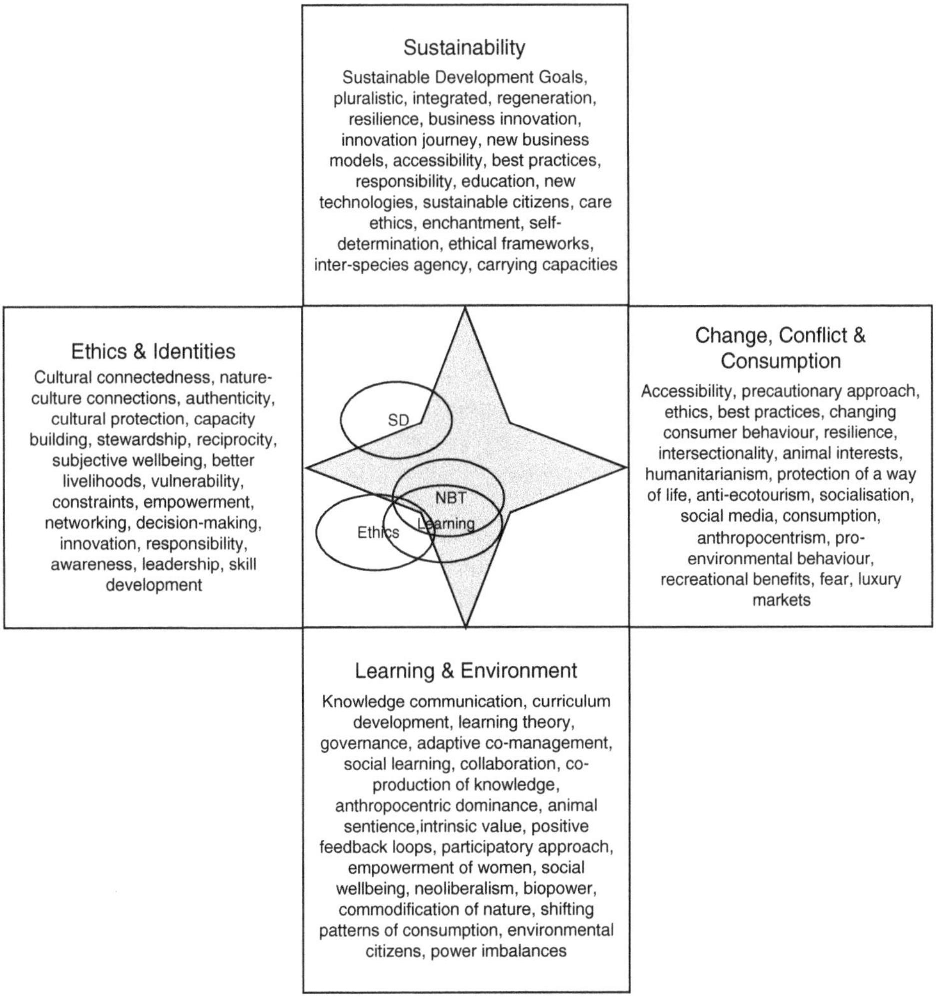

Sustainability

Sustainable Development Goals, pluralistic, integrated, regeneration, resilience, business innovation, innovation journey, new business models, accessibility, best practices, responsibility, education, new technologies, sustainable citizens, care ethics, enchantment, self-determination, ethical frameworks, inter-species agency, carrying capacities

Ethics & Identities

Cultural connectedness, nature-culture connections, authenticity, cultural protection, capacity building, stewardship, reciprocity, subjective wellbeing, better livelihoods, vulnerability, constraints, empowerment, networking, decision-making, innovation, responsibility, awareness, leadership, skill development

SD

NBT

Ethics

Learning

Change, Conflict & Consumption

Accessibility, precautionary approach, ethics, best practices, changing consumer behaviour, resilience, intersectionality, animal interests, humanitarianism, protection of a way of life, anti-ecotourism, socialisation, social media, consumption, anthropocentrism, pro-environmental behaviour, recreational benefits, fear, luxury markets

Learning & Environment

Knowledge communication, curriculum development, learning theory, governance, adaptive co-management, social learning, collaboration, co-production of knowledge, anthropocentric dominance, animal sentience,intrinsic value, positive feedback loops, participatory approach, empowerment of women, social wellbeing, neoliberalism, biopower, commodification of nature, shifting patterns of consumption, environmental citizens, power imbalances

Figure 29.1 Connecting sections and core criteria in ecotourism research

The first section of the book focused on sustainability as a core ecotourism criterion, with a primary focus on some of the macro sustainability issues that have an impact on ecotourism. Foremost among these topics is the linkage to the United Nations' Sustainable Development Goals, which have relevance to ecotourism as one of the greenest or most responsible forms of tourism. Spenceley and Rylance argued that it is not just a few of the global goals that have relevance to ecotourism (e.g., life below water, or life on land), but all 17. This chapter set the stage for viewing ecotourism in a pluralistic and integrated manner. Connected to this pluralistic nature is the belief that ecotourism should act as a model or flagship form of tourism to balance social, ecological, and environmental objectives. Such a position was championed by Day, Sydnor, Marshall, and Noakes who maintained that ecotourism businesses are often at the forefront of efforts to achieve sustainability. While regeneration was a key theme that emerged in their work, such was difficult to achieve in the absence of resilience and innovation in achieving sustainability. The innovation journey often follows a rather uniform pathway according to three periods: initiation, development, and implementation/termination (Van de Ven, 2017) that leads to desired changes in operations.

Schweinsberg and Darcy argued that transportation is one of the thorniest issues confronting ecotourism, as ecotourists often move considerable distances to satisfy their touristic needs and wants. These authors argued that transportation must continue to evolve by introducing new business models and ensuring accessibility in conforming to the universal right to travel. The idea behind newer and better business models was reinforced by Sheppard, and the idea that ecotourism and sustainability will need to be resilient in achieving best practice. Overtourism also surfaced as a formidable issue in ecotourism as ecotourists are quite literally loving some protected areas to death. Aloudat argues that tour operators have important roles as conduits of education and responsibility in the protection of natural areas based on the human connection between supply and demand. Fennell's message is that there are several disruptive influences that we have not encountered in the past, the combination of which has challenged sustainable ecotourism on several fronts. New technologies can aid in the new age of disruption in helping with issues of accessibility and as an option for sustainable citizens who wish to minimise their impact on the planet.

The second section on ethics and identities is a hybrid of two subsections that are connected by how we treat each other and the natural world. For Caton, Hurst, and Grimwood, cracks that are inherent in ecotourism must be filled by addressing colonial legacies through thinking and acting according to care ethics to 'arouse our sensory engagements with the world', and also through enchantment. Read and Grimwood continue the pursuit of the proper ethical ends for ecotourism through a pluralistic approach based on the connection between global ethics and development ethics—culturally sensitive universalism—in addressing the self-determination priorities of indigenous people. But the circle of morality does not stop with human-human relationships. Not unlike Caton et al, and Read and Grimwood, above, representing animals (wolves) demands the use of ethical frameworks like post-humanism to deconstruct the problems we encounter in thinking how best to balance profit and pleasure against the agency of wolves (Thomsen).

The identities sub-section uncovered several ethical issues that focused on the tensions that exist between culture and nature. Graci found that ecotourism can be a key component of capacity building for Indigenous people, along with cultural connectedness and authenticity, which will advance at the intersection of sound entrepreneurship and conservation. Similar themes were discussed by De Bernardi in reference to the Sámi people of northern Europe. Connection to nature and culture, and the protection of Sámi culture, through more ethical practices of cultural promotion and entrepreneurship, will sustain the ecotourism industry if authenticity is preserved.

Demand aspects of ethical and responsible stewardship by volunteer tourists was a theme discussed by Chilembwe in emphasising, like Read and Grimwood, that local to global ethical frameworks are needed to protect the interests of local people and nature. But relationships have to be reciprocal in how local people and volunteers learn and benefit from each other. The use of theoretical frameworks for better ecotourism practices was also supported by Munanura and Sabuhoro in achieving better livelihoods and subjective wellbeing for local communities. Communities are vulnerable to several internal and external constraints that impinge on their ability to realise Sustainable Development Goals. The final chapter in this section by Pirnar focused on entrepreneurship, but through the lens of female participation. Constraints and motivations to participation were discussed as well as empowerment through networking, decision-making, innovation, responsibility, environmental awareness, sharing local resources, increased skills, leadership, and management.

The change, conflict, and consumption section provided a range of different perspectives, some of which are relatively new to ecotourism research. Garrod's study on ecotourism and persons with disabilities shares with Schweinsberg and Darcy the need to make ecotourism much more accessible for people with a range of disabilities. As a neglected area of study (and practice), Garrod argues that there is an ethical imperative to adopt precautionary approaches in steadily moving towards a model of best practices through balancing economic and sustainability needs of supply and demand.

Day and Noakes argue that stemming the climate change tide will necessitate changes in consumer behaviour, along with mitigation, and anticipatory and adaptation strategies designed to make organisations more resilient in the face of often considerable vulnerabilities. Observing that climate change is intersectional is to recognise that a whole variety of different social, cultural and natural elements are at risk. How ecotourism can overcome these intersectionalities is an immediate concern. The new imperative to consider the interests of animals, especially in ecotourism, resurfaces in this section. In the face of conflict and commodification, Wise argued that humanitarian efforts are important in building new ecotourism economies and awareness in light of the dark and tragic influences that animals face in conflict zones.

Change, or the need *not* to change, was a theme explored by Pavelka. Ecotourism is sometimes rejected because it is incompatible to the values and priorities of communities, many of which are already structured around primary industries. The protection of a way of life, therefore, demands thoughtful consideration before implementation, especially for community members who have been socialised in a way that is contrary to the values of ecotourism. Socialisation was also discussed in the context of trip and post-trip accounts of ecotourists and how and why they choose to share their experiences. Social media, Sarkar contends, is playing an increasingly important role in the consumption of ecotourism, especially in how it catalyses pro-environmental behaviour and recreational benefits.

Consumption of ecotourism was also discussed from the perspective of market segments (Do, Weaver, and Lawton). Results confirmed that although Vietnam is viewed as an unconventional ecotourism market, there are parallels with Western ecotourism according to three dominant core criteria of ecotourism: nature-based, learning, and sustainability. An anthropocentrism dimension emerged in their analysis, emphasising more of a focus on larger crowding thresholds, manipulation of the natural environment, and fear of wilderness. The final chapter explored the luxury ecotourism market, which although small, is lucrative with ethical implications on hard and soft forms of ecotourism. Volo contends that expectations of luxury have evolved in line with the sustainability imperative, and this includes the use of ecotourism's core criteria in high-end ecotourism travel. These results correspond to a tradition in ecotourism theory and practice that emphasises greater value on higher-paying tourists, smaller group sizes, and longer stays at the destination.

The final section of the book pulled together chapters on learning and the natural environment. A major thrust of the ecotourism learning and education agenda circulates around guide training and interpretation. What is helpful to more deeply embed this research ontologically and pedagogically is exactly what has been provided by González-Herrera and Giralt-Escobar on theories of learning and education in ecotourism. Key areas of emphasis were provided on learning theory, knowledge communication, curriculum development, and learning synergies. While education filters through many of the chapters in this volume (see Table 29.1), other chapters dedicated to this key them would have been well received.

Switching to the broad topic of environment, the second chapter of the section by Cakar investigated environmental governance through a content analysis of 64 articles. Results indicate that there is a massive gap in the discussion on ecotourism and governance, especially in regard to overtourism hotspots. Governments are no longer taking the lead in governance, but rather pluralistic and innovative strategies are being developed that align with adaptive co-management with emphasis on social learning, collaboration and co-production of knowledge. Returning to the discussion of animals in ecotourism, Burns argues that anthropocentric dominance has a long history in tourism, but ecotourism should place us on a different trajectory towards an effective balance between animal use and animal protection in recognition of animal sentience and intrinsic value. The discussion returns to the necessity of incorporating local to global solutions in mitigating the impacts on animals.

The connection between community and biodiversity conservation is discussed in a case study on agroecotourism (heritage seeds) by Hatipoglu, Denizci, and Imamoğlu, where there is a discernible positive feedback loop was found between farming and ecotourism. Employing a participatory approach, the empowerment of women emerged as an important contributor to social wellbeing and learning in a climate of co-management and co-ownership. The theoretical richness that is a stable feature of this book is no clearer than in Lindahl Elliot's chapter on rewilding in Europe. Lindahl Elliot dug deep into the complexities of rewilding by suggesting that ecotourism and wildlife tourism can replace older forms of rural development (see Pavelka), but that there are issues of neoliberalism and biopower at hand that emphasise the commodification and control of nature for instrumental rather than intrinsic ends.

If rewilding European landscapes isn't complex enough, the Half-Earth vision discussed by Kopnina is also a tall order. But Kopnina sketches a realistic scenario on this topic, where ecotourism is positioned as a "multidimensional complex process" and ecotourists can act as environmental citizens in making difference for secies ecosystems, and humanity. There will be costs, especially for larger commercial enterprises, but shifting patterns of consumption (vegetarianism and veganism) and the combined efforts of policymakers, scientists, and conservationists are necessary in the transition.

The final chapter in the book by Stronza, Hunt, and Fitzgerald peels back the many layers, and years, of the contentious relationship that exists between conservation and tourism—a question that Budowski (1976) characterised as conflict, coexistence, or symbiosis in the early stages of tourism scholarship. A prominent deficiency in ecotourism research is the dearth of longitudinal studies that provide greater scope into governance, conservation, biodiversity, and ecosystem integrity. Future studies should focus on competing land uses (as observed in this section and others in the book), landscape or ecosystem-level analyses instead of species-level, and the social outcomes of ecotourism in areas such as environmental ethics, conservation attitudes, and changing social relations based on power imbalances.

Connectivity to the core criteria of ecotourism

The second approach to summarising the content of the 28 chapters in the book involves analysing how closely these chapters address the core themes that are often used to define ecotourism, i.e., sustainability, nature-based focus, learning and education, and ethics. A breakdown of this information is found in Table 29.1.

Results indicate that, combined, these four core criteria were mentioned in total 1702 times in the book. Sustainability was mentioned the most at 623 times (36.6% of all mentions), followed by learning and education (26.0%), nature and nature-based (25.1%), and finally ethics and morality (12.3%). This statistic serves to indicate the dominance of the first three of these criteria, as noted above, and the more recessive nature of ethics. Notwithstanding, ethics still emerges as a theme of interest in many chapters and in the field in general.

Table 29.1 Frequencies and percentages of core aspects of ecotourism in the sections

Section	Core elements of Ecotourism (N = 1702)									
	Ethics/Morality (n = 209; 12.3%)		Sustainability (n = 623; 36.6%)		Nature/NBT (n = 428; 25.1%)		Learning/Education (n = 442; 26.0%)		Total (n =1682)	
(No. of chapters)	n	%	n	%	n	%	n	%	n	%
Sustainability (n = 6)	12	5.7	257	41.3	64	15.0	38	8.6	371	22.1
Ethics/identities (n = 8)	154	73.7	178	28.6	132	30.8	67	15.2	531	31.6
CCC (n = 7)	13	6.2	53	8.5	120	28.0	54	12.2	240	14.3
Environment/learning (n = 7)	24	11.5	121	19.4	112	26.2	283	64.0	540	32.1

As we look *down* each of the four columns in Table 29.1, there are many interesting results. Of the 209 times that ethics and morality were mentioned in the book, 73.7%, not surprisingly, were mentioned in the section on ethics and identities. This was followed by a considerable drop in the other sections. The environment and learning section mentioned ethics and morality 24 times, followed by 13 times in the chapter on change, conflict, and consumption, and a mere 12 times in the sustainability section indicating a poor link between ethics and sustainability. Even though there is explicit recognition of a moral imperative tied to sustainability (Fennell, 2018) and to the UN Sustainable Development Goals more specifically (mshank, 2016), ethics is still not well referenced in ecotourism-sustainability research.

Results show that the term *sustainability* was mentioned 257 times, representing 41.3% of all mentions. The section on ethics and identities was next with 28.6% of mentions, followed by environment and learning (19.4%), and then the section on change, conflict, and consumption (8.5%). What this suggests is that authors of chapters in the ethics and identities section are far more willing to mention sustainability, than, as we saw above, authors of chapters in the sustainability section mentioning ethics despite the explicit connection between the two concepts (as noted above). One of the most revealing statistics is far less reference to sustainability by authors in the change, conflict, and consumption section, where sustainability as mentioned only 53 times in the seven chapters, comparatively speaking.

Far more balance was achieved in the column that focused on nature and nature-based in ecotourism, which, while not a dedicated section in the book, connected strongly (and not surprisingly) to the learning and environment section. Indeed, the learning and environment section represented just over one-quarter (26.2%) of all nature and nature-based tourism mentions. Another prominent finding is that the chapters in the sustainability section were far less likely to connect with nature and nature-based criterion (15.0%) than the other three sections. Again, we see less of a focus on other core criteria in chapters that are more explicitly centred on sustainability as compared to other themes.

In the fourth column on learning and education, almost two-thirds (64.0%) of all mentions of learning were in the section on environment and learning—again not surprising. But that which is less frequent is most revealing. Chapters on sustainability mentioned learning and education 38 times in the six chapters, with slightly more in the change, conflict and consumption set of chapters and more still in the ethics and identities one. It should be noted, however, that learning and education mentions were dominated by the chapter by González-Herrera and Giralt-Escobar on ecotourism and theories of learning, where over one-half (52.7%) of all mentions of education and learning occurred.

As we look *across* the rows there is a focus on how well represented the four core criteria of the definition of ecotourism are in each of the four sections of the book. It is arguable that sustainability is the most important ingredient in ecotourism because it takes into consideration so much. There is almost universal acceptance of the fact that local communities need to have control, decision-making abilities and that benefits and incentives should follow from the sound planning, development, and management in communities; while conservation of biodiversity and ecosystems is a strong feature of ecotourism. Yet, the content analysis illustrates that the focus is more squarely on the sustainability core criterion in the sustainability chapters, and less likely to discuss aspects of learning and education ethics and nature-based than other realms. Has sustainability become so unified and specific as a research theme that it demands use of its own vernacular, even when tied to ecotourism?

Chapters on ethics and identities (n = 8) can be characterised as more generalist when it comes to the use of ecotourism's core criteria, with good representation across the board (31.6% of all core criteria can be found in these chapters, but we should acknowledge that it is the largest section by number of chapters at eight). Only 67 (15.2%) of the core criteria in ethics and identities dealt with learning and education, suggesting that there may be greater scope for connecting these two criteria in future research.

It is also interesting to observe that chapters in the section on change, conflict, and consumption, what we can characterise as newer themes in ecotourism research, relied less on the conventional core criteria of ecotourism. Examples include anti-ecotourism, socialisation in ecotourism, and ecotourism as a luxury form of consumption. As these new themes continue to emerge and develop in ecotourism studies, it will be interesting to see how they further depart from these criteria or begin to more closely connect with them. Indeed, one of the main objectives of the book was to explore new terrain related to ecotourism. Arguably, this was most strongly represented in the section on change, conflict, and consumption. Finally, the last section containing seven chapters on environment and learning shows a rather skewed distribution of the four major criteria in ecotourism. Lower percentages of ethics and sustainability were included in these chapters (11.5% and 19.4%, respectively), with higher percentages for nature and nature-based and, of course, learning and education.

The star in the middle of Figure 29.1 provides a rough indication of how the four core criteria of ecotourism correspond to the four sections of the book. We observe how sustainability (SD) and ethics are restricted in their range across the chapters, and how nature and nature-based, especially, inhabits a more central position between the sections.

While the foregoing discusses what this book *is*, here, at the end, it is perhaps of some value to explain what this book is *not,* in keeping with the theme of emphasising future directions. The initial call for chapters included several topics that were not covered, including life cycle analysis, carbon-neutral destinations, justice and rights, intercultural collaboration and co-operation, feminist theory, LGBTQIA, ecotourism and the law, ecotourism and cultural change, risk management, poverty reduction, neoliberalisation, south-south cooperation, post-colonialism and post-development, and alternative energy technologies.

Ecotourism has relevance to all of these themes, and it is indeed unfortunate that these topics were not covered in the book. I argue therefore that there is considerable scope for pushing the limits of ecotourism—to fill vacant niches in the future—in keeping pace with several new areas of research in other disciplines, and to perhaps change the very essence of what ecotourism is. This work waits for us, but will no doubt benefit richly from the diversity of perspectives in the preceding pages.

References

Budowski, G. (1976). Tourism and environmental conservation: Conflict, coexistence, or symbiosis. *Environmental Conservation, 3*(1), 27–31.

Fennell, D. (2018). Sustainability ethics in tourism: The imperative next imperative. *Tourism Recreation Research*, 44(1), 117–130.

Fennell, D. A. (2020). *Ecotourism*, 5th ed. London: Routledge.

mshank. (2016). Ethics in Action: "The UN's 17 Sustainable Development Goals". Retrieved from https://www.unsdsn.org/news/2016/11/14/ethics-in-action-the-uns-17-sustainable-development-goals.

Nowaczek, A. & Smale, B. J. A. (2010). Exploring the predisposition of travellers to qualify as ecotourists: the Ecotourist Predisposition Scale. *Journal of Ecotourism*, 9(1), 45–61.

Van de Ven, A. H. (2017). The innovation journey: You can't control it, but you can learn to maneuver it. *Innovation, 19*(1), 39–42. doi:10.1080/14479338.2016.1256780.

INDEX

Note: *Italic* page numbers refer to figures and tables.

For Product Safety Concerns and Information please contact our EU
representative GPSR@taylorandfrancis.com
Taylor & Francis Verlag GmbH, Kaufingerstraße 24, 80331 München, Germany

www.ingramcontent.com/pod-product-compliance
Lightning Source LLC
Chambersburg PA
CBHW081037220326
41598CB00038B/6906

* 9 7 8 1 0 3 2 0 6 7 2 3 0 *